WILDLAND WATERSHED MANAGEMENT

WILDLAND WATERSHED MANAGEMENT

SECOND EDITION

Donald R. Satterlund
Washington State University

Paul W. Adams
Oregon State University

JOHN WILEY & SONS, INC.

New York • Chichester • Brisbane • Toronto • Singapore

Library of Congress Cataloging-in-Publication Data:

Satterlund, Donald R.

Wildland watershed management / Donald R. Satterlund and Paul W. Adams. — 2nd ed.

p. cm.

Includes bibliographical references and index.

ISBN 0-471-81154-8

1. Watershed management—United States. 2. Wilderness areas--United States. I. Adams, Paul W. II. Title.

TC409.S37 1992

333.91—dc20 91-34849

CIP

Printed and bound in the United States of America by Braun-Brumfield, Inc.

10 9 8 7 6 5 4 3 2 1

PREFACE

The second edition of Wildland Watershed Management is more than just an updated version of the first edition. It reflects the conviction that the key to maintaining or improving water yields and quality from wildland watersheds under the impact of their use and development for wood, wildlife, minerals, recreation, grazing, and other values rests upon a firm knowledge of the variable linkages of the terrestrial to the water resource system in time and in space.

Small portions of most watershed lands are tightly linked to the stream system at nearly all times (e.g., riparian zones), whereas much of the remainder is well buffered from it most of the time. However, the area of tight linkage shrinks and expands, and its location shifts over time. It is least, and usually only near perennial channels at the end of long dry seasons, and greatest and widespread into every swale and lower slope at the peak of the wet season. The extent and location of strong linkage also varies with general climate and physiography, being more persistently widespread and ill-defined in humid regions of moderate topography, but consistently limited and well-defined in steep, arid lands. These variable linkages cause even more variable water yield responses to identical impacts on the watershed, depending on impact location, extent, and timing, even to the degree that an identical modification (e.g., removal of vegetation by fire) may be beneficial, neutral, or harmful to the water resource at different times or places on a watershed.

The relationships are explained and illustrated, with methods to help the reader determine them in the field, throughout the book. The emphasis is on small headwater stream systems, where the linkage between land and water is most pronounced, and where most wildlands—forest, range, and alpine lands—are found.

This book reflects experience gained since first working in watershed management on the Fool Creek watershed experiment near Fraser, Colorado in 1950 and teaching, research, and study since then throughout the United States, but mostly in the Lake States, the Northeast, Pacific Northwest, and Northern Rocky Mountains. It also reflects the changing emphasis from water supply and flood control in the early years to protection of water quality and maintenance of a desirable environment today.

The basic organization of the second edition is unchanged, but consideration of soil erosion and sediment and water quality has been expanded as has the overall scope of watershed management issues and opportunities. The presentation is more quantitative for greater precision and conciseness, and units are expressed primarily in the SI unit (Système Internationale d'Unités) system to conform with the scientific literature.

DONALD R. SATTERLUND

Pullman, Washington

PAUL W. ADAMS

Oregon State University
January 1992

PREFACE TO THE FIRST EDITION

Watershed management is the management of all the natural resources of a drainage basin to protect, maintain, or improve its water yields. It becomes wildland watershed management when it is applied on the non-urban, non-cultivated lands that are chiefly covered by forest, range, and alpine vegetation. They are the primary source of fresh water for man throughout the world.

This book is designed to present the fundamental theory and basic practices for all who make decisions about the management of wildlands for water: the natural resources student, practicing land managers and planners, and educated citizens who ultimately determine the fate of all their natural resources.

Any book in watershed management is a work in synthesis. Though it draws on many fields, it is not a substitute for the separate study of those from which it draws. The focal point is the integration of various aspects of hydrology, ecology, soils, physical climatology, and other sciences to provide the scientific base of watershed management; to develop from this base rational procedures of applying this information to achieve desired results; and to derive guidelines for choosing acceptable management alternatives within the context of social wants and needs.

This text represents the outgrowth of more than a decade of classroom teaching of forestry, range, wildlife, geography, engineering, and other subjects in both the eastern and western United States. It reflects the benefits of association with teachers, researchers, and practicing wildland managers from many places. All have contributed substantially to the development of this book.

In addition, I wish to thank Professor Raymond A. Gilkeson, Dr. William E. Sopper, and my students, who provided valuable criticism of the manuscript, and Mrs. Maxine Andrews, who made its preparation possible. Finally, I thank Lily-Ann, Nels, Ruth, and Lisa, whose support and encouragement were unflagging.

DONALD R. SATTERLUND

Pullman, Washington
September 1972

CONTENTS

WILDLAND WATERSHED MANAGEMENT

PART I
Water, Watersheds and People

1 Water Resources and Water Problems

Water is many things to many people. It may be one of the most common substances on earth, yet it has most uncommon characteristics. It is essential to all life, but may cause widespread death and destruction. It makes our land productive, but may tear away its productivity through erosion. Water is necessary, along with sunshine and land, to provide the food to nourish our bodies; and the aesthetics of water, as clouds in the sky, turbulent brooks, or waves breaking on a rocky shore, provide a beauty that helps to nourish our souls. It may be as calm as a misty pond on a clear autumn morn, or as furious as a raging torrent.

Because water is all these things and more, it may be considered an asset or a liability. It is one of our most valuable natural resources, serving people in many different ways. It is a problem because it can destroy us or our works, impair our environment, or simply because its lack prevents us from enjoying benefits that might otherwise be available, and that we have come to expect.

It is surprising that, in the United States, water is generally perceived to be a scarce resource. Yet, as an economic resource, water can hardly be considered scarce, if scarcity is considered to be the relationship between supply and demand for a good or service as reflected in its price. Water is the cheapest of all commodities on the market. In most large cities in the United States it is delivered to any home on demand at a price of from 5 to 25 cents per ton. Not even fill dirt is available so cheaply (Stults, 1973).

Perhaps water is perceived as scarce because water problems, as with any problem, are a product of human psychology; a matter of perceptions and expectations as well as desires and needs. If we have been led to expect unlimited supplies of pure, cheap water as an inalienable right, and limits in the form of either restricted uses (e.g., limited lawn watering) or high price appear, then we perceive it as being scarce; not because the price is truly high, but because it is higher than we think it ought to be.

When we think of all the different people, with their different perceptions of, and expectations about our water resource, it is little wonder that we see problems at every hand. As populations grow and become more concentrated in urban areas, the problems grow and claim increasing attention. They extend far beyond the centers of population to our wildlands; the range, forest, and alpine lands that are home to less than 10% of the population of the United States. They extend to wildlands because wildlands are our major source of usable water. In the 11 conterminous western states, over 90% of the usable water originates from wildlands (Water Resources Council, 1968).

Water problems have become even more challenging as water resource concerns have expanded to include aquatic habitat values and the unique functions of wetlands and areas bordering water bodies (Mitsch and Gosselink, 1986; Salo and Cundy, 1987). For example, low streamflows not only may be a problem for downstream users, but also for local fish and wildlife populations.

It is not the intent of this book to cover all water problems as they exist today or appear in the future. Such broad studies are available elsewhere (e.g., Guldin, 1989; Water Resources Council, 1978). Instead, a brief survey will be presented of the nature of our water resources, problems that exist, and resource management alternatives that are available to help solve our problems. We will then review the basic scientific principles that govern the behavior of water in the environment. This will provide a foundation of understanding that is essential for a detailed examination of our primary topic: *wildland watershed management*, the management of forest and other land in a drainage basin to protect or enhance its water resources.

Our discussion largely focuses on conditions and examples in temperate North America. However, whenever possible, we emphasize those basic principles and management concepts that can be applied more broadly, and cite pertinent findings from throughout the world.

1.1 NATURAL WATER SUPPLIES

In all the solar system, Earth could be known as the watery planet. Oceans cover more than two-thirds of its surface, and freshwater ice, lakes, streams, swamps, and marshes cover more than 10% of the land. But, 97% of the world's water is salty and most of the remaining 3% consists of polar ice and glaciers. Nevertheless, the liquid fresh water on and in the earth is sufficient to cover all the land surface to a depth of nearly 100 m (see the Appendix for conversion factors for other units used in hydrology). Most of it (~98%), consists of groundwater. One-half of the remainder is stored in lakes, and the rest is stored in the soil, the atmosphere, and in rivers and streams. At any given time, there is about twice as much water in the soil as in the atmosphere, while streams and rivers contain less than one-tenth of that in the atmosphere.

Precipitation and runoff over the land areas of the world tend to be concentrated in the tropics and decrease irregularly toward the poles. Land surfaces in the southern temperate regions of the earth are somewhat wetter than in the north, but the amount of land surface is much less.

Australia is the driest inhabited continent, receiving 47 cm of precipitation and yielding only 5 cm of runoff yearly, while South America is the wettest at 163 and 93 cm, respectively. The rest all receive between 60 and 69 cm of precipitation and discharge 25–34 cm of runoff (Zubenok, 1970). Comparable figures for the United States are 76 and 26 cm, respectively. Average figures, such as those above, however, may be greatly misleading. Australia contains rain forests as well as deserts, and the range of variation is great on all the inhabited continents. Only Europe has no true desert.

To be most useful, water must be located where other elements of the environment—soils, climate, and topography—are also suitable for human occupation. Unfortunately, fresh water is often most abundant in such areas as the Amazon and Congo Basins, the mountains, and frigid and subfrigid regions that are not very hospitable to humans.

Water is a flow resource, and except for some groundwater trapped in closed basins, it is always moving. Water may be detained; in the ground, as glacial ice, or in lakes and reservoirs, but ground water seeps, glaciers flow inexorably downward and finally melt, and all water exposed to the atmosphere is subject to evaporation, moving off invisibly through the air. This flow is continuous but irregular, both in time and in space. So even areas that are usually well supplied may be subject to flood or drought at varying intervals. Water supply is variable; in space, in time, and in quality, because of the irregular flow of water through the hydrologic cycle. All three characteristics of water—its quantity, *regimen* (the distribution of flow in time) and quality—are needed to describe the water resource.

1.2 THE HYDROLOGIC CYCLE

The major source of all our water is precipitation, which falls primarily as rain or snow. The amount, kind, and seasonal distribution of precipitation is the dominant factor in the determination of water yields from any area, but other factors are also involved.

Part of the precipitation striking the ground and its vegetative cover is retained at the surface, to be evaporated back to the atmosphere. The remainder may run over the surface to the streams or infiltrate into the ground. Part or all of the water entering the ground is held by capillary forces and may be evaporated directly or absorbed by plant roots and removed to the atmosphere in *transpiration*. If more water enters the ground than can be retained by capillary forces, the excess percolates through the soil, sooner or later moving into a stream channel or into the groundwater, where it may later appear in springs or wells or sustain the flow of streams during dry periods.

The hydrologic cycle may be divided simply into four parts: *precipitation* (P), *runoff* (RO—that which becomes streamflow), *evapotranspiration* (ET—evaporation and transpiration combined), and *storage* (S) in the air, on or in the land, and in the oceans. In equation form:

$$P = RO + ET \pm \Delta S \tag{1.1}$$

Changes in moisture storage tend to balance out from year to year or over longer periods of time and can be ignored in long-run considerations of the quantity of water yield. Ignoring storage and rearranging the formula a little, it is clear that runoff is the amount of water left after evapotranspiration losses are subtracted from precipitation:

$$P - ET = RO \tag{1.2}$$

The fact that runoff is a residual is very important. It represents the amount left over after the needs for evapotranspiration are met. The evapotranspiration demand in turn, is a function of the amount of energy received and the temperature. The quantity of runoff therefore depends not on the amount of precipitation alone, but the amount by which precipitation exceeds evaporative losses. It is quantity of runoff that is usually meant when people refer to water supply.

1.3 VARIABILITY OF WATER SUPPLY IN SPACE

To understand problems of water quantity it becomes necessary to look at each of the elements of the hydrologic equation and their relationships. Using the United States as an example, precipitation averages 76 cm, evapotranspiration losses about 50 cm, and runoff is about 26 cm annually, but varies considerably from place to place.

1.3.1 Precipitation Supply

Figure 1.1 is a map of conterminous United States showing average annual precipitation. The range is very great, from more than 500 cm annually on the Olympic Peninsula of Washington to less than 10 cm in the deserts of the Southwest. The distribution and amount of precipitation depends on these factors:

1. Air mass circulation patterns (storm paths).
2. Distance and direction from large water bodies.
3. Location with respect to barriers.
4. Altitude.

Thus, there is great variability in precipitation around the western coastal mountains. Storms off the north Pacific Ocean from a southwest direction drop more than 500 cm yearly on the southwest portion of the Olympic Mountains, but near Sequim, Washington, on the northeast tip of the peninsula, annual precipitation is less than 40 cm and irrigation is used to grow crops. The distance between the two extremes is only about 70 km.

In a very general manner, the United States can be divided into six more or less uniform precipitation regions as shown in Fig. 1.2 (Humphrey, 1962).

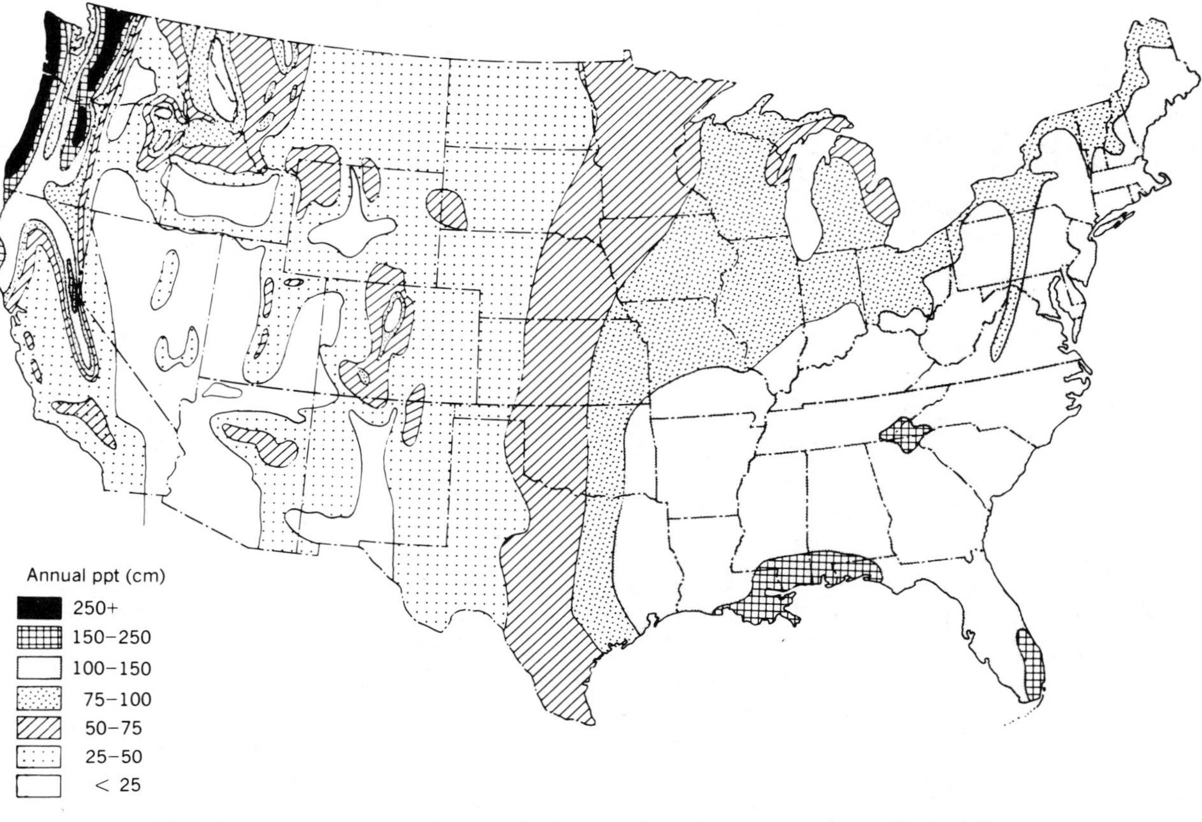

FIG. 1.1 Average annual precipitation (cm) in the conterminous United States.

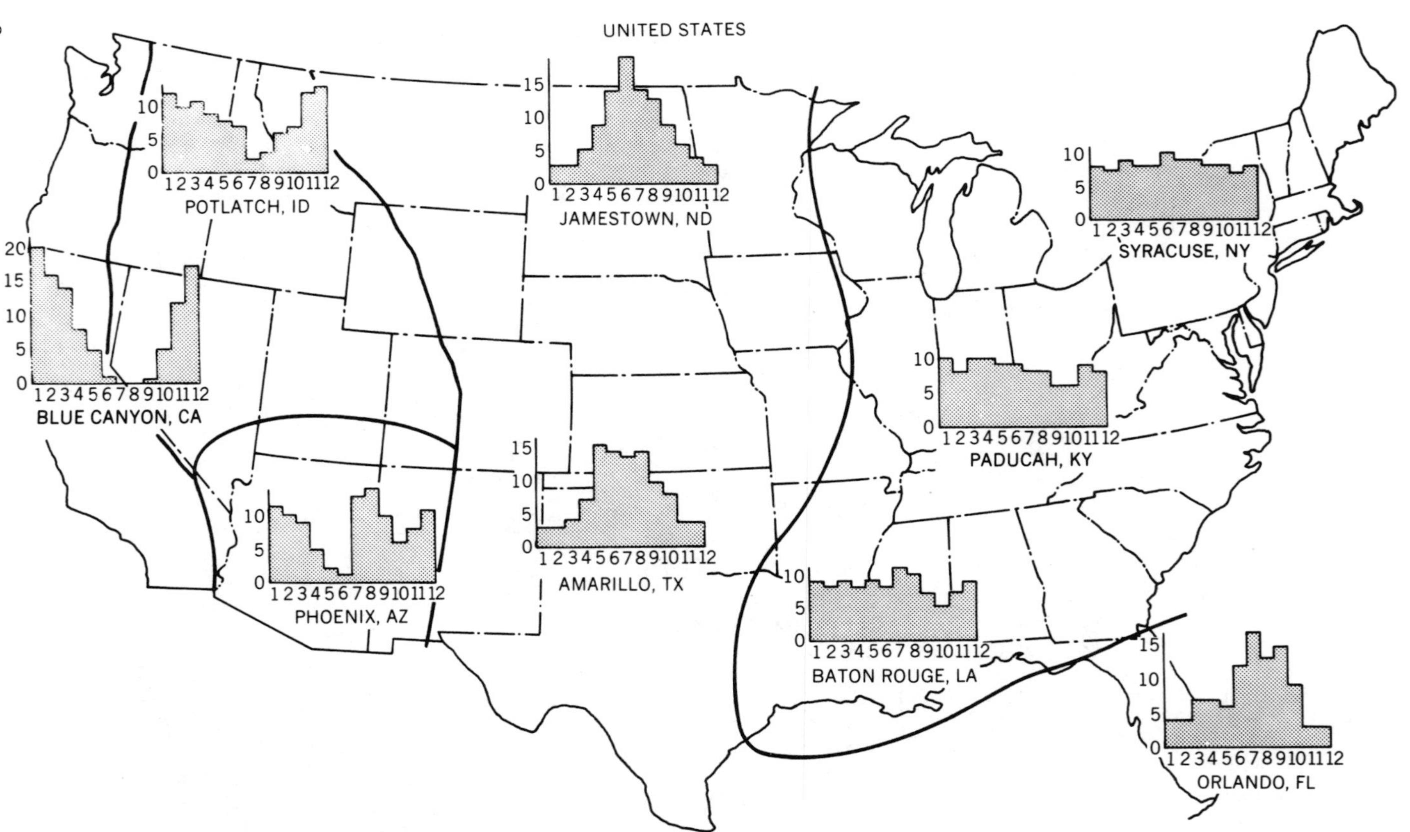

FIG. 1.2 Average monthly distribution (percent of annual precipitation) of precipitation in the United States.

1. *Eastern.* All of the United States from eastern Texas north to the upper Mississippi River and north of Florida is characterized by ample precipitation to grow crops without irrigation. Precipitation decreases from about 150 cm annually on the south coast and mountains to about 70 cm in the northern lowlands. It is distributed fairly evenly throughout the year. Snow constitutes more than 20% of the annual precipitation only in the northern tier of states. Snowpacks, containing from 10 to 50 cm, water equivalent, develop in the northern Lake States and from parts of Pennsylvania northward nearly every winter.

2. *Florida.* Florida is characterized by heavy rainfall occurring predominantly during summer and early autumn. Late fall through spring is relatively dry. Total precipitation mostly exceeds 125 cm, of which less than one-third falls in the long dry season.

3. *Great Plains.* Precipitation decreases gradually westward and northward across the Great Plains from about 100 to about 25 cm annually. Precipitation increases from its minimum in midwinter to a maximum in late spring or early summer. Although the Plains are well known for their fierce winter blizzards, snow contributes only a small portion of annual precipitation, and a continuous snowpack seldom persists for more than a few weeks except in the extreme north.

4. *Southwestern.* In the inland portion of the Southwest, precipitation varies sharply with elevation, ranging mostly between 10 and 100 cm yearly. The seasonal pattern of distribution in Arizona, southern Utah, and western New Mexico is one of prolonged, gentle winter rain or snow (depending on elevation), a very dry spring, intense local summer storms, and moderate drought in the fall. Snow may be very important in the closed forest zones of the upper elevations.

5. *Pacific Coast.* Precipitation along the Pacific Coast is heaviest in the north, exceeding 250 cm yearly in much of western Washington, decreasing rapidly in the valleys between the coastal mountains and the crest of the Cascade–Sierra Nevada Ranges, reaching less than 50 cm in the southern valleys of California. It is heaviest in winter, with maxima in December and January, tapering off into spring. Summers are dry, being nearly rainless in the south. The rains of fall begin earliest in the north and spread southward, increasing in amount and frequency with the approach of winter.

Snow is of major importance in the Pacific Coast Region. It is the dominant form of precipitation above an altitude of 500–700 m in the Olympic and northern Cascade Mountains, where snowfall may reach 20 m or more in one season. The snow line rises southward to above 1000–1500 m in coastal southern California and in the southern Sierra Nevada Mountains. Glaciers occur in the Olympics and on many peaks exceeding 2700 m in the Washington Cascades. They occur less frequently in Oregon and northern California, but snowfields persisting well into the summer are found at high elevations throughout the region.

6. *Intermountain.* Between the crest of the Rocky Mountains on the east and the crest of the Cascade–Sierra Nevada Ranges on the west, a large region receives much of its precipitation during the cool seasons of the year, although the difference between seasons is not as pronounced as on the Pacific Coast. The region lies in the rain shadow between north–south mountain ranges, and large areas at lower elevations are arid to semiarid receiving only 10–25 cm annually. Precipitation at the higher elevations ranges from about 200 cm annually in the north to about 100 cm in the south. Snow is of major importance in all the mountains and in northern valleys. Small glaciers may be found in Montana and Wyoming, and melting summer snowfields are an important source of water in the central Rockies (Martinelli, 1965).

1.3.2 Evapotranspiration Losses

The amount of precipitation returned to the atmosphere as vapor depends on the energy applied to water and the temperature. At low temperatures, only a small proportion of the energy is used to evaporate water, but at high temperatures nearly all the energy absorbed by water is used in evaporation. The primary source of energy is solar radiation and a secondary source is advected heat transferred from a warm to a cooler region.

If water were always available, then actual evapotranspiration would be the same as potential evapotranspiration and dependent only on energy supply and temperature. Thus, a map of potential evapotranspiration (Fig. 1.3) is essentially a map of available energy and temperature, and in lowlands decreases regularly from south–north as in the central section of the United States.

In the mountains, the energy supply is similar at high and low elevations; solar radiation is more intense at high elevations, but this is usually offset by greater cloudiness. However, temperatures decrease regularly with altitude and radiation is less on north-facing slopes; so isolines of potential evapotranspiration reflect the prevailing orientation and altitude of western mountains and valleys.

The greatest potential water losses are always at low elevations and in the south, whereas the least loss occurs high in the mountains and in the north. Thus, the mountains of the interior West serve as oases in the sky; not only because they receive more precipitation, but also because water losses are less.

1.3.3 Water Surplus and Deficit Regions

Where precipitation exceeds potential evapotranspiration, a *water surplus* exists that is available to regularly recharge ground water aquifers and provide runoff for streamflow. Many crops can be grown without irrigation in such regions, and they are characterized by frequent streams that flow continuously throughout the year.

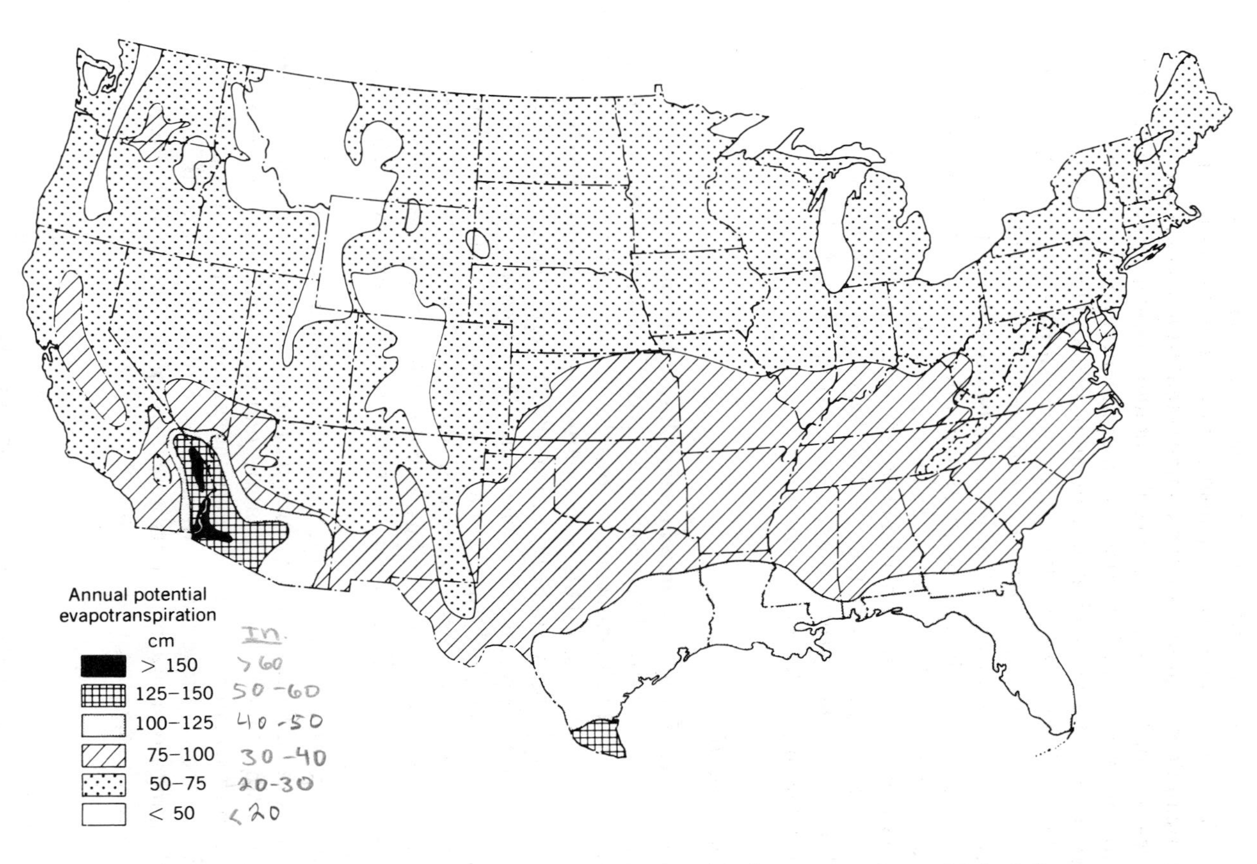

FIG. 1.3 Potential evapotranspiration (cm) in the conterminous United States.

Where annual precipitation is less than potential evapotranspiration, a *water deficit* exists. Here, less water is available than could be used by vegetation, and dryland crop production is hazardous even with such techniques as fallowing; irrigation is needed to insure many crops. The only water available for streamflow or groundwater recharge is that which moves over or through the earth's surface quickly or deeply enough to escape evaporation. It is a region of few perennial streams and intermittent runoff, most occurring during periods of heavy precipitation or snow melt and low evapotranspiration.

The map in Fig. 1.4 shows the average water surplus or deficiency regions in the United States. The nation can clearly be divided into two main parts, the "haves" and the "have nots" approximately along the 96th meridian. To the east, all lands contribute regularly to streamflow. To the west, only the higher mountains and the Pacific Coast region are water surplus lands. They are the source of most of the perennial streams of the West, and the area that generates the greatest surplus, along the coast of Washington, Oregon, and northern California, has many rivers that flow directly to the sea without ever passing near a major population center.

1.3.4 Runoff

Annual runoff (Fig. 1.5) shows a very close relationship to water surplus, which tends to confirm that theory and fact fit. Everywhere, however, runoff is greater than would be indicated by considering only precipitation and potential evapotranspiration. This is due partly to rain that falls on soils with a greater intensity than can be absorbed, and the rainfall excess flows over the surface into the streams. Also, where precipitation occurs in the winter, the supplies of energy and water are desynchronized. Much of the winter precipitation, which cannot be stored in the soil, flows off before it receives enough energy to evaporate. Thus, actual runoff always exceeds the theoretical water surplus, although the differences may be small, especially in drier regions.

In many of the deficit regions, stream flow is steadily depleted after it leaves the mountains. Small tributary streams, perennial near their source, become intermittent at lower levels, and only infrequently discharge flow throughout their entire length. Streams that drain the surrounding land are called *effluent streams*, whereas those that drain into the channel bed material, sometimes recharging ground water aquifers, are called *influent streams*. It is possible for a given reach in a stream to be effluent in wet seasons, and to be influent in dry seasons.

1.4 VARIABILITY OF WATER SUPPLY IN TIME

Areas that receive ample amounts of water on an annual basis may be subject to prolonged periods of inadequate supply and brief periods of disastrous

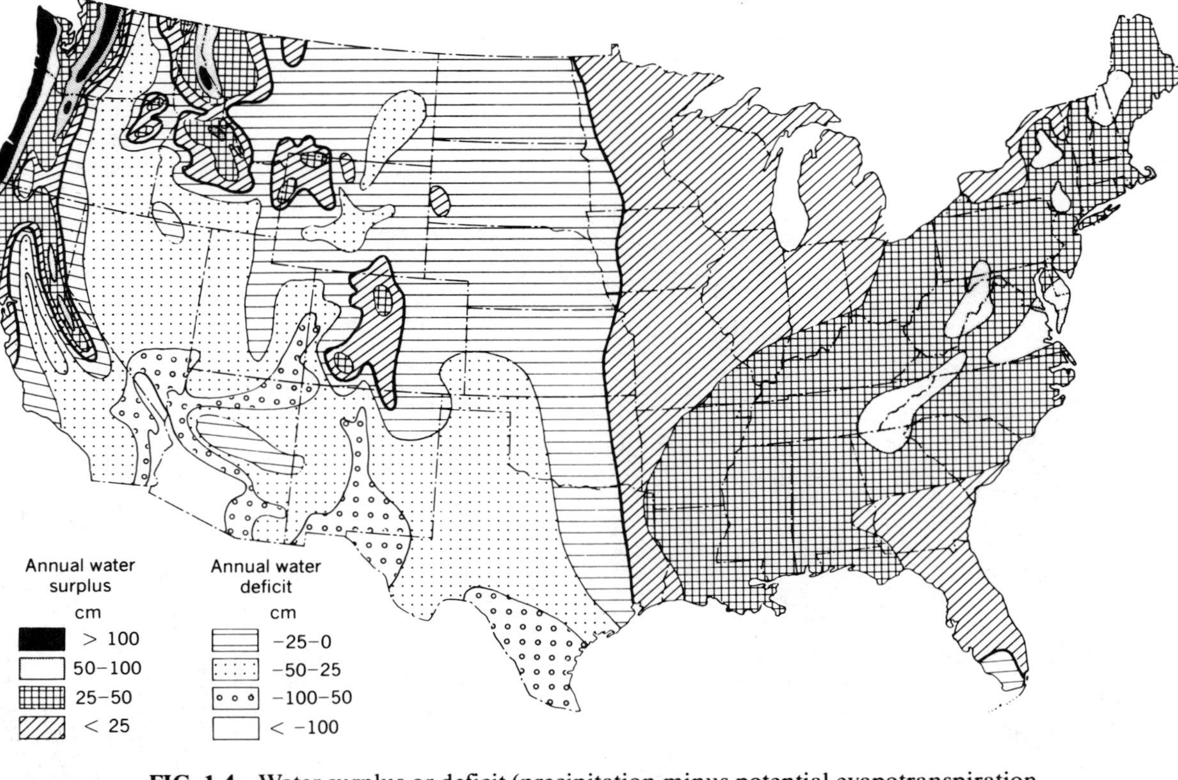

FIG. 1.4 Water surplus or deficit (precipitation minus potential evapotranspiration in cm) in the conterminous United States. The heavy line separates areas of surplus and deficit.

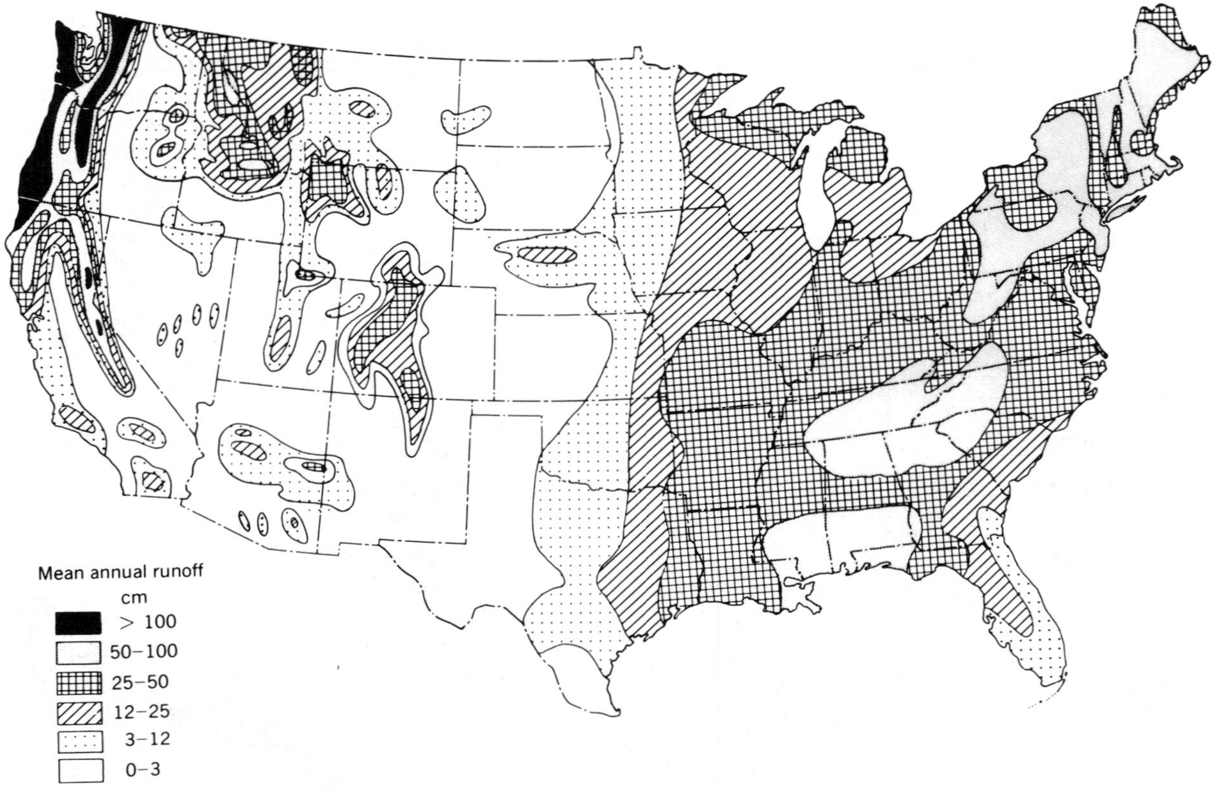

FIG. 1.5 Mean annual runoff (cm) in the conterminous United States.

excess. There are two ways of looking at *streamflow regimen*, or the distribution of streamflow in time; in terms of the long-term average pattern of flow, or in terms of the pattern during any specific short period.

The general seasonal pattern largely reflects climate; the amount, season and form of precipitation, and the amount and timing of evapotranspiration losses and snowmelt. Only large changes in moisture storage, such as recharge of depleted soil moisture in autumn or drainage of the mantle during long dry periods, modify noticeably the effect of climate on runoff. In most of the United States, streamflow peaks in late winter or spring, with low flows from midsummer through autumn. In northern regions characterized by long continuous periods of subfreezing weather, flow may decrease throughout the winter until spring melt begins. Figure 1.2 shows the average seasonal distribution of precipitation and Fig. 1.6 shows the average timing of runoff in the United States. The difference between the two figures makes quite evident the need to make allowance for summer water losses, autumn moisture recharge, and snow storage and melt.

Other factors are also important in determining the specific runoff pattern of any given stream. Geological factors of topography, surficial and bedrock deposits, can have a pronounced effect. Figure 1.7 shows the difference in flow of two rivers within the same climatic zone, but with different geology that results in different amounts, rate of intake, and release of water from the watershed mantle. The size of a river system and amount of natural surface storage is also related to uniformity of flow. Some streams, such as the St. Lawrence River, are exceptionally stable, for they are so large that tributary inflows are desynchronized, and the tremendous storage capacity of the Great Lakes buffers seasonal and other variations in flow.

The regimen in any specific period is, even to some degree in the very largest river systems, a function of the particular weather as it interacts with available storage characteristics of the watershed. During drought, streamflow is dependent on the amount of water in storage and its rate of release to the channel. During periods of heavy precipitation, the amount of storage space in the mantle, and the rate at which water can infiltrate may determine whether floods or merely high flows are produced. In the short term, rates of precipitation and storage may be more important than amounts, especially in small upstream systems where desynchronization of flow may be minimal.

When considering the need for, and the supply of water, it is well to remember that averages may be extremely misleading. It is the extremes that set the limits to a given water supply. For example, water needs are usually greatest during hot, dry periods when streamflows may be at their minimums. On the other hand, one day of flood every few years may keep a flood plain uninhabitable.

1.5 WATER QUALITY LIMITS SUPPLY

Of all natural substances on earth, water is the universal solvent. So long as most dissolved substances are in relatively limited concentration, this is no

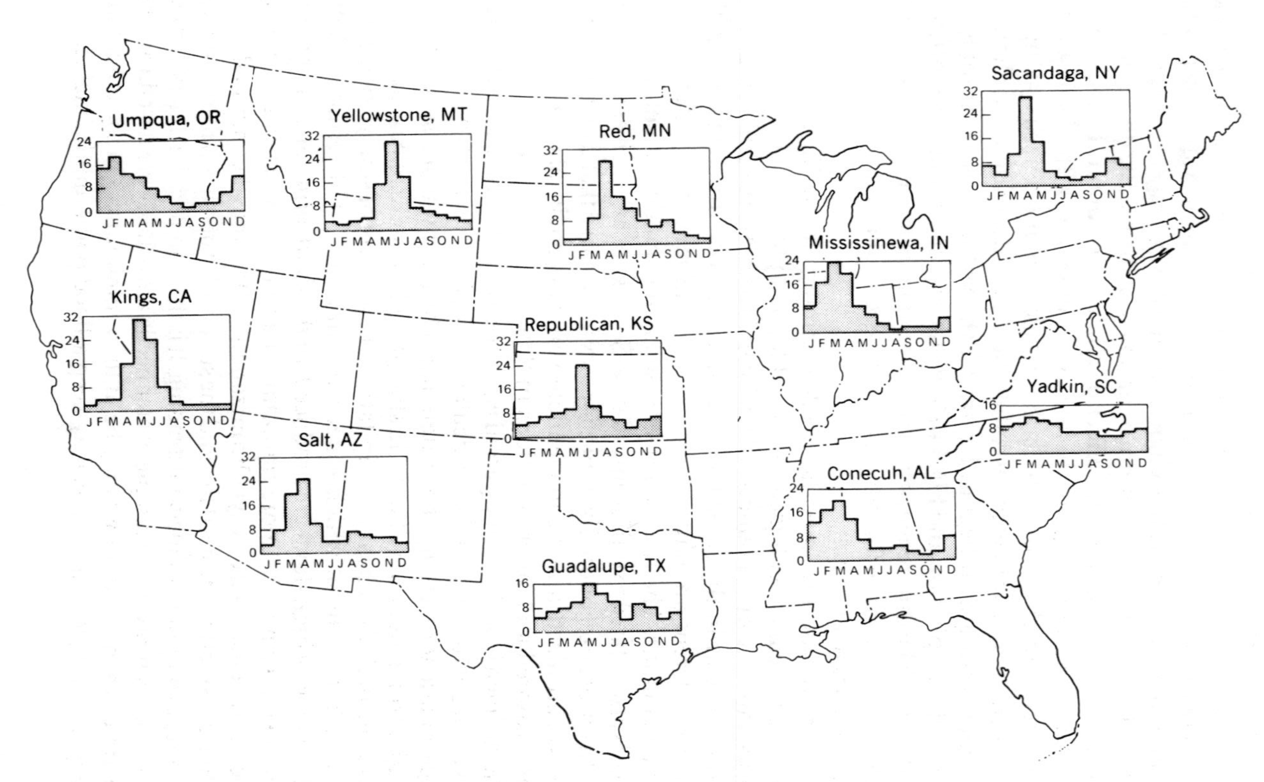

FIG. 1.6 Average monthly runoff (percent of annual flow) of selected rivers in the United States.

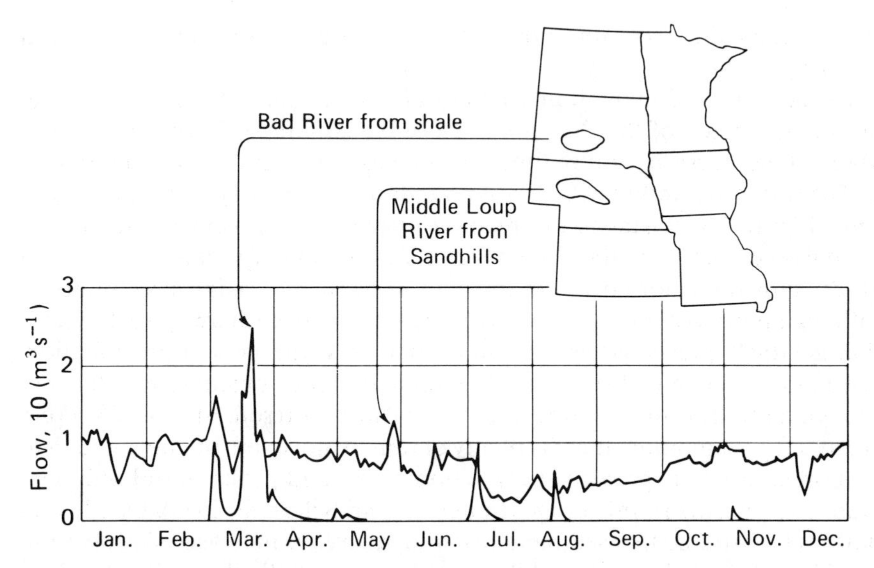

FIG. 1.7 Geology modifies flow regimen from watersheds in the same climatic area (adapted from Fig. 11, U.S. Senate, 1960).

problem. In fact, most essential plant nutrients are absorbed in ionic form from water in terrestrial as well as aquatic ecosystems. However, large concentrations of dissolved substances, particularly salts, may make water unusable for many purposes. If the oceans were fresh instead of salty, large areas of low elevation deserts in Africa, the Near East, Australia, and the Americas easily could be made to bloom. Cities near oceans would be free of water constraints and inland waters that are now diverted to supply coastal cities would be available in the interior.

The solvent properties of water may result in problems of water quality in freshwater rivers, streams, and lakes when water leaches salts from land surfaces that become concentrated when most of the water evaporated. This is why many water quality problems are greatest in arid and semiarid regions.

Another reason that water quality problems develop is that moving water literally has the power to wear down the earth and carry it away. Most of the landforms on earth were sculpted by water in its solid or liquid form or resulted from the deposition of the material that was moved. Again, the problem of erosion and sediment is not that it occurs, but is a matter of erosion rate and sediment concentration.

Where the rate of erosion exceeds the rate of soil formation, as may naturally occur in badlands, the productive capacity of the land is lost. The muddy water itself may directly injure fish, wildlife, and other aquatic organisms and slow down the rate at which a stream can purify itself from organic wastes. When sediment is deposited it may seal the surface of productive soils, fill reservoirs, and aggrade streambeds, increasing the frequency and severity of floods. And there is no question that muddy water is aesthetically repugnant to nearly everyone.

Erosion and sedimentation are natural phenomena. The Missouri River ran muddy much of the time long before the coming of European settlers. Many of the worst natural sediment problems are areas of low precipitation and irregular streamflow, which compounds the severity of water supply problems. Figure 1.8 illustrates the relation of natural sediment production to annual precipitation in the United States. Contrary to expectations, sediment production does not continue to increase with increasing precipitation. Although more water is available to erode soil and carry it away, surface erosion is largely prevented as vegetation increases with greater precipitation. Vegetation is largely absent with less than 10 cm of precipitation, but not enough rainfall occurs to seriously erode and transport soil. At about 25–35 cm of annual precipitation, there is plenty of rain to move soil, but not enough to support continuous protective vegetation cover, and peak natural sediment production occurs in this range (Langbein and Schumm, 1958). Dendy and Bolton (1976) found that sediment reaching reservoirs increased sharply from 0 to 5 cm of annual runoff and decreased exponentially thereafter as runoff reached 100 cm. High sediment yields in dry areas also reflect annual precipitation that often occurs in a few short, intense storms.

High sediment yields are also found in areas of recent tectonic uplift, as on the chaparral covered slopes of the Transverse Range of California, where massive erosion is repeatedly triggered by intense fires on the steep slopes. The highest natural sediment yields in the United States occur in the heavily vegetated north coastal region of California, however. There, highly sheared, easily weathered marine sedimentary rocks have been recently uplifted. Rapid downcutting by streams and heavy winter rains cause oversteepened slopes characterized by recurrent landslides and persistent creep that deliver large quantities of sediment to stream channels (Kelsey, 1982).

Other natural water quality characteristics may limit the usable supply of water. Although quality tends to be worse in arid and semiarid areas and those recently uplifted, such problems as color, hardness, dissolved iron, and others may exist in any region.

1.6 WATER PROBLEMS AND PEOPLE

Water becomes a problem because people are unwilling to accept limits to their activities set by natural water yield characteristics. People tend to have a near-insatiable appetite for growth and consumption that tends ultimately to press beyond those limits, whether this entails overdevelopment of flood plains of rivers or flocking to enjoy the salubrious climate of arid and semiarid regions as in the American Southwest—one of the fastest growing regions of the nation.

Human demands for products of the land; for food, fuel, and fiber, may cause adverse changes in water yield characteristics. Water itself is used in

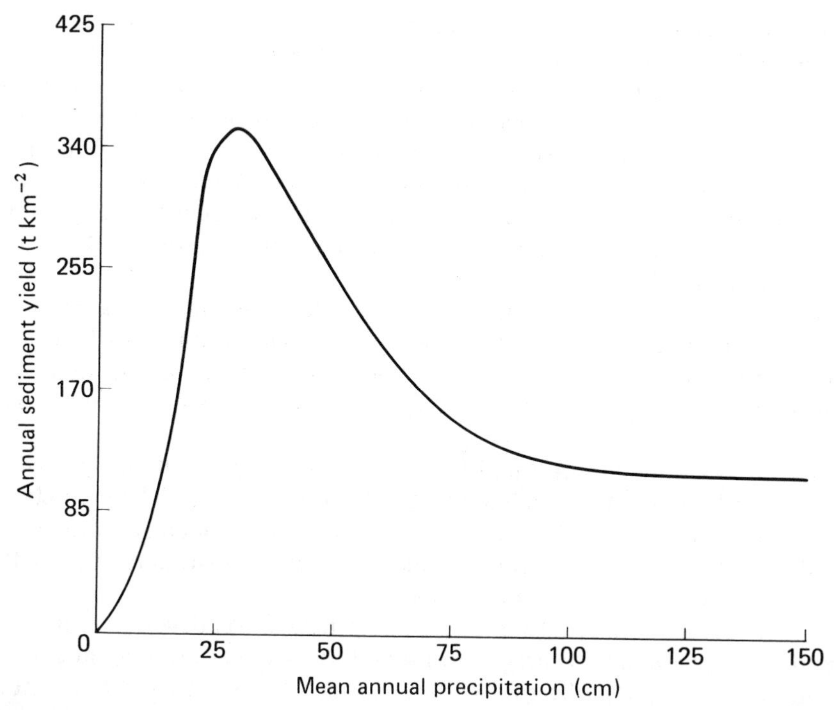

FIG. 1.8 Relationship between natural (geologic) sediment yields and mean annual precipitation. Conditions are worst under arid to semiarid climates (adapted from Langbein and Schumm, 1958).

ways that damage its capacity to supply other services or meet other needs. Some uses of water are mutually exclusive and result in conflict. For example, when water is withdrawn from a stream, it is not available for such instream uses as the production of fish and wildlife. Some uses do not preclude others, but may greatly increase costs, as when water discharged by an upstream user must be intensively treated before it can be used by those downstream. Sometimes the capacity of the water resource to serve humankind is lost through simple carelessness or ignorance.

One type of problem arises because of the human propensity to concentrate in large urban areas. When a region becomes densely settled, and the land is covered with houses, streets, factories, and office buildings, local water supplies may become contaminated, the flow regime may become more irregular, and no longer meet human needs. The supply must then be drawn from the hinterlands. When, as in eastern United States, an almost continuous metropolitan area extends from Portland, Maine to Richmond, Virginia, there is not enough hinterland to provide suitable natural supplies for everyone without conflict and cost.

Even rural areas are not immune to water problems and conflicts. Demand for water for agricultural irrigation has risen rapidly, and, in some areas, has reached capacity. While many areas have had stagnant or even declining populations, other rural areas have shown strong growth as retirees and owners of second homes have sought less stressful, natural environments. Recreational activity has increased in many areas, and water bodies are often a key focal point. Many urban residents who move to or recreate in rural areas have little knowledge of the natural processes that influence water resources, or of the effects of rural land uses such as forest management and grazing. Where basic knowledge and understanding are lacking, problems and conflicts are inevitable.

Conflict and cost are central elements of water problems, that are exacerbated when a large population deliberately chooses to live and work in regions with limited water supplies. No one should be surprised at the strenuous efforts made by the people of the American Southwest to seek water supplies as far away as the Columbia River.

People not only concentrate in areas of short supply, but also persist in occupying flood plains, which are really a part of the river. The predictable consequences are heavy losses of property, and too frequently, human life. Probably, few water problems are as thoroughly misunderstood by the average citizen as the problem of floods.

Floods are a normal part of the life of a river. As a river develops, it erodes a channel large enough to carry most flows. It does not erode a channel large enough to carry the highest flows because such flows are too infrequent to maintain the large channel that would be needed to keep the water within its banks. As a consequence, flow in a normal river fills about one-half of its channel at average flow and flow is less than average about 75% of the time. About once or twice a year the river flows about bank-full. Every few years, however, a much larger than normal storm or excessive snowmelt occurs, forcing flow over the river banks. Then the flood plain, which is an integral part of the river, carries the flow in excess of channel capacity. At rare intervals, exceptional storms generate spectacular quantities of runoff and the flood plain is deeply submerged. Annual flood damages now average over $1 billion annually, and the average annual human toll in the United States in the 1970s exceeded 167 killed (Marrero, 1979).

If humankind has had water problems in the past, it has had only a sample of the future. Though progress is evident on many fronts, much remains to be done. Populations continue to increase. Pressures upon water resources continue to grow. It is clear that few parts of any nation will have an excess of cheap, useable water. All measures of water resource management that can contribute to the solution of present or future water problems will be needed.

LITERATURE CITED

Dendy, F. E. and G. C. Bolton. 1976. Sediment yield-runoff-drainage area relationships in the United States. *J. Soil Water. Conserv.* 31:264–266.

Guldin, R. W. 1989. An analysis of the water situation in the United States: 1989–2040. USDA Forest Service, Rocky Mt. For. Range Expt. Sta., Fort Collins, CO. General Technical Report RM-177. p. 178.

Humphrey, R. R. 1962. *Range Ecology*. Ronald Press, New York.

Kelsey, H. M. 1982. Hillslope evolution and sediment movement in a forested headwater basin, Van Duzen River, north coastal California. In: Swanson, F. J., J. R. Janda, T. Dunne, and D. N. Swanston, Eds. *Workshop on sediment budgets and routing in forested drainage basins: proceedings*. Gen. Tech. Rept. PNW - 141. Portland, OR. USDA Forest Service, Pac. Northwest For. Range Expt. Sta. pp. 86–96.

Langbein, W. B. and S. A. Schumm. 1958. Yield of sediment in relation to mean annual precipitation. *EOS, Trans. Am. Geophys. Union* 39:1076–1084.

Marrero, J. 1979. Danger: flash floods. *Weatherwise*. 32:34–37.

Martinelli, J. Jr., 1965. An estimate of summer runoff from alpine snowfields. *J. Soil Water Conserv.* 29:24–26.

Mitsch, W. J. and J. G. Gosselink. 1986. *Wetlands*. Van Nostrand - Reinhold, New York.

Salo, E. O. and T. W. Cundy, eds. 1987. *Streamside management: Forestry and fisheries interactions*. Contribution No. 57. College of Forest Resources, University Washington, Seattle.

Stults, H. M. 1973. Myths—cornerstones for counterpositions. *Water Spect.* 5(4):9–15.

United States Senate. 1960. *Water resources activities in the United States*. Committee Prints. Select Committee on National Water Resources, pursuant to S. Res. 48, 86th Cong.

Water Resources Council. 1968. *The nation's water resources* (Parts 1–7). US Government Printing Office, Washington, DC.

Water Resources Council. 1978. *The nation's water resources 1975–2000*. Vol. 1 Summary. US Government Printing Office, Washington, DC.

Zubenok, L. I. 1970. Refined water balance of the continents. *Transactions of the Voyeykov Main Geophysical Observatory (Trudy GGO)* No. 263, pp. 79–82. Seen in: *Soviet Hydrology: Selected Papers* Issue 6, June 1970. pp. 516–518.

2 Water Resource Management: Approaches to the Solution of Water Problems

Water resource management includes all the myriad activities undertaken to obtain, distribute, use, regulate, treat, and dispose of water.

Throughout the United States most water resource problems are social, economic, and legal problems of competing uses and water quality rather than those of physical supply. For example, even the limited water supply of the arid Southwest could support a much larger population than at present if only 10% of the water now used for irrigation agriculture were reallocated to municipal and industrial use. But, however economical and legitimate it may be to reallocate the existing supply, the amount of water is not increased by such means. To the degree that any given use is consumptive, that water is simply not available for another use, and even many nonconsumptive uses limit the suitability of supplies for other uses. These problems are most often resolved through the political system and will be considered only as they influence management designed to modify water yields.

A major water conservation activity is to use water with greater efficiency to avoid waste, as when irrigation canals are lined to prevent seepage. Similarly, there exist many technologies, particularly in industry, by which a given supply of water may be reused again and again. However, it should be made clear that while such measures often reduce the amount of water withdrawn from the supply, they seldom reduce the amount consumed or increase the total supply. Water treatment to prevent pollution of water supplies by municipal, industrial, and other sources has long been mandated by law.

Protecting or improving water yields is another management approach. It is difficult, if not impossible, to change any one characteristic of water without changing the others, whether accomplished by engineering methods or land management. Nevertheless, alternative solutions for overcoming water yield problems can most conveniently be considered by separating the problems into three groups:

1. Problems of absolute supply (quantity).
2. Problems of regimen (temporal shortages of supply and floods).
3. Problems of quality.

2.1 PROBLEMS OF ABSOLUTE SUPPLY

Problems of inadequate absolute supply are of two major types: *offstream*, such as withdrawal or consumption for domestic, industrial, and agricultural uses, and *instream uses*, such as for fish and wildlife, recreation, hydroelectric generation, or navigation. No water resource region of the United States is free of at least local problems of absolute supply, but they are worse from the central Great Plains southward and westward across the nation (Guldin, 1989; Water Resources Council, 1978).

Where water yields are insufficient for present and future uses, four possible alternatives for obtaining additional water exist.

1. Importing water from a region where unused supplies exist is usually the first, and most popular alternative considered everywhere in the world.
2. Modification of the weather to increase precipitation is another alternative that excites the imagination and has great popular appeal.
3. Desalinization of saltwater receives strong consideration in many areas.
4. Reduction of evaporative water losses by various means so that more precipitation becomes streamflow, and more streamflow is retained for human use, is steadily receiving more attention.

Each of the above alternatives offers certain benefits and may suffer from limitations depending on the circumstances involved. It makes sense to examine each case.

2.1.1 Water Transfer

Water transfer was the first means used by civilized people to overcome adverse water supplies. Canals and aqueducts have existed in the eastern Mediterranean region since the dawn of history. So long as unused water exists in the watershed that is being tapped, the problems of water transfer are usually those of engineering and economics. Numerous examples of water supply systems that depend on diverting water from one stream system to another could be cited, among them Boston, New York, Denver, and Los Angeles.

At first, when needs are small in relation to supply, simple and inexpensive structures are often sufficient. In general, they consist of diversion and distribution systems. As demands rise, however, structures of increasing cost and complexity become necessary to capture a greater portion of the water. At some point it becomes uneconomic to obtain more water from the same source, even though unused supplies still exist. More will be said about this when problems of regimen are discussed.

Water is usually transferred from nearby sources first, but as local physical or economic limits of supply are reached, sources farther away are drawn upon. Ultimately, several limitations arise. First, costs of transporting water over ever greater distances become prohibitive for many users. Second, water may be moved great distances with little energy demand if the source is at a sufficiently high elevation that gravity may be used. However, with energy supplies for pumping becoming more limited, it is difficult to justify their use to transport large quantities of such a low value commodity as water. Finally, a point is reached where two or more users seek the same supply and conflicts develop.

A question finally arises as to whether any region has a water supply that is not used, regardless of physical supply. When instream as well as offstream uses are considered, it is difficult to conceive of an unused supply. For example, in spring, high flows from the Columbia River system to the sea are needed to trigger downstream migration of salmon and steelhead smolts, and high flows are necessary in summer to maintain suitable temperatures for returning adults. Yet, to many people outside the region, this represents an unconscionable waste of "unused" water.

Until the energy crisis of 1973, grandiose schemes such as the North American Water and Power Alliance, envisioning a giant water transfer system designed to solve the water supply problems of the entire North American continent, received considerable support (Moore, 1965). In the more sober environment of today, however, though we may still accept that what people can imagine, people can do, we question whether they should do all they can imagine.

Nevertheless, imagination is more necessary today than ever, and many proposals, such as towing Antarctic icebergs to water-short regions should not be rejected without careful analysis and possible testing.

Regardless of the problems involved, interbasin transfer of water will continue to be the dominant viable alternative for solving problems of absolute supply. However, some absolute supply problems will remain that cannot be solved by water transfer, and other alternatives should also be examined.

2.1.2 Weather Modification

Ever since Native Americans have danced rain dances, people have sought some means to make it rain. Two things are needed to directly induce precipitation.

First, vapor must condense or freeze to form a liquid or solid. However, these conditions, while necessary, are not sufficient. Witness the numerous days with clouds in the sky (clouds consist of liquid or solid water, or both) when no precipitation occurs. Precipitation also requires that the water droplets or ice crystals grow to sufficient size to fall through rising air and reach the ground.

In 1946 Vincent J. Schaefer and Irving Langmuir dropped dry ice into a cloud and observed snowfall moments later. Shortly thereafter, Bernard Vonnegut reasoned that if water vapor freezes on contact with ice, it might also freeze on other six-sided crystals that simulate ice. Tests revealed that silver iodide could cause ice crystal formation in supersaturated clouds as warm as −5 °C. This was about 10 °C warmer than the temperature where naturally occurring freezing nuclei are effective (Droessler, 1960). When ice crystals are present in the presence of water droplets, they grow at the expense of the liquid. This occurs because the vapor pressure of ice is lower than that of liquid water at subfreezing temperatures and vapor moves preferentially across the vapor pressure gradient toward the ice (see Section 5.2.1.2 for more detail). The ice crystals often grow to a size sufficient to fall out of the cloud. Therefore, silver iodide has the potential to trigger precipitation that might not otherwise occur.

This discovery promptly generated a rash of activity. Newspapers and popular magazines hailed this breakthrough as the answer to saving crops stricken by drought and filling empty reservoirs. Professional rainmakers seeded clouds throughout the world and claims of success ran wild.

Today our enthusiasm has been tempered as we learn more about the complex physical processes that result in precipitation and the difficulties of demonstrating success with weather modification experiments. Both increases and decreases have been demonstrated.

In general, most consistent success seems to come with winter orographic clouds (lifting produced by topographic features) where increases ranging from 10 to 30% have been demonstrated (Dirks, 1983). Cold-top cumulus cloud experiments have shown both increases and decreases due to seeding, but as more is learned about cloud characteristics, the rate of success may improve (National Academy of Science, 1973; Gagin and Neumann, 1981; Woodley et al. 1982). Solid evidence has been lacking of beneficial results from seeding warm cumulus clouds, where the ice phase is not significant to the precipitation process, as in tropical and semitropical regions (Cotton, 1982).

Yet, even during the period when successful seeding of cumulus clouds was taking place in Florida, fire and water shortages in the Everglades made clear a major limitation to all types of cloud seeding. If there are no clouds in the sky, precipitation will not take place, by natural or artificial means.

Furthermore, even if seeding is successful in regions where water shortages exist, a given amount of precipitation does not insure the same increase in streamflow. Rain falling on a dry, permeable soil produces no runoff until the demands of storage and evapotranspiration are met, for runoff is the residual quantity in Eq. 1.2. However, if rainfall is normally distributed, increasing rainfall should result in some increase in runoff. The increase would be greatest, however, during wet periods and in wet regions, when additional runoff is less needed (Crawford, 1965; Satterlund, 1969).

Probably the most important limitations to artificially induced precipitation are the social and legal conflicts that could arise. Among the potential conflicts are those that result from:

1. The perception that seeding robs downwind areas of their precipitation.
2. An increase in snow cover extent, depth and duration.

3. The perception that cloud seeding might cause floods.
4. The possibilities of increased erosion.

The perception that seeding may cause a decrease in precipitation in downwind areas is supported by some unsuccessful cloud seeding experiments (e.g., Neyman et al., 1969), but the notion that successful cloud seeding dries out the atmosphere, thus preventing rain downwind, is not supported by the evidence. Instead, the presence of increases may extend as much as 400-km downwind (Howell, 1977).

The potential increase in snow cover is well supported. The additional snow may clearly reduce the amount and quality of winter range for wildlife, but it would be beneficial to skiers. Avalanches might increase. It seems clear that both liabilities and benefits exist.

It is less likely that cloud seeding would cause significant flooding. There would be little incentive to seed clouds during wet periods when flooding is most likely, and seeding effects in dry weather are unlikely to be great enough to bring runoff to flood stage. Snowmelt floods are determined more by melt patterns than the amount of snow. Nevertheless, cloud seeding was thought by some persons to contribute to the Rapid City, South Dakota flood of 1972 that killed 236 people, though there was little evidence to support the contention. However, conflicts and problems can result from perceptions alone and if enough people think that something is a problem, it automatically becomes one, even if the perceptions are fallacious. This should never be forgotten.

Erosion due to increased precipitation may be increased in arid areas and in agricultural areas where bare soil is exposed. In areas of natural vegetation where precipitation exceeds 35-cm annually, increased vegetation growth should result in less erosion (see again Fig. 1.8).

To sum up, it seems clear that some weather modification is feasible now and it will become increasingly so in the future. It seems equally clear that atmospheric conditions are frequently unsuitable precisely when more precipitation is needed, which no technology can overcome. Finally, conflict and controversy will impose a social and legal limitation even where technological solutions exists. Weather modification will play an increasingly important role in solving problems of absolute supply, but it is no panacea.

2.1.3 Saline Water Conversion

Millions of sailors have learned that salt water can readily be converted to fresh water. It has been done successfully for many years and by many means. Large land-based units producing millions of liters per day from the sea exist in many places such as Kuwait; Guantanamo Bay, Cuba; Freeport, Texas, and others. In none of these installations is cost or available energy a major consideration.

Under most circumstances, however, cost and energy are major impediments to saline water conversion. Only where water is used for very high value purposes, such as domestic consumption, can high monetary or energy costs be justified. For most agricultural and many industrial uses, the value added to product output that may be attributed to water is very low. In many irrigated areas, agriculture is profitable only because the cost of water delivered to the farmer is held down by public subsidies, often to levels of only 1 to 3 cents per ton.

Energy requirements are closely associated with the methods used to desalinate water. Basically, there are three methods of separating salts from water; distillation, freezing, and use of membrane technologies such as electrodialysis and reverse osmosis. In distillation, the water is evaporated, leaving the salts behind, then the vapor is condensed. The energy requirement is very high, 2.26 MJ cm^{-3} at the boiling point. With freezing, the ice crystals consist of pure water, leaving behind an ever more concentrated brine liquid. The ice must first be washed of the brine, and it is then melted. Energy requirements are only about one-seventh of those needed for distillation. Electrodialysis and reverse osmosis separate salt from water across a semipermeable membrane by use of an electrical current or pressure, respectively. Energy requirements depend on membrane design and salt concentration, but are comparable to freezing. These techniques are particularly adapted to purification of brackish water of relatively low salt concentration.

The cost of treatment depends on both fixed capital costs and variable cost for supplies and energy. Solar stills are available, so energy costs are free, but the capital costs needed to capture the solar energy make most large installations economically infeasible. On the other hand, the availability of concentrated energy may greatly reduce capital costs. All in all, little water would be available more cheaply than the 48 cents per ton cost to obtain fresh water from brackish sources in Israel in the 1970s (Buras and Darr, 1979).

Where saline water is available, but fresh water supplies are limited, desalting is a viable alternative for use with domestic supplies and where water produces a high value added to the goods or services produced. This method of securing fresh water will continue to expand greatly in the arid areas of the world. Even so, saline water conversion probably will never supply more than a small fraction of the needs of most nations. Though a 1000 m^3day^{-1} may seem like a vast production of fresh water, it would only meet the average summer daily needs of about 1000 urban dwellers.

Desalinization is not practical in many inland areas where there is no salt water, and transfer of salt water would add to the high cost of production.

2.1.4 Reduction of Evapotranspiration Losses

About two-thirds of all the water falling on the United States returns to the atmosphere before it can reach a stream. In the drier portions of the country where absolute supply is a significant problem, as in the Interior West, average losses are closer to 90–95%. In addition, significant quantities are lost from stream channels, lakes, and reservoirs. Not all of this represents a total loss to humankind, however. Much passes as transpiration through growing crops and forests, producing goods and services as necessary as water itself. Nevertheless, large quantities are lost that provide little or no benefit, as when direct evaporation occurs from bare soils, rivers, lakes, stockponds, and reservoirs, and when weedy vegetation transpires. Even useful vegetation may transpire more water than needed for adequate growth.

Evapotranspiration is the direct consequence of the application of energy to water exposed to the atmosphere. Therefore, all methods for reducing water losses must be based on reducing the amount of energy applied to water, preventing the exposure of water to the air, or both. In other words, separate the water from the energy or the air. This is not a simple task where large amounts of water are involved, but such actions as building roofs over small reservoirs, recharging unfilled ground water aquifers, and applying chemical films to exposed water surfaces are all more or less effective.

A distinction should be made between the losses from surface storage facilities and those that take place on the land. Storage in surface reservoirs always increases evaporation, and the greater the water surface the greater the losses. At Lake Mead, on the lower Colorado River, evaporation removes the top 2 m of water each year. As more reservoirs are built, particularly in arid areas, evaporation losses inevitably increase. The application of measures to reduce evaporation from reservoirs will not increase the absolute amount of water over that existing before storage was developed, but can only hold down the increased losses. In addition, no economic methods are yet known for reducing evaporation from large reservoirs and water surfaces. Materials such as hexadecanol and other fatty alcohols that form a monomolecular film to suppress evaporation are effective only on smaller reservoirs and ponds.

On the other hand, any reduction in the amount of evapotranspiration on land surfaces should ultimately appear as increased runoff. Theoretical considerations indicate that reducing evapotranspiration rates would not be as effective in increasing runoff as would increasing precipitation (Crawford, 1965; Satterlund, 1969), but the feasibility of achieving a measure of control over evapotranspiration is much greater than that of increasing rainfall.

Furthermore, because runoff is a residual, a relatively small decrease in evapotranspiration may result in a relatively large increase of runoff. For example, if precipitation is 53 cm annually and evapotranspiration is 50 cm, runoff is 3 cm. A small reduction in evapotranspiration of only 6% would double runoff (3–6 cm). Reducing evapotranspiration by even 6% may be difficult or impossible in many places suffering water shortages, however. Much more will be said later on this subject, as it is part of the central theme of this book.

2.2 PROBLEMS OF REGIMEN

Many water problems are due to variation in the rate of streamflow from one time to another. Streamflow is usually greatest in winter and spring when human needs are least, whereas it is usually least in summer and early fall when needs are greater. Though it is usually low flows that set the limits of a given water supply for most uses, it is the high flood flows that receive the most attention. This is because high flood flows are spectacularly violent and may take human lives. Therefore, we shall consider first the problem of floods.

2.2.1 Floods

Floods occur when streamflow exceeds the capacity of the channel and overbank flow occurs. Sometimes, a flood is defined as a rate of flow that exceeds a certain threshold amount, but there are good reasons not to use this definition, as will be seen subsequently.

The rate of streamflow (Q) of any stream is a function of the cross-sectional area of the channel (a) times the average velocity (v) of flow, or

$$Q = av \tag{2.1}$$

This simple equation indicates what must be done if floods are to be prevented. Only three alternatives and their combinations exist: reduce the amount of flow, increase the channel cross section, or increase the velocity of flow so that streamflow can be retained within the channel. However, flood damage can be reduced even if floods are not controlled. We will consider flood prevention first.

2.2.1.1 Flood Prevention. Floods can be prevented only to the degree that it is possible to increase the capacity of the channel or to decrease the amount of flow. Either method may be used alone, but in practice they are often used together.

2.2.1.1.1 Increasing Channel Capacity. There are several means to increase the capacity of a stream channel to handle greater flows. The most common way is to increase the height of the banks by constructing levees or dikes. Another means is by dredging the channel to make it deeper. Building or developing a floodway, which is essentially a diversion channel to add to existing channel capacity, is also feasible in some locations, as on the lower Mississippi River around New Orleans. The channel capacity can also be increased by increasing the velocity of flow. This can be done by channel straightening, which increases the channel gradient and may reduce the resistance to flow. Removing obstacles such as fallen trees and other vegetation or debris, or smoothing channel sides and bottom to reduce friction will speed up the flow.

All of the above methods are efficient means of obtaining a high degree of protection for limited areas of high value. Alone, however, they are insufficient to provide continuous protection for long distances along a river. By confining flood flows to the river channels, they prevent storage on the flood plain. During floods, the flood plain is a natural reservoir. Water spreading out from the channel is stored as a temporary lake across the plain, and as long as it remains, it reduces the volume of the floods moving downriver. When flood plain storage is prevented by confining the flow to the channel, the volume and depth of the flood may be increased downstream. For example, it is estimated that progressive constriction of the Mississippi River above St. Louis, Missouri since 1837 caused the 1973 flood at St. Louis to be 2.9 m higher than it would have crested with the same rate of flow in the early nineteenth century (Belt, 1975). This is why defining a flood as a rate of flow exceeding some threshold rate may be misleading.

Other problems also arise from levee building. When the surface of a confined river is much higher than the surrounding areas, drainage from tributary streams and land behind the levee is impeded. There is no place for water to drain unless it is pumped up to the river. Flooding from local runoff is substituted for flooding by the main river.

In some places, building dikes is inefficient because sediment-laden streams are continually *aggrading*, or depositing sediment in the channel. The bottom of the river rises until even with high dikes, the diked channel is no larger than the original. At that time, the river has risen above the surrounding land. If seepage occurs, surrounding areas may become waterlogged, but the greater problem is levee failure during floods. Levee failure may change the course of an entire river overnight. The Yellow River of China has a long history of such catastrophes, with 26 major course changes and more than 1500 levee breaches from 602 B.C. to 1949 (Shen, 1979). No wonder it is called China's "River of Sorrows."

Channelization, or the dredging to deepen, straighten, and smooth a channel to increase flow capacity may also cause adverse effects. Gradients may be increased in the channelized reach, inducing an increase in the rate of erosion of the river and its tributaries, accompanied by aggradation of downstream reaches and increased flooding (Emerson, 1971). *Riparian ecosystems* (the unique communities bordering rivers and other water bodies) and instream habitat are often damaged, and the aesthetics of channelized reaches are repugnant to many people.

To sum up, increasing channel capacity may be a useful and efficient means of protecting local areas of high value, but cannot be relied upon for long reaches. The method usually shifts problems downstream. Shifting a problem from one place to another of lower value may be wise, but it too often is overdone, and most of the benefits are lost. Sometimes new problems are created that are worse than the old problem.

2.2.1.1.2 Reducing the Rate of Flow. Decreasing the rate of flow so that water does not cover the flood plain is the other basic means of flood control. Essentially, this method involves storing water one place instead of another—in the soil, in underground aquifers, in large reservoirs, or a system of small ones—instead of on the flood plain.

However, in some areas, and with extreme floods, it may not be physically possible to reduce flow volume enough to prevent flooding. For example, in the 1937 flood on the Ohio River, 32.64 cm of precipitation fell on a watershed covered with snow equivalent to 0.25 cm of water. Flood runoff was equivalent to a depth of about 22 cm over the entire watershed (United States Senate, 1960). There seems no question that nothing short of stopping the rain could have prevented the flood. The land quickly became saturated and could only retain about 10 cm of water. Little possibility existed for holding more. Reservoirs were quickly filled and were overtopped. Additional storage to completely fill Lake Mead, one of the nation's largest reservoirs, would have been necessary to store the excess water. Nowhere in the heavily populated Ohio River Valley is there a satisfactory location for that much reservoir storage. In fact, using all the means at our disposal, floods have not been eliminated from the Ohio, and probably never will be.

Nevertheless, flood control efforts by building reservoirs will continue, because humans simply cannot tolerate the lack of control exhibited in nature and because there is very good evidence that such flood control is effective. It works. Not everywhere, every time, or to the degree claimed by many of its advocates, but well enough to convince governments, the major providers of flood control.

The effectiveness of storage as a means of flood control depends, among other things, on the type of flood that occurs. Basically, there are two types of floods: flash floods and long-lasting, large runoff floods. The former occur mostly on relatively small watersheds (usually less than 1000 km^2) and result from relatively small volumes of intense precipitation that runs over the land so quickly that the stream channels cannot hold it within their banks. The streams quickly rise and fall, and may be back within their banks in a few days or even hours. Large volume floods may occur on any size watershed and usually result from long rains or melting snow for days on end, and there simply is not enough room in the watershed mantle to store it all. The excess travels through or over the soil to the streams and flow steadily rises until rivers are in flood.

Obviously, the long rain or snowmelt floods require great amounts of storage to control. Figure 2.1 illustrates the great volume of storage that may be needed for even a small reduction in the height (stage) of such floods.

On the other hand, small quantities of storage can result in relatively great reductions in the stage of flash floods as illustrated in Fig. 2.2. Dams and reservoirs provide more efficient protection per unit of storage when the flood crest is sharp than when it is broad.

Using storage for flood control depends on having suitable sites to store water, either on or in the land or in reservoirs. Usable sites are not always available. A watershed with thin soils will have limited storage capacity regardless of land use, and dam sites are limited by topography, geology, and human developments such as cities, farms, roads, industries, and so on. People also may be reluctant to establish artificial flood storage in areas of recreation or fish and wildlife habitat. Thus on most watersheds, flood control is incomplete for lack of suitable places to store water. At best, we should consider our efforts to be flood reduction rather than flood control.

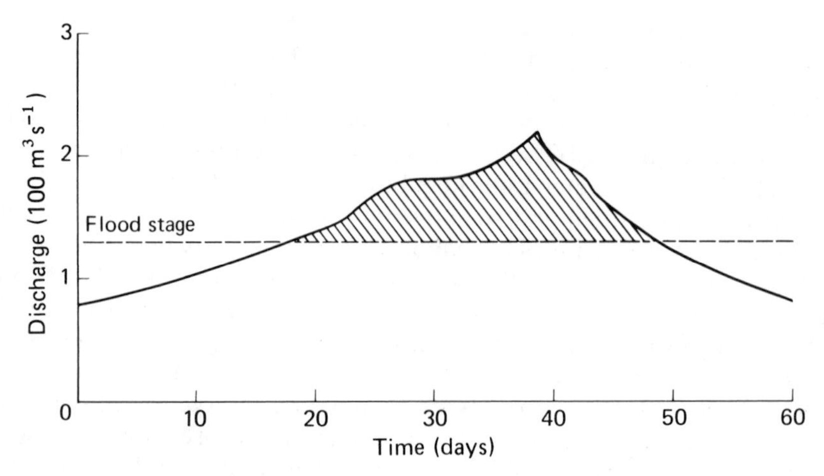

FIG. 2.1 Large quantities of water must be stored to reduce a long-rain or snow melt flood peak. Crosshatch area is proportional to necessary storage capacity.

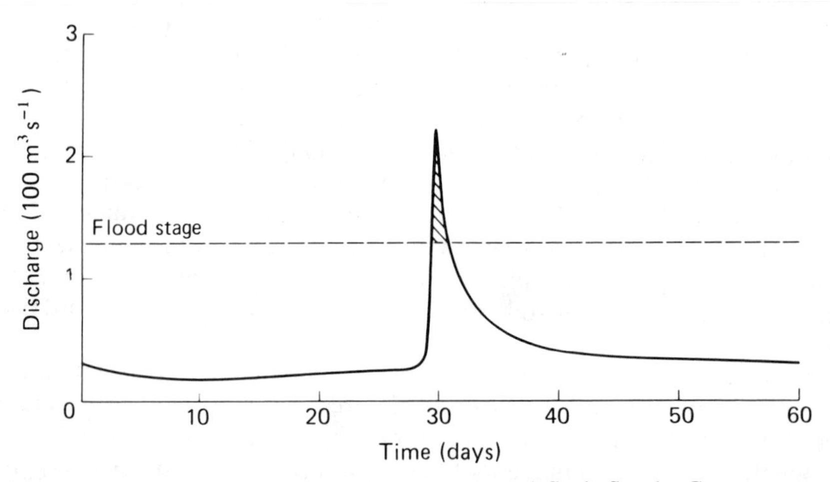

FIG. 2.2 Much less storage is necessary to control flash floods. Compare cross-hatched area of this figure with Fig. 2.1. Peak flow reduction is the same in both cases.

The question of where water should be stored is a recurrent issue whenever flood protection is considered. It also appears as big dams versus little dams or sometimes as dams versus land management. The problem was discussed at length by Leopold and Maddock (1954) and the arguments are still valid. They will not be repeated, but the essential points are as follows.

First, it seems clear that land management is not an effective means to control floods on large rivers. Most of these floods are long rain and snowmelt floods, involving extremely large volumes of water. Even the best land cover cannot increase the capacity of a saturated soil to store water. De Soto observed the Mississippi River to be in flood during his journeys in the sixteenth century, and the greatest flood of the Ohio River at what is now Pittsburgh took place before there was any significant settlement west of the Allegheny Mountains (Hoyt and Langbein, 1955). It seems doubtful that the flood control capacity of the land can be improved much over virgin conditions, and floods occurred then as now.

Second, with respect to dams, only large dams provide enough storage to control large volumes of water. Such dams must be located downstream if they are to control the runoff from a substantial portion of the contributing watershed. But, dams only protect areas downstream. They do nothing to protect upstream areas. Therefore, large dams cannot eliminate the need for small upstream dams if upstream areas are to be protected against flooding.

There are some who insist that a system of small dams in the headwaters can eliminate the need for large dams downstream. The arguments are complex, but thorough analysis has shown that a system of small dams in the headwaters does not have the same flood control capacity as a single large dam with the same amount of storage located just above the point to be protected (Leopold and Maddock, 1954).

To sum up, all methods of flood prevention are incomplete, but all are useful in the appropriate circumstances. No method can control extreme floods or those in large river systems, and all have associated adverse effects. The part that wildland watershed management can play in flood control is limited to small streams, and will be discussed in more detail in forthcoming chapters.

2.2.1.2 Other Methods of Reducing Flood Damage. Flood damage can be reduced by reducing floods, but flood control is not necessary to reduce damage. For example, there was no known flood damage when De Soto was traveling on the Mississippi River during the sixteenth century, for the local Indians adapted their use of the flood plain to avoid losses. Flood damage results because of our use of flood plains for cities, industry, farms, and many other purposes. Most damage could be eliminated if all but essential uses were eliminated. But the 7% of the United States that consists of flood plains now is too valuable not to be used. So long as the economic return from using the flood plains exceeds the losses from floods, it is to society's benefit to use them. However, it is of even more benefit if losses can be reduced.

Methods of reducing flood damages that do not require flood control include such things as (a) identification and mapping of zones subjected to varying degrees of flood hazard, (b) flood forecasting that permits warning and evacuation of people and property, (c) designing structures so they are less vulnerable to damage (so-called "flood proofing"), and (d) regulating use of the flood plain. Flood insurance also does more than just spread the risks from flood damage where participation in the program requires standards for land use and construction in flood plains, and where it discourages unnecessary use (Goddard, 1976; Muckleston, 1976).

Many problems in implementing the above measures exist. It takes detailed hydrologic analysis to delineate areas subject to flooding with any given frequency. Costs are high for nearly all the measures cited, and political will often fails when flood plain regulations are involved. Usually there are long established uses that cannot be suddenly prohibited, and many powerful interests may still profit from flood plain development even if society as a whole loses. Despite these limitations, the potential for reducing flood losses by these means is probably greater than can be obtained by flood prevention. Only in rare instances will wildland watershed management play a role in these measures.

2.2.2 Low Flows

Waste and want are the two sides of the coin of poor regimen. Floods not only cause destruction, but their waters are too often wasted, only too frequently followed by low flows and water shortages. For example, in March, 1913, the highest flood of record occurred on the Sacandaga River at Hope, New York, 906.14 $m^3 s^{-1}$. In September of the same year, flow declined to just 0.45 $m^3 s^{-1}$. An observer could barely see water moving through the rocky channel at such low flow. Very frequently, water shortages, both offstream and instream, are due to such variability in streamflow, rather than any absolute lack of water.

It seems logical that flood control should go hand in hand with increased water supply during periods of shortage. Obviously, only those methods are effective that store flood water for later release. In most instances, this means dams and reservoirs. Storage in surface reservoirs is the primary method of regulating streamflow in the world today.

There is a basic contradiction in using flood control reservoirs to store water for release during low flows. For most effective flood control, a reservoir should be kept as close to empty as possible to maximize flood storage. For most effective low flow augmentation, it should be kept full until release from storage is needed. This limitation may be overcome to the degree it is possible to predict future streamflows. For example, it is possible to predict with reasonable accuracy the seasonal snowmelt inflow to many mountain reservoirs. They may be drawn down to accommodate the predicted inflows. Similarly, summer low flow requirements are broadly predictable, and storage may be accumulated to meet those needs.

However, extreme storms or droughts are not predictable very far in advance, so some unfilled storage capacity must be retained at all times in case of unexpected flood, and drawdowns are seldom maximized lest future water needs not be met. Typically, a certain portion of storage in a multiple use reservoir is designated specifically for flood control, whereas another is designated specifically for low flow augmentation. Thus the reservoir must be of a larger capacity than would be needed for a single purpose or if high and low flows were perfectly predictable.

To many people, it appears obvious that with enough storage, all streams could be made to flow at their long-term average rate of flow. Practically speaking, such uniform flow cannot be obtained. One of the reasons is simply the lack of suitable reservoir sites of sufficient capacity, as discussed with regard to floods. For example, there is now often significant public resistance to dam building due to concerns for existing pristine areas and natural habitat.

Another reason is that, even if suitable sites are available, the amount of dependable flow to be obtained by increasing reservoir storage follows the law of diminishing returns. The first increment of storage will yield a large return in dependable flow, but later increments add less and less additional dependable flow. Additional storage becomes uneconomic long before dependable flow approaches the long-term average flow. In no major river in the United States is regulation sufficient to yield an assured flow much greater than one-half mean annual flow (Piper, 1965). In many places in the world, additional storage remains economically and physically feasible, but in many others it is not.

Another reason that dependable flow cannot reach the mean long-term flow is that evaporation losses are increased by surface water reservoirs. Increasing the amount of storage decreases the total amount of water. This ultimately tends to offset any gain in dependable flow by building reservoirs, and carried to extremes, may even reduce it. In the Cheyenne River Basin above Angostura Dam, stock ponds and reservoirs have greatly reduced the flow of the river through increased evapotranspiration from the exposed water surfaces and adjacent areas wetted by seepage (Culler, 1961). Where the Missouri River flows through the large Fort Peck, Garrison, Oahe, Fort Randall, and Gavins Point reservoirs across the Great Plains, evaporation has been estimated to amount to 20% of the streamflow (Meyers, 1962).

It may be argued that vast underground aquifers, which have been depleted so badly in southwestern United States, could easily store excess waters for later retrieval by pumping. While the vast capacity for storage may exist, the problem is the limited rate at which recharge may be induced. Even in pervious material, floods may pass by before substantial recharge takes place.

The ultimate limitation with storage of excess runoff as a means of improving water supply is that only a small part of the hydrologic cycle is available to work with in most regions. In the more water-deficient regions, total runoff is frequently less than 10% of precipitation, and excess flow may be only a small portion of total flow. Most of the water falling on the land never appears as streamflow, but returns instead to the atmosphere through evapotranspiration. If even a small portion of the massive evapotranspiration losses could be made available as streamflow during low flow periods, it would greatly improve them. Modifying evapotranspiration by wildland watershed management sometimes offers this alternative of complementing storage to improve streamflow, and will be examined in later chapters.

2.3 MODIFYING WATER QUALITY

Much water is unsuitable for many uses in its natural state and more has become unsuitable because of human-caused pollution. No simple statement of means of solving these problems is adequate, however, because of the many different kinds of water quality characteristics and because the use of water determines the quality standards that are necessary.

Where natural water quality is unsuitable for use, the major alternatives for obtaining suitable water are either development of a different supply that is suitable or treating the present supply to make it usable. The alternative of a different supply simply does not exist in many areas. It can be expected to become increasingly unrealistic everywhere in the future. Treatment to meet standards of use is, and will continue to be, the predominant means of coping with unsatisfactory natural water quality. It can be as simple as chlorination to insure health safety or may require a series of sophisticated measures before use.

If poor natural water quality is a problem, it is one that is greatly overshadowed by pollution by people. *Water pollution* may be defined as any impairment of water that lessens its usefulness or degrades its character. Any of the following may contribute to poor water quality (Wolman, 1971).

1. The presence of disease organisms or toxic materials that endanger health.
2. High biochemical oxygen demand and reduced dissolved oxygen.
3. High water temperatures.
4. The presence of solid wastes, organic or inorganic.
5. The presence of dissolved solids, including nutrients, organics, acids, and salts.
6. Sediment and turbidity.

Pollutants may be classified according to their source. *Point sources* of pollution are those that may be traced to a specific outlet, as a sewer pipe or a smokestack. *Nonpoint sources* are diffuse in origin, such as sediment and pesticides that may be carried in runoff from a farmer's field.

People have always used water as a receptacle and transport agent for ridding themselves of wastes. They recognized that most natural waters have some capacity for self-purification of certain wastes and transport others away. But increasing demands require that much of the world's water serve as a source of supply and receiver of wastes simultaneously. Contaminated water led to many outbreaks of disease.

As the epidemiology of disease became better understood, the first water pollution control measures were undertaken to protect public health. By the late nineteenth century sanitary sewers in large cities collected sewage that was treated to kill disease organisms before the effluent was released (Berger and Dworsky, 1977).

As industry, agriculture, and urban areas grew, the capacity of rivers and lakes to purify themselves of the waste loads they received came under increasing stress. The old motto that served so well when demands for self-purification were modest, "the solution to pollution is dilution," was no longer valid under increasing loads of waste and increasingly toxic and persistent chemicals. Many waters were overwhelmed; some rivers became virtual open sewers and some lakes began to die. The major principle of pollution control began to emerge: IF YOU DON'T WANT POLLUTED WATER, KEEP THE POLLUTANTS OUT.

This principle found its strongest expression in the 1972 Amendments to the Water Pollution Control Act (Public Law 92-500), which set as a national goal the elimination of discharge of pollutants into navigable waters by 1985. The act specifically bans streamflow augmentation (pollution dilution) as an alternative to adequate waste treatment, but does recognize it as useful in reducing the impact of nonpoint pollution (Randolph, 1973).

Modern emphasis on water quality problems is upon prevention of pollution. This means that solid wastes are disposed by other means such as burial in landfills or by incineration. Some, such as dried sewage sludge, have proven useful in reclaiming severely disturbed lands, such as strip mines. Dissolved solids in treated effluent, however, are much more difficult to remove. Some municipal sewage treatment plants now remove up to 95% of the biochemical oxygen demand, but most still remove only around 50%. Effluent containing dissolved plant nutrients is still commonly discharged. Acids still escape from mines and mine wastes and irrigation return flows may be laden with salts leached from the soil. Although substantial progress has been made since the early 1970s, many water quality goals have yet to be fully realized, and low flow augmentation remains a major concern (Environmental Protection Agency, 1990; Guldin 1989).

From this short discussion, it should be clear that most of the problems of and most of the solutions to poor water quality lie beyond the scope of wildland watershed management. Nevertheless, there remain many aspects of water quality, particularly nonpoint sources of sediment, solid and dissolved materials, temperature, and chemicals used in wildland management that lie partly or wholly within its domain. Methods of using wildlands and potential agents of pollution on them to maintain water quality from wildlands will be examined in detail in Chapter 12.

LITERATURE CITED

Belt, C. B. Jr., 1975. The 1973 flood and man's constriction of the Mississippi River. *Science* 189:681–684.

Berger, B. B. and L. B. Dworsky. 1977. Water pollution control. *Trans Am. Geophys. Union* 58:16–28.

Buras, N. and P. Darr. 1979. An evaluation of marginal waters as a natural resource in Israel. *Water Resour. Res.* 15:1349–1353.

Cotton, W. R. 1982. Modification of precipitation from warm clouds—a review. *Bull. Am. Meteorol. Soc.* 63:146–160.

Crawford, N. H. 1965. Some illustrations of the hydrologic consequences of weather modification from synthesis models. Paper presented at Symposium on Economic and Social Aspects of Weather Modification, Boulder, CO. July 1–3, 1965.

Culler, R. C. 1961. *Hydrology of stock-water reservoirs in upper Cheyenne River Basin*. USDI Geol. Survey Water-Supply Paper 1531-A.

Dirks, R. A. 1983. Progress in weather modification research: 1979–1982. *Rev. Geophys. Space Phys.* 21:1065–1076.

Droessler, E. G. 1960. The present status and the promise of weather modification. *Trans. Am. Geophys. Union* 41:26–31.

Emerson, J. W. 1971. Channelization: a case study. *Science* 173:325–326.

Environmental Protection Agency. 1990. National Water Quality Inventory: 1988 Report to Congress. EPA 440-4-90-003. US EPA, Office of Water, Washington, DC. p. 227.

Gagin, A. and J. Neumann. 1981. Second Israeli randomized cloud seeding experiment: evaluation of results. *Appl. Meteorol.* 20:1301–1311.

Goddard, J. E. 1976. The nation's increasing vulnerability to flood catastrophe. *J. Soil Water Conserv.* 31:48–52.

Guldin, R. W. 1989. An analysis of the water situation in the United States: 1989–2040. USDA Forest Service, Rocky Mt. For. Range Expt. Sta., Ft. Collins, CO. General Technical Report RM-177. p. 178.

Howell, W. E. 1977. Environmental impacts of precipitation management: results and inferences from project skywater. *Bull. Am. Meteorol. Soc.* 58:488–501.

Hoyt, W. G. and W. B. Langbein. 1955. *Floods*. Princeton University Press, Princeton, NJ.

Leopold, L. B. and T. Maddock, Jr., 1954. *The flood control controversy*. The Ronald Press, New York.

Meyers, J. S. 1962. *Evaporation from the 17 western states (with a section on evaporation rates, by T. J. Nordenson)*. USDI Geol. Survey Prof. Paper 272-D.

Moore, A. W. 1965. Watering a continent (the NAWAPA concept). *Proc. 9th Annual Arizona Watershed Symposium*. pp. 29–32.

Muckleston, K. W. 1976. The evolution of approaches to flood damage reduction. *J. Soil Water Conserv.* 31:53–59.

National Academy of Science. 1973. *Weather and climate modification, problems and progress*. NAS-NRC, Washington, DC.

Neyman, J., E. Scott, and J. A. Smith. 1969. Areal spread of the effect of cloud seeding at the whitetop experiment. *Science* 163:1445–1449.

Piper, A. M. 1965. *Has the U.S. enough water?* USDI Geol. Survey Water-Supply Paper 1797.

Randolph, J. 1973. New commitment for cleaner water. *Water Spect.* 5:9–13.

Satterlund, D. R. 1969. Combined weather and vegetation modification promises synergistic streamflow response. *Hydrology* 9:155–166.

Shen, H. W. 1979. Some notes on the Yellow River. *Trans. Am. Geophys. Union* 60:545–547.

United States Senate. 1960. *National water resources and problems*. Select Comm. on

National Water Resources. Comm. Print No. 4.

Water Resources Council. 1978. *The nation's water resources 1975–2000*. Vol. 1. Summary. U.S. Government Printing Office, Washington, DC.

Wolman, M. G. 1971. The nation's rivers. *Science* 174:905–918.

Woodley, W. L., J. Jordan, A. Barnston, J. Simpson, R. Biondini, and J. Flueck. 1982. Rainfall results of Florida area cumulus experiment, 1970–1976. *Appl. Meteorol.* 21:139–165.

3 Watersheds and People Through Time

3.1 RECOGNITION OF THE RELATION BETWEEN WATERSHEDS AND STREAMFLOW

It is difficult to trace the origin of the recognition of the relation of land use to the flow of water, but no reader of classical literature can long remain unaware that it lies in antiquity. Recognition certainly did not occur until civilization developed to a stage where people's activities and population were sufficient to make a sizable impact upon the nature of the land. Perhaps vague stirrings arose while people were still nomadic herdsmen, for the first concepts of property in land seem to have been applied to springs and wells (Ely and Wehrwein, 1940). However, large populations and the growth of cities awaited the development of agriculture.

Paradoxically, increased food supplies from agriculture increased, rather than decreased, the pressure on wildlands. Forests and forage in the hinterland belonged to those who exploited it, and a growing demand for timber, wool, and goat's milk cheese in the cities sent the woodcutters and shepherds high into the hills. King Solomon was said to have supplied 80,000 lumberjacks to work in the forests, and before 2900 B.C., a temple inscription announced the arrival in Egypt of 40 ships laden with timber from Lebanon (Lowdermilk, 1953). First the cedars disappeared, then the grasses and forbs, and finally, the goats killed even the thorn bushes.

> There are mountains in Attica which can now keep nothing but bees, but which were clothed, not so very long ago, with fine trees producing timber suitable for roofing the largest buildings, and roofs hewn from this timber are still in existence. There were also many lofty cultivated trees, while the country produced boundless pasture for cattle.
>
> The annual supply of rainfall was not lost, as it is at present, through being allowed to flow over a denuded surface to the sea, but was received by the country, in all its abundance, into her bosom, where she stored it in her pervious potter's earth and so was able to discharge the drainage of the heights into the hollows in the form of springs and rivers with an abundant volume and a wide territorial distribution. The shrines that survive to the present day on the sites of extinct water supplies are evidence for the correctness of my present hypotheses. (Plato—"Criteas.")

Despite the observations of Plato, the ideas of his student, Aristotle, dominated ancient Greece and much of the world for many centuries thereafter. Unfortunately, the ideas of Aristotle were completely wrong. He believed that rivers arose in deep, dark, cold caves, where air was transformed into water.

The ancient Chinese had a proverb, "To rule the mountain is to rule the river." A concept of the hydrologic cycle arose there as early as 900 B.C., but had no influence on Western thought at that early time (Nace, 1974).

So, although both China and the Mediterranean region before the birth of Christ seemed to have all the requirements to be the birthplace of watershed management—water scarcity, high demand for irrigation and municipal use, and knowledge of the effects of poor land use—they were not. Furthermore, the rules, the laws, and the customs developed by western civilization applied to the distribution and use of water, but not to the lands that were its source. Early civilization was a city phenomenon, which was unconcerned with wildlands as long as the supplies of wood and wool issued forth. The concept of property did not apply to wildlands, and without it, there could be no concept of responsibility in their use. Land-use anarchy prevailed (a condition markedly paralleled until this century in parts of the American West). Through the time of Rome, deterioration continued, and was occasionally remarked upon.

The record fell silent through the Dark Ages until 1215, when Louis VI of France issued a decree on waters and forests. Whether any positive results were derived is unknown, but in the light of later events, it seems unlikely. In 1342, a community in Switzerland reserved the first "ban" or protection forest, and from the sixteenth century onward, many such forests were proclaimed (Kittredge, 1948). About the same time, Aristotolean authoritarianism was gradually being overcome. The Paulini brothers of Venice correctly accounted for the cause of flooding and silting of the lagoons of Venice (Zinke, 1981). Bernard Palissy of France published a generally correct version of the hydrological cycle in 1563 (Nace, 1974). However, it was not until the 1670s that Pierre Perrault measured and correctly accounted for the major elements of the hydrological cycle; precipitation, evapotranspiration and runoff in the Seine River basin (Biswas, 1970).

3.2 THE AGE OF PROPAGANDA AND ACTION

The eighteenth and early nineteenth centuries were a period of ferment and contradiction in watershed management. The destruction of forests over large portions of the Alps and other mountains gave rise to destructive torrents in France, Austria, and Italy. Articles by engineers, foresters, and others on the beneficial effects of forests on climate, runoff, and springflow filled the literature. Governments on several levels passed laws requiring protection and rehabilitation of forest lands, and at the same time, often imposed such a heavy tax burden on them that they forced their destruction (Marsh, 1907).

The development of ideas and practices in the old world preceded those in the new by 50–100 years, but otherwise, they were quite parallel. In America, the forest also fell rapidly before the onslaught of the axe and plow as the new nation grew. Attention focused on the relation between forest and water, for the rivers and lakes were highways to settlement of new lands, arteries of commerce for their products, sources of power to centers of manufacturing, and the domestic supply for the populace. As early as 1811, the Commissioners of the Erie Canal reported to the New York State Legislature their fears of the effect of forest cutting on the supply of water for the canal.

> . . . Large tracts (for instance, west of the Genesee), which appear as swamps, and through which causeways of logs are laid for roads, will become dry field, when no longer shaded (as at present) by forests impervious to the sun . . .
>
> Thus the summer supply of rivers will be in part destroyed, and in part consumed, whereby their present autumnal penury must be further enhanced. (Rafter, 1905, p. 594.)

Recognition of the relation between land and water was usually followed by a period of propaganda during which causes of problems were eloquently argued and solutions proposed, usually on the basis of authority and personal opinion. The most favored measure to prevent water damage was complete prohibition of forest cutting or grazing; that of improvement was the planting of trees and grass, with or without structural measures on the land. The first extensive empirical measures of rehabilitation were taken in France, where by 1868, 190,000 acres had been reforested and 7000 acres turfed over in the Alps (Marsh, 1907).

In the United States, an early leader in creating a national awareness toward watershed management was George Perkins Marsh, a lawyer, diplomat, scholar, and US Senator from Vermont who in 1864 wrote a book called *Man and Nature*, which was later extensively revised and republished several times under the title *The Earth as Modified by Human Action*. Marsh wrote extensively on the effects of deforestation and stimulated great interest.

New York State in 1872 appointed seven Commissioners of Parks who were directed to inquire into the desirability of vesting title of the timbered regions of the Adirondacks in the State. Among the Commissioners were Verplanck Colvin and Franklin B. Hough. Colvin conducted a triangulation, topographic and hydrographic survey, and his viewpoint regarding forest cutting and water was summed up in a report to the Legislature in 1874:

> . . . unless the region be preserved essentially in its present wilderness condition, the ruthless burning and destruction of the forests will slowly, year after year, creep onward after the lumberman, and vast areas of naked rock, arid sand and gravel will alone remain to receive the bounty of the clouds—unable to retain it.

> The rocks warmed by the summer sun will, like the heated pot stones that serve the savage to boil his food in kettle of bark, throw back the rain as vapor; and the streams that now are icy cold in the shadows of the dark, damp woods, will flow, exposed to the sun, heated and impure. (Colvin, 1874, p. 116.)

In 1876, Dr. Hough, a physician from Lowville, New York, was appointed to prepare the first comprehensive report upon all aspects of forestry for the Congress of the United States. Completed in 1877, about one-quarter of the book dealt with the relations between forests, land use, climate, and streamflow. The prevailing opinions of the day can be summed up as follows:

> It is a matter of common remark, that our streams diminish as our woodlands are cleared away, so as to materially injure the manufacturing interests depending on hydraulic power, and to require new sources of supply for our state canals, and for the use of cities and large towns. Many streams, once navigable, are now entirely worthless for this use. (Hough, 1878, p. 289.)

Also in 1876, the first attempt to create national forests in America was made in the Congress. A bill was introduced that proposed the preservation of forest lands at the sources and tributaries to rivers to prevent them from becoming "scant of water." It failed to pass, but numerous bills were introduced regularly thereafter (Cameron, 1928).

The first major governmental action, however, was taken by New York State, which withdrew from sale its forest lands in the Adirondacks in 1883, and established a forest preserve and a forest commission in 1885 for the ". . . preservation of the headwaters of the chief rivers of the state. . . ."

National forests (forest reserves) were finally established in 1891 by setting aside unallocated public domain in the West, and by 1911, the Weeks Law authorized the establishment of national forests by purchase in the East. The real and fancied relationships between forest and water provided a major impetus and justification for their establishment.

Not so remarkably, the second stage of watershed management, the age of propaganda, was accompanied by an action program of forest protection and preservation in the new world as it had been in the old. It was an age of authoritative opinion, and simple answers to complex questions that were seized upon by a public largely born and raised in the country, but increasingly cramped into the smokey, dirty cities of the industrial East. The remembered virtues of the good old days, of cool streams and sylvan hills, the cleanliness and quiet of the forest, rendered any argument in support of forests, whether valid or not, acceptable.

Everyone, it seemed, had an opinion on the relationship between forests and waters, and early research on forest influences apparently consisted chiefly of opinion polls, which strongly supported the view that forest cutting damaged streamflow (Zon, 1927; Lull, 1963). Opinions did differ, however, and increasingly, people of weighty reputation clashed. The efficacy of the forest in regulating streamflow was questioned by those who pointed out that floods and low flows had occurred before the destruction of the forest, and such flows were a result of the weather, not the woods. The only way to settle the question was by scientific research, which began near the turn of the century, but developed slowly until after World War II.

3.3 SCIENTIFIC BEGINNINGS

In the United States, the first records of federal research effort appear in 1890, when the forestry appropriation for fiscal 1891 and 1892 directed Dr. B. E. Fernow, director of the Bureau of Forestry, to engage in rain-making experiments. Fernow protested the assignment in vain, and apparently the study was conducted, though no results are available. By 1908, the first forest experiment station was established in Arizona, followed thereafter by stations in Colorado, Idaho, Washington, California, and Utah. Forest watershed research, however, really began in 1910 with a cooperative study conducted by the US Forest Service and the Weather Bureau at Wagon Wheel Gap, Colorado (Cameron, 1928).

The Wagon Wheel Gap study became the classical watershed experiment wherein two adjacent undisturbed watersheds were selected, and after a period of calibration (consisting of measuring streamflow from both for a sufficient period, so that the flow from one watershed could be predicted by knowing the flow of the other) one was deforested, and the change in streamflow observed. While the Wagon Wheel Gap study was being conducted, the forest-streamflow issue seemed to fade away. In 1927, however, major flooding struck the lower Mississippi with losses of life and property that were unprecedented, and the controversy of forest and streamflow began anew. In 1928, the final report of the Wagon Wheel Gap study was issued, and while it confirmed that the forest reduced the high flows of spring, the forest also somewhat reduced the low flows of summer (Bates and Henry, 1928). The same year, the McNary–McSweeney bill was passed, expanding the forest experiment station system.

During the years of the Great Depression, the economic dislocation of the country, floods, dust storms, and erosion led to a series of action and research programs in soil and water conservation. The U.S. Forest Service released a major review of watershed concerns and recommended programs that included fire protection, reforestation, grazing management, and research (USDA Forest Service, 1933). The Soil Erosion Service (later the Soil Conservation Service) and the Tennessee Valley Authority were created in 1933. The great floods on the Ohio and Mississippi River systems were followed by the Flood Control Act of 1936, which recognized land treatment as a necessary adjunct to structures for flood control. The Civilian Conservation Corps planted trees, built roads, check dams, and contour furrows and seeded grass on eroding forest and range lands. Research was broadened and accelerated by many agencies. Led by the paper written by Hoover (1944), the studies initiated in the 1930s were reported in increasing numbers and the era of science in watershed management began.

A parallel pattern of climatological research took place from the late nineteenth century until the end of World War II. It moved from descriptive, empirical studies to application of the basic physical laws of energy exchange that drive the hydrologic cycle (Miller, 1969; Caborn, 1973).

But much of watershed management is still characterized by the overthrust of ideas beyond their time. The theory that forests increase runoff and regulate floods is simple and attractive, and superficially self-evident. It can also be extravagantly overdone, and hence public pressures for certain types of wildland use (or nonuse) continue to be based on some theories that are correct, some that are applicable in degree, and some that are completely erroneous.

Wise use and development of wildland resources rests partly upon public understanding of the basic physical, biological, and social relationships of watershed management. To the degree that these are lacking, decisions may be made that are inconsistent with or obstruct the ability of our wildlands to serve our water resource needs. The chapters that follow present the principles and some of the results of scientific research on the relation of wildlands to water, which will help to create this understanding.

LITERATURE CITED

Bates, C. G. and A. J. Henry. 1928. Forest and streamflow at Wagon Wheel Gap, Colorado. Final Report *Mo. Weather Rev. Suppl.* 30:1–79.

Biswas, A. K. 1970. *History of hydrology*. North-Holland, Amsterdam.

Caborn, J. M. 1973. Microclimates. *Endeavor.* 32:30–33.

Cameron, J. 1928. *The development of governmental forest control in the United States*. The Johns Hopkins Press, Baltimore, MD.

Colvin, V. 1874. *Report on the topographical survey of the Adirondack wilderness of New York for the year 1873*. N.Y. Senate Document No. 98. Albany.

Ely, R. T., and G. S. Wehrwein. 1940. *Land economics*. Macmillan, New York.

Hoover, M. D. 1944. Effect of removal of forest vegetation upon water yields. *Trans. Am. Geophys. Union* 6:969–975.

Hough, F. B. 1878. *Report upon forestry*. U.S. Government Printing Office, Washington, DC.

Kittredge, J. 1948. *Forest Influences*. McGraw-Hill, New York.

Lowdermilk, W. C. 1953. *Conquest of the land through seven thousand years*. U.S.D.A., Soil Conservation Service, Ag. Information Bull. No. 99.

Lull, H. W. 1963. Forest influences research by questionnaire. *J. For.* 61:778, 780, 782.

Marsh, G. P. 1907. *The earth as modified by human action*. A last revision of "Man and Nature," 1864. Charles Scribner's Sons, New York.

Miller, D. H. 1969. Develc pment of the heat budget concept, *Yearbook, Assoc. Pacific Coast Geographers*. Oregon State University Press. 30:123–144.

Nace, R. L. 1974. History of hydrology—a brief summary. *Nature Resour.* X:2–9.

Rafter, G. W. 1905. *Hydrology of the State of New York*. New York State Museum Bulletin 85.

USDA Forest Service. 1933. Watershed and other related influences and a watershed protective program. From "A national plan for American forestry." Senate Document No. 12, Separate No. 5. US Government Printing Office, Washington, DC.

Zinke, P. J. 1981. Cumulative impacts on watershed processes and soil productivity. pp. 22–35. In: R. B. Standiford and S. I. Ramacher, Eds. *Cumulative effects of forest management on California watersheds*. Spec. Pub. 3268, Div. of Ag. Sciences, University of California, Berkeley, CA.

Zon. R. 1927. *Forests and water in the light of scientific investigation*. US National Waterways Commission Final Report. Appendix V. Senate Document 469, 62d Congress, 2d Session.

PART II
The Wildland Hydrologic System

INTRODUCTION

The natural elements of the wildland hydrologic system are the atmosphere; soil, subsoil and rock of the upper mantle; the biota, and water. These elements are tied together into a system by the flow of water and energy.

The dominant matter that flows through the system is water, from the vapor in the air to streamflow into the sea (Fig. II.1). Dissolved, suspended, or entrained materials also move with water through parts of the system, and are responsible for many of the system's (and water's) desirable or undesirable characteristics. Most mineral nutrients in terrestrial biota are carried to and within them dissolved in water. The productivity of aquatic ecosystems is largely determined by the amount and kind of dissolved gases and mineral materials. Water quality may be altered by sediments that are torn from the surface by moving water or transported in it, sometimes resulting in deterioration of the water and the surface from which it is removed or upon which it is deposited. Gases and particulates found in the atmosphere may be dissolved or entrained in precipitation, leading to such concerns as acid rain.

Water is unique in that it is not only the dominant kind of matter that flows through the system, but it also represents a major energy flow as well. It is primarily water that ties the flow of energy and matter together. Vapor that rises into the air not only represents the flow of matter, but also of energy by virtue of its physical state and by virtue of its increased height. The vast energy manifested by the awesome power of a hurricane is derived by the condensation or freezing of rising water vapor. The kinetic energy of falling rain and flowing water and ice has provided the power to sculpt most of the face of the earth. In many parts of the world, flowing water is used to generate most of the electric energy to run factories, provide transportation, and furnish heat and light for buildings.

Part II will consider the flow of water and energy through the atmosphere, terrestrial and surface water system. It will focus primarily on the wildland environment; the flow of water through plants and the upper mantle, into the atmosphere and the stream channel. The flow of water in the air and in stream channels will be considered chiefly as it influences or is influenced by the terrestrial wildland ecosystem, or improves our understanding of the land phase of the hydrologic cycle.

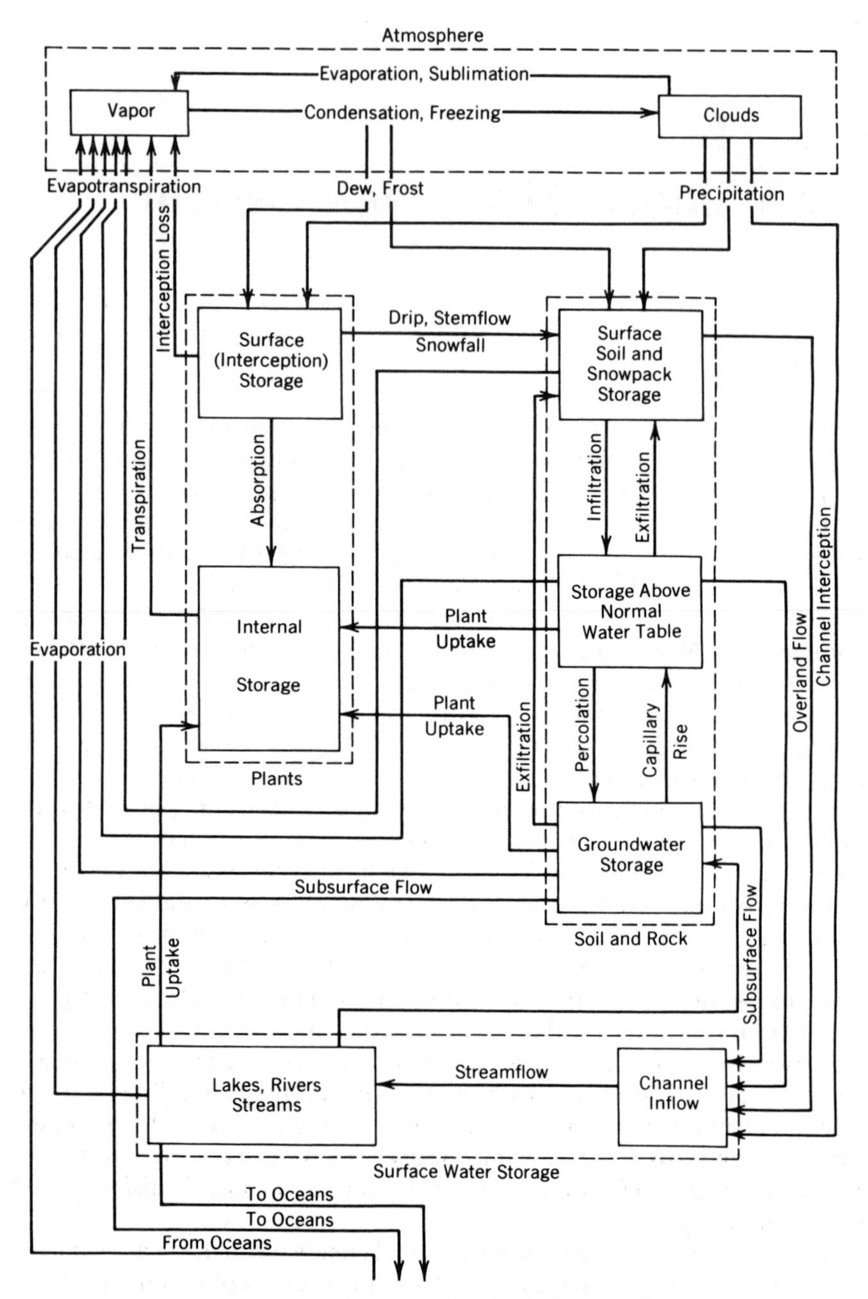

FIG. II.1 Flow chart of the hydrologic cycle.

The energy status of water determines its physical state as solid, liquid, or gas. It also determines how it moves into, through and out of plants, soil and rocks, and erodes the earth's surface. Energy exchange at the earth's surface as it influences evapotranspiration and snowmelt will be considered, as will the energy status of water as it influences its storage and movement in soil and plants. Finally, the principles of erosion as influenced by moving water and gravity, will also be covered.

4 How Runoff Is Generated by a Watershed

The water issuing from a watershed reflects the integrated net effect of all the watershed characteristics that influence the hydrologic cycle. Two aspects of water yields, quantity and regimen, are completely expressed by the hydrograph of the stream as it leaves the watershed. A *hydrograph* of a stream is simply a continuous graphical representation of its discharge or flow over time.

A primary objective of watershed management is to maintain or improve water yields as expressed by the hydrograph. To do so requires an understanding of how a watershed functions in the delivery of water to its outlet.

4.1 STREAMS AND WATERSHEDS

A *watershed* is defined by the stream that drains it. It is simply the area that collects and discharges runoff through a given point on a stream. The term is often used synonymously with *drainage basin* or *catchment*.

There are many different classifications of streams. They may be grouped according to whether they gain or lose water by drainage, the persistence of their flow, their position in the tributary system, their drainage pattern, or their density.

A stream that is fed by inflow into the channel is called an *effluent stream*. It transports drainage from the land. All streams are effluent at their source and many are effluent throughout their length. Thus, streamflow usually increases in a downstream direction as more and more drainage is accumulated. However, some sections of a stream may lose water by drainage from the channel into ground water systems or into the surrounding soil where it is lost by evapotranspiration. Such streams are called *influent streams*. The downstream portions of many streams in arid regions are influent, and some stream sections in humid regions may be influent during dry periods or where they cross highly permeable formations such as porous limestones. Most small stream systems (fourth order or less, see paragraph below) are effluent throughout their length most of the time.

Streams may also be classified by the persistence of their flow as perennial, intermittent, or ephemeral. *Perennial streams* exhibit essentially continuous flow. They flow year round, although during extreme drought, surface water may not be visible if the channel bottom consists of permeable material.

Intermittent streams flow during wet seasons, but dry up during a portion of all normal years. Continuous flow may last from a few weeks or months each year to nearly year round. In wet years they may appear to be perennial. *Ephemeral streams* flow only in direct response to precipitation or snowmelt. They are the most common type of stream in arid regions, but occur in all climates.

Perennial streams always have well-defined channels, as do many intermittent streams, but in humid regions ephemeral channels and some intermittent channels may be poorly defined and are evident only as swales or depressions. However, in arid and semiarid regions all channels tend to be more clearly defined whether their streamflow is perennial or ephemeral.

Streams are also classified by stream order. A *stream order* identifies the position in a hierarchy of tributaries occupied by a stream segment. Any clearly defined channel without tributaries usually is designated as order 1 or as a *first-order channel*. Where two first-order channels join, they form a channel of order 2; where two second-order channels join, a channel of order 3 is formed, and so on (Strahler, 1968). The number of stream segments of each order in a drainage tends to form an inverse geometric sequence. Thus, most stream channels are order 1 and most of the watershed area drains directly into first-order channels (Horton, 1945). Figure 4.1 shows a fourth-order stream system, and lists the number of stream channels of each order and the

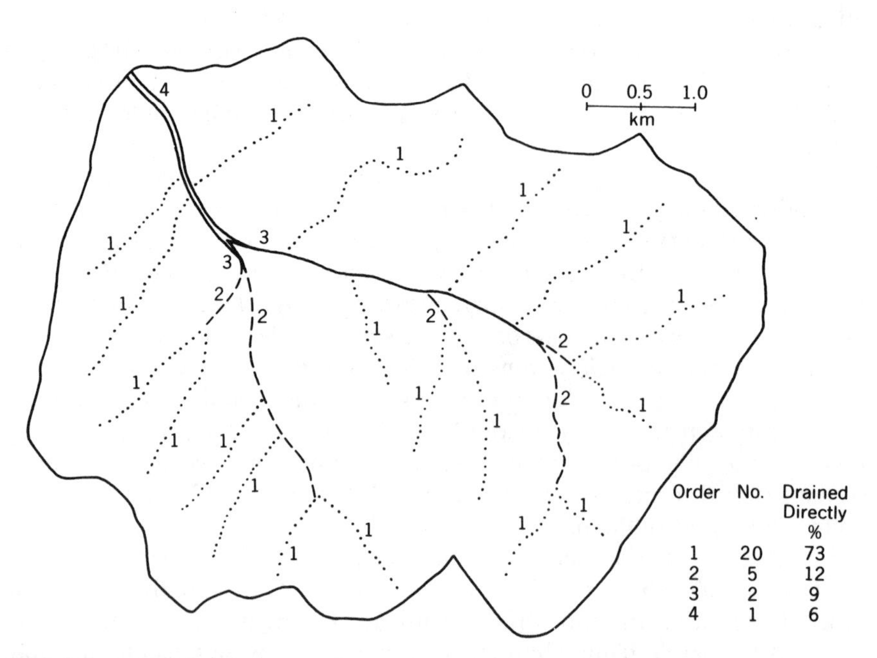

Order	No.	Drained Directly %
1	20	73
2	5	12
3	2	9
4	1	6

FIG. 4.1 A fourth-order, dendritic stream system. Most of the watershed drains into first-order tributaries. The number of stream segments of each order tend to form an inverse geometric sequence.

percentage of area draining directly into each. Obviously, the direct effect of land management is greatest on small stream systems, considered herein to be fourth order or less.

Drainage patterns are of several kinds. The most common are dendritic, parallel, trellis, and radial. Figure 4.1 exhibits a generally dendritic pattern (treelike form where the trunk is the highest order channel), but the first-order tributaries in the upper portion of the figure show a parallel pattern. Trellis patterns are dendritic in form, but tributaries join higher order segments at nearly right angles. Trellis patterns are typical of folded belts of rock. Radial patterns result from drainage of conical or domelike landforms (Paine, 1981).

Drainage density is the ratio of total channel length of all stream orders within a basin to the basin area. In the United States, density ranges from about 2 to 800 km km^{-2} and depends on such factors as climate, topography, and soils or bedrock. Low densities occur where relief is low, soils and rocks are highly permeable, and vegetation is dense. Maximum values are found on badlands where vegetation is lacking, slopes are steep, precipitation is intense, and weak impermeable clays are exposed (Strahler, 1968).

4.2 SOURCES OF RUNOFF

There are two basic types of streamflow; storm flow and base flow. *Storm flow* appears in the channel in direct response to precipitation and/or snowmelt, whereas *base flow* sustains streamflow during interstorm or subfreezing periods. However, in downstream, high-order channels of large stream systems, the distinction becomes blurred, for it may take several days or even weeks for storm flow peaks in low-order tributaries to travel down to the main stem. The routing of channel flow downstream will be discussed after flow from the watershed into the channel is considered.

For many years, hydrologists considered stormflow to result from surface runoff or overland flow, and base flow to be derived from groundwater, based largely on Horton's (1933) pioneering early work on infiltration. More modern studies reveal no such clear distinction among the sources of storm flow and base flow.

A review of the flow chart of the wildland hydrologic system (Fig. II.I) reveals that water reaches the channel system by several routes. Storm flow reaches the channel as channel interception, overland flow, or subsurface storm flow.

Channel interception is precipitation that falls directly into the water that is already in a stream channel. It becomes runoff immediately.

Overland flow or *surface runoff* is derived from precipitation or snowmelt that does not infiltrate the soil. It flows over the surface as a thin film or in small rills before reaching a channel.

Subsurface storm flow or *interflow* comes from rain or snowmelt that infiltrates the soil. It either flows rapidly through highly permeable portions of the mantle or displaces existing water into a channel. The latter process is somewhat analogous to the movement of water through a full garden hose; any water that enters the end attached to the faucet must displace water already in the hose out the open end (Harr, 1976). Displacement can also occur in moist, but unsaturated soils (Abdul and Gillham, 1984). The displacement process is sometimes referred to as *translatory flow* (Hewlett and Hibbert, 1967). Both processes can occur simultaneously.

A part or all of the water entering unsaturated soil may be retained by molecular forces on the surface of soil particles or in soil pores. Only that portion that is not retained can contribute to storm flow or deep percolation. In many storms, only a small portion of infiltrated water contributes to streamflow, if at all.

Base flow is the water that drains from the land to sustain streamflow during dry periods. It has two basic sources. *Groundwater flow* moves out of an underground reservoir of saturated material, the upper surface of which is called a *water table*. The material in which groundwater is stored, and through which it moves is called an *aquifer*. Sometimes a saturated layer may be held temporarily near the land surface by an impermeable or slowly permeable layer, and saturation is discontinuous with depth. Under such conditions, a *perched water table* is said to exist.

Another source of base flow is drainage from unsaturated zones. Slow, continuous drainage of unsaturated soil can be sufficient to sustain streamflow for long periods in the absence of a water table (a water table may exist immediately adjacent to the channel), particularly in steep, mountainous country (Hewlett, 1961b).

It should be made clear that whether a given inflow to the channel represents storm flow or base flow depends on the time it enters the channel in relation to the time of rain or snowmelt, not the precise route that it follows. Although channel interception and overland flow are so rapid that they must almost always cause storm flow, water that moves through the soil and rock mantle cannot easily be separated into each class.

Water draining from areas with perched water tables may contribute to both storm flow and base flow if it reaches the channel during or soon after rain or snowmelt and then continues to drain for a long time. Similarly, unsaturated drainage is most rapid at the high moisture contents prevailing during storms and snowmelt. It then slows markedly as the moisture content decreases, but drainage may continue at very slow rates for weeks or months (Hewlett, 1961a). Many intermittent streams reflect such a pattern of channel inflow.

On any given watershed, one type of runoff may be converted to another, as when overland flow from an impermeable area infiltrates as it flows over a more permeable one. It may then be stored or become subsurface storm flow or base flow. Subsurface flow may be forced to the surface and become overland flow before it reaches a channel.

To summarize, streamflow may result from several sources of flow into a channel. In equation form:

$$RO = CI + OF + SSSF + BF \tag{4.1}$$

where RO is channel runoff, CI is channel interception, OF is overland flow, $SSSF$ is subsurface storm flow, and BF is base flow.

A given hydrograph may contain the contribution of any or all sources of runoff, except that channel interception cannot occur by itself. There must be water in the channel from some other source for it to fall into. The character of the hydrograph is determined by the character of the storm or snowmelt and the timing of channel inflow, modified by the changes that occur as flow travels down the channel to the point of measurement. The effect that each has on the hydrograph may be better understood if each factor is examined separately.

4.2.1 Channel Interception

Channel interception is often a minor factor in the hydrograph when the area of exposed water surface in stream channels is very small compared to the area of the watershed or when storms are light. For example, a watershed that contains 3 km of stream channel of 1-m average width per square kilometer has a water surface of only 0.3% of the total area. With most storms, channel interception would be unimportant, and is often ignored. However, as the channel surface area increases to 1% or more of the watershed area, it rapidly becomes a factor of importance. This is because of two characteristics.

1. All or most of the rain falling in the channel appears on the hydrograph.
2. Channel interception appears on the hydrograph faster than any other type of runoff, the only lag being the time it takes to travel downstream to the point of measurement.

Channel interception appears on the hydrograph soon after rain begins and ends soon after the storm is over, the exact time lag depending on the time it takes to route flow from the various parts of the channel system past the gaging station. Losses from channel interception are negligible because of the rapid movement of water out of the watershed; there is little time for it to be depleted by evaporation. Evaporation rates tend to be low under the cloudy, humid conditions that prevail during and just after a storm.

The surface water area in a channel network is a dynamic feature of watersheds and it naturally follows that the importance of channel interception varies. For example, following a long dry spell, only perennial stream segments contain surface water; then channel interception might be unimportant. Toward the end of a large storm during the wet season, or during snowmelt, every swale and depression may contain its own ephemeral stream, and the amount of exposed water surface may be an order of magnitude (10 times) or more greater. Blyth and Rodda (1973) found that the length of flowing streams in a small (1856 ha) watershed in southeast England varied by a factor of 5 over a 1-year period. Considering that average width must have at least doubled, it is easy to see how channel interception can become a significant source of runoff during long storms in wet seasons, even in semiarid regions.

The stream system shown on most maps is not representative of the active channel system during all flow conditions. Channel widths are rarely given and most maps of humid regions represent perennial and major intermittent flow conditions unless they are of very large scale.

4.2.2 Overland Flow or Surface Runoff

Precipitation falling outside the stream channel is stored on the watershed surface, infiltrates into the soil, or runs off. Only that portion that runs off over the surface is *overland flow* or *surface flow*. The two terms are used synonymously in this book.

There is always some precipitation that is stored on the watershed surface even if the surface is impermeable. For example, it may take several minutes after rain has begun before drip falls from the eaves of a house or water begins to flow down a hard surfaced road. On vegetated surfaces, several millimeters of rain may be stored before substantial quantities reach the ground. Once on the ground, additional water may be stored in surface depressions and the litter.

In addition to surface storage, most water reaching the ground soaks into the soil. Many storms generate no runoff at all except for negligible amounts of channel interception; all the water reaching the ground is stored, either on the surface or in the ground. Overland flow will not occur until surface storage and infiltration requirements have been met. The portion of precipitation that becomes overland flow is sometimes termed *precipitation excess*. In equation form:

$$P - S - I = P_e = OF \tag{4.2}$$

where P is precipitation, S is surface storage, I is infiltration, e is excess, and OF is overland flow or surface runoff. This type of overland flow is often called *Hortonian surface runoff*, after R. E. Horton, one of the dominant figures in early scientific hydrology.

Overland flow is generated more consistently on some parts of a watershed and at some times rather than at others. Rain or snowmelt on impermeable surfaces, such as rock outcrops, roads, and densely frozen soils, generates surface runoff soon after the storm or melt begins. Similarly, sites with a low capacity to store and transmit additional subsurface moisture, such as shallow soils, moist swales or waterlogged areas will be frequent source areas of surface runoff after they become saturated.

Sometimes, water *exfiltrates*, that is, it is forced out of the ground by hydrostatic pressure to become surface runoff. Exfiltrated water is sometimes called *return flow*. Areas of exfiltration are often termed *seeps*. Surface runoff from seep areas or other places where the soil is saturated to the surface is sometimes called *saturation overland flow* to distinguish it from Hortonian surface runoff (Dunne and Black, 1970 a).

The surface need not be impermeable or saturated in order that surface runoff be generated. Whenever the amount of water exceeds the surface storage capacity and it is applied at a rate that exceeds the capacity of the soil to absorb it, Hortonian surface runoff will be generated. On the upland areas of many watersheds covered by vegetation, the infiltration capacity greatly exceeds any precipitation rate that occurs, and hence, surface runoff does not take place. However, where infiltration capacities are limited, as on heavy clay soils, on sites where vegetation is sparse or when soils are frozen, surface runoff may be common (Baker and Mace, 1976; Hibbert 1976). Soils exposed or compacted by logging, grazing, fire, or other disturbances may also generate surface runoff.

Overland flow is the primary source of runoff in the ephemeral streams of most arid regions. It almost always results from high intensity rains that exceed infiltration capacities, or from impermeable surfaces such as rock outcrops. Precipitation that enters the soil in arid regions rarely appears as runoff, for it is stored near the surface and is lost by evapotranspiration.

Regardless of how surface runoff is generated, it appears rapidly on the stream hydrograph. It flows over the surface initially as a thin film, and is impeded in its movement by friction, the stems and litter of vegetation, and natural surface roughness. Surface irregularities soon cause it to gather in rills after moving a short distance, forming *rivulets* (literally: little streams) where resistance to flow is reduced, and it moves quickly to a channel.

Some surface runoff may appear on the hydrograph almost immediately after the beginning of a storm, and it usually ends soon after rain stops. The amount of lag is determined by the velocity of flow over the surface, the distance water must flow before reaching a channel, and the distance and velocity of flow through the channel system to the point of measurement.

On watersheds, where surface runoff occurs frequently, a dense channel network tends to develop, especially in arid and semiarid regions. The *time to peak* or *concentration* storm flow (above base flow) is characteristically short with short overland flow distances and an efficient channel transport system. Where overland flow is rare, as in well-vegetated regions, stream densities are lower and longer overland flow distances and greater flow resistance cause a greater lag in the appearance of surface runoff on the hydrograph. Slope steepness also influences the time of concentration.

4.2.3 Subsurface Storm Flow or Interflow

Hydrologists studying storm flow in forested basins of the Coweeta Hydrologic Laboratory in the southern Appalachians discovered in the early 1940s that overland flow could not account for all the storm runoff being produced. They proposed that infiltrated water moving rapidly through the soil must be responsible (Hewlett and Hibbert, 1967; Hibbert, 1976).

If more water infiltrates into the soil than can be retained in the soil pores, the excess tends to percolate vertically downward. Should the downward flux of water exceed the *permeability* (capacity to transmit a fluid) of a lower layer, a saturated layer will form, or a portion of the downward flux will be diverted laterally.

There are basically two types of subsurface storm flow. Translatory flow displaces water already in the mantle by entry of water into the surface. Soils exhibiting translatory flow may be either saturated or unsaturated, but are usually quite moist. Water can also flow fast enough to appear as storm flow down large pores (*macropores*) or shrinkage joints under the force of gravity without displacing existing soil water (Flugel and Schwarz, 1983). Movement through macropores can occur through soil layers that are far below saturation, but water entry into the macropores usually takes place in a zone that is saturated or nearly so. A combination of the two types of flow can also occur, a process sometimes referred to as *partial displacement* (Thomas and Phillips, 1979). Details of water movement and storage in soils will be discussed more thoroughly in Chapter 6.

Field evidence indicates that inhomogeneities in the soil may be the triggering mechanism in generating subsurface storm flow. They consist of permeability variations that allow shallow saturated conditions to build up. The *hydraulic head* (water with positive pressure and energy potential) may then be propagated through the interconnected filled pore system as translatory flow. In other cases the hydraulic head in the saturated zone may force water into structural and biotic macropores that permit very rapid drainage even through unsaturated layers (Bevin, 1981; McCaig, 1983). The hydraulic head may increase rapidly, caused by a large rise in the water table when a small amount of infiltration converts the *capillary fringe* (saturated zone above the water table held by capillary tension) into water with a positive head (Sklash et al., 1986).

Although soils do not have to be saturated to produce subsurface storm flow, a zone of saturation always exists immediately adjacent to, and slopes toward the channel. Water cannot flow against a hydraulic gradient into the channel. The saturated zone need not extend to the soil surface, however.

Subsurface storm flow does not always flow directly into a channel. Some of it may exfiltrate, where it is converted to overland flow. Where exfiltration occurs, it insures that all rain reaching the surface is converted to runoff. Saturation overland flow may therefore be comprised of a subsurface storm flow component and a direct precipitation component and can occur at a rate faster than the rainfall rate.

Unlike channel interception and overland flow, which always appear as storm flow in small stream systems, water flowing through the soil is identifiable as subsurface storm flow only by the timing of its appearance on the hydrograph. Only that portion of streamflow that can be clearly associated with a given storm or snowmelt period can be considered storm flow. Both translatory flow and flow through macropores can continue long enough that they may be considered as base flow. Therefore, some method to separate storm flow from base flow must be applied.

The simplest technique, developed by Hewlett and Hibbert (1967) is to use a line of arbitrary slope (they found 0.55 L s^{-1} km^{-2} h^{-1} works well in the eastern United States) starting at the time of hydrograph rise (beginning of *rising limb*), extended until it intersects with the *falling* or *recession limb* (period of declining flow after peak) of the hydrograph. All flow above the line represents storm flow; the rest is considered base flow.

Another technique is based on the assumption that although both storm flow and base flow decline in the same manner (the logarithm of flow tends to decline linearly with time), they do so at different rates. Thus, if flow is plotted on the appropriate paper, two straight lines should result, with the point of their intersection identifying when storm flow ends and base flow begins. A similar result, based on the same assumption, can be obtained by calculating a *recession coefficient*, or the ratio of discharge at any given time on the recession limb of the hydrograph to discharge at some fixed time earlier. For many peaks, this ratio will increase until storm flow ends and then becomes essentially constant for base flow depletion.

All these methods are illustrated in Fig. 4.2. In this particular case, there is little difference in the results; when the line of arbitrary slope is used, storm flow ends about 11 h after the storm peak. Plotting the flow on semilog paper yields storm flow cessation about 1 h earlier. The recession coefficient becomes essentially constant about 13 or 14 h after the peak. The volume of storm flow is about the same for all three. Differences among methods are sometimes greater than in Fig. 4.2, but all methods are somewhat arbitrary and none can be considered particularly better than the others.

4.2.4 Base Flow

It is evident from the previous discussion that there is no clear point at which storm flow ends and base flow begins. Neither can the two be distinguished on the basis of the route followed by the water through the soil to the stream channel. At one time, base flow was generally thought to result from the slow drainage of ground water from an aquifer. However, even this distinction was tenuous where perched temporary groundwater formed at shallow depths and contributed readily to storm flow.

Base flow may be sustained for long periods of time by drainage from unsaturated zones as well as from saturated zones. For example, if drainage reduced the moisture content of a layer of soil 1-m deep over an area of 1 km^2 by just 1% over a 30-day period, it would sustain an average base flow of 3.85 L s^{-1}. Hewlett (1961b) has shown that such drainage does occur.

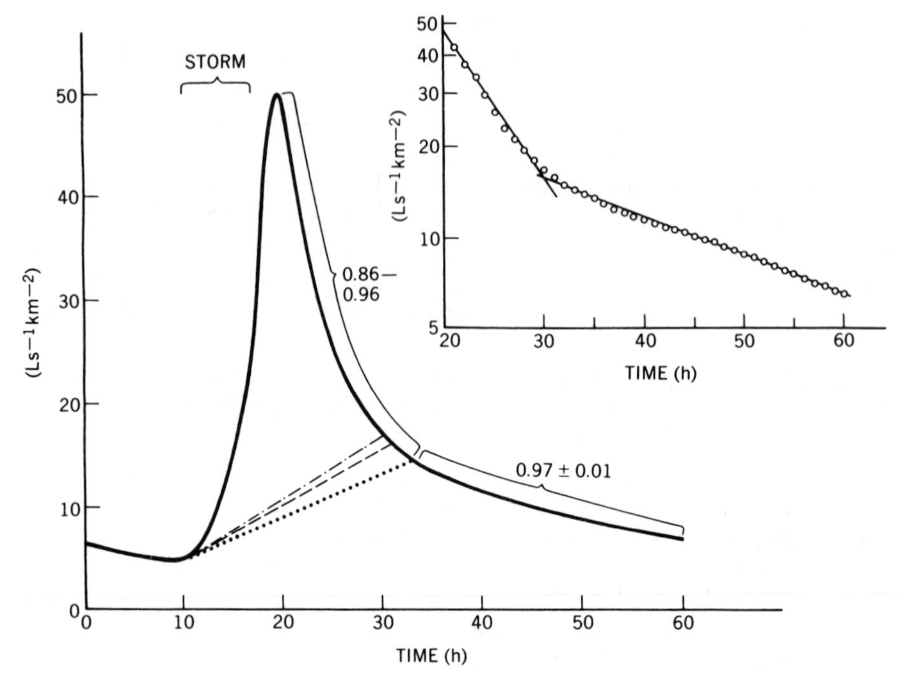

FIG. 4.2 Hydrograph separation into storm flow and base flow by three different methods. Dashed line is an arbitrary slope of 0.55 L s^{-1} km^{-2} h^{-1}, dotted line slopes to point where the recession coefficient becomes essentially constant, and the dot–dash line is derived from plotting the recession leg on semilog paper (insert figure).

The soil moisture or groundwater that supplies base flow is not always replenished by every storm or period of snowmelt that generates storm flow, as may be implied from some hydrograph separation techniques. This is particularly the case when storm flow is comprised of channel interception or Hortonian overland flow. However, base flow is usually increased subsequent to substantial storm flow or whenever subsurface storm flow is a clear component of storm flow.

4.2.5 The Hydrograph of Stream Systems

The nature of any hydrograph at a given location in a stream system is the integrated result of two factors: (a) the amount and timing of flow into each channel segment, and (b) translations in flow from each segment as it travels downstream to the point of measurement.

The amount and timing of inflow to each channel segment depends on the precipitation amount and timing and the flow path to the channel. As most water drains directly to first-order channels (Fig. 4.1), flow paths from the land tend to be short. Within most small stream systems, first-order watersheds tend to respond similarly to rainfall, so the pattern of inflow tends to reflect local rainfall characteristics. The same cannot be said for snowfall, however. Snowmelt may vary sharply with changes in slope direction and with elevation, particularly in mountainous topography.

Once water enters any channel, it flows downstream. The time of travel to the gage is a function of flow velocity and distance to the gage. It may be assumed that average velocity is approximately the same from each first-order tributary to the gage in most small stream systems, unless the basin is quite heterogeneous. Therefore, if inflow into first-order channels is uniform, the shape of the downstream hydrograph will largely reflect the frequency distribution of distances from each first-order channel to the gage (Rogers, 1972).

All sharp peaks from any channel segment tend to flatten as they move downstream. Flattening of the peak results primarily from changes in dynamic storage within the channel as the stream surface (or *stage*) rises and falls. For any given stream segment at any given period, successively more water enters storage with a rising stage. Since outflow equals inflow minus changes in storage, less water flows out than enters the segment in a given time. Therefore, part of the peak is delayed in its appearance in the next segment downstream. Thus the peak is both attenuated and delayed. Figure 4.3 shows how both distance and translation modify the hydrograph downstream.

The farther downstream in any stream system, the less important the discharge from any single tributary becomes in determining the nature of the hydrograph. The relative amount of flow contributed by the tributary declines rapidly downstream, its shape is dampened, and its identity is ultimately masked completely. Thus, the nature of inflow into the channel system is important in determining hydrograph characteristics only in low-order streams. It naturally follows that wildland management, which can affect primarily inflow into the channel system, would be most likely to have an important effect only on flow in small stream systems.

4.3 WATERSHEDS SHOW A VARIABLE RESPONSE TO PRECIPITATION

In the early 1960s, John D. Hewlett and his associates at Coweeta introduced a concept that may be considered one of the most important advances in modern hydrology; the *variable source area* concept of watershed runoff (Hewlett, 1961a). This concept of the runoff process has also been referred to as the partial source area and dynamic contributing area concept, but both are essentially the same as variable source area, which seems to be the term becoming established in the literature. Mathematical and physical models and field studies have established the essential validity of this newer concept of the runoff process (Hewlett, 1961b; Hewlett and Hibbert, 1963; Whipkey,

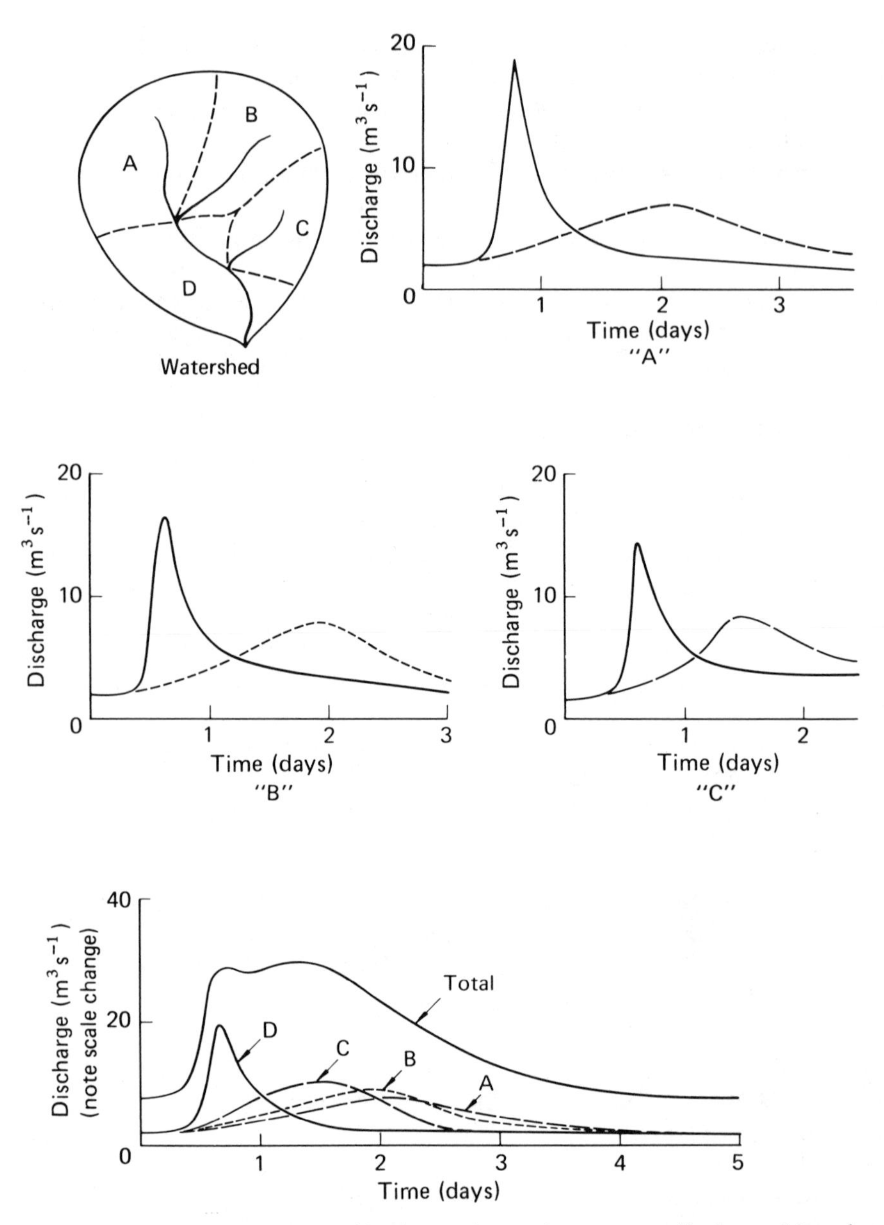

FIG. 4.3 Effect of translation of hydrographs on downstream discharge. Note that contribution of area *D* is not translated. Change in hydrograph shape is due to different lag times from tributaries to the mouth (desynchronization) and change in the shape of tributary contribution as it moves down the channel.

1965; Horton and Hawkins, 1965; Betson and Marius, 1969; Weyman, 1970; Dunne and Black, 1970 a, b, 1971; Black, 1970; Freeze, 1972, 1980; Scholl and Hibbert, 1973; Stephenson and Freeze, 1974; Engman and Rogowski, 1974; Martinec, 1975; Zalesskiy, 1976; Lee and Delleur, 1976; Beasley, 1976; Palkovics and Petersen, 1977; de Vries and Chow, 1978; Feller and Kimmins, 1979; Mosley, 1979, 1982; Bouma, 1981; Bevin, 1982; Bevin and Germann, 1982; Smith and Hebbert, 1983; Bevin and Wood, 1983; and others).

Essentially, the variable source area concept states that most of the time only a portion of any watershed actively generates runoff in response to precipitation and/or snowmelt (hence: partial source area). However, the portion generating runoff changes, growing during a storm or snowmelt and shrinking after the end (hence: variable source area or dynamic contributing area). The concept will be explained in more detail below.

Prior to a storm or snowmelt, water is not distributed uniformly over a watershed. For example, after a long dry spell, soil moisture on most of the watershed has been depleted by evapotranspiration and drainage. A free water surface may exist only in the perennial stream channel. A small area next to the channel contains a saturated zone, fed by slow, deep drainage from the larger pores of moist, but unsaturated soils from above. Farther from the channel, drainage of large pores has been completed (temporary or *detention storage* is exhausted), although the smaller pores retain moisture against the force of gravity by capillary or matric forces (*retention storage* space is filled). At greater distances above the channel, evapotranspiration has depleted retention storage from an increasing portion of the root zone. Near the watershed divide, nearly all retention storage space is depleted of moisture (Fig. 4.4).

At the beginning of a storm, only a small area including and adjacent to the perennial channel can contribute to channel inflow, in the absence of surface runoff due to precipitation excess. Elsewhere, detention and retention storage space is recharged as precipitation enters the soil. In the soils in which saturated or unsaturated drainage is occurring at the beginning of the storm, recharge by precipitation may induce translatory flow immediately. However, some subsoil zones away from the stream channel exhibiting drainage may be separated from the surface by soil in which retention storage space has been depleted. Drainage from these subsoils does not respond to precipitation until retention storage space in the soil above has been filled, unless gravity flow takes place directly through the macropore system.

As the storm continues, a larger and larger area contributes to runoff. The stream system expands as it receives more water and intermittent and ephemeral channels become active. Swales and other topographic hollows may receive moisture draining from above as well as direct precipitation. If they are unable to transmit it all, a saturated layer may develop and rise toward the surface. Overland flow may begin as areas become saturated with water, mostly in hollows and swales, moist areas near expanding channels and areas with thin soils (Fig. 4.4). On the drier upland areas, most of the infiltration initially recharges depleted soil moisture zones and is retained, but as the

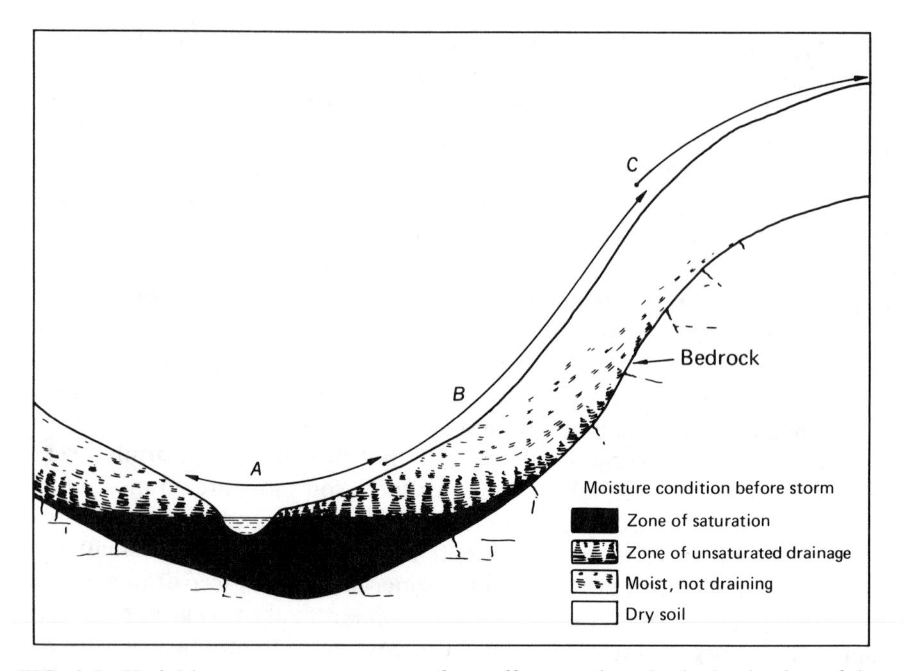

FIG. 4.4 Variable source area concept of runoff generation. At the beginning of the storm, Area *A* immediately begins to generate runoff. As the storm continues and retention storage capacity of the soil is exceeded, the area generating runoff expands as shown by *B*. Area *C* does not contribute to runoff if the storm ends before recharge of retention storage is completed.

storm continues, the mantle may begin yielding subsurface drainage over a larger and larger area as a greater and greater portion of the retention storage capacity is satisfied (Fig. 4.5). Whether drainage on upland areas well removed from the channel contributes to storm flow or base flow depends primarily on whether translatory flow takes place, the steepness and permeability of the mantle, and the length of the storm.

If a storm is short, only a small part of the watershed is likely to be effective in generating runoff, either as storm flow or base flow. Ultimately, the entire watershed may generate storm flow if the storm lasts long enough and antecedent conditions have been wet. Once all the watershed contributes, the rate of runoff approaches the average rate of precipitation (Rothacher et al., 1967). Although rare in the eastern United States (Hewlett et al., 1977), streamflow generation from most or all of a watershed is common in winter in the coastal forests of the Pacific Northwest during long storms. Harr (1977) found that storm flow averaged 38% of precipitation for seven winter storms that were ". . . low to moderate size at best."

The variable source area concept does not apply just to subsurface storm flow. There seem to be few exceptions to the process, regardless of the path followed by rainfall until it enters the channel. Clearly, channel interception parallels the expansion and contraction of the surface area of any stream system. Even with relatively uniform rates of precipitation, Hortonian surface runoff seems to be generated more frequently and in greater amounts near the channel in arid and semiarid areas. This is apparently due to the lesser opportunity for infiltration near the channel (Lane et al., 1978; Pilgrim et al., 1978; Yair et al., 1980). With nonuniform infiltration capacities over a watershed surface, any change in precipitation rate will be accompanied by a change in the area generating Hortonian overland flow. Saturated overland flow is also clearly related to changes in area where the zone of saturation intersects the watershed surface.

When a storm ends, a reverse process of shrinkage of the watershed area generating runoff begins. Channel interception stops immediately. Hortonian surface runoff is increasingly absorbed before it can reach a channel, but exfiltration may continue to generate decreased amounts of saturation overland flow. Water yielded by translatory flow and rapid gravity flow through macropores is not replaced by infiltrating water, and the contributing area shrinks as downslope drainage cannot replace all water lost to the channel. As the most mobile water drains, storm flow stops and only base flow continues at a declining rate. Meanwhile, water is depleted from retention storage by evapotranspiration, creating new storage space that must be largely replenished before substantial subsurface flow can be generated by the next storm. Finally, the system may decline until only perennial channels, fed by drainage from small moist areas, remain. When another storm begins, the process begins anew.

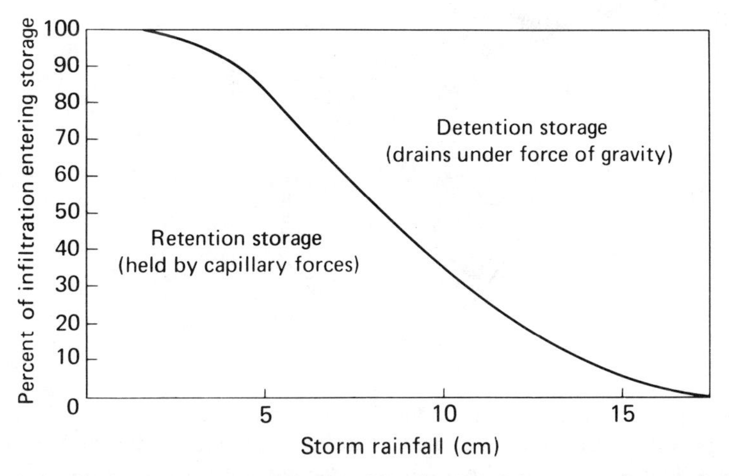

FIG. 4.5 Change in proportion of infiltration entering retention and detention storage on a watershed as the storm rainfall accumulates.

The pattern of runoff generation varies from storm to storm, partly because each storm is different. In winter, precipitation may fall as snow, which is retained in storage until it melts. But even if all storms were uniform rain storms, variation would still exist due to differences during interstorm periods that lead to initial differences in the amount and distribution of moisture within the watershed. Both the amount of drainage and the amount of evapotranspiration depend on the length of the interstorm period. Evapotranspiration rates also vary with the weather. Despite all the variation, the runoff response of a watershed varies in a recognizable pattern with season (Fig. 4.6).

In most temperate areas, summer is a season of moisture depletion, soils dry rapidly, and the rate at which contributing areas expand in response to rain is relatively slow. A like situation prevails in the tropics where there is a dry season. With the cooler weather of autumn, or in the tropics with the onset of the wet season, recharge of storage begins to predominate over depletion, soils

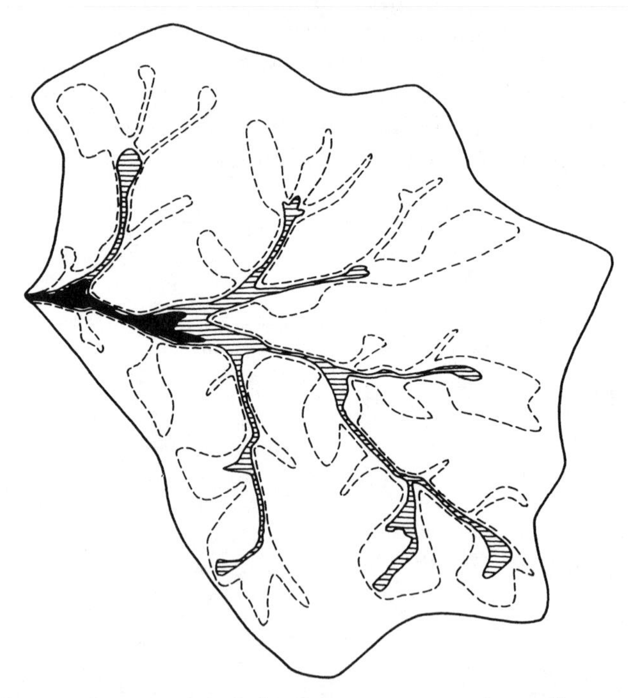

FIG. 4.6 Pattern of seasonal variation in source areas of runoff generation. Solid, source of perennial streamflow; hatched, source of late winter, spring, and early summer intermittent flows; dotted lines encompass source areas generating ephemeral flow during wet seasons. The entire watershed may be active in generating runoff for a few days during long storms during wet seasons or during the snowmelt season.

remain moist longer, and the watershed responds more quickly and completely to each succeeding storm. Where winter precipitation is in the form of rain, low water losses may permit recharge to become and remain nearly complete, and large portions of a watershed yield copious quantities of precipitation as streamflow. In the moist tropics, heavy rains in the wet season keep watersheds highly charged with moisture all the time. In spring, or within the tropics with the end of the wet season, depletion begins to dominate again, and the watershed responds to precipitation with less and less runoff.

Where a snowpack develops, streamflow tends to lag until melt. If autumn rains are insufficient to recharge storage before snows begin, and freezing conditions persist, streamflow may decline to its yearly low in late winter in northern regions. In spring, the first snowmelt is used to complete recharge and only after recharge is largely completed does streamflow rise substantially unless frozen soils prevent infiltration. In many regions, the period of recharge may be very brief; only a few days in spring. Since the snow line in mountainous regions moves upward from the stream channel with time, in spring a watershed may respond to rain by expanding its source areas from moist channel areas upward and downward from the melting snowpack zone simultaneously.

Because the calendar year does not coincide with hydrologic seasons, hydrologic years frequently differ from calendar years. In the United States, the Geological Survey uses a *water year* running from 1 October to 30 September, which conforms approximately with the end of the season of soil moisture depletion, when storage is least. Some hydrologists use a hydrologic year that begins in spring, near the start of the growing season because soils tend to reach a more consistent moisture content at the moist than at the dry end of the moisture scale.

Variations in the manner in which watersheds respond to precipitation by generating streamflow also occurs over longer periods of time. During droughts, only a small portion of precipitation appears as streamflow; most goes to replenish depleted retention storage. During periods of above-normal percipitation, parts of a watershed that seldom contribute to streamflow become consistent generators of runoff. The drier the region, the smaller the portion of a watershed that is consistently effective in generating streamflow. The remainder may contribute so little and so infrequently that it may be misleading to call the entire area a watershed at all. In many northern and mountainous areas of western United States where a snowpack develops, the only time that runoff is consistently generated from deep, well-drained upland soils is during and just after snowmelt each spring.

Perhaps this is why some hydrologists prefer the term "catchment," for it can be seen, on the basis of the variable source area concept, that the term watershed can often be a misnomer. Nevertheless, the term watershed is so deeply entrenched in American literature that it will be retained in this book. But, the word should be used with a clear understanding of its limitations as a description of the runoff process.

4.4 IMPLICATIONS OF VARIABLE SOURCE AREA THEORY TO WATERSHED MANAGEMENT

Since the time of Plato it has been known that watersheds and streams are linked into one interrelated system. The variable source area concept reveals, however, that all parts of a watershed are not equal when it comes to generating streamflow. Some portions of the land are more tightly linked to the stream system than others. It is not just a simple matter of distance from a perennial channel; some areas away from the channel may be more consistent runoff generators than others nearby. Source areas represent the interface between the land system and the stream system, regardless of location, and hence they are key areas with respect to watershed management.

Any management activity on a source area is far more likely to exert a clear and rapid impact on the water resource, for good or for ill, than the same action taken elsewhere on the watershed. Drainage water from roads, discharged onto source areas, is almost certain to reach a stream, and may carry whatever entrained or dissolved materials are contained within it. Harvesting trees on source areas should lead to lesser recharge requirements before streamflow is generated. Pesticides applied on source areas might contaminate the stream system, whereas those applied elsewhere might be quite harmless to the water resource. Bren and Turner (1980) show how timber harvest on source areas can change streamflow regimen, making it less stable.

The importance of source areas to both land and water systems makes their recognition a primary requirement for rational watershed management. The lack of recognition of source areas in the design and conduct of many older watershed experiments seriously limits their application beyond the watershed upon which they were conducted.

One key to the recognition of source areas is that they tend to remain, or quickly become highly charged with water. Riparian areas are usually source areas, but source areas go beyond riparian zones. They include the more moist sites and others, such as shallow soils, that require little recharge of retention storage before they begin generating runoff, by whatever route. They may be identified by many techniques. Many early studies relied upon direct measurements of flow into trenches cut into slopes (Whipkey, 1965; Dunne and Black, 1970 a; among others). Piezometers and wells were used to observe groundwater and measure hydraulic gradients (Holbo et al., 1975). Seeps and saturated soils extending to the surface may be observed directly (Dunne et al., 1975; Bevin, 1978).

Topography is closely related to the distribution of moist source areas. Water moves downward and concentrates where slopes converge (Anderson and Burt, 1978; O'Loughlin, 1986). However, soil water convergence may not always match surface topography (Anderson and Kneale, 1980; Heerdegen and Beran, 1982).

Soil properties are closely related to processes of infiltration and drainage. Coloration and mottling, the presence or absence of pans, texture, structure, and depth may all be used to infer locations of source areas (England and Holtan, 1969). Subsurface geologic characteristics may determine the flow paths of subsurface water and may be useful for locating source areas (Burroughs et al., 1965; Yamamoto, 1974).

Plants may also be used to indicate the location of source areas. Many species are closely associated with the presence of saturated areas and the persistence of readily available soil moisture. For example, in the northern Rocky Mountains, soils in the *Thuja plicata-Athyrium filix-femina* habitat type are almost always wet (Daubenmire and Daubenmire, 1968). Any plant species that consistently indicates wet or moist sites would be useful in this regard (Pole and Satterlund, 1978; Roberts et al., 1980; Gurnell, 1981; Comeau et al., 1982; Winkler and Rothwell, 1983).

Source areas characterized by Hortonian overland flow may be identified by erosion patterns where soils are not heavily vegetated (Gleeson, 1953). Rill and gully densities tend to be high where surface runoff is common. In humid areas where vegetation is more dense, it is more difficult to detect Hortonian surface runoff source areas.

Recognizing source areas of watershed runoff is an essential early step in all watershed management planning. The manager should use every tool available to insure that these critical areas linking the land and water systems are properly identified.

LITERATURE CITED

Abdul, A. S. and R. W. Gillham. 1984. Laboratory studies of the effects of the capillary fringe on streamflow generation. *Water Resour. Res.* 20:691–698.

Anderson, M. G. and T. P. Burt. 1978. Toward more detailed field monitoring of variable source areas. *Water Resour. Res.* 14:1123–1131.

Anderson, M. G. and P. E. Kneale. 1980. Topography and hillslope water relationships in a catchment of low relief. *J. Hydrol.* 47:115–128.

Baker, M. B., Jr., and A. C. Mace, Jr. 1976. Factors affecting spring runoff from two forested watersheds. *Water Resour. Res.* 12:719–729.

Beasley, R. S. 1976. Contribution to subsurface flow from the upper slopes of forested watersheds to channel flow. *Soil Sci. Soc. Am. J.* 40:955–957.

Betson, R. P. and J. B. Marius. 1969. Source areas of watershed runoff. *Water Resour. Res.* 5:574–582.

Bevin, K. 1978. The hydrological response of headwater and sideslope areas. *Hydrol. Sci. Bull.* 23:419–437.

Bevin, K. 1981. Kinematic subsurface stormflow. *Water Resour. Res.* 17:1419–1424.

Bevin, K. 1982. On subsurface stormflow: Predictions with simple kinematic theory for saturated and unsaturated flows. *Water Resour. Res.* 18:1627–1633.

Bevin, K. and P. Germann. 1982. Macropores and water flow in soils. *Water Resour. Res.* 18:1311–1325.

Bevin, K. and E. F. Wood. 1983. Catchment geomorphology and the dynamics of runoff contributing areas. *J. Hydrol.* 65:139–158.

Black, P. E. 1970. Runoff from watershed models. *Water Resour. Res.* 6:465–477.

Blyth, K. and J. C. Rodda. 1973. A stream length study. *Water Resour. Res.* 9:1454–1461.

Bouma, J. 1981. Soil morphology and preferential flow along macropores. *Agric. Water Manage.* 3:235–250.

Bren, L. J. and A. K. Turner. 1980. Hydrologic output of small forested catchments: Implications for management. *Aust. For.* 43:111–117.

Burroughs, E. R. Jr., S. P. Hughes, B. L. Hicks, and H. F. Haupt. 1965. Geophysical exploration in watershed studies. *J. Soil Water Conserv.* 20:3–7.

Comeau, P. G., M. A. Comeau, and G. F. Utzig. 1982. A guide to plant indicators of moisture for southeastern British Columbia, with engineering interpretations. Land Mgt. Handbook No. 5, ISSN 0229-1622. Ministry of Forests Pub. H 18-81064. Victoria, B.C. Canada.

Daubenmire, R. and J. B. Daubenmire. 1968. Forest vegetation of eastern Washington and northern Idaho. Wash. Agric. Expt. Sta. Tech. Bull 60. Pullman, WA.

de Vries, J. and T. L. Chow. 1978. Hydrologic behavior of a forested mountain soil in coastal British Columbia. *Water Resour. Res.* 14:935–942.

Dunne, T. and R. D. Black. 1970a. An experimental investigation of runoff production in permeable soils. *Water Resour. Res.* 6:478–490.

Dunne, T. and R. D. Black. 1970b. Partial area contributions to storm runoff in a small New England watershed. *Water Resour. Res.* 6:1296–1311.

Dunne, T. and R. D. Black. 1971. Runoff processes during snowmelt. *Water Resour. Res.* 7:1160–1172.

Dunne, T., T. R. Moore, and C. H. Taylor. 1975. Recognition and prediction of runoff-producing zones in humid areas. *Hydrol. Sci. Bull.* 20:305–327.

England, C. B. and H. N. Holtan. 1969. Geomorphic grouping of soils in watershed engineering. *J. Hydrol.* 7:217–225.

Engman, E. T. and A. S. Rogowski. 1974. A partial area model for storm flow synthesis. *Water Resour. Res.* 10:464–472.

Feller, M. C. and J. P. Kimmins. 1979. Chemical characteristics of small streams near Haney in southwestern British Columbia. *Water Resour. Res.* 15:247–258.

Freeze, R. A. 1972. Role of subsurface flow in generating surface runoff. 2. Upstream source areas. *Water Resour. Res.* 8:1272–1283.

Freeze. R. A. 1980. A stochastic-conceptual analysis of rainfall-runoff processes on a hillslope. *Water Resour. Res.* 16:391–408.

Flugel, W. A. and O. Schwarz. 1983. Oberflachenabfluss and Interflow auf einem Braunerde-Pelosol Standort im Schonbuch; Ergebnisse eines Beregnungsversuchs (surface runoff and interflow on a Braunerde–Palosol site at Schonbuch; results of a rain test). *Allg. Forst. U.J. Ztg.* 154:59–64.

Gleeson, C. H. 1953. Indicators of erosion on watershed land in California. *Trans. Am. Geophys. Union* 34:419–426.

Gurnell, A. M. 1981. Heathland vegetation, soil moisture, and dynamic contributing area. *Earth Surface Processes Landforms* 6:553–570.

Harr, R. D. 1976. Hydrology of small forest streams in western Oregon. USDA Forest Service Gen. Tech. Rept. PNW-55. Pacific Northwest For. and Range Expt. Sta., Portland, OR.

Harr, R. D. 1977. Water flux in soil and subsoil on a steep forested slope. *J. Hydrol.* 33:37–58.

Heerdegen, R. G. and M. A. Beran. 1982. Quantifying source areas through land surface curvature and shape. *J. Hydrol.* 57:359–373.

Hewlett, J. D. 1961a. Watershed management. pp. 61–66. In: Annual Report for 1961. USDA Forest Service. Southeastern For. Expt. Sta. Asheville, NC.

Hewlett, J. D. 1961b. Soil moisture as a source of base flow from steep mountain watersheds. USDA Forest Service. Sta. Pap. 132. Southeastern For. Expt. Sta., Asheville, NC.

Hewlett, J. D. and A. R. Hibbert. 1963. Moisture and energy conditions within a sloping soil mass during drainage. *J. Geophys. Res.* 68:1081–1087.

Hewlett, J. D. and A. R. Hibbert. 1967. Factors affecting the response of small watersheds to precipitation in humid areas. pp. 275–290. In: W. E. Sopper and H. W. Lull, Eds. *Int. Symp. For. Hydrology.* Pergamon, New York.

Hewlett, J. D., G. B. Cunningham, and C. A. Troendle. 1977. Predicting stormflow and peakflow from small basins in humid areas by the R-index method. *Water Resour. Bull.* 13:231–253.

Hibbert, A. R. 1976. Percolation and streamflow in range and forest lands. pp. 61–72. In: H. F. Heady, D. H. Faulkenborg, and J. P. Riley, Eds. *Watershed management on Range and Forest Lands*, Utah Water Research Laboratory, Utah State University, Logan, UT.

Holbo, H. R., R. D. Harr, and J. D. Hyde. 1975. A multiple-well, water-level measuring and recording system. *J. Hydrol.* 27:199–206.

Horton, J. H. and R. H. Hawkins. 1965. Flow path of rain from the soil surface to the water table. *Soil Sci.* 100:377–383.

Horton, R. E. 1933. The role of infiltration in the hydrologic cycle. *Trans. Am. Geophys. Union* 14:446–460.

Horton, R. E. 1945. Erosional development of streams and their drainage basin: hydrophysical approach to quantitative morphology. *Bull. Geol. Soc. Am.* 56:275–370.

Lane, L. J., M. H. Diskin, D. E. Wallace, and R. M. Dixon. 1978. Partial area response on small semiarid watersheds. *Water Resour. Res.* 14:1143–1158.

Lee, M. T. and J. W. Delleur. 1976. A variable source model of the rainfall runoff process based on the watershed stream network. *Water Resour. Bull.* 12:1029–1036.

Martinec, J. 1975. Subsurface flow from snowmelt traced by tritium. *Water Resour. Res.* 11:496–498.

McCaig, M. 1983. Contributions to storm quickflow in a small headwater catchment—the role of natural pipes and soil macropores. *Earth Surface Processes Landforms* 8:238–252.

Mosley, M. P. 1979. Streamflow generation in a forested watershed, New Zealand. *Water Resour. Res.* 15:795–806.

Mosley, M. P. 1982. Subsurface flow velocities through selected forest soils, South Island, New Zealand. *J. Hydrol.* 55:65–92.

O'Loughlin, E. M. 1986. Prediction of surface saturation zones in natural catchments by topographic analysis. *Water Resour. Res.* 22:794–804.

Paine, D. P. 1981. *Aerial photography and image interpretation for resource management*. J. Wiley. New York.

Palkovics, W. E. and G. W. Petersen. 1977. Contribution of lateral soil water movement above a fragipan to streamflow. *Soil Sci. Soc. Am. J.* 41:394–400.

Pilgrim, D. H., D. D. Huff, and T. D. Steele. 1978. A field evaluation of subsurface and surface runoff. II. Runoff processes. *J. Hydrol.* 38:319–341.

Pole, M. W. and D. R. Satterlund. 1978. Plant indicators of slope instability. *J. Soil Water Conserv.* 33:230–232.

Roberts, S. W., B. R. Strain, and K. R. Knoerr. 1980. Seasonal patterns of leaf water relations in four co-occurring forest tree species: Parameters from pressure volume curves. *Oecologia* 46:330–337.

Rogers, W. F. 1972. New concept in hydrograph analysis. *Water Resour. Res.* 8:973–981.

Rothacher, J., C. T. Dyrness, and R. L. Fredericksen. 1967. Hydrologic and related characteristics of three small watersheds in the Oregon Cascades. USDA Forest Service. Pac. Northwest For. Range Expt. Sta. Portland, OR.

Scholl, D. G. and A. R. Hibbert. 1973. Unsaturated flow properties used to predict outflow and evapotranspiration from a sloping lysimeter. *Water Resour. Res.* 9:1645–1655.

Sklash, H. G., H. K. Stewart, and A. J. Pearce. 1986. Storm runoff generation in humid headwater catchments. 2. A case study of hillslope and low-order stream response. *Water Resour. Res.* 22:1273–1282.

Smith, R. E. and R. G. B. Hebbert. 1983. Mathematical simulation of interdependent surface and subsurface hydrologic processes. *Water Resour. Res.* 19:987–1001.

Stephenson, G. R. and R. A. Freeze. 1974. Mathematical simulation of subsurface flow contributions to snowmelt runoff, Reynolds Creek Watershed, Idaho. *Water Resour. Res.* 10:284–294.

Strahler, A. N. 1968. Quantitative geomorphology. pp. 898–912. In: R. W. Fairbridge, Ed. *Encyclopedia of Geomorphology*. Halsted Press.

Thomas, G. W. and R. E. Phillips. 1979. Consequences of water movement in macropores. *J. Environ. Quality* 8:149–152.

Weyman, D. R. 1970. Throughflow on hillslopes and its relation to the stream hydrograph. *Int. Assoc. Sci. Hydrol. Bull.* 15:25–33.

Whipkey, R. Z. 1965. Subsurface stormflow from forested slopes. *Int. Assoc. Sci. Hydrol. Bull.* 10:74–85.

Winkler, R. D. and R. L. Rothwell. 1983. Biogeoclimatic classification system for hydrologic interpretations. *Can. J. For. Res.* 13:1043–1050.

Yair, A., D. Sharon and H. Lavee. 1980. Trends in runoff and erosion processes over an arid limestone hillside, northern Negev. Israel. *Hydrol. Sci. Bull.* 25:243–255.

Yamamoto, T. 1974. Seismic refraction analysis of watershed mantle related to soil, geology and hydrology. *Water Resour. Bull.* 10:531–546.

Zalesskiy, F. V. 1976. Flash flood formation in permafrost regions. *Sov. Hydrol.: Selected Papers* 15:95–97.

5 Atmospheric Moisture and Precipitation

Precipitation, or the lack of it, is the dominating factor in the hydrology of any region. It is the immediate source of all water entering the land phase of the hydrologic cycle. The general distribution of precipitation has been presented earlier, but watershed managers require a more detailed knowledge of precipitation characteristics that set the framework within which they work.

5.1 ATMOSPHERIC MOISTURE

Water is always present in the atmosphere. It exists there in all three physical states; gas (water vapor), liquid (rain and cloud droplets), and solid (snow, ice and crystals). At any given time, most of the water is present as vapor; but it also exists as clouds; composed of water droplets, ice crystals, or a mixture of both.

5.1.1 Water Vapor and Humidity

Water enters the atmosphere as vapor, where it mixes with other gases in the air, but retains its separate characteristics.

Water can exist in three physical states, as determined by temperature and its free-energy status. The lowest energy state is that of ice, which is characterized by strong, rigid molecular bonding in the form of hexagonal crystals. Molecular motion is vibratory in solids, but if enough energy is added, motion increases until sufficient momentum develops that the rigid crystaline bonds are broken and the molecules are free to move, exchanging bonding forces with a succession of others, becoming *liquid.* Molecular motion is nonuniform, and an individual molecule with a higher than average velocity may break completely free to escape into the air, becoming a vapor molecule (Schroeder and Buck, 1970). The process of changing directly from a solid to the gaseous state or vice versa is called *sublimation*.

Most vapor is derived from liquid water, for it takes less energy to break liquid bonding, and average molecular momentum is greater in liquids, which are usually at a higher temperature than solids. The process of changing from a liquid to a gas is called *evaporation*. *Vaporization*, which means becoming a gas, may be used for either sublimation or evaporation.

Evaporation and sublimation are cooling processes. *Temperature* represents the average rate of molecular activity of a substance. The rate of motion is nonuniform, and it is the most rapidly moving molecules that break free of others to become vapor. Hence, the average rate of motion of the remainder is less, and the temperature falls unless additional energy is supplied.

Water vapor molecules in the air move in random directions at varying speeds. If moist air is over water, some of the molecules reenter the water surface and become part of the liquid. This process is called *condensation*. At subfreezing temperatures, water vapor may be transformed directly into ice, a process often called *frost formation* or *freezing* to indicate the direction of change of sublimation. Frost formation and condensation are heating processes, as gas molecules have a higher average velocity than those in liquids or solids, thus their addition raises the average rate of motion, increasing the temperature.

When the average rate of molecular movement from a liquid to the air is the same as that from the air back to the liquid, the air is said to be *saturated* with respect to water vapor; that is, its vapor density is at a maximum for that temperature. Water vapor is like other gases, and exerts pressure. The partial pressure of water vapor, or *vapor pressure*, is the contribution made by water to total atmospheric pressure. If temperature is constant, vapor pressure varies directly with vapor density. Vapor pressure is expressed in pascals (Pa) in Systême Internationale (SI) units. One atmosphere of air pressure at sea level is 101.325 kPa, of which as much as 3 kPa may be due to water vapor in warm moist air.

The capacity of air to hold water vapor, or its maximum vapor density, is a function of temperature at constant air pressure. Since pressure of a gas varies with density at a given temperature, the saturated vapor pressure has the same functional relation with temperature. When air is at its maximum vapor density the temperature of the air is at its *dew point*, for further cooling would cause dew to form on surfaces in contact with the air. The vapor pressure of ice or water is always the same as the saturated vapor pressure of overlying air at the same temperature. Table 5.1 shows the maximum vapor density and saturated vapor pressure of air at sea level for the range of temperature common to most of the earth. Separate tables should be used for each 200–600-m change in altitude, as the density of air influences its ability to hold vapor.

As it takes more energy for vapor molecules to sublimate from a solid than to evaporate from a liquid, the saturated vapor pressure of ice (and air in contact with ice) is slightly lower than that of liquid water at all temperatures below 0 °C. The slight difference is highly significant to the precipitation process, as will be seen shortly. Note also that over the range of −10–30 °C, saturated vapor pressures of water approximately double for each 10 °C increase in temperature.

Humidity is a general term referring to the water vapor content of air. It may be expressed in several ways, one of which is vapor pressure. Another way is by

TABLE 5.1. Relation of Saturated Vapor Pressure and Density of Air at Sea Level Over Water and Ice to Temperature (Dew Point)

t(°C) (dew point)	Saturated Vapor Pressure (Pa)		Maximum Vapor Density (g cm^{-3})	
	Ice	Water	Ice	Water
−40	13		0.11	
−35	23		0.20	
−30	38		0.34	
−25	63	81	0.56	0.71
−20	103	126	0.91	1.07
−15	165	191	1.37	1.61
−10	260	287	2.14	2.36
−5	402	422	3.25	3.41
0	610	610	4.85	4.85
5		872		6.80
10		1227		9.40
15		1705		12.83
20		2338		17.30
25		3167		23.05
30		4243		30.38
35		5623		39.63
40		7376		51.32
45		9583		65.76
50		12334		83.64

noting the *dew point*, the temperature at which the air would be saturated and dew forms on exposed surfaces. Vapor density is still another means, and is usually called *absolute humidity*. All of the above are presented in Table 5.1.

The most common expression of humidity is that of *relative humidity*; the ratio of the actual density (or pressure) of water in the air to the maximum vapor density (or saturated vapor pressure) at the same temperature. This ratio is multiplied by 100 when relative humidity is expressed as a percentage. In equation form:

$$RH = (e/e_s)100 \tag{5.1}$$

where RH is relative humidity in percent (%), e is vapor pressure or density, and e_s is saturated vapor pressure or density at the same temperature.

For most purposes in watershed management, relative humidity measurements are among the least useful but the most readily available. They must usually be converted to vapor pressures or absolute humidity before they can be used to express the amount of water in the air or rates of evaporation or condensation.

5.1.2 Sources of Moisture in the Atmosphere

Water enters the atmosphere as vapor, a result of evaporation from both land and water surfaces. As oceans cover most of the earth's surface, they are the most important moisture source. Land surfaces are also important, however. For example, in the conterminous United States, the average daily vapor input is about 38 gigatons (Gt; 1 Gt is equivalent to 1 km^3 of water). About 26 Gt represent inflow from over oceans; 17 from the Gulf of Mexico and Atlantic, and 9 from the Pacific. About 12 Gt represents evaporation from land and internal water surfaces (Court, 1974).

The temperature of the oceans strongly influences the moisture supplied to the atmosphere and transported over adjacent land surfaces. The relatively cold currents along the west coasts of southern California, Chile, and southwestern Africa are associated with arid climates over the land. The relatively warm Japan Current and Gulf Stream provide plentiful moisture over northwestern North America and northern Europe, respectively.

Evaporation from land surfaces also varies. Where large supplies of energy and water are available, as in tropical rain forest regions, the contribution to the atmospheric vapor content can be quite high, and may nearly equal that of the oceans on a unit area basis. Where energy or water supplies are low, the land contribution to the atmospheric vapor supply is always low, as in the arctic and subarctic regions and desert zones throughout the world.

5.2 PRECIPITATION

Precipitation results when the water vapor in the air condenses or freezes and is deposited on the earth's surface. It may take many forms; snow, rain, hail, sleet, dew, frost, rime, and many others. The number of recognized forms of solid precipitation alone ranges up to 80, depending on the classification system (La Chapelle, 1969). Other characteristics of spatial distribution, intensity, and frequency are important to the practicing watershed manager. Many are related to precipitation systems and processes.

5.2.1 Precipitation Processes

Most precipitation occurs when moist air is lifted to high altitudes where cooling causes water vapor to condense or freeze, causing clouds. As water vapor is an invisible gas, the presence of visible water in the air always indicates liquid or solid water, or both.

Lifting causes cooling because as air ascends, the pressure on it decreases and it expands. Expansion against the remaining pressure constitutes work, and requires energy. The energy expended is heat energy, and the effect is to cool the air. (In descending air the process is reversed. Air is compressed and warmed.) These temperature changes are called *adiabatic* because there is no loss or gain of energy to the surrounding air. The heat energy is converted to mechanical energy of expansion, and vice versa.

Cooling can also occur by loss of heat to surroundings, as when air contacts a cold surface, or when two air masses of different temperature mix, cooling one and warming the other. However, lifting is the dominant mechanism of atmospheric cooling that induces precipitation.

5.2.1.1 ***Lifting Mechanisms.*** Certain storm characteristics are determined by the manner in which moist air is lifted to a height where precipitation can develop. The ability of the atmosphere to produce and sustain lifting forces depends on dynamic forces that initiate movement and the stability of the air mass being lifted. Air is *stable* if buoyancy forces oppose vertical motion; it is *unstable* if they tend to accelerate vertical motion, and it is *neutral* if vertical motion is neither aided nor opposed. Stability is a function of the temperature of an air parcel in relation to surrounding air at the same altitude. Air warmer than its surroundings is less dense, and is unstable; that colder is stable, and at the same temperature, it is neutral. Most precipitation occurs during unstable conditions. Lifting is initiated by four dynamic forces: orographic, frontal, convective, and convergent motion.

Orographic lifting is caused when a moving air mass is forced over a topographic barrier. The north–south oriented mountains of the United States, combined with the generally westerly air flow, cause orographic lifting to be a major source of precipitation in the Coast Ranges, the Cascade–Sierra Nevada, Rocky, and Appalachian Mountain chains. At times, warm moist air masses that move into southern United States from the Gulf of Mexico release precipitation as they rise over the escarpment of the Edwards Plateau in Texas, the Ozark highlands, Cumberland Plateau, and others. Some of the heaviest rainfall in the world occurs in the foothills of the Himalaya Mountains north of the Bay of Bengal as storms move inland from the sea.

Behind a topographic barrier, a pronounced *rain shadow* may occur as subsiding (sinking) air masses warm and reverse the condensation process, preventing further precipitation.

The orographic effect may be very important in steep mountainous country on watersheds only a few hectares in extent. It must be considered wherever a watershed shows topographic variation.

Frontal lifting produces precipitation when a moist air mass is forced upward by passage of frontal systems. When one air mass displaces another, the less dense (usually warmer) one is forced upward. The line of contact between two air masses is called a *front*. A *warm front* occurs when a warmer air mass displaces a colder one, and a *cold front* is the reverse. The warmer air mass may be well below freezing in winter.

A warm air mass displacing a cold one tends to more or less gently override it, depending on the humidity of the air. If the air is fairly dry, the atmosphere tends toward stability and only high clouds with little precipitation may occur.

If the air is moist, the cloud layer thickens from cirrus into stratus clouds, yielding slow, steady precipitation as the surface front approaches. Snow will fall if the warmer air is below freezing.

In most cases, precipitation accompanying warm fronts is light to moderate in intensity, whatever its form. It may last as long as several days and cover a broad area. Winds during storms are generally light.

A cold front, since the cold air mass is more dense, tends to displace and lift the warmer air mass more sharply and violently. The rapid lifting causes precipitation to be heavier, more localized, and of shorter duration. A narrow band of high winds, heavy showers, and towering clouds, called a *squall line*, often marks the passage of a cold front.

Convective lifting occurs when moist air is unstable and strong heating of the surface causes expansion and lifting of the air. Once lifting is initiated, the air continues to rise as condensation releases heat that prevents cooling to the temperature of the surrounding air at the same altitude. Cumulus clouds may develop into cumulonimbus clouds with their tops towering thousands of meters above the cloud base if the air is very moist and unstable. Cloud tops are often composed of ice crystals, even in the tropics. Raindrops may be held up by the rapidly rising air until they become large enough to fall through the updraft. Sometimes hail is formed when rain drops are repeatedly carried upward and freeze. Finally, the rain or hail falls through the rising warm air, cooling it so that its buoyancy may be reversed. An extremely heavy rain, accompanied by a cold downdraft, covering only a small area and lasting for only a short time, called a *cloudburst*, may then occur.

Convective storms are usually warm-weather phenomena, but may occur in cool seasons if instability is pronounced. They tend to occur in the afternoon and early evening when temperatures are highest (Osborn, 1983).

Convergence is air flowing into a low-pressure area from the higher pressure air surrounding it. The inflowing air displaces upward the air of lower pressure and density. Most precipitation is associated with low-pressure systems.

Sometimes two or more of the above types of lifting may be combined. Convective processes in warm moist air masses may be accelerated by the passage of a cold front. Particularly violent storms may occur all along the squall line, and cover a much larger area than would be covered by either type of storm alone. Fronts are almost always associated with low-pressure cells. In some mountain areas, convective–orographic storm paths may develop that lead to more consistent and heavier rainfall in some areas than in others.

5.2.1.2 Formation of Precipitation. Two requirements must be met before precipitation can fall from the sky: (a) condensation or freezing of vapor and (b) growth of the water droplets or ice crystals sufficient for them to fall through rising air to reach the ground. Dew or frost can be deposited directly on the surface through condensation or freezing of vapor, but although they may be biologically important, they are seldom hydrologically significant.

As soon as moist air is cooled to the dew point, condensation or freezing will take place if there is a suitable surface available. Dew or frost form readily on many surfaces. But in the atmosphere, extensive surfaces do not exist. If the air contained no liquid or solid particles, condensation or freezing would be extremely difficult. Pure air may become highly *supersaturated* (i.e., vapor density exceeds the saturated vapor density for a given temperature). Water molecules in a gas do not readily cohere when they collide, and only after a water droplet reaches a critical size of some 10^9 or more molecules will it sustain itself. Since molecular velocity depends on temperature, the exact critical size is a function of temperature (Miller, 1971). Spontaneous condensation or freezing may require supersaturation of as much as four times saturation.

Most of the time there are numerous particles in the air to provide surfaces or condensation nuclei upon which vapor molecules can adhere, and condensation and freezing takes place at only moderate levels of supersaturation. These particles are composed of dust, salts, and other natural and human-made materials.

Cloud droplets may be very small, and essentially pure water in small droplets will not freeze until they are greatly supercooled. (Condensation nuclei are too small to have a significant effect on purity.) Many clouds contain liquid water at temperatures lower than $-20\ °C$. They are often composed of a mixture of small liquid droplets and ice crystals during much of the year. They are held suspended in the air by rising air currents, their maximum fall velocity (*terminal velocity* is the fall velocity at which acceleration due to gravity is offset by air resistance and becomes constant), often being less than 1 km h^{-1}. Therefore, in order for precipitation to reach the ground, the particles must grow to such size that their terminal velocity exceeds the upward velocity of the air, and they are large enough that they do not evaporate before reaching the ground. Condensation alone is seldom sufficient for the required growth. Most growth occurs in two ways.

1. The first growth method is by coalescence of particles. Particles in motion in the cloud may collide and cohere to one another, inducing growth. As the larger particles are formed, they may fall through a rising stream of smaller ones and collect them. Many of the large snowflakes in winter are aggregations of many individual crystals bound together by rime, or tiny water droplets that freeze upon contact with the falling snowflake.
2. The second is growth of ice crystals by vapor transfer from water droplets. Many clouds contain a mixture of ice and water droplets. At the same temperature, ice has a lower vapor pressure than liquid water (Table 5.1). Therefore, vapor moves from the liquid to the ice, and ice crystals grow at the expense of the droplets until they are large enough to fall. Further growth may be due to a combination of vapor transfer and coalescence. The ice crystal growth mechanism is the basis for many kinds of cloud seeding.

The rising air beneath the cloud base is usually warmer and unsaturated, so rain or snow falling into it may vaporize before reaching the ground, particularly in summer during light convective storms. The streaklike sheet of water falling from the clouds that fails to reach the ground is called *virga*.

5.2.2 Precipitation Characteristics

Many characteristics of precipitation are important to the watershed manager. Among them are form, amount and spatial distribution, rate or intensity, and frequency of extreme rates and amounts.

5.2.2.1 ***Form.*** The two major forms of precipitation are rain and snow. Form is important because rain is mobile and either infiltrates or runs off, with only minor surface storage. Snow, however, may remain at the surface for weeks or months. Snow is much more difficult to measure than rain. It also has great differences with regard to energy exchange.

Minor forms of precipitation include *sleet*, rain that freezes into ice pellets as it falls to the ground; *hail*, ice pellets formed by repeated melting and refreezing of raindrops in the convective currents within the clouds; and *glaze*, formed by supercooled water that freezes upon contact with the surface. *Dew* and *frost* are formed directly from vapor condensing or freezing upon the surface. Once on the ground, these forms of precipitation are little different from rain or snow.

An important form of precipitation in certain localities is *cloud drip* (or *fog drip*) and *rime*, which are deposited directly on vegetation engulfed by clouds. Rime consists of crystals of supercooled water that freeze upon contact. In some coastal and mountainous areas cloud drip or rime constitute a substantial contribution to the hydrological cycle (Lovett et al., 1982; Harr, 1982; Berndt and Fowler, 1969; among others).

5.2.2.2 ***Amount and Spatial Distribution.*** It is not possible to know precisely how much precipitation falls on any watershed over a period of time. There are always two types of errors involved in determining watershed precipitation: measurement errors and sampling errors.

Measurement errors always occur because the deposition of rain or snow is modified by the presence of a gage. Whenever a gage is inserted into a airstream it causes turbulence that carries rain or snow past the gage opening. If the gage opening is at ground level, it may receive extra precipitation as rain splash or drifting snow. Most gages have their opening at a height of about 1 m and are deficient in catch.

Measurement errors can be minimized by proper gage exposure and by shielding the gage. Gages should be exposed in areas protected from excess wind and with no obstacles within a 45° angle from the vertical. In windy regions, shields may be necessary to break the flow of wind over the gage.

Snow is more difficult to measure accurately than rainfall. Many types of shields have been developed to try to insure a representative catch, but even at a single point, gage catch is subject to large errors due to wind (Sturges, 1986). In many cases, snow depth is measured on the ground and converted to water at the ratio of 10 cm of snow to 1 cm of water, but as the actual ratio varies from about 20:1 to 4:1, this method yields only an approximation. A more accurate point measure is obtained by taking a snow core of known diameter and weighing or melting it to determine the water equivalent. Measuring snow on the ground does not overcome the effect of wind. Protected areas usually accumulate more than falls directly, whereas open areas accumulate less. Most estimates of snowfall are approximations at best.

Sampling errors arise because precipitation is not uniformly distributed over a watershed and only a limited number of relatively small gages are used. In the United States, the standard gage has an opening of about 325 cm^2. Gage numbers range from about 1 per 250 km^2 in southern New England to 1 per 8500 km^2 in parts of the Southwest. Additional gages would be desirable in many western states (Chang, 1981), with gage number, type, and location being particularly important in mountainous areas (Corbett, 1967; Molnau et al., 1980).

Precipitation estimates often require extrapolation from a single gage outside a watershed, or interpolation from two widely separated gages. However, where several gages are found in or near a watershed, the amount of precipitation can be estimated by several techniques. The simplest is to take the arithmetic mean of the gage catch. If the watershed is relatively flat and the gages uniformly distributed the method is as accurate as any other; but these conditions are seldom met.

Where gage distribution is nonuniform the *Thiessen method* may provide a better estimate (Fig. 5.1*a*). This method weights gage catch according to the spacing and distribution of gages. Using a map of a watershed and gage locations, it requires that a straight line be drawn between each gage and its neighbors. A perpendicular bisector is extended from each of the lines until it meets those from the other lines radiating out from a given gage, forming a polygon about it. The polygon encompasses the area represented by the gage. The catch is weighted by the proportional area it represents, and watershed precipitation is the weighted sum of gage catch.

Isohyets are lines of equal precipitation drawn by interpolation or extrapolation from the catch in each gage (Fig. 5.1*b*). They should take basin characteristics such as topography into account, and may provide a more accurate representation of precipitation distribution than other methods when elevation differences are pronounced and gages are few. Precipitation estimates are obtained by weighting the proportional area between two adjacent isohyets and multiplying by their mean value, summing over the entire watershed. Computer techniques for locating isohyets are available (Wei and McGuinness, 1973), but are based only on interpolation of catch between gages.

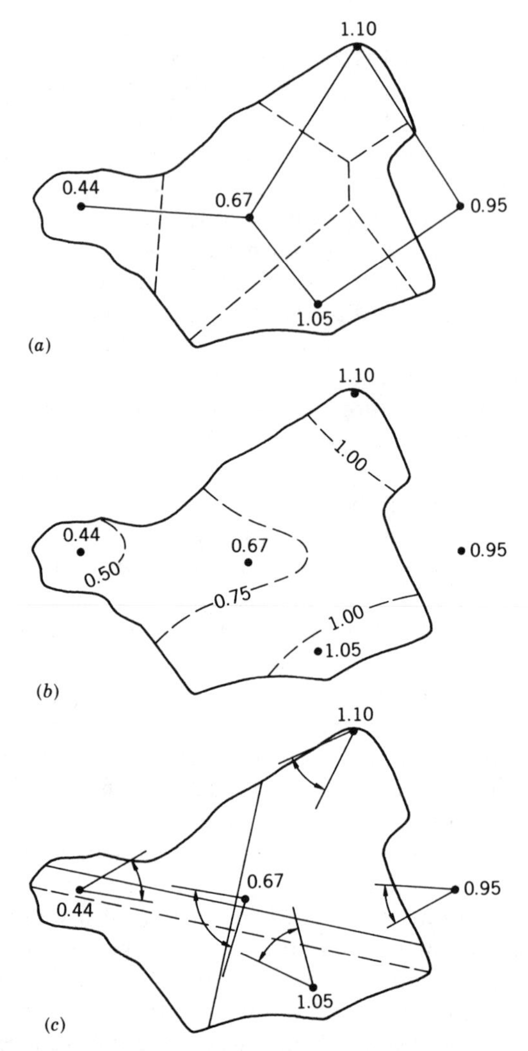

FIG. 5.1 Estimating watershed precipitation by weighting gage catch. (*a*) Thiessen polygons (dashed lines). Catch of each gage is weighted by the proportion of watershed area encompassed within the polygon. (*b*) Isohyets (dashed lines). Precipitation for the area between each pair of lines is the mean of each. (*c*) Two axis method showing longest axis (dashed line), major and minor axes (solid lines), and acute angles subtended by each gage. Weight of each gage catch is proportional to the subtended angle.

The *two axis* method (Bethlahmy, 1976) offers another method of weighting gage catch to estimate watershed precipitation. Weighting is based on gage location with respect to the center of the watershed rather than area (Fig. 5.1*c*). On a map of the watershed, a line drawn along the longest axis of the basin is fitted with a perpendicular bisector, which is extended to each boundary. It is called the *minor axis*. The minor axis in turn is bisected by a line parallel to the longest axis, extended to the boundaries, which is called the *major axis*. The intersection of the two axes is the approximate center of the watershed. At each gage, two lines are drawn toward the farthest point on each axis, creating an acute angle at each gage. Gage catch is weighted by the angle it subtends as a fraction of the sum of subtended angles for all the gages. Court and Bare (1984) tested this method against others and found that it was much easier to use than Thiessen means and isohyets, and yielded similar estimates.

Precipitation ordinarily increases with elevation in amounts that vary with the position of the gage in relation to storm paths, by region, and by location with respect to other topographic features. Precipitation does not follow the details of rough topography, but reflects only the broad outlines. It is controlled more by the location and spacing of the higher projecting features than lower protected ones. Thus, Fenn Ranger Station, Idaho, in the narrow Selway River valley at an elevation of 482 m, but with nearby ridges reaching from 1500 to 2000 m, receives nearly 70% more precipitation than Grangeville, Idaho, located to windward at an elevation of 1023 m on a broad plateau with no large nearby ridges. Precipitation amounts in wider valleys (5–10 km wide and greater, depending on depth and orientation) may exhibit typical rain shadow effects. In the Olympic Peninsula of Washington, annual precipitation increases by 15–30 cm per 100 m increase in elevation on the windward southwestern slopes, but decreases at nearly twice that rate in the rainshadow to the northeast (Meteorology Committee, 1969).

Even on watersheds with more gentle topography, storm rainfall may be highly variable from one part to another. Many convective storm cells are less than 10 km^2 and the area of intense rain may be even smaller.

5.2.2.3 Intensity and Frequency. The rate at which rain falls is an important hydrologic parameter, especially as it affects storm hydrographs and erosion. The design of drainage facilities: bridges, culverts, ditches, and other water control devices, requires knowledge of the rainfall intensities that can be expected to occur with any given frequency. In watershed management, snowfall rates are less important, but they may be extremely important for avalanche warning and control and highway clearing.

In the United States, short-term precipitation rates are measured by recording rain gages, based on either weighing or "tipping bucket" mechanisms. For periods longer than 2 days, nonrecording gages may provide satisfactory data. Rates are usually expressed in depth per unit of time, except for periods shorter than 1 h. Thus rainfall may be 200 mm in 6 days, 121 mm in 24 h, or 51 mm in 1 h; but 114 mm h^{-1} for 20 min if 38 mm falls in one 20-min period.

Most hydrologically important intensities are the higher extremes that occur only at infrequent intervals, but most rain falls at low intensities. Cooper (1967) found that the proportion of total rain falling at or above a given rate follows an exponential relation for rates above 38 mm h^{-1} and amounts above 2.5 mm. In equation form:

$$R = ae^{-bI} \qquad 5.2$$

where R is the proportion of rain falling at or above a given intensity I, (mm h^{-1}), e is the base of the natural logarithm (2.718 . . .), and a and b are the constants; $a = 0.13$ and $b = 0.0264$ for his data.

These rates were based on the amounts and intensities that fall during uniform bursts and not on mean rainfall rates observed in 2 min, 15 min, or other fixed time period. According to Eq. 5.2, less than 1% of the rainfall in his study fell at rates exceeding 100 mm h^{-1}.

Small showers are seldom hydrologically important regardless of intensity. Over one-half the summer storms in the eastern United States produce less than 1 cm of rain, and in areas such as the plains of Montana, 70% of summer showers yield under 2 mm (Weaver, 1985).

The watershed manager must rely on historical data to determine the frequency of occurrence of extreme rates of rainfall. Short records are of little value, for they are likely to contain too few of the rare events to be meaningful. Rainfall intensity data, for durations from 20 min to 24 h and return periods of various lengths have been published by the U. S. Weather Bureau (now the National Weather Service) in a series of Technical Papers (U.S. Weather Bureau, 1961a,b; 1962, 1963). Other data sources may be found locally (Haines, 1977), and guidelines for the summary of such data are available (Finklin, 1983).

The *return period* of a storm with rainfall equal to or exceeding a given intensity for a given duration is the average interval in years between such events over a long period of time. The reciprocal of the return period provides an estimate of its probability in any given year. Thus if a return period for a storm of, say, 100 mm h^{-1} for 20 min is 10 years, then the probability of a storm having that intensity or greater is 1:10 or 0.1 in any given year. In equation form:

$$p = 1/T_r \qquad 5.3$$

where p is the probability of a storm of equal or greater intensity in any year, and T_r is the return period, in years.

The probability of a storm of a given intensity (or greater) occurring at least once over a given span of years is

$$P_n = 1 - q^n \qquad 5.4$$

where n is the number of years, and $q = \dfrac{(T_r - 1)}{T_r}$

5.2.3 Effect of Forests on Precipitation

For many years, the idea that forests increase local precipitation has persisted. Clearly, forests are found in regions of high rainfall and are absent in areas of little rainfall. But, does the rain cause the forests or the forests cause the rain? This question was considered by Kittredge (1948), who summarized the information available to his time, and more recently by Chang and Lee (1974) and Hamilton (1986).

There is no way that the question can be answered with absolute certainty on a worldwide or large regional scale, but fears that deforestation will greatly change climate seem not to be supported by historical evidence. Changes in the world's forests over the past thousand years have been great; those in climate have not (Bryson, 1974). Increased "greenhouse gases" due to human activity appears to have the greatest implications for climate change (Harrington, 1987), although debate continues over whether or not significant "global warming" will be the result (Wheeler, 1990).

On a local scale, it is clearly evident that forest vegetation can greatly influence the amount of precipitation reaching the ground from rime and cloud drip. It also seems evident that tall trees can increase topographic barrier elevations by up to 50 m, possibly influencing orographic precipitation.

On the whole, it seems that the idea that forests importantly affect precipitation is rejected by most meteorologists, except where fog drip and rime occur frequently.

LITERATURE CITED

Berndt, H. W. and W. B. Fowler. 1969. Rime and hoarfrost in upper-slope forests of eastern Washington. *J. For.* 67:92–95.

Bethlahmy, N. 1976. The two-axis method: a new method to calculate precipitation over a basin. *Hydrol. Sci. Bull.* 21:379–385.

Bryson, R. A. 1974. A perspective on climatic change. *Science* 184:753–760.

Chang, M. 1981. A survey of the U.S. national precipitation network. *Wat. Resour. Bull.* 17(2):241–243.

Chang, M. and R. Lee. 1974. Do forests increase precipitation? *W. Va. For. Notes*. No. 2, 1974, Morgantown, WV. pp. 16–20.

Cooper, C. F. 1967. Rainfall intensity and elevation in southwestern Idaho. *Water Resour. Res.* 3:131–137.

Corbett, E. S. 1967. Measurement and estimation of precipitation on experimental watersheds. pp. 107–129. In: *Proceedings of the International Symposium on Forest Hydrology*, 1965, Pennsylvania State University National Science Foundation, Washington, DC.

Court, A. 1974. Water balance estimates for the United States. *Weatherwise* 27:252–255, 259.

Court, A. and M. J. Bare. 1984. Basin precipitation estimated by Bethlahmy's two axis method. *J. Hydrol.* 68:149–158.

Finklin, A. I. 1983. Summarizing weather and climatic data—a guide for wildland managers. USDA For. Serv. Gen. Tech. Rep. INT-148. Intermountain For. Range Exp. Sta., Ogden, UT.

Haines, D. A. 1977. Where to find weather and climatic data for forest research studies and management planning. USDA For. Serv. Gen. Tech. Rep. NC-27. North Cent. For. Exp. Sta., St. Paul, MN.

Hamilton, L. S. 1986. Towards clarifying the appropriate mandate in forestry for watershed rehabilitation and management. pp. 33–51. In: *Strategies, Approaches and Systems in Integrated Watershed Management.* Food and Agriculture Organization of the United Nations, Rome, Italy.

Harr, R. D. 1982. Fog drip in the Bull Run municipal watershed, Oregon. *Water Resour. Bull.* 18:785–789.

Harrington, J. B. 1987. Climatic change: a review of causes. *Can. J. For. Res.* 17(11):1313–1339.

Kittredge, J. 1948. *Forest Influences*. McGraw-Hill, New York.

La Chapelle, E. R. 1969. *Field guide to snow crystals*. University of Washington Press, Seattle, WA.

Lovett, G. M., W. A. Rieners, and R. K. Olson. 1982. Cloud droplet deposition in subalpine balsam fir forest: hydrological and chemical inputs. *Science* 218:1303–1304.

Meteorology Committee. 1969. *Climatological handbook, Columbia Basin states. Precipitation.* Vol. 2. Pac. Northwest River Basins Commission. Vancouver, WA.

Miller, A. 1971. *Meteorology.* 2nd ed., Charles E. Merrill Publishing Co., Columbus, OH.

Molnau, M., W. J. Rawls, D. L. Curtis, and C. C. Warnick. 1980. Gage density and location for estimating mean annual precipitation in mountainous areas. *Wat. Resour. Bull.* 16(3):428–432.

Osborn, H. B. 1983. Timing and duration of high rainfall rates in southwestern United States. *Water Resour. Res.* 19:1036–1042.

Schroeder, M. J. and C. C. Buck. 1970. *Fire Weather.* USDA, Forest Service. Agric. Handbook 360. U.S. Gov. Print. Off., Washington, DC.

Sturges, D. L. 1986. Precipitation measured by dual gages, Wyoming-shielded gages, and in a forest opening. pp. 387–396. In: D. L. Dane, Ed., *Cold regions hydrology*, Proc. of a symposium, Am. Water Resour. Assoc., Bethesda, MD.

U.S. Weather Bureau. 1961a. Rainfall-frequency atlas of the United States for durations from 30 minutes to 24 hours and return periods from 1 to 100 years. Tech. Pap. No. 40. GPO, Washington, DC.

U.S. Weather Bureau. 1961b. Generalized estimate of probable maximum precipitation and rainfall-frequency data for Puerto Rico and Virgin Islands for areas to 400 square miles, durations to 24 hours, and return periods from 1 to 100 years. Tech. Pap. No. 42. GPO, Washington, DC.

U.S. Weather Bureau. 1962. Rainfall-frequency atlas for the Hawaiian Islands for areas to 200 square miles, durations to 24 hours, and return periods from 1 to 100 years. Tech. Pap. No. 43. GPO, Washington, DC.

U.S. Weather Bureau. 1963. Probable maximum precipitation and rainfall-frequency data for Alaska for areas to 400 square miles, durations to 24 hours, and return periods from 1 to 100 years. Tech. Pap. No. 47. GPO, Washington, DC.

Weaver, T. 1985. Summer showers: their sizes and interception by surface soils. *Am. Midl. Nat.* 114:409–413.

Wei, T. C. and J. L. McGuinness. 1973. *Reciprocal distance squared method. A computer technique for estimating areal precipitation*. ARS-N.C.8. USDA, ARS, North Central Region, Peoria, IL.

Wheeler, D. L. 1990. Scientists studying "the greenhouse effect" challenge fears of global warming. *J. For.* 88(7):34–36.

6 Soil Moisture Movement and Storage

Soils and subsoils may be viewed as a reservoir, into which water received at the surface may infiltrate and either be stored or percolate through, or both. The soil surface is a critical hydrologic interface, as the path followed by water reaching the surface may determine the nature of the hydrograph and erosion from the watershed. Moisture stored in the soil influences the rate and amount of evapotranspiration and drainage.

Wherever people use the land for any purpose, they intentionally or unintentionally modify surface and subsurface properties, changing water yield characteristics. Thus, if we wish to maintain or improve water yields while using the land, we must understand how moisture moves into, through, and is stored in the soil and subsoil.

6.1 BASIC PRINCIPLES

6.1.1 Laws of Water Movement

Water moves into and through soils whenever the forces acting upon it are unequal. It then moves in the direction of the greatest inequality of forces, from the higher toward the lower force. The movement of water constitutes *work*, and *potential* is defined as the work necessary to move a unit mass from one place to another. The total potential energy of water in soil is the ability to do work, and consists of three potentials:

$$\Psi_t = \Psi_p + \Psi_o + \Psi_z \tag{6.1}$$

where Ψ_t is the total potential energy, Ψ_p is the pressure potential, Ψ_o is the osmotic potential, and Ψ_z is the gravity potential.

Pressure potential is of two kinds, positive and negative. Positive pressures (sometimes called *pressure head potential* or *hydraulic head*) represent the force exerted by the weight of a column of water, expressed in terms of the height of the water surface above the point of measurement. Negative pressures (called *capillary* or *matric potential*) occur in unsaturated soils where adhesive forces between soil and water and the cohesive forces within water produce a *tension* or *suction* force. Zero pressure occurs at the water table, where saturation is complete, and no water column exists above the point of measurement. All pressures are measured with atmospheric pressure (101 kPa) as the reference point. Pressure head potential acts downward, but capillary potential can act in any direction.

Gravity potential is due to the force of gravity and always acts downward. *Osmotic potential* results from different concentrations of solutes in water. It is important primarily in internal plant–water relations (Kramer, 1969).

The rate of moisture movement in saturated soil can be expressed by the Darcy equation (Childs, 1967; Klock, 1972):

$$Q = -K \text{ grad } \phi \tag{6.2}$$

where Q is the volume flux of water per unit of cross-sectional area (cm^3 cm^{-2} s^{-1}), K is the saturated hydraulic conductivity (cm s^{-1}), and grad ϕ is the hydraulic gradient or change of head per unit distance (cm cm^{-1}). Also,

$$K = k\rho g/\eta \tag{6.3}$$

where k is the intrinsic permeability of the soil (cm^2), ρ is the density of water (g cm^{-3}), g is the acceleration of gravity (cm s^{-2}), and η is the viscosity of water (N-s cm^{-2}). Intrinsic permeability depends mostly on pore amount, size distribution, and continuity; the more large, connected pore space, the greater the permeability. As the viscosity of water approximately doubles as temperature is reduced from 25 to 0 °C, it can be seen that conductivity would be reduced by about one-half.

The Darcy equation can also be applied to unsaturated soils in a modified form in which K is a function of the volumetric moisture content (θ), as

$$q = -K(\theta) \text{ grad } \phi \tag{6.4}$$

$K(\theta)$ decreases very rapidly as θ decreases from saturation levels, because as θ decreases, large pores empty first. Thus, there is a rapid decline in available conducting area of pores per unit cross-sectional area of soil, and because resistance to movement is inversely proportional to the square of the pore radius through which water moves. Volume flow in pores increases as the fourth power of the radius when both factors are combined (Dixon, 1971). Conductivity may decline by a factor of 10^6 as a soil dries from saturation to the permanent wilting point at about −1.5 MPa (Philip, 1969). Figure 6.1 illustrates the relation between hydraulic conductivity and moisture content.

Capillary rise from a water table in an open capillary is an inverse function of the radius of the pore, and is directly related to the surface tension of water and the *wetting angle*, or the contact angle between the water meniscus and the pore wall. In equation form:

$$h = \frac{2\gamma}{r\rho g} \cos a \tag{6.5}$$

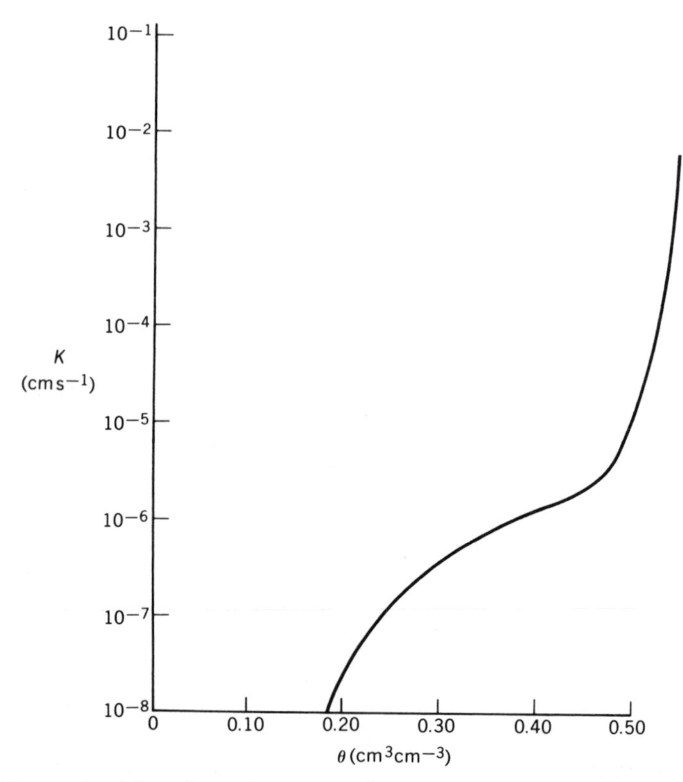

FIG. 6.1 Example of the relation between soil conductivity (K) and volumetric moisture content (θ) over the range from saturation to the permanent wilting point.

where h is the height of rise (cm), γ is the surface tension (N cm^{-1}), r is the radius of the capillary (cm), ρ is density of water (g cm^{-3}), g is acceleration of gravity (cm s^{-2}), and a is wetting angle. In most soils the wetting angle is approximately zero, but some soils are water repellent and the wetting angle can reach 90° or more. Figure 6.2 illustrates capillary rise (or depression) as capillary radius is varied. As surface tension decreases with increasing temperature by about 5% from 0 to 25 °C, capillary rise decreases with increasing temperature (De Bano, 1981; Klock, 1972).

6.1.2 Factors Influencing Water Movement and Storage

Water, like any other substance, interacts with its environment, and the nature of that environment influences those interactions. The movement and storage of water therefore reflects the nature of soils and subsoils. The variables in Eq. 6.1–6.5 help us to identify which soil characteristics are important and how they interact. The grad ϕ is arbitrarily established as a unit gradient. Several of the other elements in the equations are essentially constant, and their slight variation over the range of normal conditions can be ignored, as the density of water (ρ), the acceleration of gravity (g), the gravity potential (Ψ_z), and for most soils, the osmotic potential (Ψ_o). Surface tension (γ) and viscosity (η) of water both vary with temperature sufficiently that soil temperature should be taken into account. The wetting angle (a) of water with the pore wall reflects soil chemical characteristics that indicate water attracting or repellent properties, commonly called *wettability*. Hydraulic conductivity (K) is essentially a function of intrinsic permeability (k) and volumetric moisture content (θ), both of which are related to the amount, size, and distribution of pore space and its wetting and drying at a given temperature.

6.1.2.1 Pore Space and Water Content. For the purposes of hydrology, soils and subsoils can be thought of as consisting of a matrix of solid particles, containing pore spaces that are filled with air or water. The amount, continuity, and size distribution of the pore space and its water content are the dominant factors in water movement and storage.

6.1.2.1.1 Types of Pore Space. Most soils have two types of pore space; that within the aggregates and that between them. Pore space within aggregates predominantly reflects soil particle size distribution and arrangement,

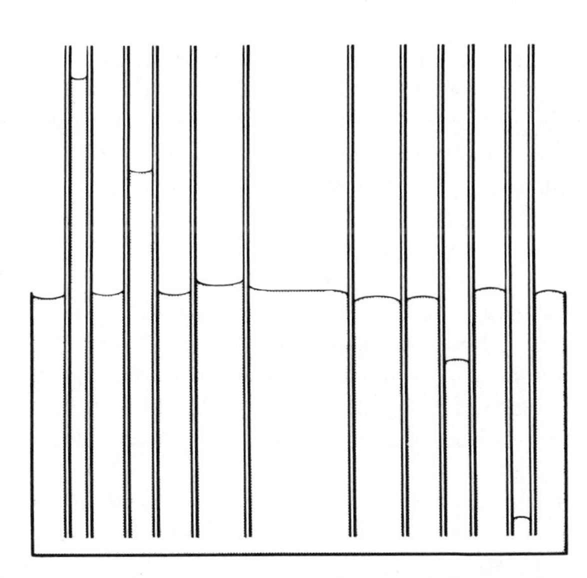

FIG. 6.2 Diagrammatic representation of capillary forces in water. The capillary tubes on the left are hygrophilic, those on the right, hygrophobic. The direction of the capillary force is toward the center of the circle whose arc is described by the meniscus. The amount of force primarily is a function of the contact angle of the meniscus, surface tension and the radius of the tubes, and is reflected by the distance the water level is raised or depressed.

while pore space between aggregates reflects the size and shape of the aggregates. The former is sometimes called *textural porosity* and the latter *structural porosity*.

A third type of void space may develop in soils containing certain types of surface active organic colloids and clays that shrink when dry and swell when moist. In these soils the amount and size distribution of pores vary with moisture content. When very dry, large cracks may develop and pore amount and size is at its maximum. When very wet, both the amount and size of pores are at a minimum (Childs, 1967). Some swelling clays such as bentonite are virtually impermeable when saturated, because pore size and amount is so small.

In addition to the above, large pores may be created in soils by biological activity, both plant and animal. Decaying roots often leave nonstructural channels in the soil, as do animals such as insects, earthworms, and large burrowing mammals such as badgers. Smaller organisms are important as well, both for the channels they create and their activity in creating structural aggregates.

6.1.2.1.2 Porosity and Soil Water Classification. The amount of pore space determines the amount of water that a soil can hold at saturation. Below the water table, in the *zone of saturation*, the amount of porosity is determined primarily by texture. Above the water table in the *zone of aeration*, where the moisture content may fluctuate over a wide range, the amount of porosity is a function of all the different kinds of porosity.

The total porosity of a soil is often estimated from its *bulk density*; the mass of dry soil per unit bulk volume, or its apparent specific gravity. Bulk density is determined by weighing an oven dry sample of known volume. As the specific gravity of individual mineral soil particles averages about 2.65, total porosity may be approximated by the following equation:

$$\xi_t = 1 - D_b/2.65 \tag{6.6}$$

where ξ_t is soil porosity and D_b is bulk density (g cm^{-3}). For example, if bulk density is 1.2 g cm^{-3}, then total porosity is 0.55 (or 55% when multiplied by 100 for percent expression). Most soils have bulk densities ranging from about 0.5 for loose forest soils high in organic matter to about 1.8 for heavily compacted soils or clayey subsoils. Bulk density normally increases with depth until the deep subsoil is reached.

Total porosity determines the maximum water content at saturation, but soils within the zone of aeration are seldom saturated. Most soils contain some entrapped air. Saturated or nearly saturated soils drain rapidly at first, but drainage becomes extremely slow as capillary tensions reach −10 to −30 kPa. Capillary potential can be calculated from Eq. 6.5, where h represents the capillary potential. Soil pores having an equivalent diameter greater than 0.01–0.03 mm will drain under the force of gravity. The soil moisture content when saturated soils are allowed to drain until they reach a given tension, usually within the range of −10 to −30 kPa, is called the *field capacity* of that soil. Most soils that are wetted beyond field capacity and are free to drain will reach field capacity in from 12 to 24 h.

Water that drains readily from the soil under the force of gravity is called *gravitational* or *free water*. The pores from which it drains are called *noncapillary pores* or *macropores*, but some investigators set an arbitrary tension of −0.1 kPa, or an equivalent pore diameter of about 3 mm as the lower boundary for macropores (Germann and Bevin, 1981; Bevin and Germann, 1982). Most noncapillary pores are structural pores, shrinkage cracks, or biological channels, except where soils are uniformly coarse in texture. Most soil pores are very irregular and many portions of large diameter do not drain under the force of gravity. This is because the smallest neck determines the capillary forces of the meniscus as it reaches that point, so drainage will continue only until the meniscus reaches a sufficiently small neck that capillary tension prevents further drainage (Fig. 6.3).

The water content of any given soil at a given moisture tension may be different, depending on whether the soil is being wetted or draining, due to pore irregularity. Once water has drained past a large void below a narrow neck, refilling of the void can occur only at much lower tensions, as the meniscus must move upward through the larger diameter void. Thus, at a given tension, drying soils hold more water than soils being wetted. The inequality in moisture content–energy relations depending on whether the soil is being wetted or draining is called *hysteresis*. Soil moisture content values such as field capacity are always determined under conditions of drainage.

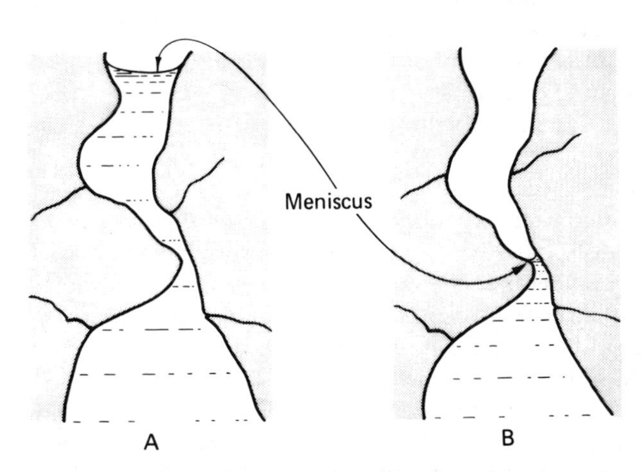

FIG. 6.3 Drainage of an irregular large pore filled with water (A) continues only until meniscus reaches a small neck (B). As the meniscus recedes downward, the capillary force increases with the decrease in pore diameter. The diameter of the pore at the location of the meniscus determines the capillary force with which water is retained, regardless of the pore diameter elsewhere.

Gravitational water is merely detained before it drains from the soil, so it represents *detention storage.* The amount of detention storage space in the soil is the approximate volume of noncapillary porosity; approximate because saturation of upland soils is rare—there is usually some entrapped air that does not have time to dissolve in the water before it drains. The volume of capillary pores can be determined by weighing soil of known bulk volume at field capacity, drying and reweighing to find the volumetric moisture content:

$$\theta = \rho w/V \tag{6.7}$$

where θ is the volumetric moisture content ($cm^3\ cm^{-3}$), ρ is the density of water ($g\ cm^{-3}$), w is the weight of water at field capacity (g), and V is the soil bulk volume (cm^3). Noncapillary pore space is then

$$\xi_n = \xi_t - \theta \tag{6.8}$$

where ξ_n is the noncapillary porosity, an ξ_t is the total porosity, and θ is the volumetric moisture content at field capacity.

Water retained in capillary pore space represents water in *retention storage.* Plants can extract part of the water held in retention storage, for most plant roots can exert tensions of about 1.5 MPa before they become permanently wilted. Some can exert tensions up to 4 MPa or more before wilting, but the *permanent wilting point* is usually set as the soil moisture content when drying causes tensions to reach 1.5 MPa. Therefore, plants extract water from pores down to 0.001 or 0.002 mm in equivalent diameter. The amount of pore space that holds water available to plants (*available water*) can be determined by subjecting a soil of known field capacity to a pressure of 1.5 MPa, determining the volumetric moisture content at the wilting point by Eq. 6.7 and subtracting from the moisture content at field capacity.

Soils at the permanent wilting point may continue to dry by evaporation until virtually all free water is removed. When all water is removed, the soil is *oven dry*, as this state is usually obtained by drying a sample at 105 °C for 24 h in an oven. Air at low relative humidity and high temperature can exert suction forces reaching to 150 MPa, and the air dry moisture content may become virtually nil. The water retained in soils between the permanent wilting point and oven dry is called *hygroscopic water.* The amount of space holding hygroscopic water is the volumetric moisture content at the wilting point.

Most field soils vary in moisture content from somewhat below complete saturation at *field maximum moisture content* to air dry soils with a moisture content somewhere between the permanent wilting point and oven dry at *field minimum moisture content.* They also vary widely in the amount of water they hold at a given tension, because of differences in texture and structure. Coarse textured soils, such as sands, typically have less total porosity than well-aggregated soils of finer texture. The range of values in volumetric moisture content at saturation, field capacity, and the permanent wilting point in relation to texture is shown in Fig. 6.4.

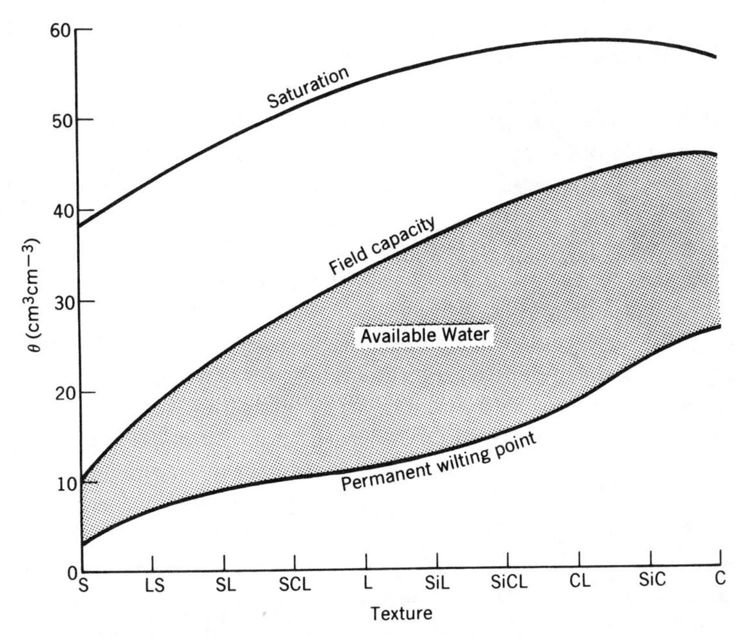

FIG. 6.4 The approximate volumetric moisture content (θ) of mineral soils of various textures at saturation, field capacity, and permanent wilting point. The difference between field capacity and wilting point generally represents plant available water. S = sand; LS = loamy sand; SL = sandy loam; SCL = sandy clay loam; L = loam; SiL = silty loam; SiCL = silty clay loam; CL = clayey loam; SiC = silty clay; C = clay.

Many factors determine the moisture storage capacity of a watershed in addition to the pores in the soil fines. Nonporous rocks and stones in the soil limit space available to hold water. Rocky talus slopes and weathered and fractured bedrock may be hydrologically active. For example, water may be held in a film about 0.1 mm thick and as small droplets at contact points in boulder fields, talus deposits, and cobble-covered ephemeral stream beds in deserts (El Bushi and Davis, 1969). Significant quantities of available soil moisture may be contained in rock fragments found in skeletal soils of southwest Oregon, particularly in smaller and well-weathered pieces (Flint and Childs, 1984a). Bedrock characteristics may limit or enhance moisture storage, depending on porosity and fracturing (Megahan, 1973). The quantity of active water storage capacity (difference between field maximum and minimum) is referred to as the *hydrologic depth* of the watershed mantle.

Hydrologic depth is determined by the above factors and the depth of the mantle itself. For example, a coarse sandy soil 1 m in depth is hydrologically shallower than a well-aggregated silt loam of the same depth because of the greater moisture holding capacity of the latter. Nonporous rocks in soil reduce hydrologic depth, but weathered or fractured bedrock may increase it.

Hydrologic depth may also be limited by a relatively static water table that limits active moisture storage capacity.

6.1.2.1.3 Conductivity and Soil Pores. Each soil has a unique conductivity depending on pore amount, size distribution, and arrangement. Soils having large amounts of large, continuous pores have very high saturated conductivities and may have very high unsaturated conductivities provided water is supplied to the large pores as during infiltration (Dixon, 1971; Germann and Bevin, 1981; Megahan and Clayton, 1983). Mosely (1982) measured subsurface flow velocities averaging 0.3 cm s^{-1} through macropores in a tussock grassland and several forest soils but more recent work (Pearce et al., 1986) raises doubts on the occurrence of such rapid throughflow velocities. A key requirement for substantial macropore flow is that somewhere the soil must be near saturation so the water can enter macropores; once having done so, movement readily takes place through unsaturated zones.

In the absence of additional infiltration or a saturated zone, macropores are quickly drained. Then soil moisture moves only through the finer capillary pores, and conductivity often shows an abrupt decrease when field capacity is reached. As the proportion of macropore porosity decreases, saturated conductivity tends to decrease in all soils, even though total porosity remains high; but unsaturated conductivity may be higher than in soils with many macropores. In general, well aggregated soils of all kinds and those of coarse texture exhibit greater conductivity than those with poor structure or fine texture.

6.1.3 Soil Wettability

Most soils are *hygrophylic* (literally: water loving) in that they readily attract water, but some soils are *hygrophobic* (literally: water hating) in that they resist wetting. In extreme degree, water may be repelled by some substances as shown in Fig. 6.2. Such soils are not readily wetted.

Water movement can be severely limited in water repellent soils, as capillary potential is reduced or reversed. In general, water moves more slowly into water repellent dry soils and out of water repellent wet soils than in soils that attract water.

Water repellency in soils seems to be related to nonpolar chemical substances found mostly in organic materials. Soil particles may be coated by hygrophobic materials that leach from plant litter, are deposited by microbes, or are distilled by fire and condensed on cooler underlying soils (De Bano, 1981).

Coarse textured soils tend to become nonwettable more easily than fine textured soils, as coarse particles have much less surface per unit volume or mass. Therefore only a small amount of hygrophobic material is needed to coat particle surfaces enough to induce repellency.

Many different plants seem to be associated with water repellency, including fungal mycelia, grasses, trees, and shrubs. However, water repellency is most pronounced in burned areas. Apparently, fire distills hygrophobic substances and some of it migrates along the sharp vapor pressure gradients to condense in the cool soil beneath. However, some fires intensify and others destroy water repellency. Temperature during the fire seems to be the critical factor; heating at 260 °C produces an extremely repellent condition if hygrophobic substances are available, but beyond 371 °C repellency decreases and as temperatures increase further, repellency is destroyed (De Bano and Krammes, 1966).

6.1.4 Temperature

Water in the soil in liquid form exists at temperatures from somewhat below 0 °C in small pores to about 30 °C, except near the surface where it may be warmer. It also exists in solid form as soil frost at temperatures below 0 °C.

When frozen, water in the soil is immobile, but liquid water may move into and through frozen soils (Kane and Stein, 1983). The effect of frost on liquid water movement depends on frost structure and density. *Concrete frost* consists of lenses of solid ice, and is essentially impermeable. *Honeycomb* and *granular frost*, which terms indicate their structure, may increase the amount and size of pore space in the soil, actually enhancing moisture movement (Megahan and Satterlund, 1961). During freezing, water moves to the freezing front, and since water expands about 9% upon freezing, it may open cracks and fissures and heave soil (Heidmann and Thorud, 1976). Sometimes soil frost forms a series of vertical columns called *needle ice* or *stalactite frost*; the latter term being a misnomer, as stalactites grow downward, not upward as needle ice does.

Both surface tension and viscosity of liquid water vary with temperature. All other factors being equal, the saturated hydraulic conductivity of a soil would approximately double as the temperature increased from 0 to 25 °C, and volumetric moisture content at field capacity would be reduced by as much as 2% (Klock, 1972). This may be enough to sustain base flow all summer long from deep soils recharged only during snowmelt.

6.2 INFILTRATION

Infiltration is the process by which water enters and moves through the watershed surface. It is strictly a surface phenomenon, but it may be influenced by subsurface conditions. Water either passes through the soil surface or it does not. If not, small quantities may be stored in surface irregularities and as an adhesive film, but the rest becomes surface runoff. The surface of any watershed is thereby a critical factor in the disposition of rain or snow.

Two groups of factors control the infiltration rate into the watershed surface: (a) the soil factors that determine how rapidly water can be absorbed and (b) the rate at which water is applied from rain or melting snow. The first group determines the *infiltration capacity*, or the rate at which water can be absorbed, but does not necessarily determine the rate at which it will infiltrate. The *infiltration rate* is always less than or equal to the infiltration capacity, depending on the rate at which water is applied. Infiltration capacity is seldom limiting, for most precipitation or snow melt occurs at rates much lower than infiltration capacity. Thus infiltration capacity is significant only at excessive rainfall or snowmelt rates.

Nevertheless, most attention in watershed management is focused on the factors that control infiltration capacity, and with considerable justification, for it is often the extremes that are most important in hydrology, and many of these factors are readily modified by land use. On the other hand, the factors that determine the rate at which water reaches the surface from rain or melting snow are less susceptible to control.

6.2.1 Changes in Infiltration Capacity with Time

The infiltration capacity of a given soil is not constant. If water is applied to a dry soil, the rate of intake is rapid at first, but declines rapidly and then levels off, reaching an essentially constant rate. However, if the surface is nonwettable, the initial low infiltration capacity may increase to a peak before declining or leveling off (Fig. 6.5).

Among the factors responsible for the changes in infiltration capacity are (a) sealing of the soil surface; (b) decreases in capillary potential as moisture content increases; (c) swelling of soil colloids; (d) decrease in amount, size, and continuity of solid pores with depth; and (e) entrapment of soil air. Finally, the infiltration capacity of a saturated soil is limited by drainage rates as defined by Eq. 6.2, for water can enter a saturated soil only as fast as water in the soil is displaced.

The relative importance of each factor to infiltration capacity varies. Where soils are bare, surface sealing may dominate, but sealing is usually unimpor-

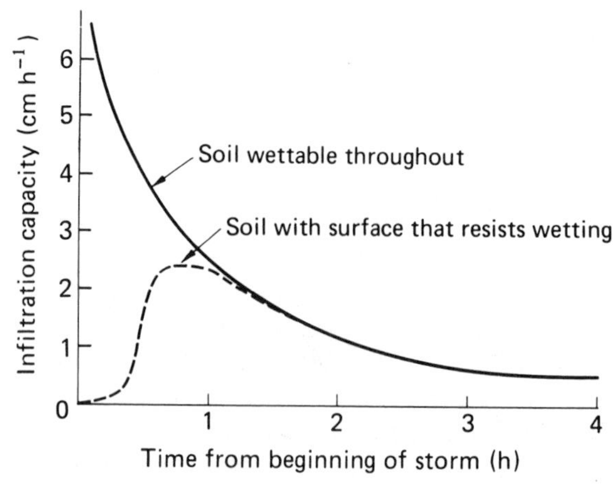

FIG. 6.5 Common trend of infiltration capacity during a single storm.

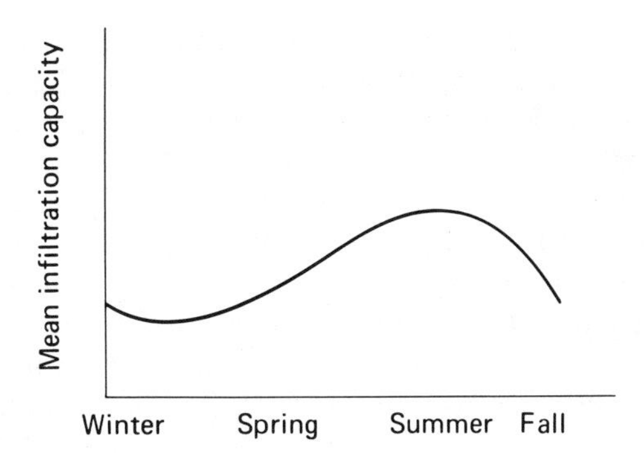

FIG. 6.6 General seasonal trend of infiltration capacity.

tant under a good cover of litter or low plants. Once any soil layer becomes saturated, capillary forces disappear, along with pore meniscuses, and this may cause a flow increase in nonwettable soils and a decrease in wettable ones. Some soils are much more susceptible to shrinkage and swelling than others, depending on the type and amount of colloidal material they contain. However, many cracks and fissures are closed more by material carried into them along with surface water than by swelling.

After rainfall stops and water is lost from the soil by drainage and evapotranspiration, the infiltration capacity recovers toward the state existing before the storm. However, it is seldom restored to precisely the same point even though the soil moisture content becomes the same. Progressive, seasonal changes occur in most soils due to the addition and decomposition of organic matter, the growth of plants and the activities of soil fauna, grazing, and differences in climate. The general pattern of seasonal changes in infiltration capacity is shown in Fig. 6.6. Maximums usually occur in late summer or early autumn and minimums in late winter or early spring, but exceptions have been noted. Infiltration capacities increased into fall on forested sites in southwestern Oregon (Johnson and Beschta, 1981), and compacted soils may be loosened by frost action and be most permeable in spring.

6.2.2 Effects of Land Use on Infiltration Capacity

The direct effects of land use are mostly changes in the amount and kind of vegetative cover and soil exposure and compaction associated with various activities. These activities have modified soil texture, structure, wettability, moisture content, frost type and occurrence, depth and nature of organic horizons, and soil biota. That the effect on infiltration rates has not been greater is because in most places, original infiltration capacities so far exceeded those necessary to absorb high intensity rain, that a considerable margin of safety existed.

But the margin of safety certainly can be exceeded, with risks increasing under the impact of continuing, ever more intensive use. Soil exposure leading to erosion and compaction are particularly critical land use impacts that affect infiltration capacity.

Erosion modifies infiltration by removing upper permeable soil layers, providing a source of fine particles for sealing surface pores and reducing the capacity of the soil to support protective vegetative cover and soil litter. Erosion is a consequence of reduced infiltration rather than an original cause, but once the cycle of accelerated erosion has begun, feedback effects may multiply the adverse consequences. Erosion will be considered in a separate chapter.

Soil compaction may also critically affect infiltration capacity by reducing the amount, size, and continuity of surface pore space. *Compaction* occurs when forces (loads, pressure, or vibration) at the soil surface pack the solid particles and aggregates together, making the soil more dense by reducing pore spaces. A reduction in pore space also may be caused by washing of fine particles into pores where they collect. Bulk density often is used as an index of compaction, although a variety of approaches have been employed (Flint and Childs, 1984b; Gifford et al., 1977; Howard and Singer, 1981).

There are many factors that cause compaction (Greacen and Sands, 1980). The exposure of soil is a common cause. The reasons are many. The very act of exposing soil implies removal of overlying litter and other organic material, the decomposition of which supports soil organisms active in creating a loose soil structure, and which is the source of cementing materials that help to bind aggregates together. As the organic material remaining in the soil is mineralized, organism populations decrease and aggregate development declines. Many of the cementing agents, organic gels, and the like, are themselves broken down and deaggregation occurs easily. Bare soils exposed to the atmosphere also exhibit a more severe environment for organisms than well-vegetated soils. Temperature and moisture extremes are more pronounced, such as freezing and thawing, alternate wetting and drying, and extreme heating. It is little wonder that the amount of organic matter on and in a soil is usually well correlated with its infiltration capacity, except when the organic matter is water repellent.

The exposure of bare soil to raindrop impact is an even more important factor in compaction. Raindrops have a double effect. They exert a considerable force when striking the earth, which moves soil particles closer together. More important, this force may break down aggregates and carry small particles of clay and silt into larger pores as water infiltrates. These small particles may fill and clog the pores until the amount and size of the pore space in the surface soil is very seriously reduced (Mc Intyre, 1958; Skaggs, 1971; Morin et al., 1981). On soils covered with litter and vegetation, the force of rain is largely dissipated. The topic of rainfall impact is discussed more thoroughly in Chapter 10 (Erosion and Sediment), which is closely related to the problems of infiltration and the energy of moving water.

Soils need not be bare and exposed to be compacted. Many soils have a low resistance to the forces of compaction, regardless whether the force is the impact of falling rain, the vibration of a bulldozer track, the boot of a hiker, or the hooves of cattle or elk.

An excellent review of soil compaction on forest and range lands was presented by Lull (1959), most of which is still valid. He summarized key principles of compaction as follows:

The amount of compaction will depend on the degree to which the stress applied to the soil overcomes the resistance the soil offers to compaction.

The resistance that the soil offers depends on its moisture content, texture, structure, density, and organic content.

As this resistance is overcome, the effect is to pack individual soil particles closer together and to crush soil aggregates, thus, reducing pore space.

Resultant additional soil materials per unit volume increase the resistance of the soil to deformation to a point where resistance and stress are in equilibrium and no further compaction occurs.

Soils that have the greatest range in particle-size (i.e., medium textured soils) compact to greatest densities, finer particles filling the voids between coarser particles.

The less dense the soil, the greater opportunity for compaction.

The greater the organic content, the smaller the maximum compaction.

Soil freezing tends to compact soil by breaking down water-stable aggregates and tends to loosen compacted soils.

Duration of compaction depends largely on the stresses the soil undergoes by swelling and shrinking from changes in moisture content and temperature.

Compaction increases bulk density, reduces total pore space by the same proportion, reduces noncapillary (large) pore space by a greater amount, and has its greatest effect on infiltration and percolation.

It follows from the principles of compaction that any land use that substantially reduces production of organic material will increase the tendency of soil to compact. In most cases, removal of vegetation is associated with direct compaction from equipment or animals, and an increase in the amount of soil moisture may compound the problem by promoting *soil puddling* (loss of soil structure at high moisture contents). The converse is also true. Increases in vegetation and organic matter decrease the tendency of soil to compact. Furthermore, they may help to restore compacted soils to their original state. Gaiser (1952) indicated that a 51-year-old white oak stand developed over 10,000 highly permeable root cavities per hectare through a compact clay. The cavities were filled with loose organic matter or mixtures of organic matter and soil from the activity of soil organisms. Macropores may persist for long periods. Aubertin (1971) found at least a third generation live root in one old root channel. Persistence of channels was greatest in silt and silty clay loams and least in coarse textured soils.

With regard to compaction and its subsequent restoration, the time span for each may be quite different. A single trip with heavy machinery over a moist soil may be enough to compact it severely, but recovery processes may take years. Studies in southern United States show recovery from compaction in from 8 to 18 years (Hatchell and Ralston, 1971; Dickerson, 1976), whereas decades may be needed in the Pacific Northwest (Froehlich and McNabb, 1983; Froehlich et al., 1985), and it may take a century or more in arid regions (Webb and Wilshire, 1980; Iverson et al., 1981.)

A continuous cover of vegetation alone may not be enough to prevent compaction. Old pastures on heavy soils of abandoned farms in New York State are often more highly compacted than abandoned cultivated fields, and their recovery may be much slower. The effect is probably due to grazing and trampling in wet weather, for summer rain is frequent. Whereas fields were not cultivated when wet, cattle grazed continuously throughout the growing season.

Compaction also influences the type of soil frost that develops in regions of frozen soils, as does the vegetation type and density and quantity of litter on the soil. Concrete frost typically forms in compacted soil containing little organic matter. On the other hand, frost formation in loose soil with large pores and much organic matter is more frequently of a porous granular or honeycomb form, which may enhance infiltration (Megahan and Satterlund, 1961). Litter also provides insulation that helps to keep the soil from freezing as deeply or completely as bare soils.

Fire, both wildfire and prescribed burns, may also influence infiltration by modifying soil wettability, as discussed in Section 6.1.3. Sites susceptible to formation of nonwettable soils should be burned at temperatures that limit development of nonwettable conditions or be protected from fire.

6.3 SOIL MOISTURE REDISTRIBUTION AND DRAINAGE

Water infiltrating through the surface is either stored or continues to move through the soil. Unless soil macropores are open to the surface, a saturated layer usually develops before water enters the noncapillary pore system where it may move into and through unsaturated zones. Part of the water in structural pores always tends to move into the finer textural pores of the aggregates where it is retained by capillary forces after the larger pores drain. The net effect is a redistribution of moisture within the soil and subsoil column as shown in Fig. 6.7.

In moist soils, a distinct wet front may not develop as water moves through the soil profile, nor does the soil need to become saturated at any level. Nixon and Lawless (1960) observed that passage of moisture through recently wetted permeable profiles caused but a relatively slight increase in moisture content

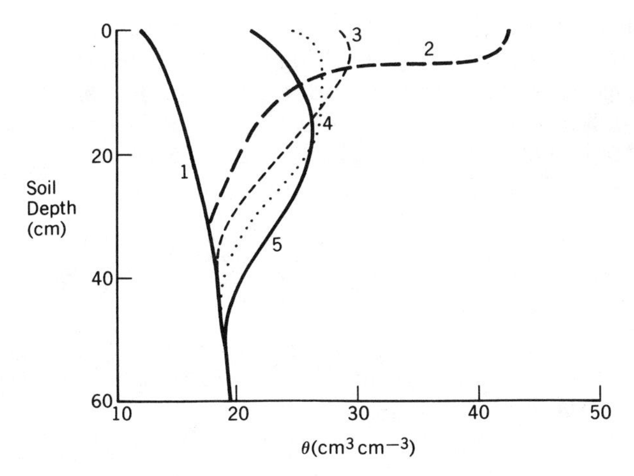

FIG. 6.7 Distribution of moisture (expressed as volumetric content θ) in a soil profile before, during, and after a storm. (1) Before storm. (2) End of storm. (3–5) One, two, and three days after end of storm, respectively.

at each depth, and the additional increment of rainfall was rapidly distributed rather uniformly throughout.

Land use has a much lesser effect on subsurface properties that influence soil moisture redistribution, storage, and drainage than on surface properties. Quite obviously, anything that influences evapotranspiration has a strong influence. However, this subject is deserving of separate treatment. Our concern here is upon practices that influence the soil and subsoil.

Few rapid or pronounced changes in subsurface properties result directly from land use except for compaction that may extend to as much as 30 cm or more deep. Most changes are gradual or slight. Nevertheless, they may be of great hydrologic importance in the long run.

The most important effects that land use has on subsurface soil characteristics are those that change pore size, quantity, and arrangement. They come about directly, as by the development of root channels by vegetation, or indirectly through organic matter production and biotic activity in the soil that may modify structural development. Here the type of vegetation can be important. Much of the organic matter produced by grasses develops underground as roots, whereas most of the organic matter produced by trees develops above ground. Similarly, there are vast species differences in root type, depth, and distribution. Different types and quantities of soil fauna are associated with different plant communities. It would require a thorough exposition of the ecology of many plants and animals that is beyond the scope of this presentation to discuss these many factors in detail. But any form of land use that results in the incorporation and decomposition of large amounts of organic matter deep in the soil will generally maintain or enhance desirable subsurface conditions for the movement and storage of moisture.

Under most circumstances, however, land use effects on subsurface soil conditions are not likely to be controlling in a watershed management program. Land use practices that affect evaporative withdrawal and soil surface conditions generally outweigh them. Furthermore, they are strongly interrelated, and only seldom will a practice that enhances surface conditions fail to maintain or enhance desirable subsurface conditions.

LITERATURE CITED

Aubertin, G. M. 1971. Nature and extent of macropores in forest soils and their influence on subsurface water movement. USDA Forest Service Res. Pap. NE - 192. Northeastern For. Expt. Sta., Upper Darby, PA.

Bevin, K. and P. Germann. 1982. Macropores and water flow in soils. *Water Resour. Res.* 18:1311–1325.

Childs, E. C. 1967. Soil moisture theory. *Adv. Hydrosci.* 4:73–117.

De Bano, L. F. 1981. Water repellent soils: a state of the art. USDA Forest Service. Gen. Tech. Rept. PSW - 41. Pac. Southwest For. Range Expt. Sta., Berkeley, CA.

De Bano, L. F. and J. S. Krammes. 1966. Water repellent soils and their relation to wildfire temperatures. *Int. Assoc. Sci. Hydrol. Bull.* 11:14–19.

Dickerson, B. P. 1976. Soil compaction after tree-length skidding in northern Mississippi. *Soil Sci. Soc. Am. Proc.* 31:565–568.

Dixon, R. M. 1971. Infiltration role of large soil pores: a channel system concept. pp. 136–147. In: *Biological effects in the hydrological cycle,* Proceedings. Purdue University, West Lafayette, IN.

El Bushi, I. M. and S. N. Davis. 1969. Water-retention characteristics of coarse rock particles. *J. Hydrol.* 8:431–441.

Flint, A. L. and S. Childs. 1984a. Physical properties of rock fragments and their effect on available water in skeletal soils. pp. 91–103. In: D. M. Kral, Ed. *Erosion and productivity of soils containing rock fragments.* Soil Sci. Soc. Am. Spec. Publ. No. 13, Madison, WI.

Flint, A. L. and S. Childs. 1984b. Development and calibration of an irregular hole bulk density sampler. *Soil Sci. Soc. Am. J.* 48(2):374–378.

Froehlich, H. A. and D. H. McNabb. 1983. Minimizing soil compaction in Pacific Northwest forests. pp. 159–192. In: E.L. Stone, Ed., *Forest Soils and Treatments Impacts.* Proc. Sixth North Am. Forest Soils Conf. University of Tennessee Conferences, Knoxville, TN.

Froehlich, H. A., D. W. R. Miles, and R. W. Robbins. 1985. Soil bulk density recovery on compacted skid trails in central Idaho. *Soil Sci. Soc. Am. J.* 49(4):1015–1017.

Gaiser, R. N. 1952. Root channels and roots in forest soils. *Soil Sci. Soc. Am. Proc.* 16:62–65.

Germann, P. and K. Bevin. 1981. Water flow in soil macropores. I. An experimental approach. *J. Soil Sci.* 32:1–13.

Gifford, G. F., R. H. Faust, and G. B. Coltharp. 1977. Measuring soil compaction of rangeland. *J. Range Manage.* 30:437–460.

Greacen, E. L. and R. Sands. 1980. Compaction of forest soils. A review. *Aust. J. Soil Res.* 18:163–180.

Hatchell, G. E. and C. W. Ralston. 1971. Natural recovery of surface soils disturbed in logging. *Tree Planters Notes* 22(2):5–9.

Heidmann, L. J. and D. B. Thorud. 1976. Controlling frost heaving of ponderosa pine seedlings in Arizona. USDA Forest Service Res. Pap. RM - 172. Rocky Mt. For. Range Expt. Sta., Ft. Collins,CO.

Howard, R. F. and M. J. Singer. 1981. Measuring forest soil bulk density using irregular hole, paraffin clod, and air permeability. *For. Sci.* 27:316–322.

Iverson, R. M., B. S. Hinckley, R. M. Webb, and B. Hallet. 1981. Physical effects of vehicular disturbances on arid landscapes. *Science* 212:915–917.

Johnson, M. G. and R. L. Beschta. 1981. Seasonal variation in infiltration capacities of soils in western Oregon. USDA Forest Service Res. Note PNW - 373. Pac. Northwest For. Range Expt. Sta., Portland, OR.

Kane, D. L. and J. Stein. 1983. Water movement into seasonally frozen soils. *Water Resour. Res.* 19:1547–1557.

Klock, G. O. 1972. Snowmelt temperature influence on infiltration and soil water retention. *J. Soil Water Conserv.* 27:12–14.

Kramer, P. J. 1969. Plant and soil-water relations: a modern synthesis. McGraw-Hill, New York. pp. 482.

Lull, H. W. 1959. *Soil compaction of forest and range lands.* USDA, Washington, DC, Misc. Pub. No. 768.

Mc Intyre, D. S. 1958. Permeability measurements of soil crusts formed by raindrop impact. *Soil Sci.* 85:185–189.

Megahan, W. F. 1973. Role of bedrock in watershed management. *Proc. Irrig. Drain. Div. Am. Soc. Civ. Eng. Conf.* held at Ft. Collins, CO. pp. 449–470.

Megahan, W. F. and J. L. Clayton. 1983. Tracing subsurface flow on roadcuts on steep, forested slopes. *Soil Sci. Soc. Am. J.* 47:1063–1067.

Megahan, W. F. and D. R. Satterlund. 1961. Winter infiltration studies on abandoned and reforested fields in central New York. *Proc. Eastern Snow Conf.* pp. 121–132.

Morin, J., Y. Benyamini, and A. Michaeli. 1981. The effect of raindrop impact on the dynamics of soil surface crusting and water movement in the profile. *J. Hydrol.* 52:321–335.

Mosely, M. P. 1982. Subsurface flow velocities through selected forest soils, South Island, New Zealand. *J. Hydrol.* 55:65–92.

Nixon, P. R. and G. P. Lawless. 1960. Translocation of moisture with time in unsaturated soil profiles. *J. Geophys. Res.* 65:655–661.

Pearce, A. J., M. K. Stewart, and M. G. Sklash. 1986. Storm runoff generation in humid headwater catchments. 1. Where does the water come from? *Water Resour. Res.* 22:1263–1272.

Philip, J. R. 1969. Theory of infiltration. *Adv. Hydrosci.* 5:215–289.

Skaggs, R. W. 1971. Infiltration of water into the soil profile. pp. 123–135. In: *Biological effects in the hydrological cycle.* Proceedings, Purdue University, West Lafayette, IN.

Webb, R. H. and H. G. Wilshire. 1980. Recovery of soils and vegetation in a Mohave desert ghost town, Nevada, USA. *J. Arid Environ.* 3:291–303.

7 Energy Exchange

Water is unique in that it simultaneously represents both the greatest flow of matter and a major flow of energy through terrestrial ecosystems. More than any other substance on earth, it ties together the cycle of matter and energy through its great mass and latent heats of vaporization, fusion, and sublimation. It takes large amounts of energy to evaporate water and melt or sublimate ice; energy that is absorbed, moves with, and is released by water moving through the hydrologic cycle.

Energy is the capacity to do work. There are many forms of energy, all of which can be converted from one kind to another. The important forms of energy in hydrology are radiant energy, heat or thermal energy (which may be either sensible or latent), and mechanical energy (either potential or kinetic). In this chapter, radiation, sensible heat, and latent heat will be discussed in detail. Potential energy was discussed in Chapter 6 and will be considered again in Chapter 8. Kinetic energy will be introduced, but is applied mostly in relation to erosion, which is the subject of Chapter 10.

Radiant energy is a form of electromagnetic energy that occurs when an atom becomes excited, and for a brief instant one of its electrons leaves its orbit about the nucleus and just as quickly returns. The brief electron shift is accompanied by the emission of energy. The energy so emitted is propagated across space in a manner that appears to act both as undulatory waves and as a stream of tiny particles called *photons*.

Sensible heat is heat of molecular motion contained in a substance that determines its temperature. It is called sensible heat because its relative presence can be detected by anyone's unaided senses according to whether the substance is hot or cold. A common measure of the amount of sensible heat is the *calorie* (cal), the amount of energy that is required to raise the temperature of 1 g of water by 1 °C at a temperature of 15 °C. It is equal to 4.1855 joules (J).

Latent heat is heat contained in a substance by virtue of its physical state. As pointed out in Section 5.1.1, the three physical states of water (solid, liquid, and gas) have widely different energy states at the same temperature, and large quantities of energy are needed to raise water from a lower to a higher state.

Mechanical energy has two forms: *kinetic energy*, the energy of motion, and *potential energy*, energy due to the position in the earth's gravitational field or differences in pressure status, as with soil moisture or water vapor in air. All forms of energy are convertible into all other forms, and the common unit of energy in the SI system is the joule, which is the work done when a force of 1 newton (N) displaces a point through a distance of 1 m. A *newton* is the force that, when applied to a mass of 1 kg, will give it an acceleration of 1 m s^{-1}.

Radiant energy was, until recently, commonly expressed in units of langleys (ly); a *langley* is equal to 1 cal cm^{-2} or 4.1855 J cm^{-2}. Radiant energy is now frequently expressed in watts per square meter (W m^{-2}). A *watt* is the SI unit of power equal to 1 J s^{-1}, and 1 ly min^{-1} is equal to 697.3 W m^{-2}.

In this book, we primarily use SI units. However, other units commonly are found in the literature, such as langleys for radiant energy and calories for sensible and latent heat. The serious student will need to learn the conversions in the Appendix because the widespread conversion to the SI system in scientific literature did not occur until the 1980s, and earlier publications used other systems.

Radiant and heat energy is transferred in three major ways: radiation, conduction, and convection. *Radiation* is the movement of electromagnetic energy through space as waves or discrete packets (*quanta*) of photons. Its movement is not dependent on the presence of matter; it moves best through empty space.

Conduction is the transfer of sensible heat through matter by means of intermolecular contact, without transfer of the matter itself. It always moves from the warmer to the colder point.

Convection is the transfer of either sensible or latent heat by the movement of the substance containing the heat. These movements occur only in liquids and gases. Convection is sometimes referred to as *eddy diffusion*, in that whereas diffusion is on a molecular scale, eddy diffusion is concerned with the motion of larger masses of matter. Eddy motion may be caused by thermally induced buoyancy or by the deflection of flow moving across a rough surface.

In the atmosphere, *advection* is a term used with reference to horizontal transport of heat as air masses move from one area to another. Figures 7.1 and 7.2 illustrate the modes of energy transfer during day and night that will be discussed in the sections that follow.

7.1 RADIANT ENERGY

Essentially all of the energy available on earth comes from the sun in the form of solar radiation. The sun is an intensely hot body, with internal temperatures exceeding 100000 K[1] and with a surface temperature that approximates 6000 K.

7.1.1 Radiation Intensity

All matter radiates energy at an intensity proportional to the fourth power of its absolute temperature (K). The sun radiates throughout the electromagnetic

[1]The kelvin (K) is the SI unit for temperature. Units are the same as on the Celsius scale, but zero is 273.16° lower. K = °C + 273.16. Kelvin is also known as absolute temperature.

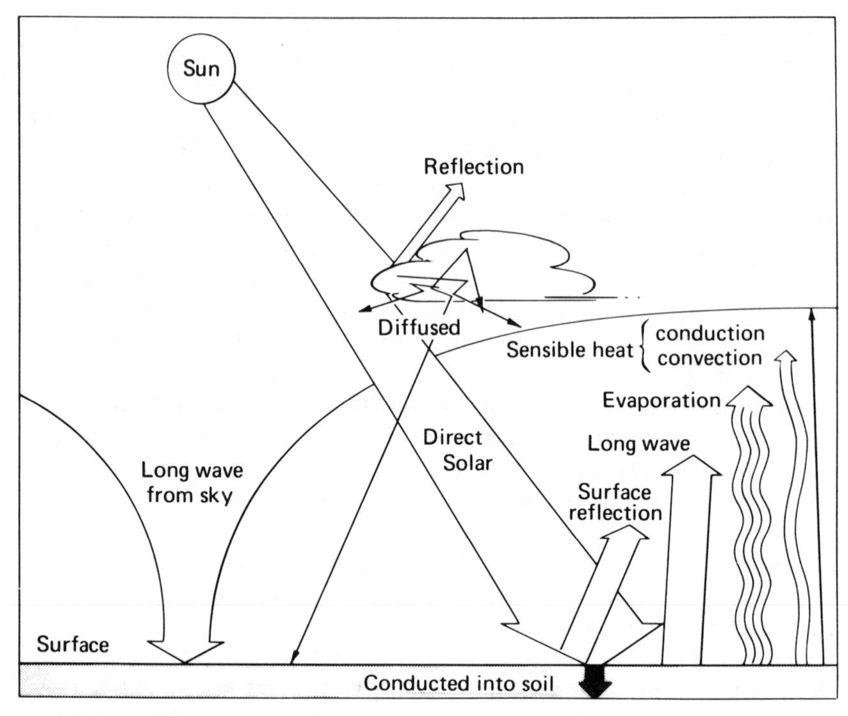

FIG. 7.1 Sources and disposition of energy at noon on a summer day.

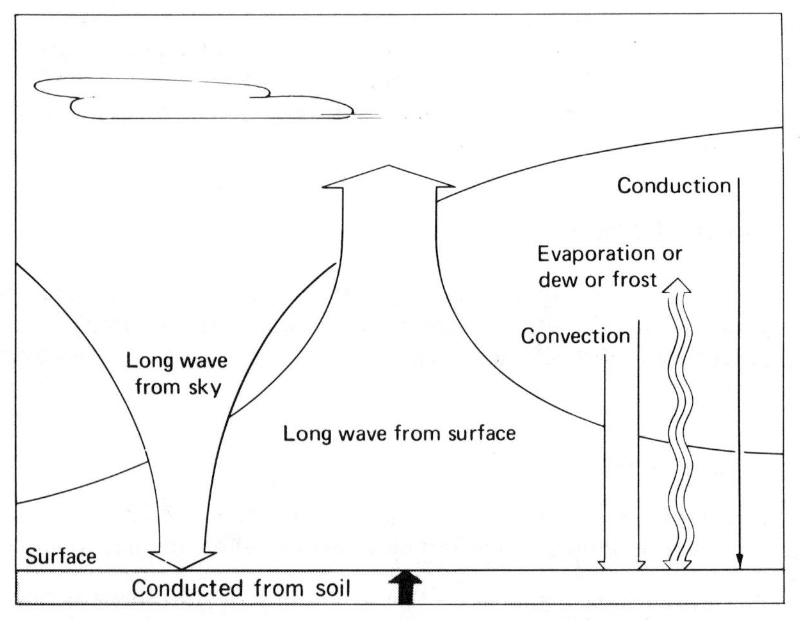

FIG. 7.2 Sources and disposition of energy at night.

spectrum, ranging from the very short wavelengths in the ultraviolet (UV) region, into the visible and infrared (IR), to long wavelength radio waves. However, most matter radiates imperfectly, being better at some wavelengths than others. Perfect radiators emit energy throughout the spectrum only according to their temperature and are called *blackbodies*. Imperfect radiators are called *gray bodies*, and the rate of their total radiation is less than would occur from a blackbody at the same temperature. The *emissivity* of a body is the ratio between its actual radiation emission and the radiation emitted by a blackbody at the same temperature, expressed as a decimal. A blackbody has an emissivity of 1.0.

The intensity of radiation in the total spectrum is related to temperature and emissivity by the Stefan–Boltzmann radiation law whereby

$$R = \varepsilon\sigma T^4 \tag{7.1}$$

where R is the total intensity of radiation (W m^{-2}), σ is the Stefan–Boltzmann constant (5.67×10^{-8}), ε is the emissivity, and T is the temperature (K) (Gates, 1962). Radiation from the sun outward into space exceeds 73 MW m^{-2} from its entire surface. At 15 °C (288.16 K), near the average temperature of the surface of the earth, a blackbody would radiate 390 W m^{-2}. Over the range in temperatures from −45 °C (228.16 K) to 50 °C (323.16 K), which encompasses most of the temperature range commonly found on the earth's surface, blackbody radiation changes by a factor of 4, from about 154 to 617 W m^{-2}.

7.1.2 Quality of Radiation

The spectral quality of radiation is also a function of the temperature of the radiating body. Although some energy is radiated at all wavelengths by any blackbody, the peak of the radiation curve follows the Wien law,

$$\lambda = \frac{2897}{T} \tag{7.2}$$

where λ is the peak wavelength in micrometers (μm) and T is the temperature in kelvins (Reifsnyder and Lull, 1965). Thus, at temperatures common to the earth's surface, the peak wavelength ranges from about 9 to 13 μm, well into the IR portion of the spectrum. Solar radiation peaks in the visible part of the spectrum are at about 0.47 μm. Figure 7.3 illustrates the intensity of solar and terrestrial radiation in relation to wavelength. Most of the sun's radiation occurs at wavelengths of 3.5 um or less and is often referred to as *shortwave radiation* or *solar radiation*. Radiation from the surface of the earth and its cover is mostly of longer wavelengths and is generally referred to as *longwave radiation*. Sometimes longwave radiation is called *terrestrial radiation* or *atmospheric radiation* if the matter emitting it is the earth or sky, respectively. Occasionally longwave radiation is termed *infrared (IR) radiation*, but IR simply means "beyond red,"

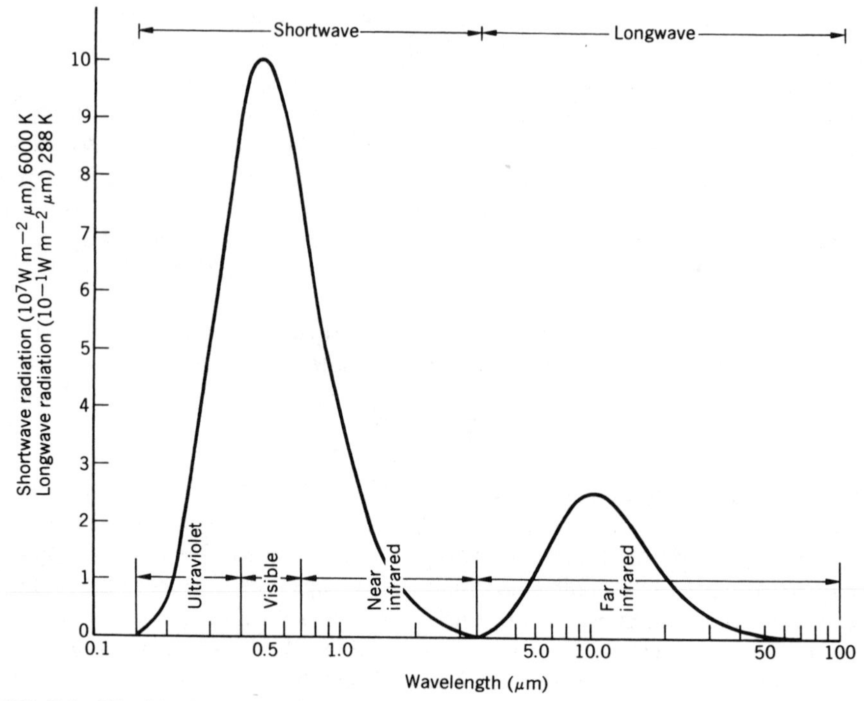

FIG. 7.3 Blackbody radiation spectra at approximate sun (6000 K) and earth (288 K) temperatures. Note different scales for each.

and occurs in both the short- and longwave parts of the spectrum. Sometimes the terms *near IR* and *far IR* are used to distinguish between the short- and longwave portions, respectively.

Certain properties of absorption, reflection, and emission of radiation are related to wavelength. These differences are very important in understanding certain characteristics of the hydrologic cycle and will be discussed in future sections, particularly in Chapters 8 and 9.

7.1.3 Radiation Reaching the Surface of the Earth

7.1.3.1 Solar Radiation. The radiant energy emitted by the sun is radiated in all directions into space, and the intensity of radiation reaching any object is inversely proportional to the square of its distance from the radiating body. The earth, at an average distance of about 150 million km from the sun (slightly less in the northern hemisphere winter and slightly more in summer), intercepts only an extremely small fraction of the solar radiation. The amount that would reach the earth's surface in the absence of an atmosphere is 1376 W m^{-2} on a plane perpendicular to the sun (Hickey et al., 1980). This quantity is called the *solar constant* and represents the maximum flux of direct beam solar radiation available at the surface of the earth.

Not all of this radiation reaches the surface of the earth, however. Air is never perfectly transparent to solar radiation, even on clear days. Most of the UV wavelengths are absorbed by the ozone layer of the atmosphere and other wavelengths are selectively absorbed by oxygen, water vapor, and carbon dioxide. Dust, smog, clouds, and other particulates cause further absorption, scattering, and reflection back to outer space. On clear, dry days at high elevations (about one-quarter of the earth's atmosphere is below 2400 m) as much as 95% of the solar radiation may reach the surface, whereas less than one-half may reach the surface under a heavy Los Angeles smog. Only 10% or less of that reaching the top of the atmosphere may reach the ground under a dense cloud cover. Variations in atmospheric conditions result in highly variable solar irradiation from place to place and day to day as the weather changes. On the average, only about 47% that strikes the top of the atmosphere reaches the surface annually in the northern hemisphere (Gates, 1962).

Outside the tropics there is an increasingly pronounced seasonal rhythm of irradiation toward the poles that results from the declination of the earth's axis by 23.45° from its plane of rotation about the sun. Thus, the position of the sun in relation to the equator (the *solar declination*) varies ±23.45° from the vertical from summer to winter, causing the angle at which the noonday rays strike a horizontal surface to vary by 46.9° annually. At Pullman, Washington (latitude 46.75° N), the sun reaches a maximum elevation above the horizon of only 19.8° on the first day of winter and 66.7° on the first day of summer. Noon solar altitude is a function only of solar declination and latitude and can be determined for any date as

$$A = 90° + d - l \tag{7.3}$$

where A is the solar altitude above the horizon, d is the solar declination, and l is latitude, all in degrees. The approximate solar declination for many dates may be found in Satterlund (1977) or Lee (1978) and can be interpolated for any date.

The energy in direct beam solar radiation is a function of the sine of the angle of incidence, so at Pullman, in the absence of an atmosphere, a unit horizontal surface would receive about 2.7 times as much energy at noon in summer as in winter. When the lower solar altitude throughout the day and shorter daylength of winter is considered, the daily potential energy in summer is about 4.3 times greater.

On any given watershed or slope facet of a watershed, the steepness and direction of the surface may greatly affect the amount of radiation received. Again at Pullman, any north slope steeper than 19.8° (36%) would receive no direct solar radiation at all on the first day of winter, but all south slopes would receive more than horizontal surfaces. The approximate potential direct beam solar irradiation can be determined for many slopes in different places on earth by means of readily available tables (Fons et al., 1960; Frank and Lee, 1966; Buffo et al., 1972). Methods of determining the mean slope of a watershed or its parts have been presented by Lee (1978). Equations for calculating irradiation in other places are presented in all the above cited papers.

The lower energy reaching the surface in northern hemisphere winters is further reduced by the greater distance that low angle beams must travel through the atmosphere, with an attendant increase in reflection and absorption of the radiation, and, in many regions, by the greater cloudiness in winter than in summer. At Spokane, Washington (latitude 47.62° N), the average solar radiation in December was only 36.3 W m^{-2} and in July was 322 W m^{-2}, during the period 1955–1964. Solar radiation received in the winter months of December, January, and February was just slightly more than one-fifth of summer radiation receipt in June, July, and August (Phillips, 1965).

The effect of the atmosphere on irradiation of slopes is to reduce the differences among them. On a clear day in winter when steep north slopes receive no direct beam radiation, they still receive radiation scattered by the atmosphere and reflected from nearby surfaces. Differences between south and north slopes are great, but with increasing cloudiness they disappear and under a heavy overcast all slopes are bathed in near equal amounts of scattered solar radiation which has no directional component (Geiger, 1959).

7.1.3.2 ***Longwave Atmospheric Radiation.*** Although solar radiation is the ultimate source of nearly all the energy on earth, it represents directly only a portion of the radiant energy received on any particular watershed. Longwave radiation is also important in energy exchange. Clouds and water vapor are especially important emitters of longwave radiation from the atmosphere. Incoming longwave radiation can be estimated over a wide range of atmospheric conditions with the following equation, modified from Satterlund (1979):

$$R_l = \sigma T^4[1 + (0.08)(1-kC)]\{1 - [\exp - (e_o{}^{T/2016})](1 - kC)\} \qquad (7.4)$$

where R_l is the incoming longwave radiation (W m^{-2}), σ is the Stefan-Boltzmann constant (5.67×10^{-8}), T is the air temperature (K), C is the cloud cover as a decimal, e_o is the vapor pressure in 10^2 Pa, and k is an empirical coefficient having a value of 0.90–0.95, depending on cloud density and altitude. Clear sky emissivity ranges mostly from 0.75 to 0.85, increasing with humidity. Emissivity approaches 1.0 as low altitude cloud cover becomes complete.

Daily incoming longwave radiation from the atmosphere usually exceeds the flux of solar radiation except under cool, dry skies in late spring or early summer. For example, under clear skies with a mean air temperature of 15 °C and a relative humidity of 40%, longwave radiation is 308 W m^{-2}; at the same temperature with 75% relative humidity and 50% cloud cover, it is 364 W m^{-2}. Even in winter, with a temperature of 0 °C, relative humidity of 60 and 50% cloud cover, incoming longwave radiation is 274 W m^{-2}.

Longwave atmospheric radiation differs from solar radiation in other respects. It bathes the surface day and night, varying in amount only as air temperature, humidity, and cloudiness change. Furthermore, since the atmosphere envelops a watershed from horizon to horizon, as does a continuous cloud layer, atmospheric longwave radiation lacks the directional feature of solar beam radiation and bathes all slopes nearly uniformly.

7.1.4 Radiation Exchange at the Earth's Surface

Radiation reaching the surface of the earth goes through processes similar to those that occur in the atmosphere, namely, absorption, reflection, and transmission. Transmission is essentially limited to a narrow layer from the top of the tallest vegetation to the ground surface or bottom of shallow lakes and streams and will be henceforth ignored except in relation to snow. The earth's surface is highly variable in character, with great differences in the disposition of radiant energy.

7.1.4.1 Absorption of Radiation. Radiant energy reaching the earth's surface is either reflected or absorbed. Therefore, the reflectivity of a surface, or its *albedo*, is of primary importance in determining the energy available to do work. Reflected radiation does not enter into any reactions unless it is later absorbed, but merely changes its direction of movement. The albedo of various substances varies with wavelength of the incident radiation, but is usually expressed as a percentage of the entire solar spectrum. (The term is not used in regard to longwave radiation, as most substances that cover the earth's surface have similar absorbance and emittance characteristics to longwave radiation and approach blackbody efficiency.) The albedos of common materials covering watersheds are given in Table 7.1.

TABLE 7.1. Albedo of Watershed Surfaces

Surface Cover	Albedo (%)
Water	5–10
Bare soil (light colored, dry)	20–35
(dark colored, moist)	8–15
Grass (short, green, dry)	25–35
(short, green, wet)	15–20
(tall, cured)	25–30
(tall, green)	15–20
Marsh and Bogs	15–20
Forests (spruce, dense, no snow)	5–10
(spruce, dense, snow)	20–25
(mixed conifer-hardwood, in leaf)	10–15
(hardwoods, in leaf)	15–20
(hardwoods, winter, snow)	35–45
Snow (fresh)	80–95
(old)	40–70

Several things stand out from a consideration of albedo, foremost the great variability. It varies by vegetative type and even by species; for example, in the range shown for forests, dark species such as spruce or hemlock have lower, and light species such as aspen and birch have higher albedos. However, as the human eye is sensitive to only about one-half of the solar radiation reaching earth and albedo varies with wavelength, visual estimation of albedo can be somewhat misleading. Note that wet surfaces have lower albedos than dry ones, so they absorb more solar energy. Note too, the extremely high and changeable albedo of snow.

Sometimes, figures for the albedo of single leaves are given in the literature. In comparison with the albedo of an entire tree or stand of vegetation, these figures are invariably high, for they do not take into account the opportunities for absorption of radiation reflected from one surface to another in the many surfaces of a full stand of vegetation. Multiple reflection and absorption always reduces the albedo of a deep stand of vegetation as compared to a single surface (Miller, 1955).

Absorption is the complement of albedo, so the amount of radiation absorbed by any surface is

$$R_a = R(1.00 - r) \tag{7.5}$$

where R_a is the radiation that is absorbed, R is the incoming radiation, and r is the reflectivity or albedo of the surface expressed as a decimal. Thus, a reduction of the albedo of fresh snow from 0.9 to 0.8 as it begins to melt results in a doubling of the amount of shortwave radiation that is absorbed.

Upon absorption, radiant energy is always converted to some other kind of energy, usually sensible or latent heat. It cannot be stored as radiant energy. Only absorbed energy is available to do the work of heating the environment and driving the hydrologic cycle.

7.1.4.2 ***Emission of Radiation.*** Since all matter radiates energy in accordance with Eq. 7.1, much of the radiation that is absorbed by the earth's surface is reradiated back toward space. Such radiation is continuous and as most natural substances on the earth's surface (including snow) radiate nearly as blackbodies at normal temperatures, outgoing longwave radiation is substantial in amount. At the temperature of melting snow, it is 315 W m^{-2}, whereas a hot surface of 50 °C may emit 617 W m^{-2}.

Not all outgoing radiation reaches space, however. At any given temperature all matter absorbs the same wavelengths as it emits, so much of the outgoing longwave radiation is absorbed by the oxygen, carbon dioxide, and water vapor of the atmosphere even when skies are clear. It is ultimately reradiated in all directions by the atmosphere. When the sky is covered by clouds it absorbs nearly all the outgoing radiation, and cooling of the earth's surface by loss of outgoing radiation is largely offset by radiation from the base of the clouds back to the surface.

Ultimately, all radiation received by the earth is lost back to outer space. If less were lost than received, the earth would become steadily hotter, whereas if more were lost it would become steadily colder. Since in the long run, the temperature of the earth and its atmosphere is fairly constant, incoming and outgoing radiation must be in balance for the earth as a whole.

7.1.4.3 ***Net Radiation and the Radiation Balance.*** It is one thing for the earth as a whole to exhibit a near-perfect radiation balance, but it is quite another thing for any given place of the surface of the earth to do so. Tropical regions consistently receive more radiation than they emit, and polar regions usually emit more than they receive. In temperate regions, the yearly budget may be roughly in balance, but late winter to early summer is a time of surplus and late summer to early winter a time of deficit. The difference between incoming and outgoing radiation is termed *net radiation*.

Net radiation represents the quantity of radiation that can be converted to other forms of heat energy to make ecosystems function; warming the atmosphere and surface, melting snow, evaporating water, and providing energy for photosynthesis. Photosynthesis, though vital to all life, accounts for only a minor fraction of net radiation and will be neglected as a component of the radiation balance.

Net radiation at any point can be determined by summing all the sources of incoming radiation and subtracting all the sources of outgoing radiation over a period of time. It may have a positive, zero, or negative value. When negative, some other form of energy is needed to supply the requirement for the excess outgoing radiation. When positive, the excess is stored in the matter that absorbs it, or transferred elsewhere, largely in the form of currents of warm air or water vapor. A representative radiation balance in a large, open, horizontal field on a summer day in a temperate region is shown in Table 7.2, and the diurnal pattern is illustrated in Fig. 7.4.

TABLE 7.2. Representative Radiation Balance for a Grassy Field on a Clear Summer Day in the Temperate Zone

Source	Amount ($W\ m^{-2}$)
Incoming	
Shortwave (solar)	332
Longwave (atmosphere)	330
Total	662
Outgoing	
Albedo (reflected shortwave) at 25%	83
Longwave emitted (mean blackbody temperature of 293 K)	418
Total	501
Net	161

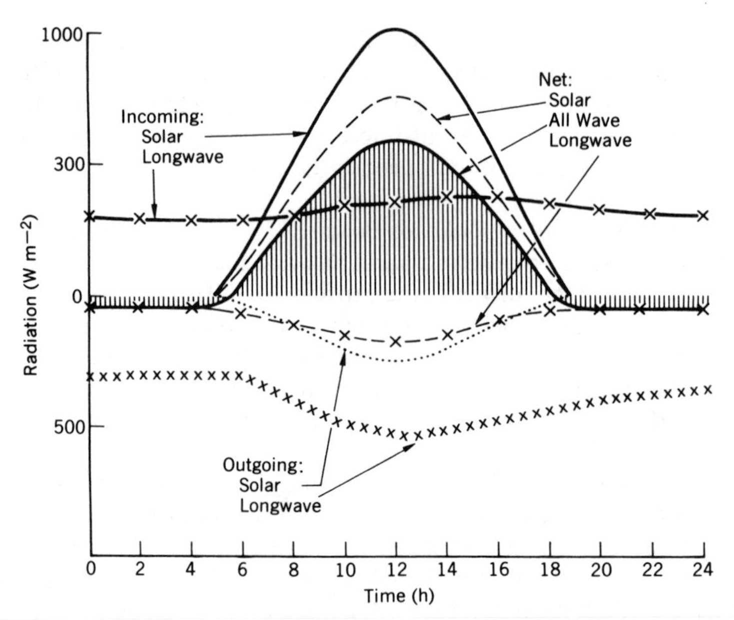

FIG. 7.4 Diurnal pattern of the radiation balance for a grassy field on a clear summer day in the temperate zone.

7.1.4.4 Shadows and View Factors. The radiation balance of any point is greatly influenced by the nature of its surroundings. Except in large, open, horizontal fields, part of the sky is obscured by hills, mountains or trees, or other tall vegetation. These elements of the landscape may block or reflect the sun's beams, and act as near-blackbody sources of longwave radiation while reducing incoming longwave radiation from the atmosphere.

Whether a given point is in the shade or the sun at any given time depends on the position of the topography or vegetation relative to that of the sun. The sun's position in the sky (its altitude and direction) from any point at any time may be calculated from equations such as those presented in Satterlund (1977). It can be approximated by the use of sun path diagrams such as are found in Buffo et al. (1972) or Brown (1973) and illustrated in Fig. 7.5. If the topography or vegetation in the direction of the sun extends to a greater angular altitude than the sun, the point is in the shade; if less, it is in the sun.

Estimates of the amount of shade cast by forest stands can be developed using the methods of Wilson and Petzold (1973) for uniform stands, and of Satterlund (1983) for randomly spaced stands.

The proportion of sky visible from a point on the surface, relative to a complete hemisphere, is the *view factor* of that point. It is important because longwave radiation exchange to and from any point on a surface takes place with everything that can be seen from that point. Radiation exchange with a clear sky may be quite different than with objects blocking the sky from view. This is because the sky may have an emissivity of much less than 1.0, so the *radiant temperature* of the sky (its equivalent blackbody temperature) may be much colder than that of surface features that radiate nearly as blackbodies. For example, if clear sky emissivity is 0.8 and air temperature is 10 °C, its radiant temperature is only −5.3 °C, and longwave radiation exchange from a surface at the same temperature as the sky will be strongly negative. Therefore, the sur-

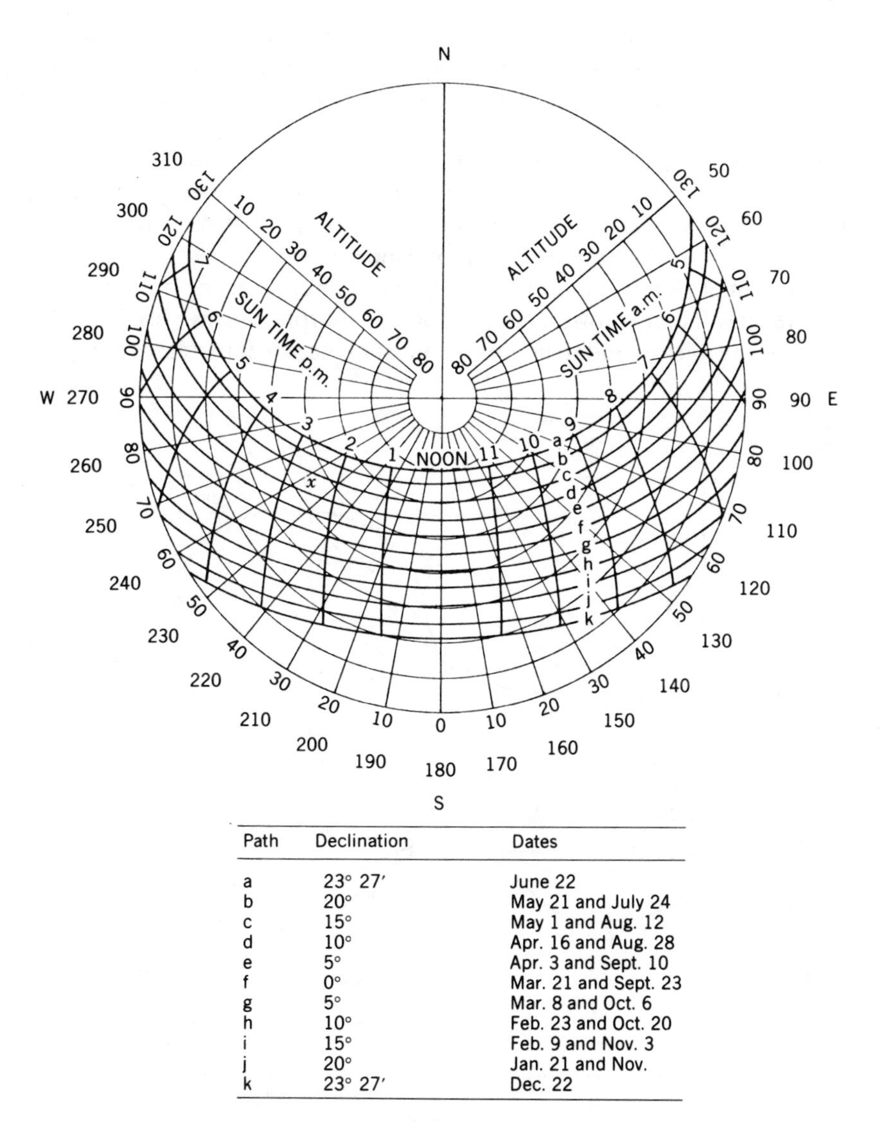

Path	Declination	Dates
a	23° 27′	June 22
b	20°	May 21 and July 24
c	15°	May 1 and Aug. 12
d	10°	Apr. 16 and Aug. 28
e	5°	Apr. 3 and Sept. 10
f	0°	Mar. 21 and Sept. 23
g	5°	Mar. 8 and Oct. 6
h	10°	Feb. 23 and Oct. 20
i	15°	Feb. 9 and Nov. 3
j	20°	Jan. 21 and Nov.
k	23° 27′	Dec. 22

FIG. 7.5 Sun path diagram for 45° N. latitude. The sun's position from any point at this latitude on April 24 and August 20 at 1430 sun time is 44° above the horizon at an azimuth of 237° and is shown by an x on the diagram (after Brown, 1973).

face will tend to cool. If little sky is visible, as within a forest, the surface temperature may remain near air temperature, but where the view factor is large, as in the center of a large open field, frost may form.

The view factor in the center of a circular opening may be found from the equation

$$V = \sin^2 [\tan^{-1}(r/h)] \tag{7.6}$$

where V is the view factor, r is the radius of the opening, and h is the height of the objects around the perimeter of the opening (Reifsnyder and Lull, 1965). As the view factor represents the relative portion of sky, its complement $(1.0 - V)$ represents the relative proportion of surface objects in view.

Equation 7.6 can be modified and solved numerically to yield the view factor for many conditions other than the center of a circular opening that may be of interest in watershed management (McAdams, 1954; Waggoner and Reifsnyder, 1961; Cochran, 1969; and Wilson and Petzold, 1973). A map of the distribution of view factors within an irregular forest opening is shown in Fig. 7.6. The view factor is lowest adjacent to the sharply concave borders where openings are narrow and trees obscure most of the sky. It increases along con-

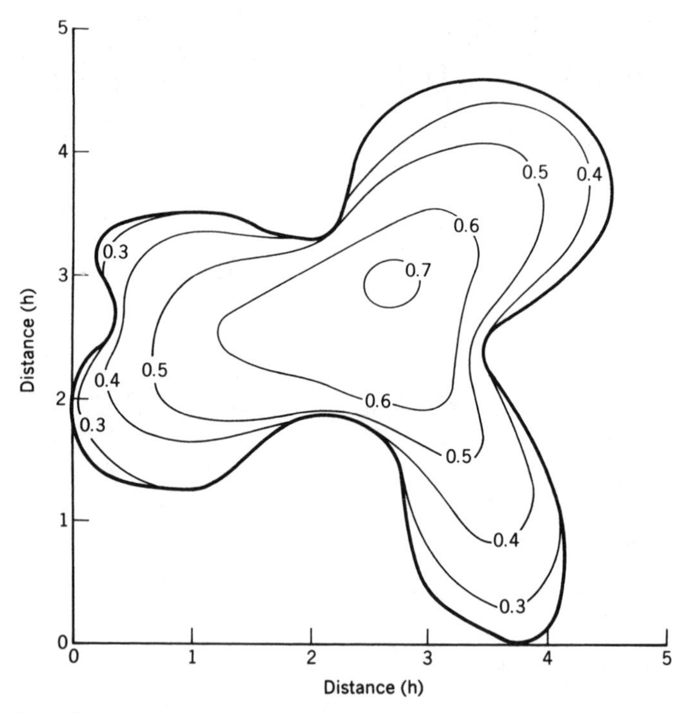

FIG. 7.6 Distribution of view factors within an irregular opening on a horizontal surface surrounded by trees of height (h).

vex borders where the opening widens and becomes greatest outward into the center of the opening.

7.2 THERMAL ENERGY

When a molecule of a substance absorbs radiant energy it becomes excited, or raised to a higher energy level. Any one or more of several things may happen to excited molecules. They may simply increase their average rate of motion, which raises the temperature of the substance. Such activity represents an increase in the form of energy called sensible heat.

If the excited molecules overcome molecular bonding forces, the substance may melt or vaporize. This change in thermal energy manifested by a change in physical state is a latent heat conversion and may be reached with no change in temperature. Latent literally means "hidden," and latent heat cannot be sensed by our ordinary senses as heat.

Water that absorbs radiant energy usually converts part of it to sensible heat and part to latent heat, simultaneously.

7.2.1 Sensible Heat

Heat energy is the total molecular energy of a substance. It is a function of the number, kind, and structure of the molecules and of their average rate of activity. *Temperature* is one characteristic of sensible heat represented by the average rate of motion, or the relative hotness or coldness of a substance; it represents the kind, not the amount of sensible heat. The amount of sensible heat is expressed in calories or joules, as mentioned earlier.

The total amount of heat in a substance is rarely of interest to anyone, but the change in amount of heat stored under different conditions may be of widespread interest. The change in amount of heat contained in a substance for a given change of temperature is

$$\Delta H_S = cM(\Delta T) \tag{7.7}$$

where ΔH_S is the change in sensible heat content (J), c is the specific heat, M is the mass (g), and ΔT is the change in temperature (K or °C).

The *specific heat* of a substance is the amount of heat necessary to raise the temperature of a mass of 1 g by 1 K. It is usually expressed as a decimal of the specific heat of water, which has a value of 1.0 cal g^{-1} or 4.1855 J g^{-1}, the highest of any common natural substance. The specific heat is a function of the chemical composition and physical structure of a substance. It varies with temperature, but within the temperature ranges that exist naturally on most watersheds the variation is negligible and specific heat may be treated as a constant. The specific heats of water and other common substances are given in Table 7.3.

TABLE 7.3. Specific Heats of Some Common Substances on Watersheds, with Comparative Figures for Lead and Iron

Substance	Specific Heat ($J\ g^{-1}$ at 15 °C)
Lead	0.13
Iron	0.46
Dry mineral soil	0.75–0.92
Air	1.00
Peat and humus	1.88
Ice, snow	2.05
Water, liquid	4.19
Leaves, fresh (range varies with moisture content)	3.35–3.77
Twigs, bark, wood (moisture 50–150% of dry weight)	2.55–3.22
Moist sandy soil (moisture 10–20% of dry weight)	1.17–1.51
Moist loam soil (moisture 20–35% of dry weight)	1.05–2.09
Wet organic soil (moisture 100–200% of dry weight)	2.93–3.35

Substances having a high specific heat have a greater *heat capacity* than those with low specific heats, for they can store more heat in a given mass for a given increase in temperature.

An understanding of specific heat aids us in understanding the nature of the commonly measured factor of temperature. Many persons mistakenly consider temperature as representing an amount of heat; so much for each degree. This cannot be true, for Table 7.3 and Eq. 7.7 show that the amount of heat that would raise the temperature of 1 g of pure water by 1 K would raise the temperature of the same amount of dry soil by 5 K, of iron by 9 K, and of lead by more than 33 K!

Nevertheless, if (and only if) specific heat and mass are held constant, then the temperature of a substance is directly related to its heat content. Therefore, if heat is added to a substance, its temperature will rise in proportion to the increase in heat content. Hot substances of the same specific heat and mass always contain more heat than cold ones, but different substances of the same mass and temperature may contain vastly different amounts of heat, as in the case of lead and water.

Temperature is an important characteristic of heat in other respects. It always influences the rate and direction of heat transfer, whether by radiation, conduction, or convection. Hot materials have a higher energy flux, or rate of exchange with their environment than cold ones, as is evident from Eq. 7.1 and the laws of conduction and convection. Temperature also determines the spectral quality of radiation and influences the rate of biochemical reactions.

7.2.2 Latent Heat

As brought out in Section 5.1, water exists freely in three physical states: solid, liquid, and gas. It readily changes phase as it gains or loses energy. The energy associated with physical phase change is latent heat.

There are three possible phase changes and three associated latent heats. Latent heats vary somewhat with temperature and pressure for each phase change, however. Changes of water from solid to liquid (melt) and liquid to solid (freezing) are characterized by the latent heat of fusion in the amount of 335 J g^{-1} at 0 °C. Changes from liquid to gas (evaporation) and gas to liquid (condensation) are characterized by the latent heat of vaporization. As evaporation and condensation occur over a wide range of temperature and pressure, the latent heat of vaporization is similarly varied, ranging from about 2.51 kJ g^{-1} at 0 °C and 2.45 kJ g^{-1} at 20 °C to 2.26 kJ g^{-1} 100 °C. Change from a solid to a gas (sublimation) and from a gas to a solid (frost formation) takes place mostly at temperatures near 0 °C and are characterized by the latent heat of sublimation. It equals the sum of the latent heats of fusion and vaporization, or 2.85 kJ g^{-1}.

A large portion of the total energy absorbed on most watersheds is converted to latent heat. To explain how much is partitioned into latent and sensible heat we must examine how these forms of heat are transferred away from the surface of the watershed.

7.3 CONDUCTION OF HEAT

Conduction is the transfer of sensible heat through matter by direct molecular contact without movement of the matter itself. Heat always moves from the hotter to the colder point. Whenever energy is absorbed by a dry surface, the surface temperature is increased, setting up a temperature gradient. Sensible heat flows along the gradient at a rate determined by the conductivity of the substance and its temperature gradient, or

$$C = K \operatorname{grad} T \tag{7.8}$$

where C is the heat transfer (J s^{-1}), K is the conductivity (J s^{-1} cm^{-1} cm^{-2} $°C^{-1}$), and grad T is temperature gradient (°C cm^{-1}). The approximate conductivity of common materials found on a watershed is shown in Table 7.4.

In general, conductivity is greatest where molecular contact per unit volume is greatest, as in dense solids and liquids, and least where molecules are highly dispersed, as in diffuse gases. Thus, snow is a much poorer conductor than ice because molecular density is so much less in snow. Insulating power or resistance to conductive heat transfer is the reciprocal of conductivity, and still air is a good insulator, as are most highly porous materials. As molecular density decreases, so does conductivity, which becomes zero in a complete vacuum.

Highly porous materials, such as loose litter or dry soil, conduct heat very slowly. Slow rates of energy transfer are not always negligible, however. Even small rates of energy transfer can accumulate to considerable sums over a long period of time. Thus, enough heat may be conducted downward to dry soils to the wilting point or lower to a depth of a meter during a long period without

TABLE 7.4. Approximate Heat Conductivity of Various Watershed Materials (10^{-3} J s^{-1} cm^{-1} cm^{-2}°C) at Ordinary Temperatures

Material	Conductivity (K)
Rocks	20.9–23.3
Ice	16.7
Water	5.1
Soil, moist	4.7–8.1
Soil, dry	1.4–3.5
Snow (varies with density)	0.6–2.3
Forest humus, dry	0.6
Air	0.2

precipitation (Colman, 1953). Federer (1966) demonstrated that the slow heat conduction upward from cold, but unfrozen ground under a snowpack is sufficient to account for winter streamflow in the White Mountains of New Hampshire during long periods of subfreezing weather, by melting the snowpack at the groundline.

7.4 CONVECTIVE HEAT TRANSFER

Convective heat transfer is the most complex and least-understood mechanism of energy exchange on the watershed. Yet, it is of great importance, for both sensible heat and latent heat exchange between watershed surfaces and the atmosphere is primarily by convection. In the long run, virtually all the energy of net radiation absorbed by a watershed surface is converted to sensible and latent heat and transferred into the atmosphere.

The basic theory of convection is well presented by Thom (1975) and Lee (1978). Since convection is heat transfer by means of mass movement of the substance containing the heat, some transfer mechanism between the surface and the atmosphere must operate to get the heat from the surface into the air. Sensible heat crosses the still boundary layer between the surface and the free atmosphere mostly by conduction and latent heat crosses primarily by molecular diffusion of water vapor.

If conduction and diffusion were the only factors operative over great distances, heat transfer would be very slow. But the boundary layer that must be crossed by these mechanisms is very thin, of the order of millimeters or even less. Very steep temperature and vapor pressure gradients may develop between the surface and moving air, in inverse correspondence to the increasing speed of air movement as the distance above the boundary increases. The turbulent air movement itself may result from buoyancy induced by heating of layers of air near the surface, or *thermal convection*. Turbulent motion may so be caused by the deflection of the general winds of the atmosphere as they move across rough surfaces, or *forced convection*. Usually, convective transfer in the daytime involves both phenomena, but at night surface heating is absent; the lower layers of air often become calm, and convection stops. If night winds occur, forced convection can take place.

Three characteristics of the atmosphere and surface are dominant factors in convective heat transfer:

1. The wind speed of free air above the surface.
2. The temperature and vapor pressure gradients between the surface and air.
3. The roughness of the surface.

The relations between wind, temperature, and vapor pressure to height above a smooth and rough surface are shown in Figs. 7.7–7.9.

The rate of convective heat transfer is directly proportional to the temperature and vapor pressure gradients between the surface and the free air. Thus, a doubling of either variable would double the rate of transfer. Wind speed and roughness interact to determine the resistance of the air to the movement of momentum, sensible heat or latent heat to or from the free air and the underlying surface. However, for a given net energy supply to the surface, a feedback mechanism tends to operate so that increasing wind speed causes a decrease in the temperature (or vapor pressure) gradient. Total convective heat transfer from any surface is determined largely by the energy supply to the surface, and the dryness of the air. Wind speed and temperature and vapor pressure gradients adjust to the energy available.

The roughness of the surface is expressed in terms of a roughness parameter, symbolized Z_0. It is a function of the height and irregularity of elements that project above the earth's surface, such as vegetation. Rough surfaces deflect air moving across them, causing eddying and efficient mixing of surface air with the air above, thus causing more efficient transfer of heat and vapor upward. The taller and rougher the surface, the greater the roughness parameter. Roughness parameters for various surfaces that might be encountered on a watershed are given in Table 7.5. Variation is greater than four orders of magnitude from smooth snow to tall forests. The effect of increasing roughness is to cause a logarithmic increase in the rate of convection.

As wind speed reaches zero, convective transport stops and sensible heat transfer in air becomes dependent on the very inefficient process of conduction. Vapor transfer becomes dependent on diffusion. This happens frequently on clear nights when net radiation is negative and a still, stable layer of cold air develops near the surface. If the air is moist, diffusion of vapor through the still air layer may occur, but the amount of dew deposited over a period of several hours is small, (Stone, 1963; Hornbeck, 1964; Fritschen and Doraiswamy, 1973; Hicks, 1983) and may be evaporated quickly in the morning once convective processes become active again.

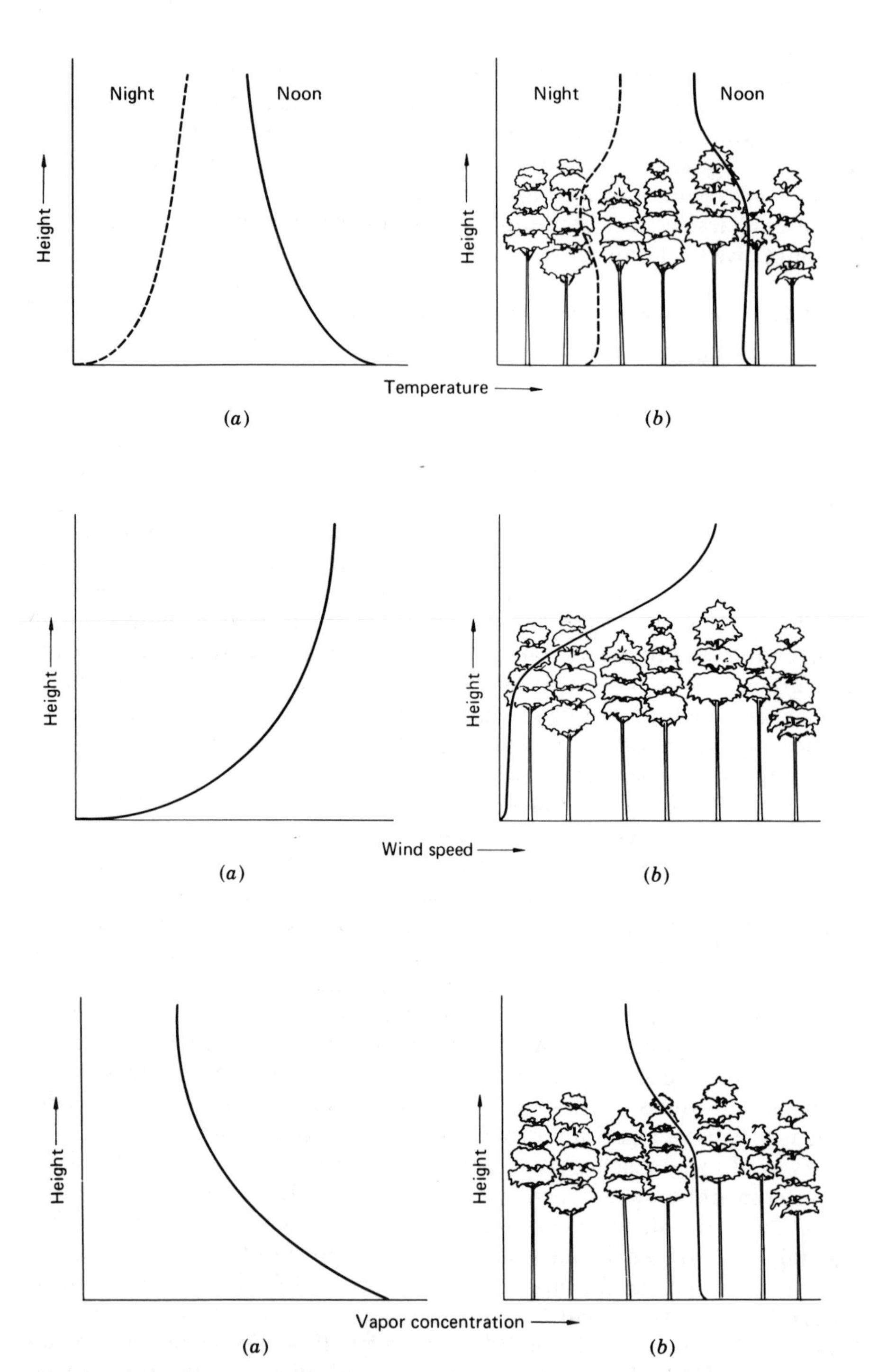

FIG. 7.7–7.9 Curves of temperature, wind, and water vapor in relation to active surfaces. (*a*) Smooth bare field and (*b*) forest.

TABLE 7.5. Roughness Parameters for Various Watershed Covers

Cover	Z_0 (cm)
Level, open snowfield	0.25–0.50
Short grass	20
Tall grass	30
Tall shrubs, brush	40–100
Conifer forest	100–290

Source: Sverdrup, 1936; Geiger, 1959; Byers, 1959; Baumgartner, 1956.

7.5 ADVECTIVE ENERGY RELATIONS

The sensible and latent heat that is transferred to the air in one place may be transported elsewhere in response to local or general atmospheric circulation patterns. Areas to the east of large warm water bodies, such as in North America and Europe, may have mild winters despite their northerly latitudes. The effects of advection may extend far inland.

Warm, moist air masses off the Pacific Ocean strongly influence winter and spring energy relations of watersheds on the slopes of the Cascade and Sierra Nevada Mountains inland to the Rocky Mountains of western United States and Canada. Snowbelts along the Great Lakes reflect latent heat advected to leeward shores. The Gulf of Mexico and Atlantic Ocean are sources of advected energy moving into southern and eastern United States and as far north as Canada.

Local advection is common and occurs during all seasons, as when heated air from dry western valleys moves into moister and cooler surrounding mountain forests. On a smaller scale, it may move from a dry southwesterly slope across a cool, moist riparian forest in a nearby valley bottom. It occurs also on a still smaller scale as in the movement of heat from a leaf in direct sunlight to one in the shade of the crown interior.

The importance of advected heat as a source of energy has been greatly underestimated in the past, particularly in humid areas, but more recent studies indicate it is frequently an appreciable factor in the energy balance. One study in Nebraska indicated that advection provided from 15 to 50% of the energy consumed in evapotranspiration. Both regional and local sources contributed, with regional advection being three or more times as important as local advection (Brakke et al., 1978), but local advection may be much more important elsewhere.

Advected energy is transferred to the watershed surface by several means. Warm air moving over a cool surface transfers heat directly by convection. Similarly, a warm, moist air mass with a vapor pressure greater than that of melting snow may transfer latent heat by direct condensation on the snowpack. In clear weather, the amount of energy that can be supplied is limited largely by the temperature of the air, except for small amounts represented by dew or frost formation on the surface.

Much advected energy is represented by latent heat that is converted to sensible heat in the atmosphere by condensation and reaches the surface as long-wave radiation from clouds, as convective transfer of sensible heat, or directly in the mass of falling precipitation. Most condensation that takes place in the atmosphere is due to the adiabatic cooling of rising air. The adiabatic *lapse rate* (change of temperature per unit change in altitude) of dry air is 1 °C per 100 m.

When considerable moisture is present in rising air, it cools at the dry adiabatic lapse rate until the air cools to or below the dew point and condensation begins. Condensation releases the latent heat of vaporization, which slows the rate of cooling of the rising air. This *wet adiabatic lapse rate* varies, depending on the rate of condensation and whether the condensed moisture falls from the rising air or is carried upward with it. On the average, it is about 0.6 °C per 100 m.

When moisture laden winds rise over a mountain range, considerable latent heat may be converted to sensible heat, accompanied by high precipitation on the windward side and low humidities and high temperatures as it descends to the leeward. On a snow covered range, such as the Cascade Mountains in winter, additional latent energy exchange may take place without adding to the sensible heat content of the air when the latent heat of condensation is used to melt snow by direct condensation on the snowpack surface. Over a winter season, large quantities of advected heat, both latent and sensible, may be available to melt snow in coastal mountains.

As air descends on the leeward side of the mountains, it warms adiabatically. Since much of the moisture has been removed by condensation and precipitation on the windward side, the air is dry and warms at the dry adiabatic lapse rate. Large quantities of sensible heat are therefore available to sublimate or melt snow or evaporate water. The descending warm, dry winds are often called *chinooks* in western North America and *foehn winds* in many places. "Snow eating" chinook winds may extend inland across the Rocky Mountains well northward into Canada, and may cause sublimation of as much as 88 mm (water equivalent) of snow during January–March from snowpacks above the treeline or in large openings (Golding, 1978; Swanson, 1980). Longley (1967) estimated that chinooks are present on a third of the days during December–February along the Rocky Mountain front in parts of Alberta. At Rapid City, South Dakota, on 13 January, 1913, the temperature rose from −27.2 °C at 0800 to 8.3 °C at 2200 during a strong chinook (Blair, 1948). Figure 7.10 illustrates the orographic effects on advected energy.

7.6 THE ACTIVE SURFACE AND ENERGY EXCHANGE

A key aid in integrating and understanding energy exchange on a watershed is the "active surface" concept developed by Rudolf Geiger (1959). The *active sur-*

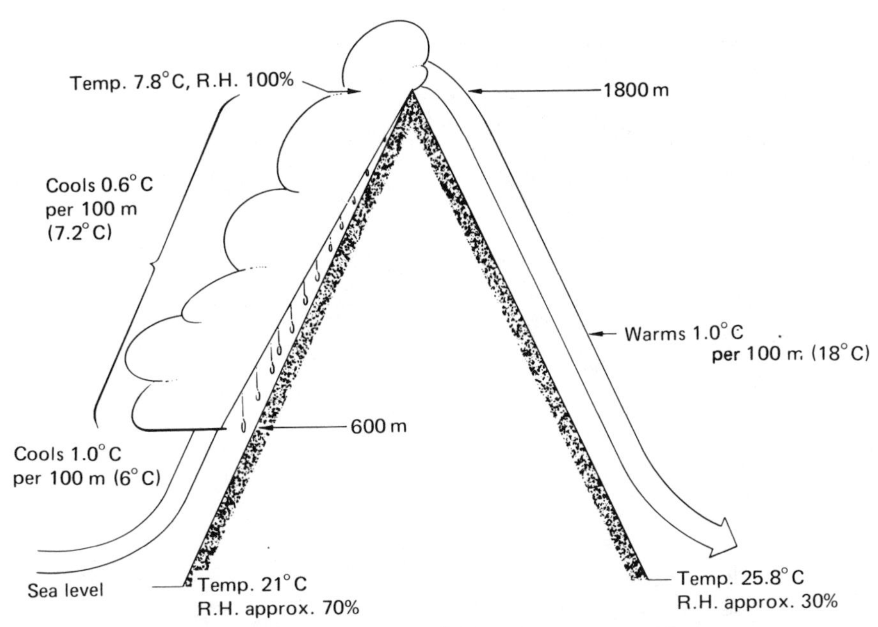

FIG. 7.10 Orographic effects on advected heat. Note the changes both in temperature and relative humidity. They are due to the conversion of latent to sensible heat during condensation and the removal of moisture by precipitation as the air mass ascended the mountain barrier.

face is the surface where most energy exchange takes place and climatic processes are modified: where radiation is received, absorbed, and transformed to other kinds of energy; where sensible heat is transferred to the atmosphere and into the ground; which initially catches precipitation and is the source of moisture for evapotranspiration; which provides frictional resistance to the wind and roughness to cause mixing of the lower atmosphere. It is primarily by modifying the active surface and the soil immediately below that the watershed manager can modify energy exchange and water relations on the watershed.

The characteristics of the active surface of a watershed are a function of fixed and variable properties. Fixed properties include factors of latitude and slope, which determine the potential irradiation received by any surface, as illustrated in Fig. 7.11 (Frank and Lee, 1966). The effect of the active surface is also influenced by its position and extent. In the free atmosphere, temperatures decrease with altitude in a more or less regular manner according to adiabatic relations. Thus, if the dry adiabatic lapse rate of 1 °C per 100 m holds, the air at 1500 m above a plain at sea level would be 15 °C cooler than air temperatures at the surface. However, if there is an extensive plain at 1500-m elevation, we would not expect its temperature to be so much lower, for adiabatic (without gain or loss of heat) relations no longer apply. The active surface of the plain absorbs solar radiation, converts it to sensible heat, and

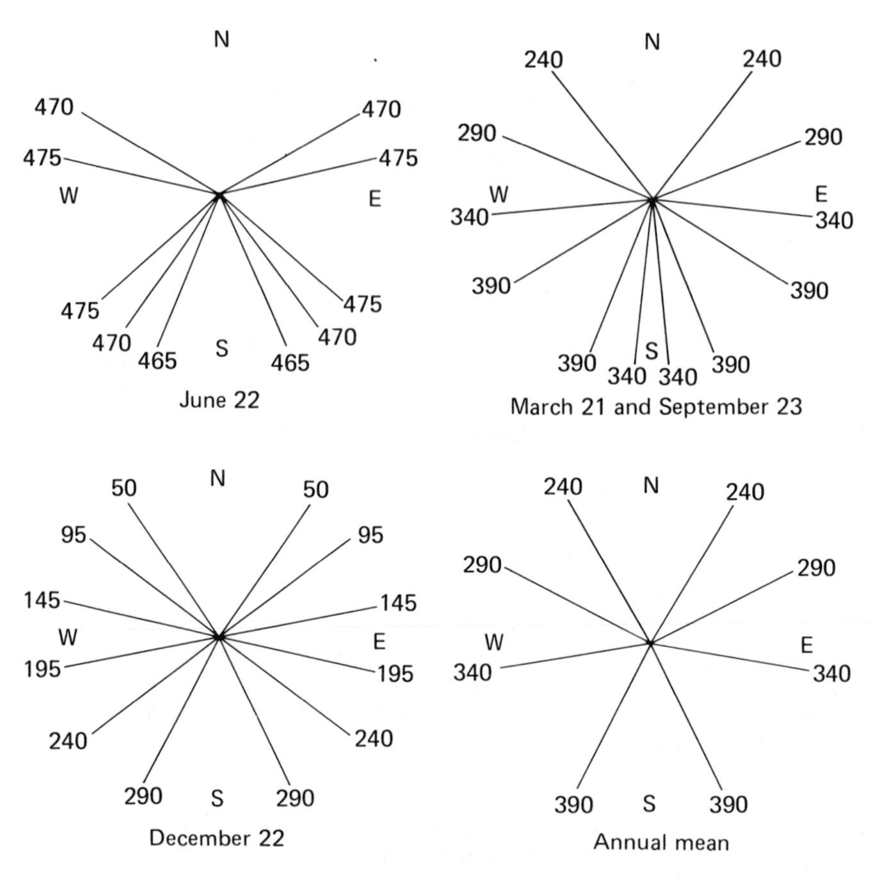

FIG. 7.11 Daily potential direct-beam solar radiation (W m^{-2}) for slopes of 40% inclination and various aspects at latitude 40° N. Variation by slope direction is least in June and greatest in December. Diffuse radiation and cloudiness tend to reduce the variation shown here.

transfers it to the atmosphere. Hence, temperatures on the high plateaus of Wyoming frequently reach 35 °C or more in summer.

Yet, at the summit of Mt. Washington, in the White Mountains of New Hampshire at a similar elevation, temperatures in summer seldom exceed 25 °C, although obviously its surface is also an active surface. The difference between it and equivalent plateau elevations in Wyoming is due to the extent and continuity of the surface. Mt. Washington can be considered more or less as a probe extending into the atmosphere. The characteristics of the air moving across it have been determined by their source high above the active surface of the much lower lands that surround it. There is simply not enough surface on the upper part of the mountain itself to greatly modify the large quantities of air moving across it, and hence the air retains the climatic characteristics of an atmosphere well above the active surface (Fig. 7.12). The upper elevations of all mountains exhibit this phenomenon to a greater or lesser degree, depending largely on differences in total relief and the surface area at each elevation. Complex wind patterns in mountains may tend to obscure otherwise obvious relationships.

The nature of the active surface is also a function of variable features such as soil moisture, thermal conductivity, albedo, roughness, and others (Stoutjesdijk, 1977, 1980). Some of the variable features may be modified by changing vegetation and land use, whereas others such as those that change in response to weather, are not easily controlled. The effects of the active surface can be seen more clearly by considering energy exchange processes there in more detail.

7.7 THE ENERGY BALANCE OF AN ACTIVE SURFACE

The energy balance of any surface can be written

$$R_n + H_s + LE + G + M = 0 \tag{7.9}$$

Where R_n is the net radiation, H_s is the sensible heat exchanged with the atmosphere, LE is the latent heat exchanged with the atmosphere, G is the heat exchanged with the ground or vegetation mass, and M is the metabolic heat of photosynthesis or respiration. Heat transfer may be either toward or away from the surface. If we ignore metabolic heat, which is negligible in the long run on most watersheds, and rearrange the equation, it becomes

$$R_n = H_s + LE + G \tag{7.10}$$

In the long run, there is no net change in the heat content of the soil or vegetation below the active surface, so the value of G becomes zero, and drops out.

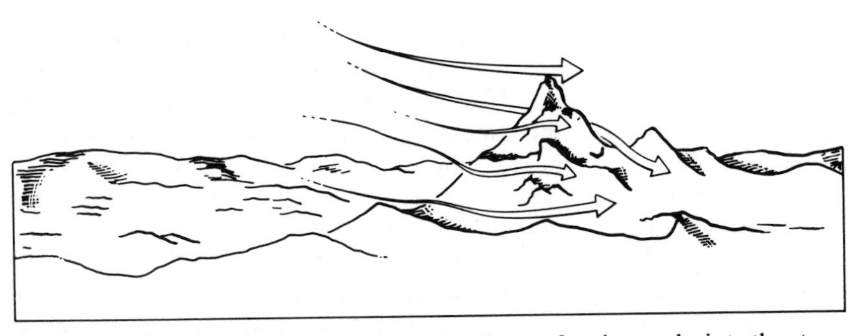

FIG. 7.12 A mountain high above the surrounding surface is a probe into the atmosphere. Atmospheric conditions on the upper slopes reflect the source of air high above the surrounding active surface.

Thus, it becomes clear that virtually all net radiation that is absorbed by a surface is transformed over the long run into sensible or latent heat that is transferred into the atmosphere.

The process that determines the rate of sensible and latent heat transfer to the atmosphere is dominantly that of convection. Sensible heat and latent heat are transferred independently, but simultaneously, according to the temperature and vapor pressure gradients that exist. Over the long run, the total outgoing energy flux cannot exceed the net radiation flux. Thus,

$$R_n = H_s + LE \tag{7.11}$$

where

$$H_s = f(\text{grad } T, \mu, Z_o) \tag{7.12}$$

where grad T is the temperature gradient, μ is wind speed, and Z_o is the roughness parameter, and

$$LE = f(\text{grad } e, \mu, Z_o) \tag{7.13}$$

where grad e is the vapor pressure gradient.

The relative proportion of sensible to latent heat transferred to the atmosphere can be expressed as:

$$\text{BR} = H_s/LE \tag{7.14}$$

where BR is the *Bowen ratio*. According to Eqs. 7.12 and 7.13, convective transfer varies with temperature and vapor pressure gradients, thus the Bowen ration can be determined from

$$\text{BR} = C\,\frac{T_s - T_a}{e_s - e_a} \tag{7.15}$$

where C is a coefficient representing the mixing ratio of water vapor in dry air, T_s is the surface temperature, T_a is the air temperature, e_s is the surface vapor pressure, and e_a is the vapor pressure of the air. Air temperature and vapor pressure are measured at the same elevation above the surface. A high Bowen ratio (>1.0) means that net radiation is being partitioned more into sensible heat than latent heat, whereas a low ratio indicates that net radiation is transformed predominantly into latent heat of vaporization.

When air over a moist surface is near saturation, as with humid air over water or irrigated short grass, the Bowen ratio, and hence relative partitioning of net radiation into sensible and latent heat, varies with temperature because of the exponential relation of saturated vapor pressure and temperature.

Then, latent heat transfer may be obtained by multiplying net radiation by a weighting factor (Pruitt, 1974)

$$LE = R_n (\Delta/\Delta + \gamma) \tag{7.16}$$

where Δ is the slope of the saturated vapor pressure curve at a given temperature (Pa/°C) and γ is the psychrometric constant at the same temperature. The value of γ varies slightly with pressure and temperature, mostly between 63 and 68, averaging about 66 at typical growing season temperatures near sea level. The approximate partitioning of net radiation under these moist conditions yields a Bowen ratio of 1.0 at about 6 °C, decreasing to about 0.35 at 25 °C; values for a wide range of temperatures are given in Table 7.6.

When the air over a moist surface departs from a near-saturated condition, Eq. 7.16 must be modified to take the greater vapor pressure gradient into account so that

$$LE = R_n (\Delta/\Delta + \gamma) + h(e_s - e_a) \tag{7.17}$$

where h is a transfer coefficient that takes wind speed and roughness into account, and e_s and e_a are vapor pressures of the surface and air, respectively (Pruitt, 1974).

TABLE 7.6. Approximate Partitioning of Net Radiation Under Humid Conditions for Various Temperatures, Near Sea Level and at an Elevation of 1000 m

	Bowen ratio		H_s/R_n X 100		LE/R_n X 100	
Elevation (m):	0	1000	0	1000	0	1000
Temperature (°C)						
−40	29.86	26.86	97	96	3	4
−35	19.58	17.59	95	95	5	5
−30	12.91	11.59	93	92	7	8
−25	8.75	7.86	90	89	10	11
−20	5.91	5.32	86	84	14	16
−15	4.17	3.74	81	79	19	21
−10	2.83	2.54	74	72	26	28
−5	2.02	1.81	67	64	33	36
0	1.44	1.30	59	57	41	43
5	1.07	0.96	52	49	48	51
10	0.79	0.71	44	42	56	58
15	0.60	0.54	37	35	63	65
20	0.46	0.41	31	29	69	71
25	0.35	0.32	26	24	74	76
30	0.27	0.25	21	20	79	80
35	0.21	0.19	18	16	82	84
40	0.17	0.16	15	13	85	87

Bowen ratios tend to vary diurnally and seasonally. During the low temperatures just after sunrise and in the winter, a greater proportion of net radiation tends to be transformed into sensible heat and exchanged with the atmosphere and less latent heat (evaporation) exchange occurs. In the higher temperatures of afternoon and in summer, the situation is reversed, provided water is available for evaporation.

7.7.1 Negative Bowen Ratio and the Oasis Effect

Situations sometimes occur when the direction of sensible and/or latent heat movement is from the atmosphere to the active surface. Such situations are normal when net radiation is negative, as on clear nights when outgoing radiation from a cool surface of high emissivity exceeds incoming radiation from warmer air that has a lower radiant temperature. Dew and frost are manifestations of downward transfer of latent heat and inversions occur during downward sensible heat transfer. Such heat movement downward from the atmosphere is given a negative sign in Eqs. 7.11 and 7.14. If net radiation is negative and both sensible and latent heat are transferred downward from the atmosphere, then the Bowen ratio remains positive. A negative Bowen ratio may occur when a surface is cooler than air, so that sensible heat moves downward along the temperature gradient, but the vapor pressure of the air is less than that of the cooler surface, which is common when the humidity of the air is low and the surface is moist. Then, latent heat exchange is upward from the surface into the air; the sign for sensible heat is negative and that for latent heat is positive, yielding a negative Bowen ratio. It is not possible to have a negative Bowen ratio by having latent heat move downward while sensible heat moves upward, for this implies that the warmer surface has a lower saturated vapor pressure than the cooler air, which cannot be, as saturated vapor pressure is an increasing function of temperature.

The "oasis effect" is a common condition that is typified by a negative Bowen ratio. Many situations occur where a moist area is adjacent to a dry one. Over dry surfaces, net radiation can only be converted to sensible heat (for there is no water to evaporate), which causes heating of the air. When this hot, dry air moves over a cooler, moist surface, sensible heat is transferred downward to the surface in accordance with the temperature gradient, whereas vapor moves upward along the vapor pressure gradient. The rate of evaporation may then exceed the rate of net radiation by a large amount (Rosenburg and Verma, 1978). The additional energy converted to latent heat comes from sensible heat transferred from the atmosphere.

While a desert oasis is the classical example illustrating the oasis effect, it is also common in irrigated areas, around springs and other moist areas in dry weather. It is not necessary that the source area for advected sensible heat be completely dry to have an oasis effect, and it can even be moist. It is only necessary that a relatively dry atmosphere be warmed to a higher temperature than that of some nearby moist surface so that the cooler moist surface has a higher vapor pressure than the warmer air. Such situations are ubiquitous in uneven topography, where one slope receives more net radiation than another, where one is drier than another or where well-drained upland soils are dissected by moist stream courses. Even in humid regions, advected sensible heat may contribute appreciably to evapotranspiration. Data for the growing season in Wisconsin by Tanner and Pelton (1960) indicated a net sensible heat gain by rough vegetation. They concluded that the surface roughness of moist vegetation is very important, particularly in arid regions and during the winter months in humid regions where advected sensible heat is large compared to net radiation. Pearce et al. (1980) indicated the large advective energy supply available to evaporate intercepted water in forests in Great Britain.

7.7.2 Energy Exchange and Topography

The topography of any given watershed is one of the most important fixed characteristics determining its energy receipt and partitioning. Ultimately, watersheds will require classification at least partly on the basis of their energy exchange characteristics if they are to be managed effectively. For example, potential solar beam irradiation of any site is determined solely by its latitude, slope, and location with respect to barriers that might shade it (Satterlund, 1981). Tabular data of potential solar beam irradiation for most slopes on watersheds between 30° and 60° latitude are readily available (Fons et al., 1960; Frank and Lee, 1966; Buffo et al., 1972). Lee and Baumgartner (1966) classified more than 350 km^2 of Bavarian mountainous land on the basis of potential insolation. The location of shadows from barriers can be determined using the method of Satterlund (1977).

Several studies have shown that even such simple classifications are suitable for explaining a large portion of the variation among mountain watersheds with respect to annual water yield (Lee, 1964) and snowmelt (Hendrick and Filgate, 1971; Bethlahmy, 1973).

The relation between net radiation and evapotranspiration is strongly affected by temperature as explained previously. Thus, other factors equal, evapotranspiration decreases with increasing altitude because at the lower temperatures of higher elevations, relatively more net radiation is partitioned into sensible heat. Similarly, one would expect that nearby east and west facing slopes would exhibit different moisture regimes even though precipitation and net radiation were the same. The portion of net radiation partitioned into latent heat energy may increase by as much as 35% as temperatures increase from near 10 °C at dawn to 30 °C in late afternoon. Thus, east slopes, which receive most of their net radiation in the morning, tend to be more moist than west slopes that receive most of their net radiation in the afternoon.

Furthermore, different slopes tend to receive different proportions of their total energy as radiation and sensible heat. In the northern hemisphere, south slopes receive more radiant energy than north slopes. They also tend to be a source of sensible heat into the atmosphere. North slopes receive less radiant energy, and tend to be sinks (absorbers) for sensible heat, which may comprise a large portion of their total energy income. Rough surfaces are efficient exchangers of sensible heat with moving air. Canopy temperatures of subalpine forests on different slopes showed little difference during daylight hours at the Fraser Experimental Forest, Colorado though clouds and thunderstorms caused considerable, short-lived local variation (Kaufmann, 1984). The quantitative significance of the above relationships to watershed management remains largely unknown, but they are certainly related to different soils and vegetation, and greater water yields on north and east slopes than on their counterparts. They are also likely to be responsible for the greater increase in water yields following timber harvest on north and east slopes.

It should be clear from the foregoing that the fixed physical characteristics of a watershed, together with the local climate, set certain limits to energy receipt and partitioning. Within those limits, variable characteristics of land use and vegetation type and density influence the actual energy exchange that takes place.

7.7.3 Edge Effects

Air moving over a surface with any given set of energy exchange characteristics, such as albedo, roughness, slope, moisture availability, and so on, tends to move toward equilibrium with it. Atmospheric conditions and energy exchange may become relatively homogeneous and simple over wide areas where the surface is uniform, as in high-pressure cells over tropical seas, arctic tundra, or arid plains.

Where adjacent surfaces differ from one another, however, the moving air tends toward a different equilibrium over each, carrying characteristics developed over one surface to another. The disequilibrium is greatest where the two surfaces meet, and the sharper the difference between adjacent surfaces and the closer equilibrium has been attained with the previous surface, the greater the disequilibrium will be with the next. A good example of a sharply different surface is the desert oasis previously described, or the shore of a desert reservoir. This condition of greater or lesser disequilibrium at each boundary of unlike surfaces may be called an *edge effect*.

The edge effect not only results in sharply increased evaporation where moist areas lay behind dry ones, but evaporative and heat stress may be reduced in the lee of moist areas. Thus, in a grassy opening surrounded by tall trees, the windward edge may be protected for some distance into the opening, and the edge toward the sun is shaded by the trees. Evapotranspiration is reduced on the sheltered edge, and snowmelt may be slower. If the opening is large, the center is much more exposed to wind, heat, and sun.

The scale of the different surfaces may modify the degree of edge effect. For example, the edge effect may be very strong at the shore of a large lake surrounded by dry land, but less so if many small lakes are scattered uniformly over the land surface. The effect may be virtually absent when water droplets alternating with similarly sized dry spots are dispersed over a wide area, as any air mass moving over it would tend to come into equilibrium with the average condition. Similarly, edge effects may be particularly great where smooth, low surfaces are adjacent to tall rough ones. The rate of change of air mass characteristics is greatest at the boundary of two different surfaces and diminishes the farther it moves over homogeneous ones.

Where watershed surfaces are composed of a mosaic of different vegetation types, soils, slopes, and elevations, it is extremely difficult to establish "typical" energy exchange characteristics for any given site or for the watershed as a whole. These problems make it difficult to extrapolate research from one area to another and from small watersheds to large river basins. Pattern is therefore an important consideration in energy exchange that must be taken into account, but our ability to do so with great accuracy is limited. Heterogeneity breeds a complexity that we have not yet learned to resolve.

LITERATURE CITED

Baumgartner, A. 1956. Untersuchungen über den Warme- und Wasserhaushalt eines jungen Waldes. Berichte des Deutshcen Wetterdienstes. Berlin. 5:4–53. (C.S.I.R.O. translation 3760 by Dr. E. Pichler, Melbourne, 1958.)

Bethlahmy, N. 1973. Water yield, annual peaks and exposure in mountainous terrain. *J. Hydrol.* 20:155–169.

Blair, T. A. 1948. *Weather elements.* 3rd ed. Prentice-Hall, New York.

Brakke, T. W., S. B. Verma, and N. J. Rosenburg. 1978. Local and regional components of sensible heat advection. *J. Appl. Meteor.* 17:955–963.

Brown, J. M. 1973. Tables and conversions for microclimatology. USDA Forest Service. Gen. Tech. Rept. NC-8. North Central For. Expt. Sta., St. Paul, MN.

Buffo, J., L. J. Fritschen, and J. L. Murphy. 1972. Direct solar radiation on various slopes from 0 to 60 degrees north latitude. USDA Forest Service. Res. Paper PNW-142. Pac. Northwest For. Range Expt. Sta., Portland, OR.

Byers, H. R. 1959. *General meteorology*. McGraw-Hill. New York.

Cochran, P. H. 1969. Lodgepole pine clearcut size affects minimum temperature near the soil surface. USDA Forest Service. Res. Paper PNW-86. Pac. Northwest For. Range Expt. Sta., Portland, OR.

Colman, E. A. 1953. *Vegetation and watershed management.* Ronald Press, New York.

Federer, C. A. 1966. Sustained winter streamflow from groundmelt. USDA Forest Service. Res. Note NE-41. Northeastern For. Expt. Sta., Upper Darby, PA.

Fons, W. L., H. D. Bruce, and A. McMaster. 1960. Tables for estimating direct beam solar irradiation on slopes at 30° to 46° latitude. USDA Forest Service. Pac. Southwest For. Range Expt. Sta., Berkeley, CA.

Frank, E. C. and R. Lee. 1966. Potential solar beam irradiation on slopes: tables for 30° to 50° latitude. USDA Forest Service. Res. Paper RM-18. Rocky Mt. For. Range Expt. Sta., Ft. Collins, CO.

Fritschen, L. J. and P. Doraiswamy. 1973. Dew: an addition to the hydrologic balance of Douglas-fir. *Water Resour. Res.* 9:891–894.

Gates, D. M. 1962. *Energy exchange in the biosphere.* Harper and Row, New York.

Geiger, R. 1959. *The climate near the ground.* (Translated by M. H. Stewart et al.) Harvard University Press, Cambridge, MA.

Golding, D. L. 1978. Calculated snowpack evaporation during chinooks along the eastern slopes of the Rocky Mountains in Alberta. *J. Appl. Meteor.* 17:1647–1651.

Hendrick, R. L. and B. D. Filgate. 1971. Application of environmental analysis to watershed snowmelt. *J. Appl. Meteor.* 10:418–429.

Hickey, J. R., L. L. Stowe, H. Jacobwitz, P. Pellegrino, R. H. Maschoff, F. House, and T. H. Vander Haar. 1980. Initial solar irradiance determinations from Nimbus 7 cavity radiometer measurements. *Science* 208:281–283.

Hicks, B. B. 1983. A study of dewfall in an arid region: an analysis of Wangara data. *Quart. J. R. Met. Soc.* 109:900–904.

Hornbeck, J. W. 1964. The importance of dew in watershed-management research. USDA Forest Service. Res. Note NE-24. Northeastern For. Expt. Sta., Upper Darby, PA.

Kaufmann, M. R. 1984. Effects of weather and physiographic conditions on temperature and humidity in subalpine watersheds of the Fraser Experimental Forest. USDA Forest Service. Res. Pap. RM-251. Rocky Mt. For. Range Expt. Sta., Ft. Collins, CO.

Lee, R. 1964. Potential insolation as a topoclimatic characteristic of drainage basins. *Int. Assoc. Sci. Hydrol. Bull.* 11:27–41.

Lee, R. 1978. *Forest microclimatology.* Columbia University Press, New York.

Lee, R. and A. Baumgartner. 1966. The topography and insolation climate of a mountainous forest area. *For. Sci.* 12:258–267.

Longley, L. W. 1967. The frequency of winter chinooks in Alberta. *Atmosphere.* 5:4–16.

McAdams, W. H. 1954. *Heat transmission.* 3rd ed. McGraw-Hill, New York.

Miller, D. H. 1955. Snow cover and climate in the Sierra Nevada, California. *Univ. Calif. Pub. Geog.* 11.

Pearce, A. J., J. H. C. Gash, and J. B. Stewart. 1980. Rainfall interception in a forest stand estimated from grassland meteorological data. *J. Hydrol.* 46:147–163.

Phillips, E. L. 1965. *Washington climate for these counties: Adams, Lincoln, Spokane, Whitman.* Agric. Ext. Serv., Washington State University, Pullman, WA.

Pruitt, W. O. 1974. Factors affecting potential evapotranspiration. pp. 82–101. In: E. J. Munke, Ed. *Biological effects in the hydrological cycle.* Proceedings, Third Int. Seminar for Hydrology Professors, 18–30 July, 1971. Purdue University, West Lafayette, IN. UNESCO.

Reifsnyder, W. E. and H. W. Lull. 1965. Radiant energy in relation to forests. USDA, Forest Service. Tech. Bul. No. 1344.

Rosenburg, N. J. and S. B. Verma. 1978. Extreme evaporation by irrigated alfalfa: a consequence of the 1976 midwestern drought. *J. Appl. Meteor.* 17:934–941.

Satterlund, D. R. 1977. Shadow patterns located with a programmable calculator. *J. For.* 75:262–263.

Satterlund. D. R. 1979. An improved equation for estimating long-wave radiation from the atmosphere. *Water Resour. Res.* 15:1649–1650.

Satterlund, D. R. 1981. Climatic classification of watersheds. pp. 33–42. In: D. M.

Baumgartner, Ed. *Interior west watershed management.* Proceedings of a symposium, Cooperative Extension, Washington State University, Pullman, WA.

Satterlund, D. R. 1983. Forest shadows: how much shelter in a shelterwood? *For. Ecol. Manage.* 5:27–37.

Satterlund, D. R. and J. E. Means. 1978. Estimating solar radiation under variable cloud conditions. *For. Sci.* 24:363–373.

Stone, E. C. 1963. The ecological importance of dew. *Quart. Rev. Biol.* 38:328–342.

Stoutjesdijk, P. 1977. High surface temperatures of trees and pine litter in the winter and their biological importance. *Int. J. Biometer.* 21:325–331.

Stoutjesdijk, P. 1980. The range of micrometeorological diversity in the biological environment. *Int. J. Biometeor.* 24:211–215.

Sverdrup, H. V. 1936. Eddy conductivity of air over a smooth snowfield. *Geofysiske Publikesjoner* 11(7):1–69.

Swanson, R. H. 1980. Surface wind structure in forest clearings during a chinook. pp. 26–30. In: *Proc. 48th Ann. Meeting Western Snow Conf.*, Laramie, WY.

Tanner, C. B. and W. L. Pelton. 1960. Potential evapotranspiration by the approximate energy balance method of Penman. *J. Geophys. Res.* 65:3391–3413.

Thom, A. S. 1975. Momentum, mass and heat exchange in plant communities. pp. 57–109. In: J. L. Monteith, Ed. *Vegetation and the atmosphere.* Vol. 1. Academic, London.

Waggoner, P. E. and W. E. Reifsnyder. 1961. Difference between net radiation and water use caused by radiation from the soil surface. *Soil Sci.* 91:246–250.

Wilson, R. G. and D. E. Petzold. 1973. A solar radiation model for sub-arctic woodlands. *J. Appl. Meteor.* 12:1259–1266.

8 Water Losses

Precipitation falling on a watershed may be held in storage or leaves it as vapor or liquid flow. That which is evaporated is generally considered to be lost from further use. To be sure, some water is lost from further use by other means, as by incorporation into the dry matter that comprises the biomass of terrestrial ecosystems, but the amount is negligible. Baker (1950) calculated that the water required to form the wood produced annually by a productive forest is equivalent to only 1 mm of precipitation. With such exceptions, essentially all water losses are evaporative in nature, whether occurring as direct evaporation from open water surfaces, interception of precipitation, transpiration by plants, sublimation from blowing and drifting snow, or evaporation from soils. Horton (1973) compiled much of the earlier literature related to such water losses from wildlands.

Under natural conditions, water losses largely reflect the evaporative characteristics of the climate of any watershed as determined primarily by supplies of energy and water. Losses may be modified by changing surface conditions so that evaporative processes are modified, but climate largely sets the limits of response to most changes on wildland watersheds.

8.1 EVAPORATION AS A CLIMATIC CHARACTERISTIC

Evaporation as a climatic characteristic has long been of utmost interest in agriculture, forestry, and ecology, as well as in water resources. Numerous methods of measuring evaporation or estimating the potential evaporating power of the atmosphere have been developed. Two common principles are in use:

1. Direct measurement of evaporation loss from pans, lysimeters and lakes, and reservoirs.
2. Estimates based on climatic elements upon which evaporation depends or with which it is associated.

Nearly all estimates of potential evapotranspiration are based on the assumption that evaporation from a free water surface or short vegetation well supplied with water is primarily a function of atmospheric conditions. Potential evapotranspiration includes all forms of evaporative water loss to the atmosphere. Direct measurements are made by observations of water loss under standard conditions, whereas estimates are based on known physical laws, usually expressed in somewhat simplified equation form.

8.1.1 Direct Measurements

Evaporation pans are containers with a free water surface exposed to the air in which the water level is measured at intervals. The difference between successive measurements represents evaporation for the interval between measurements. If precipitation occurs between measurements, the amount is added to the antecedent measurement. Sometimes, scales are used to weigh the pan and the weights are converted to water depths. Continuous records may be obtained by use of floats or scales with a pen attached to a clock-driven chart.

Lysimeters are earth filled containers, either bare or vegetated, maintained in a constantly moist condition, that are weighed at intervals or continuously. In case of heavy precipitation, they are fitted with drains from which the excess water is caught and measured so a complete account of input and loss may be made.

Evaporation measurements may be made from lakes, ponds, and reservoirs by measuring changes in water level, provided surface area is known for each level and inflow and outflow are known. However, accurate measurements may be difficult to obtain, for inflow includes not only measured streamflow, but also precipitation, springs, and possible surface runoff, while outflow includes possible seepage as well as surface streamflow. If evaporation is small in relation to inflow and outflow, large errors in the evaporation term may result.

In all of the above methods, the basic continuity equation is used:

$$I - O \pm \Delta S = 0 \tag{8.1}$$

where I is input, O is output, and ΔS is the change in storage. Evaporation is represented by output, and if output does not include liquid flow, it is usually determined as a difference between input and change in storage. If output includes liquid flow, then it must be separated into each phase.

Evaporation pans are the simplest and most widespread means of directly measuring evaporation. There are many different types of pans, and each may exhibit slightly different evaporation under the same conditions. In the United States, most stations use a standard pan 122 cm in diameter and 25.4 cm deep.

8.1.2 Estimates of Potential Evapotranspiration

Estimates of potential evapotranspiration are usually based either on convection transfer theory, whereby the evaporation rate depends on the vapor pressure gradient, or on the principle of conservation of energy and the energy balance. The first is expressed by the equation

$$E = c(e_s - e_a) \tag{8.2}$$

where E is the evaporation, c is a coefficient dependent on wind speed, barometric pressure, roughness, and others; e_s is the saturated vapor pressure of the evaporating surface; and e_a is the vapor pressure of the free air. The second is expressed by the equation

$$LE = R_n - G - H_s \tag{8.3}$$

where LE is the latent heat of vaporization expressed as the equivalent depth of water, R_n is the net radiation, G is the heat stored in the soil or water, and H_s is the sensible heat transferred to the air. To convert net radiation to depth of water, it is useful to remember that 1 ly equals 1 cal cm^{-2}. Also, 1 g of water is 1 cm^3, or an area of 1 $cm^2 \times 1$ cm depth. Then, if the latent heat of vaporization is x cal g^{-1}, x ly are equivalent to a depth of 1 cm. Conversion to J cm^{-2} or W m^{-2} may be made by applying the appropriate constant (see Appendix).

Both of the above relationships of Eqs. 8.2 and 8.3 are somewhat simplified, and many modifications have been developed to make them more precise. It is not the purpose of this section to discuss the principles of estimating evapotranspiration and all the various methods of doing so in great detail. The interested student can refer to several works that will provide a fairly complete historical review in Rohwer (1931), Thornthwaite and Holzman (1942), US Geological Survey (1954), Munn (1961), and Brutsaert and Stricker (1979). However, several methods of estimating potential evapotranspiration have come into widespread use in wildland management and some discussion is merited.

Probably the simplest well-known potential evapotranspiration model is one developed by Hamon (1963);

$$PET = C\,D\,P_t \tag{8.4}$$

where PET is the potential evapotranspiration (mm day^{-1}), D is the possible hours of sunshine in units of 12 h, P_t is the saturated absolute humidity (g m^{-3}) at the mean daily temperature, and C is an empirical coefficient (0.00026). This is basically an energy balance model whereby D serves as a proxy for energy supply and P_t serves both as an additional proxy for energy supply and as a weighting factor for partitioning into latent heat. The only data required for use are latitude, from which day length can be determined, and mean daily temperature observations.

Earlier, Thornthwaite (1948) developed a more complex model based on similar simple data and the same energy balance reasoning.

$$PET = 1.6(10\,T/I)^a \tag{8.5}$$

where T is the temperature (°C), I is a heat index calculated on the basis of an annual summation of monthly indexes based on mean temperature, and a is an empirical constant. Both the Hamon and Thornthwaite models yield comparable results. They are widely used because the only climatic data needed, air temperatures, are widely available. However, the basic assumption that the partitioning of energy is dependent only on temperature may be greatly in error under conditions of inadequate water supply. Figure 8.1 illustrates how net radiation and partitioning varies with surface water supply.

A more realistic model was developed by Penman (1956) that essentially combined the vapor pressure gradient Eq. (8.2) with the energy balance Eq. (8.3). The use of the vapor pressure gradient as a variable reduces dependence on the assumption of energy partitioning based strictly on temperature and also allows the effect of advected sensible heat to be taken into account. The basic Penman equation is

$$PET = (\Delta/\gamma R_n + E_a)/(\Delta/\gamma + x) \tag{8.6}$$

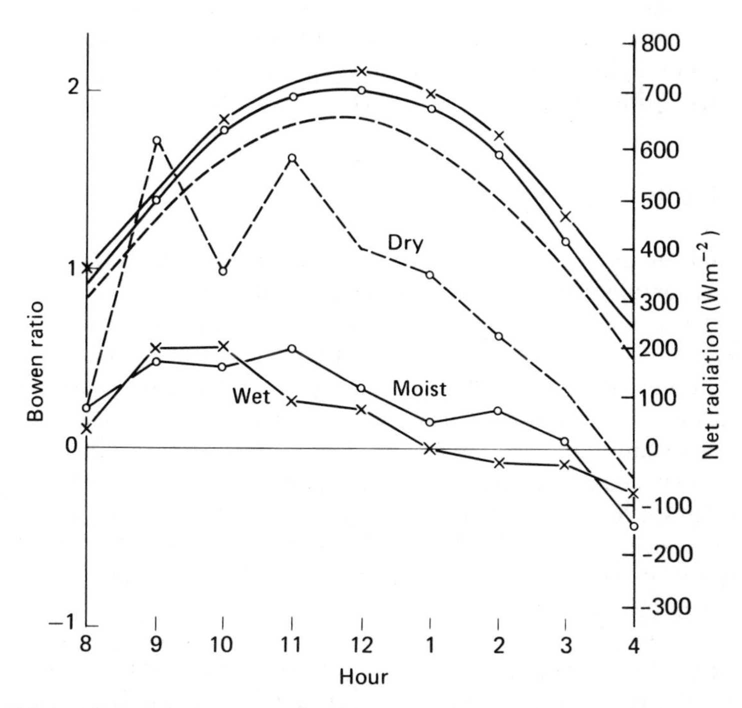

FIGURE 8.1 Change in Bowen ratios from morning to afternoon under wet, moist, and dry soil conditions and comparable solar radiation regimes. Lower irregular graphs show Bowen ratios; upper smooth graphs indicate corresponding net radiation. Note reduction in net radiation as surface dries. This would be expected from albedo changes and increased surface temperatures (adapted from Figs. 2–4, Fritschen and Van Bavel 1962).

where Δ is the slope of the saturated vapor pressure at the temperature involved (Pa/°C), γ is the psychrometric constant at the same temperature, R_n is net radiation, x is 1 for open water and >1 for green plants, and E_a is an expression for the drying power of the air based on wind speed and vapor pressure gradient. At 10, 20, and 30 °C, the values of Δ/γ are 1.3, 2.3, and 3.9, respectively, indicating that the radiation term is weighted more heavily than the vapor pressure term in seasons when temperatures are higher.

As models are made more realistic, data requirements for their use increase. To use the Penman model, the following observations are needed: (a) The duration of bright sunshine or net radiation, (b) mean air temperature, (c) mean vapor pressure, and (d) mean wind speed. These are much more than is needed for the simpler models and use tends to be limited by a shortage of necessary data.

Perhaps the greatest limitation of all potential evapotranspiration models is that evapotranspiration is not strictly a climatic phenomenon. The nature of the evaporating surface, heat storage in the substrate, and biological responses to water availability may result in water loss rates that are greater or less than might be expected from open water or short vegetation well supplied with water under the same climatic conditions (Stewart, 1983). Nevertheless, potential evapotranspiration models often prove useful in comparing the hydrology of one area with another, provided their limitations are kept in mind.

8.2 INTERCEPTION

Precipitation falling on a vegetated watershed either strikes the surface of the vegetation or falls directly through holes in the canopy to the ground. If vegetation is dense, some of the early precipitation is caught and retained by the foliage, branches, and stems. As precipitation continues, the amount retained may increase until parts of the plant can hold no more, and drip begins and water flows down the stems (or, with snow, masses slide off), striking lower layers or reaching the ground below. The portion of precipitation held by plants is exposed to evaporation, while that reaching the ground may be distributed in space quite differently from the rather uniform manner in which it fell. The process of catching and redistributing precipitation by vegetation back to the atmosphere and the ground is called *interception*.

Interception has a twofold significance. Most important, it may substantially change the amount of precipitation reaching the ground; in some forests by about as much as 40% of precipitation measured in the open. It also changes the spatial distribution of precipitation, especially snow, over the ground surface. Moisture is concentrated at the base of plants where flow down the stem is heavy and beneath drip points or around the crown periphery and in openings between crowns by falling snow. These changes may importantly modify both the amount and timing of water that ultimately becomes streamflow.

8.2.1 Rainfall Interception

The interception process begins as rain is caught by the vegetation. A person standing under a tree at the beginning of a shower can remain dry for a period of time, because, except for some raindrops that fall directly through the canopy (*throughfall*), most are retained by the plant surfaces above (entering *interception storage*). As more plant surfaces become wetted, however, more and more of the rain falls off or runs down the plant to the ground. Thereafter, *drip* from the canopy adds to the throughfall, and *stemflow* moves down the stems, reaching the ground. If precipitation continues at a rate exceeding evaporation of water from plant surfaces, the *interception storage capacity* of the plant becomes satisfied. Thereafter, water reaches the ground (*net precipitation*) so long as the rainfall above the canopy (*gross precipitation*) falls at a greater rate than evaporation occurs. When the rain stops, drip and stemflow stop soon thereafter and the water in interception storage is lost to the atmosphere by evaporation. *Interception loss* is the water returned to the air and is therefore a function of the interception storage capacity and the rate of evaporation during the storm. It is also the difference between gross and net precipitation.

Some small storms stop before the interception storage capacity is filled. Then, the interception loss approaches the gross precipitation, except for the raindrops that fall directly through the crown canopy. In equation form interception loss is

$$I = CP \quad [\text{where } P \leqslant (S + E)] \tag{8.7}$$

where I is the interception loss, C is a coefficient of catch efficiency, P is the rainfall, S is the interception storage capacity, and E is evaporation loss during the storm. Sometimes the rate of evaporation may exceed the rate of precipitation for long periods during low intensity storms. Then the interception storage capacity remains unfilled until the precipitation rate increases (Massman, 1980). For larger storms, where the interception storage capacity is full when rainfall ends,

$$I = S + E \quad [\text{where } P > (S + E)] \tag{8.8}$$

The amount of interception loss in any given storm is therefore related to (a) the vegetation characteristic of interception storage capacity, (b) the rate of rainfall, (c) the rate of evaporation during the storm, and (d) the duration of the storm.

Although interception loss is commonly represented by regression equations of the form

$$I = a + bP \tag{8.9}$$

where a and b are the empirical regression coefficients, it should be obvious that so long as the plant surface remains continuously wet and the amount of precipitation is greater than combined storage capacity and loss by evaporation during the storm, there is no direct causal relationship between the amount of precipitation and the interception loss. Nevertheless, hundreds, if not thousands of rainfall interception studies have been made in forests and shrublands throughout the world, with results expressed mostly in the form of Eq. 8.9. The results are surprisingly consistent within given regions and vegetation types, and show high accuracy. This suggests the empirical coefficients may have real physical meaning; that the coefficient a may represent interception storage capacity, whereas the slope coefficient b represents the average rate of evaporation during the storm. The accuracy of most results further suggests that there is a consistent relationship between the average precipitation intensity and the average rate of evaporation during storms for any given climatic region (Gash, 1979).

Rutter et al. (1971, 1975) and Massman (1983) developed physically based computer models of rainfall interception that offer clear insight into the interception process. They are the most rigorous method of estimating interception loss available, but they require detailed meteorological data that are seldom available, which greatly limits practical application. The major factors influencing interception losses are discussed below.

8.2.1.1 Interception Storage. Interception storage is primarily a characteristic of the vegetation; it represents the capacity of the vegetation to store water as a film over the plant surface. In theory, storage capacity should also be influenced by temperature, as temperature modifies the viscosity and surface tension of the water film, and thus its thickness before drip or flow begins.

The major plant characteristics influencing interception storage capacity are the amount and nature of the plant surface. The amount of surface determines the maximum area of the water film and the nature of the surface determines its continuity and thickness.

The primary surface of most plants consists of the leaves, with the remainder consisting mostly of stems, but includes fruits, flowers, and other plant parts. The surface area of vegetation is commonly expressed in terms of a *leaf area index* (LAI), which is the ratio of the total surface area of the leaves to the land surface area occupied by the vegetation. The LAI of vegetation ranges from less than 0.5 m^2 m^{-2} to 12 m^2 m^{-2} (Marshall and Waring, 1986). Thus, a similarly broad range of interception storage capacities might be expected; values in one extensive survey ranged from 0.25 to 9.14 mm (Zinke, 1967). Amounts are least for grasses and sparse forbs, increasing for shrubs and leafless hardwoods, and reach their maximum for tall, dense conifers.

The nature of plant leaves is highly varied, from dense woolly mats to smooth, water-repellent, waxy surfaces. Water is also retained on the bark of tree stems, and bark characteristics are similarly varied. Bark characteristics not only help to determine interception storage capacity, but along with stem form, they also determine the amount of stem flow.

Interception storage starts to build up as soon as rainfall begins. Providing the rate of rainfall substantially exceeds evaporation from the surface, it builds up rapidly. Storage may be depleted during a storm if rainfall intensities diminish below the rate of evaporation, and in small, low intensity storms, storage capacity may never become satisfied, even though the amount of rain substantially exceeds it. When a storm ends, water remaining in storage is sooner or later lost to evaporation, unless a new storm begins before the process can be completed.

8.2.1.2 Evaporation of Intercepted Water. Evaporation of intercepted water occurs at a rate determined by the available energy supply, vapor pressure gradient, and aerodynamic characteristics of the surface. It may be expressed by a modified Penman equation (Monteith, 1965):

$$PET = \frac{\Delta R_n + \rho c\,(e_s - e_a)/r_a}{L(\Delta + \gamma)} \tag{8.10}$$

where *PET* is the potential evaporation, Δ is the slope of the saturated vapor pressure–temperature curve (Pa/°C) at the prevailing temperature, R_n is net radiation, ρ is the density of air (g cm^3), c is the specific heat of air, e_s is vapor pressure of water on the leaf surface, e_a is the vapor pressure of the air, r_a is the *aerodynamic resistance* of the surface (s cm^{-2}), a function of canopy roughness and wind speed, L is the latent heat of vaporization (J cm^{-1} depth), and γ is the psychrometric constant. The rate of potential evaporation by vegetation exceeds potential evaporation from an open water surface to the degree that aerodynamic resistance (r_a) over the vegetation is less than that over water as a result of greater roughness of the vegetation.

When the complete surface of the vegetation is covered by a film of water, evaporation occurs at its potential rate. However, when the surface film is incomplete, the rate of loss may be considered to be proportional to the degree of film coverage. Thus, if one-half of the surface were dry, evaporation would occur only at one-half of the potential rate. It is not necessary that the interception storage capacity be completely filled to have a continuous film of water over the surface, but the relative degree of canopy saturation is a good approximation of the proportion of wetted surface (Rutter and Morton, 1977).

Interception loss is sensitive to rainfall rate only when the rate is insufficient to maintain the surface in a completely wetted condition. Otherwise, loss is a direct function of storm duration for any given evaporation rate.

Evaporation of intercepted water begins as soon as the first drops strike the plant and continues until the plant is dry. Rates tend to be greatest near the beginning of the storm, during conditions of high energy availability as in summer, as soon as surfaces become completely wetted, and may exceed 0.5 mm h^{-1} (Gash, 1979). Rates averaged 0.37 mm h^{-1} throughout the year at night in New Zealand, indicating that evaporation of intercepted water is largely driven by advected energy rather than net radiation. Rates of loss may be several times greater than from open water due to the efficient convective exchange with the rough forest surface that reduces aerodynamic resistance to sensible heat and vapor transfer (Stewart, 1977; Pearce et al., 1980).

8.2.2 Interception of Snow

The principles of snow interception are the same as those of rain, but several important differences exist due to the nature of snow and the climatic conditions under which snow occurs.

Interception storage of snow develops differently from interception storage of rain. The first raindrops tend to adhere to plant surfaces and form a thin film spreading throughout. The first snowflakes (unless they are wet) tend to bounce or slide off the surface and fall through all but the smallest spaces, where they may be caught and bridge across small openings. On coniferous trees, continuous snowfall provides more and more bridges across larger gaps, thus increasing the size of the surface upon which more flakes can come to rest. As the snow load builds up, the gaps are filled and the platform surface increases more slowly. Heavier snow loads on flexible branches bend them downward, and at some point, the tree can hold no more snow. Thereafter, the excess slides or falls from the canopy as rapidly as it falls from the sky (Satterlund and Haupt, 1967).

Snow held in the canopy after snowfall ends may later reach the ground, falling *en masse*, as drip after melt, or by being blown from the trees. More than 80% of the snow held in the canopy after storms subsequently reached the ground during two winters in northern Idaho (Satterlund and Haupt, 1970).

Snow generally remains in interception storage for much longer periods than rain. In the Sierra Nevada of California, intercepted snow may be present in the crown canopy of conifer forests more than 50% of the snow season (Kittredge, 1953) and in upstate New York more than 60% of the season (Lull and Rushmore, 1961; Satterlund and Eschner, 1965). It may persist almost continuously throughout the snow season in far northern regions.

Evaporation rates from intercepted snow are much lower than from intercepted rain, which is evidenced by the extreme persistence of snow in the crowns before it is lost. Hoover and Leaf (1967) were unable to detect any effect of forest cutting on snow reaching the ground in a Colorado subalpine forest and Satterlund and Haupt (1970) reported only about 5% of total snowfall was lost to interception in northern Idaho. The lower rate of evaporation from snow results from the smaller supply of energy in winter, the much lower vapor pressure gradients in subfreezing weather, and the sliding of most snow from the crowns as soon as melt begins.

Snow interception has little meaning where snow completely covers short vegetation such as grasses and low shrubs.

8.2.3 Cloud Drip and Rime

Clouds and fog frequently envelop forests in coastal regions and mountains and occur with greater or lesser frequency nearly everywhere (Hardwick,

1973). They may or may not be accompanied by precipitation. In many cases, the individual water droplets are so small they remain suspended rather than failing to the ground. Under such circumstances, the ground in the open may be dry, but under forests, a steady drip of water soaks the ground.

The clouds of water droplets moving through the vegetation are deposited on the leaves and stems where they coalesce (*not* condense, for liquids have already condensed) into larger droplets that drip to the ground. This phenomenon is also a part of the interception process, but in this case it results in a gain instead of a loss of water, and net precipitation exceeds gross precipitation. This form of precipitation is called *cloud drip* or *fog drip*, or less frequently; *occult precipitation, horizontal precipitation,* or *negative interception*.

At air temperatures below freezing, the sharp curvature of the surface of tiny water droplets depresses their freezing point, and they may remain in a supercooled state to temperatures of −30 °C. Upon deposition, the surface flattens and they may crystallize almost immediately, giving rise to an adhesive deposit called *rime*. Deposits of rime build up snowlike as much as 30 cm or more thick on trees, the load sometimes becoming so heavy that trees break or fall from the weight. Rime can also build up directly from a mixture of ice crystals and liquid water droplets. *Hoarfrost* may also be deposited from water vapor directly onto cold surfaces.

Once on the vegetation, cloud drip and rime are like any other source of water or snow in interception storage. Some evaporates back to the air, drips from the leaves, flows down the stems, or if rime, falls or slides off to the ground when melting reduces the adhesive bonds with the plant. Rime is seldom blown off by the wind.

Cloud drip and rime represent a potentially important contribution to the water balance of many catchments, as indicated by measurements of up to 1524 mm annually (Isaac, 1946; Oberlander, 1956; Ekern, 1964; Kerfoot, 1968; Berndt and Fowler, 1969; Gary, 1972 ; Azevedo and Morgan, 1974; Shevelev, 1977; Harr, 1982; Lovett et al., 1982; and Hindman et al., 1983).

Studies of the deposition process have indicated the important variables involved in the formation of cloud drip and rime. The mass flux of water transport to a unit area of surface depends on (a) water droplet density in air, (b) windspeed, (c) time, (d) droplet size, (e) aerodynamic roughness, and (f) the dimensions of the surface upon which deposition occurs. Deposition increases with all factors except surface dimensions of the object upon which deposition occurs. The larger the surface, the less efficiently it collects cloud drip or rime (Fowler and Berndt, 1971; Merriam, 1973; Shuttleworth, 1977). Forest tree leaves and needles are efficient collectors of cloud drip, but as deposits of rime build up, catch efficiency decreases, limiting further collection until the rime falls. Therefore, temperature during fog events importantly affects the water flux to the ground by determining whether fog drip or riming takes place.

One effect of cloud drip and rime is to offset interception losses from normal precipitation. In areas of frequent fog and dense, rough vegetation, interception may result in a net gain in moisture. In one study in western Oregon, the reduction in moisture added by fog drip that resulted when two watersheds were patch cut caused a decrease in water yield after harvest, rather than an expected increase (Harr, 1980, 1982). As the frequency and duration of fog enveloping forests tends to increase with elevation away from coastal regions, the likelihood of net interception gains increases with elevation in most mountainous regions. At intermediate elevations, interception may prove to be a neutral factor in the water balance, and at lower elevations, it may result in a net loss.

8.2.4 Redistribution of Precipitation Reaching the Ground

Precipitation reaching the ground may be distributed quite differently in space than if it fell on the canopy above. The variation in spatial distribution of throughfall and drip can be seen in data by Leonard (1961), who used 216 gages randomly located over 0.6 ha of northern hardwood forest. The coefficient of variation in catch ranged from 58% in a storm with 1 mm of gross precipitation to 15% in a storm with 26 mm of gross precipitation. Table 8.1 illustrates his data. Helvey and Patric (1965) calculated that 46 rain gages would be necessary to hold the standard error under 5% of mean throughfall and drip in eastern hardwood forests in summer for storms that yielded less than 4.8 mm of throughfall and drip. That storm size includes more than one-half the number of summer storms over much of the eastern United States, although a considerably smaller proportion of summer precipitation.

The redistribution of precipitation falling as snow is often visually evident following a storm (Fig. 8.2). Small openings in the forest frequently exhibit heavy catches where wind-blown snow from the canopy is deposited. The heavy catch in the openings is probably due to eddy motion across the tree tops, which carries snow into the opening (Hoover and Leaf, 1967; Swanson, 1980; Gary, 1980). In forests, the catch is least at the base of the stem. Under

TABLE 8.1. Variation of Throughfall and Drip

	Throughfall and Drip				
Storm Size (mm)	Average (mm)	Maximum (mm)	Minimum (mm)	Standard Deviation	Coefficient of Variation (%)
1.0	0.5	1.5	0.0	0.3	58
3.0	1.8	7.1	0.3	0.8	42
3.3	2.5	4.1	0.8	0.8	25
6.4	4.8	7.9	2.0	1.3	24
7.1	4.8	7.9	3.3	0.8	17
10.2	8.4	11.9	3.6	1.5	17
25.9	21.3	29.0	13.2	3.3	15

Source: Based on 216 measurements per storm (Leonard, 1961).

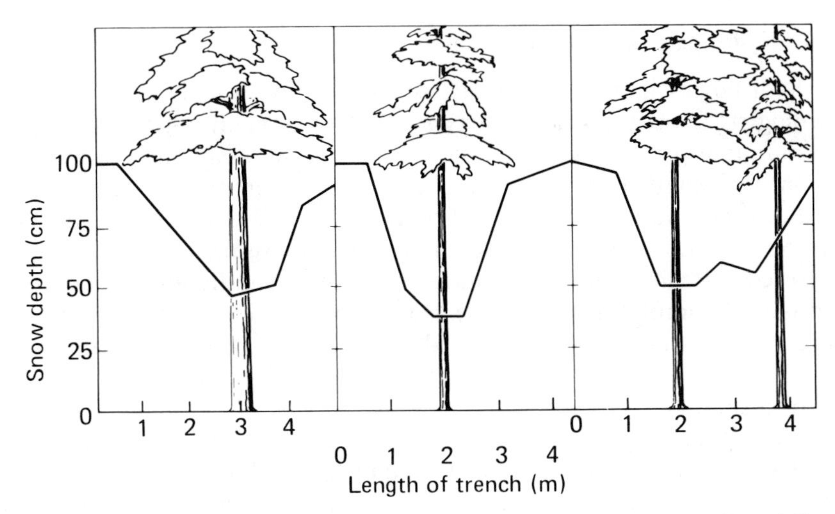

FIGURE 8.2 Typical snow profile under conifers (adapted from Lull and Rushmore, 1961).

conifers, the snow load on branches bends the tips down and snow slides off to be concentrated at the crown periphery.

Stemflow from rain may greatly concentrate water at the base of the stem. In large storms, the concentration of water over a small area may be as much as seven times the amount of gross precipitation (Leonard, 1961) and tends to follow large roots deep in the soil (Voigt, 1960). Stemflow varies greatly in amount with storm size, species characteristics such as form, size, bark characteristics (of woody vegetation), and the position of the individual crown in relation to surrounding vegetation.

Stemflow begins on smooth-barked species such as beech after as little as 1.3 mm of rainfall (Voigt, 1960; Leonard, 1961), but it may not start on rough-barked species until 23 mm of rain has fallen (Helvey and Patric, 1965). Even with evergreen species such as Douglas-fir and western hemlock, more precipitation is needed in summer than in winter to initiate stemflow (Rothacher, 1963). During small storms and with rough-barked species, stemflow seldom occurs. Stemflow is unimportant in snow interception, and in modifying the distribution of water in grasses and other low vegetation where a large number of stems are well distributed in space.

The significance of precipitation redistribution in space lies (a) in its effect of causing sampling problems of the amount of net precipitation and soil moisture, and (b) in causing concentrations of soil water that may enhance movement through soils to stream channels. It has been proposed that a large part of the increase in water yield in some snow zone forests, when small openings are created by cutting, is due to snow redistribution rather than reduced interception losses (Hoover, 1974).

8.2.5 Hydrologic Significance of Interception

The hydrologic significance of interception is now widely recognized by hydrologists. Interception losses may account for as much as 42% of gross precipitation in forests (Gash et al., 1980). On the other hand, interception may account for as much as 46% of water reaching the ground in forests where cloud drip and rime are common (Lovett et al., 1982). The gains from cloud drip and rime are readily understandable, but high loss rates, which appear to far exceed potential evapotranspiration rates and net radiation for periods as long as a full month (Sharma, 1984), require explanation. Furthermore, once energy is used to evaporate intercepted water, it is not available for transpiration, so why are interception losses simply not offset by equivalent reductions in transpiration?

With respect to energy supply, rainfall interception loss rates from forests are often in the range of 0.2–0.5 mm h^{-1}, equivalent to 50–125 J cm^{-2} h^{-1}, even in cloudy weather, in winter, and at night when net radiation is slight or nil. The temperature of wet canopies approximates the wet bulb temperature, and as the atmosphere is seldom saturated with vapor at the surface even during rainstorms, the surface temperature of the canopy is less than the air temperature during rainfall. Thus, the wet canopy is a sink for advected sensible heat that is derived from the air (Rutter, 1975; Pearce et al., 1980). However, as aerodynamic resistance (r_a) is least in tall, rough vegetation such as forests, and greatest with smooth surfaces, the rate of advected sensible heat transfer to a wet surface diminishes rapidly as the surface becomes smoother. Thus, rates of interception loss from low, smooth vegetation such as grasses may be little different than evaporation from open water.

Much of the energy used to evaporate intercepted water would not otherwise be available for transpiration. Leaves characterized by a dry surface usually approximate air temperatures when net radiation is near zero, and are neither a sink nor a source for sensible heat exchange with the air. Furthermore, much of the total interception loss takes place at night and in seasons when plants are dormant. Transpiration then is nil, so interception losses could not be offset by savings in transpiration in any case.

In daytime, when net radiation is positive, rates of water loss from interception still exceed those of transpiration. First, the net radiation of leaves with wet surfaces is greater than those with dry surfaces because (a) the albedo of wet leaves is lower, and (b) they emit less longwave radiation as a consequence of their lower temperature. Second, in transpiration vapor must move from within leaves through stomates, and overcome the resistance offered by the stomates themselves. Thus, the Penman–Monteith equation of evaporation from the canopy surface (Eq. 8.10) becomes

$$PET = \frac{\Delta R_n + \rho c\,(e_s - e_a)/r_a}{L(\Delta + \gamma)\,(1 + r_c/r_a)} \qquad (8.11)$$

where r_c represents the total *canopy resistance* to movement of water from the interior to the surface of the leaves. As r_c may be an order of magnitude greater than r_a in forests, transpiration rates would be much less than interception loss rates even if energy supplies were equal. Evaporation from a wet forest canopy may be three or more times greater than transpiration from the same surface (Stewart, 1977; Singh and Szeicz, 1979; Wronski, 1984). Where r_c is similar in magnitude to r_a, as it is in short grasses, interception losses should be more fully offset by transpiration savings, provided the plants are physiologically active (Rutter, 1975).

Similar arguments can be made regarding snowfall interception in contrast to evaporation of snow on the ground (Satterlund and Eschner, 1965). In winter, water transport in vegetation is restricted by frozen stems and the slow rate of water absorption by roots in cold soils (Kramer, 1942). Stomates are seldom open and vegetation is usually dormant. Hardwoods are often leafless.

Energy supplies to snow in trees are much greater than to snow on the ground. The albedo of intercepted snow is much less than that of a snowpack on the ground (Leonard and Eschner, 1968). The roughness of a snow-covered canopy could increase the advected energy supply to it by more than 10 times the amount available to an open, snow covered field. These differences, resulting primarily from the complex surface geometry of intercepted snow, are sufficient to account for considerably greater losses than from a snowfield on open ground.

8.3 TRANSPIRATION

The fixed characteristics of the site (climate, physiography, and soils) exert primary control over the amount of transpiration from a given watershed by controlling the amount of energy and water that is available. Within these limits, vegetation characteristics and land use exert secondary control. Natural vegetation tends to develop in response to site to the point where transpiration is the maximum that can be supported without endangering survival of the dominant species (Roberts, 1983). On wet and mesic sites, losses approach their maximum potential rates. In more xeric situations, or on wet or mesic sites where land use prevents the full development of natural vegetation, actual evapotranspiration losses are less than their potential.

Both the aerial and below-ground characteristics of plants influence the transpiration process. The aerial portions represent the major active surface where energy exchange occurs and the root system absorbs the water that is conducted upward to be transpired from the leaves. Water moves from the soil to the roots, through the plants to the leaves and into the air along a gradient of decreasing energy potential. At each step, it must overcome the resistance to water movement through the soil–plant atmosphere continuum. Thus, transpiration may be expressed as

$$T = f\frac{\Psi_{\text{soil}} - \Psi_{\text{root}}}{r_{\text{soil}} + r_{\text{root}}} + \frac{\Psi_{\text{root}} - \Psi_{\text{leaf}}}{r_{\text{stem}}} + \frac{\Psi_{\text{leaf}} - \Psi_{\text{air}}}{r_{\text{canopy}} + r_{\text{air}}} \tag{8.12}$$

where Ψ is potential and r is resistance. The energy potential of the air may be determined from its vapor pressure or relative humidity as

$$\Psi_{\text{air}} = RT \ln(e_a/e_s) \tag{8.13}$$

where R is the gas constant (0.4615 MPa K^{-1}), T is the temperature (K), e_a is the vapor pressure of the air, and e_s is the saturated vapor pressure at air temperature. Over a relative humidity range from 99 to 47%, the water potential of the atmosphere ranges from about −1 to −100 MPa, while the potential in the soil ranges mostly between 0 and −1.5 MPa, and in the leaves from −0.2 to −2 MPa.

The major resistance to water movement through the system is in the vapor phase from the leaf into the air (Rutter, 1975). However, small changes in the internal water status of plants may trigger physiological responses that greatly change canopy resistance as stomates open or close (Molz, 1981). Let us examine the factors that influence the rate and amount of transpiration.

8.3.1 Energy Absorption by Transpiring Vegetation

All other factors equal, transpiration would increase directly with energy supply to the vegetation canopy. Therefore, the vegetative characteristics that influence net energy availability are likely to influence transpiration losses. Plant characteristics influencing energy exchange have already been discussed. They include differences in albedo and roughness as they are influenced by species, height, density, and dormancy of vegetation.

Thus, Baumgartner (1967) showed that the radiant energy supply for cultivated crops, grass, and bare soil was about 5, 10, and 40% less, respectively, than that of a coniferous forest. Moore (1976) found that daily net radiation above a *Pinus radiata* forest was from 12 to 24% greater than above a nearby grassland because of the lower albedo and surface temperature of the forest. Tajchmann (1971) showed net radiation to be 20 and 16% greater over Norway spruce than alfalfa and potatoes, respectively, even though incoming shortwave radiation was 12% less than over the fields. Lowe et al. (1965) showed how albedo increased as corn matured and green leaves became dry.

Ordinarily, one might expect that vegetation on poleward facing slopes would have less solar radiation available for transpiration than that on slopes facing the equator. In at least some instances, for example, in southern California, the radiation absorbed by the vegetation canopy was greater on the poleward slope. Two factors were deemed responsible: (a) the vegetation canopy on the poleward slope was deeper and more dense, and (b) the sun's rays follow a longer, more oblique path through the canopy. Thus, even though less solar radiation was received on the poleward slope, a much higher proportion was absorbed there, and much more solar radiation reached the ground on the equator-facing slope (Miller and Poole, 1980).

8.3.2 Water Movement from the Plant canopy into the Air

Transpiration moves from plants into the air primarily via the stomates in the leaves. Minor quantities may be lost through lenticels in the stems, which serve the same functions as stomates, or when stomates are closed, by movement through the epidermis. Stomates of most plants close in response to darkness, internal water deficits that result in loss of turgor in the leaves, and either excessively high or low temperatures (Van Hylckama, 1969; Fahey, 1979), but the mechanisms triggering closure are complex and exceptions may occur.

Even when stomates are open, there may be considerable resistance (*stomatal resistance*) to the passage of water vapor from internal leaf cell surfaces, which may be considered to have a vapor pressure at saturation for a given leaf temperature, through the stomatal pore to the air. Measurements of stomatal resistance by various means have shown wide variation by species under similar moisture regimes (Shepard, 1972; Rutter, 1975; Roberts et al., 1980, 1981; Idso, 1983). In general, there is considerable overlap in life forms (grasses, forbs, shrubs, and trees), but the higher values tend to be found in trees, particularly pines. There is also large variation in resistance when stomates are closed, but epidermal resistance is usually at least one order of magnitude greater than when stomates are open.

Seasonal changes in stomatal resistance also occur as leaves mature. The lower values occur in young, green leaves and the greater in old, mature, or senescing leaves (Gee and Federer, 1972; Roberts et al., 1980). The availability of soil moisture also influences stomatal resistance, as will be discussed separately.

In communities, stomatal resistance must be replaced by canopy resistance, which represents the resistances in parallel of several layers of leaves. Canopy resistance is approximately

$$r_c = r_s/LAI \tag{8.14}$$

where r_c is canopy resistance ($\mathrm{s\,cm^{-1}}$), r_s is stomatal resistance ($\mathrm{s\,cm^{-1}}$), and *LAI* is leaf area index ($\mathrm{m^2\,m^{-2}}$). Clearly, the higher the leaf area index, the lower the canopy resistance. Canopy resistance should include not only overstory forest vegetation, but also that in the understory, which may contribute significantly to transpiration losses of the community (Cline and Campbell, 1976; Tan and Black, 1976; Roberts et al., 1980; Kelliher et al., 1986).

Several features of communities influence transpiration rates through their effects on canopy and air resistance to vapor transfer to the atmosphere. Other factors equal, tall irregular communities should exhibit greater rates of water loss because of lesser air resistance than shorter, smoother vegetation. Similarly, dense communities with higher *LAI* values could be expected to show higher water losses (Miller, 1982). Xeric species such as pines and oaks tend to higher stomatal resistances. It is no accident that the tallest, densest communities develop where moisture is seldom limiting, whereas short vegetation having incomplete coverage of the ground surface and a strong component of annuals grow on sites where drought is frequent and prolonged. Within a given climatic region, the nature of the climax community should provide a good index of relative transpiration losses.

8.3.3 Moisture Movement into and within Plants

The water available for transpiration from most plants is limited to that stored in the root zone of the soil. Important exceptions occur where ground water is available to the root system and in forest communities where storage of moisture available for transpiration in tree stems may exceed 20% of the root zone storage capacity and provide enough water to sustain transpiration for a week or more (Satterlund, 1959; Running et al., 1975).

When soil moisture is supplied to the plant at a slower rate than it is transpired, an internal water deficit develops, resulting in a decrease of leaf water potential and an increase in the potential gradient between soil and leaf. If insufficient water is absorbed at the higher gradient, leaf turgor is lost and stomates begin to close, increasing canopy resistance and reducing transpiration losses. Sammis and Gay (1979) showed a direct relation between leaf conductance (the inverse of resistance) and leaf water potential of creosotebush.

Transpiration for most plants essentially stops when soil moisture declines to the permanent wilting point, though some species can continue to extract soil moisture at potentials of −4 MPa or lower. However, the hydraulic conductivity of most soils drops sharply when soil moisture falls below field capacity, and transpiration of many species is reduced when soil moisture potential falls to only −0.1 to −0.2 MPa (Lopushinsky and Klock, 1974).

The ability of plants to absorb and transmit water at a rate sufficient to prevent stomatal closure and a reduction in transpiration depends on several factors, including: (a) the potential rate of transpiration; the loss rate when stomatal resistance is at its minimum; (b) the internal resistance of the plant roots, stems, and leaf cells; (c) the density and distribution of the root system; (d) the amount and distribution of moisture in the root zone; and (e) the unsaturated hydraulic conductivity characteristics of the soil. These factors often interact. For example, when the potential rate of transpiration is low, even a sparse root system in a soil with only a little available water and low hydraulic conductivity may absorb water fast enough to meet demands. However, only a dense, well-distributed root system in a moist soil with good hydraulic conductivity can meet high transpiration demands (Denmead and Shaw, 1962). In some cases, internal resistance in deciduous hardwoods may limit transpiration even when the trees have a well-distributed root system in wet soils when transpiration stress in high (Federer, 1979, 1982).

Most of the time when soils are moist, they do not limit transpiration. One would therefore expect little difference in losses from similar communities whether soils were shallow or deep. However, as the soils dried, the available moisture remaining in the shallow soil would decrease more rapidly than in the deep soil and transpiration rates would be reduced sooner and to a greater degree from the community on the shallower soil. Then, differences in community transpiration on such sites would be a function of the frequency and degree of soil moisture recharge. With frequent rains sufficient to recharge losses, there should be little difference, but as the interval between periods of recharge increased, the differences in losses would increase.

Similar results might be expected from communities that are shallow or deep rooted on deep soils. However, shallow rooted species that fail to utilize all the available water in a deep profile during long dry periods are likely to be successfully invaded by deep rooted species. The trend toward full occupancy of the site is a firm principle of ecological succession.

The soil moisture at which absorption becomes limiting varies with the unsaturated hydraulic conductivity of each soil. In general, a greater proportion of the available water in coarse soils is readily released at a given tension than in fine soils. Thus, available water in sandy soils may be depleted to a greater degree than that in clayey soils before transpiration rates are reduced (Zahner, 1967).

Plants differ greatly in the depth to which rooting occurs. Rooting depth also depends on soil characteristics, being shallow for all species where penetration is limited by dense layers, fragipans or bedrock, and deepest in open, well-aerated soils. In general, trees root more deeply than shrubs and shrubs more deeply than grasses, with forbs showing high variability. Important exceptions occur, however, for some grasses may root deeper than some trees. Many life forms appear to exhibit the greatest rooting depths in semiarid climates (Hellmers et al., 1955, Lewis and Burghy, 1964; Davis, 1978). Root growth of grasses may be greatly reduced by removal of tops by grazing animals, with the heavier grazing associated with the smaller root systems. The rooting depth of annuals is highly dependent on stage of growth, reaching a maximum at maturity. The deepest roots of trees appear to be those of mesquite (*Prosopsis juliflora*) found at a depth of more than 53 m in Arizona (Phillips, 1963).

8.4 DIRECT EVAPORATION FROM SOIL

The basic principles of the evaporation process can be applied directly to soils. Several factors, such as the position of the active surface in relation to the soil surface, litter interception, the transport of liquid water to the soil surface and the intensity of drying of surface soils require elaboration.

Unless a soil is bare, the active surface ordinarily stands somewhat above the soil surface. This is particularly true of tall, dense forests, and in some degree even for vegetation lower in stature and less dense. The effect of a raised active surface is to reduce the supply of energy available for direct evaporation from the soil. Ritchie (1972) indicates that evaporation from a wet soil under a canopy is reduced in direct proportion with the reduction in net radiation by the canopy. Andreychik (1976) found soil evaporation to be inversely related to crown density in forests, and Miller (1982) found it to decrease with an increase in leaf area index of chaparral.

Many hydrologists consider evaporation from the forest floor and litter as a special case of interception loss (Helvey and Patric, 1965), but the forest floor is considered as a part of the forest soil in this treatment. With a forest floor representing all stages of decomposition from a freshly fallen leaf to well-decomposed organic material mixed with mineral soil, any line drawn between soil evaporation and litter interception must be somewhat arbitrary. Probably most of the evaporation from the soil under forests occurs from the upper soil horizons that contain most of the organic material (Andreychik, 1976).

There are two basic phases of soil drying; an energy limited phase when the soil surface is wet, and a drying phase when evaporation is limited by available moisture. Some investigators include a third, transition phase between these two (Gardner and Hillel, 1962; Idso et al., 1979; Novak and Black, 1982).

While the soil surface is wet, evaporation takes place at its potential rate. It may be very rapid under high energy conditions. Removal of moisture in the surface films and pores increases the curvature of pore water surfaces and sets up a tension gradient, causing moisture to move upward from regions of lower tensions. Even with very great hydraulic gradients, the rate of upward movement in fine pores and thin water films is very slow. Continued evaporation lowers the water level in the soil pores well below the surface as the soil dries more rapidly than surface water can be replaced by liquid flow. Thereafter, water movement to the surface must occur in the vapor phase, with resistance increasing with the length of the diffusion pathway through the soil (Stewart, 1984).

The effect on evaporation is manifold. The drying of the surface increases the albedo, and hence reduces the energy supply. Soil surface temperatures increase with drying as more energy is partitioned into sensible heat and because dry soil has a much lower heat conductivity than moist soil. Net radiation is further reduced by the increased outgoing longwave radiation from the hotter surface, and more energy is lost as higher temperatures accelerate thermal convection of sensible heat to the atmosphere. Within the soil, effective vapor pressure gradients are reduced by additional pore resistance as depth to the liquid increases. Furthermore, heat flow to the liquid decreases with the decreasing rate of conduction through drying soils to greater depths. A rapidly developing cycle that increasingly tends to limit soil evaporation takes place, and evaporation declines quickly to a very slow rate before the soil is dried to any substantial depth. For example, Keen (1928) showed that evaporation essentially ceased from a saturated coarse sand in less than 1 week, though a water table remained only 36 cm below the surface. Evaporation continued for a longer period and the water table fell to greater depths in fine sand and heavy loam.

Evaporation decreases even more rapidly when the soil exhibits free drainage or contains absorbing roots, for water transport decreases rapidly as unsaturated soils dry. Where soils are covered by litter or mulches, evaporation often shows a sharp reduction as the litter or mulch dries, and there is considerable reason to believe that litter interception losses may be offset by savings in soil evaporation in areas where the litter is not rewetted frequently (Bristow et al., 1986). With frequent wetting, there is probably little difference in losses from bare or litter-covered soils.

Evaporation from an exposed surface soil may dry the soil much more completely than transpiration. Plant roots are able to exert tensions of only −1.5 to −4.0 MPa on soil moisture, equivalent in evaporating power to a relative humidity exceeding 98% at temperatures normal at a small depth below the surface. In most soils, a visible film of moisture coats soil particles, causing a darkening of color even at the permanent wilting point. On the other hand, when soils are bare, temperatures may exceed 60 °C near the soil surface with relative humidity reduced to 10% or less, creating evaporative forces exceeding −345 MPa. Direct evaporation can therefore dry surface soil almost to oven dryness.

Herein lies the root of a misconception that has plagued watershed management from its beginnings, and which is still strong in the public mind today: namely, that forests and other vegetation increase stream flow because they prevent the excessive drying of the soil. To paraphrase Mark Twain, the wonderful thing about science is that you can take one small fact, and from it, receive such a vast return in speculation. And here was a fact any superficial observer could readily see—the upper layers of bare soils dry out more completely than the upper mineral soil layers in the forest. Therefore, forests prevent the soil from drying.

The only flaw in this reasoning is that it is superficial. Although the surface layers of a bare soil may indeed dry almost to oven dryness, once the surface dries, it acts as a barrier between thermal energy and water at deeper levels. And thereby is accomplished that which must be done to prevent further water losses to the atmosphere—keep the energy away from the water. Vegetation, on the other hand, serves as a mechanism to conduct water from throughout the root zone to the energy absorbed in the canopy, drying soils to great depth. The lesser degree of drying near the surface is more than offset by the much greater depth to which drying may occur.

On shallow soils and in arid climates where rainfall seldom penetrates far beneath the surface, little difference between water losses from vegetated and bare areas should be evident (Cable, 1980). Although a dense cover might theoretically prevent the soil from drying so thoroughly as it would if bare, the situation is academic. Such dry sites cannot support a dense cover that would protect it from drying.

When soils are covered by snow, most evaporation takes place at the snow surface rather than from the soil. However, a slight vapor pressure gradient might exist from warmer underlying soil that would tend to move some soil water upward into the snowpack, even from frozen soils.

Soil freezing also causes soil moisture redistribution upward to the frozen layer. The soil moisture content at the surface may exceed saturation as freezing water expands and creates additional pore space (Krumbach and White, 1964; Harlan, 1973). Nightly freezing and daily thawing of moist bare soils may keep the surface nearly saturated over long periods, with upward water movement detected from as deep as 1.2 m. Evaporation losses may then greatly exceed those from similar soils that remained unfrozen. The greater moisture losses from the diurnally frozen soils apparently were due to the constant moisture available at the surface, the lower albedo of the surface and the greater vapor pressure gradient from the wet soil surface than that of the unfrozen soil that became air dry early in the rainless period (Anderson, 1946).

Water losses during the energy limited phase of soil evaporation are analogous to losses from interception storage, for in each case the amount of stored water readily available for loss is small. In summer, the frequency of wetting is a most important factor in determining total losses, for evaporation decreases rapidly as the surface dries. In winter, water losses from direct evaporation tend to be energy limited in humid areas and under a snowpack. In humid areas evaporation is effective in removing water only from the surface 25 cm or so of the soil profile, and when frequently rewetted, only from lesser depths. In more arid areas, where there may be long periods without precipitation, the rate of evaporation may be low, but the drying effect may extend to depths of 1 m or more.

In much of the world, snow is the primary active surface from which water losses occur during a large part of the year. Snow on the ground may be picked up by the wind, and much of the blowing and drifting snow never reaches the ground again. However, the many unique properties of snow make it desirable to discuss its hydrologic relations in a separate chapter (see Chapter 9).

LITERATURE CITED

Anderson, H. W. 1946. The effect of freezing on soil moisture and on evaporation from a bare soil. *Trans. Am. Geophys. Union* 27:863–870.

Andreychik, M. F. 1976. Water evaporation from the soil in the main types of forests in the Belorussian Pols'ye. *Forestry* 2:18–26 (seen in: *Sov. Hydrol.: Selected Papers* 15:250–255).

Azevedo, J. and D. L. Morgan. 1974. Fog precipitation in coastal California forests. *Ecology* 55:1135–1141.

Baker, F. S. 1950. *Principles of silviculture.* McGraw-Hill, New York.

Baumgartner, A. 1967. Energetic bases for differential vaporization from forest and agricultural stands. pp. 381–390. In: W. E. Sopper and H. W. Lull, Eds. *Proc. Internatl. Symposium on Forest Hydrology.* Pergamon Press, Oxford.

Berndt, H. W. and W. B. Fowler. 1969. Rime and hoarfrost in upper-slope forests of eastern Washington. *J. For.* 67:92–95.

Bristow, K. L., G. S. Campbell, R. I. Papendick, and L. F. Elliott. 1986. Simulation of heat and moisture transfer through a surface residue-soil system. *Agric. For. Meteorol.* 36:193–214.

Brutsaert, W. and H. Stricker. 1979. An advection-aridity approach to estimate actual regional evapotranspiration. *Water Resour. Res.* 15:443–450.

Cable, D. R. 1980. Seasonal patterns of soil water recharge and extraction on semi-desert ranges. *J. Range Manage.* 33:9–15.

Cline, R. G. and G. S. Campbell. 1976. Seasonal and diurnal water relations of selected forest species. *Ecology* 57(2):367–373.

Davis, E. A. 1978. Root systems of shrub live oak in relation to water yield by chaparral. *Hydrol. Water Resour. Arizona Southwest* 7:241–248.

Delfs, J. 1955. Die Niederschlagszuruckhaltung in Walde (Interception). Mitteilungen des Arbeitskreises "Wald und Wasser" von Dr. W. Friedrich-Koblenz. No. 2.

Denmead, O. T. and R. H. Shaw. 1962. Availability of soil water to plants as affected by soil moisture content and meteorological conditions. *Agron. J.* 54:385–390.

Ekern, P. C. 1964. Direct interception of cloud water on Lanaihale, Hawaii. *Soil Sci. Soc. Am. Proc.* 28:419–421.

Fahey, T. J. 1979. The effect of night frost on the transpiration of *Pinus contorta* ssp. *latifolia. Oecol. Plant.* 14:483–490.

Federer, C. A. 1979. A soil-plant-atmosphere model for transpiration and availability of soil water. *Water Resour. Res.* 15:555–562.

Federer, C. A. 1982. Transpirational supply and demand: plant, soil, and atmospheric effects evaluated by simulation. *Water Resour. Res.* 18:355–362.

Fowler, W. B. and H. W. Berndt. 1971. Efficiency of foliage in horizontal precipitation. pp. 27–33. In: *Proc. 39th Ann. Meeting, Western Snow Conf.*

Fritschen, L. J. and C. H. M. Van Bavel. 1962. Energy balance components of evaporating surfaces in arid lands. *J. Geophys. Res.* 67(13):5179–5185.

Gardner, W. R. and D. I. Hillel. 1962. The relation of external evaporative conditions to the drying of soils. *J. Geophys. Res.* 67:4319–4325.

Gary, H. L. 1972. Rime contributes to water balance in high-elevation aspen forests. *J. For.* 70:93–97.

Gary, H. L. 1980. Patch clearcuts to manage snow in lodgepole pine. pp. 335–346. Proc. of the 1980 Watershed Management Symposium, Irrig. Drain. Div., Am. Soc. Civil Eng., NY.

Gash, J. H. C. 1979. An analytical model of rainfall interception by forests. *Quart. J. R. Met. Soc.* 105:43–55.

Gash, J. H. C., I. R. Wright, and C. R. Lloyd. 1980. Comparative estimates of interception loss from three coniferous forests in Great Britain. *J. Hydrol.* 48:89–105.

Gee, G. W. and C. A. Federer. 1972. Stomatal resistance during senescence of hardwood leaves. *Water Resour. Res.* 8:1456–1460.

Hamon, W. R. 1963. Computation of direct runoff amounts from storm rainfall. Int. Assoc. Sci. Hydrology, Symposium on Surface Waters, Pub. No. 63:52–62.

Hardwick, W. C. 1973. Monthly fog frequency in the continental United States. *Mo. Weather Rev.* 101:763–766.

Harlan, R. L. 1973. Analysis of coupled heat-fluid transport in partially frozen soil. *Water Resour. Res.* 9:1314–1323.

Harr, R. D. 1980. Streamflow after patch logging in small drainages within the Bull Run municipal watershed, Oregon. USDA Forest Service Res. Pap. PNW-268. Pac. Northwest For. Range Expt. Sta., Portland, OR.

Harr, R. D. 1982. Fog drip in the Bull Run municipal watershed, Oregon. *Water Resour. Bull.* 18:785–789.

Hellmers, H., J. S. Horton, G. Juhren, and J. O'Keefe. 1955. Root systems of some chaparral plants in southern California. *Ecology* 36:667–678.

Helvey, J. D. 1974. A summary of rainfall interception by certain conifers of North America. pp. 103–113. In: E. J. Monke, Ed. *Biological effects in the hydrological cycle.* Proc. 3rd Int. Sem. for Hydrol. Professors. 18–30 July, 1971. Purdue University, W. Lafayette, IN, UNESCO.

Helvey, J. D. and J. H. Patric. 1965. Canopy and litter interception of rainfall by hardwoods of eastern United States. *Water Resour. Res.* 1:193–206.

Hindman, E. E., R. D. Borys, and P. J. DeMott. 1983. Hydrometeorological significance of rime ice deposits in the Colorado Rockies. *Water Resour. Bull.* 19(4):619–624.

Hoover, M. D. 1974. Snow interception and redistribution in the forest. pp. 114–122. In: E. J. Monke, Ed. *Biological effects in the hydrological cycle.* Proc. 3rd Int. Sem. for Hydrol. Professors. 18–30 July, 1971. Purdue University, W. Lafayette, IN. UNESCO.

Hoover, M. D. and C. F. Leaf. 1967. Process and significance of interception in Colorado subalpine forest. pp. 213–224. In: W. E. Sopper and H. W. Lull, Eds. *Internatl Symposium on Forest Hydrology*. Pergamon Press, Oxford.

Horton, J. 1973. Evapotranspiration and water research as related to riparian and phreatophyte management: an abstract bibliography. USDA For. Serv. Misc. Pub. 1234. US Govt. Printing Office, Washington, DC.

Idso, S. B. 1983. Stomatal regulation of evaporation from well-watered plant canopies: a new synthesis. *Agric. Meteorol.* 29:213–217.

Idso, S. B., R. J. Reginato, and R. D. Jackson. 1979. Calculation of evaporation during three stages of soil drying. *Water Resour. Res.* 15:487–488.

Isaac, L A. 1946. Fog drip and rain interception in coastal forests. USDA Forest Service Res. Note 34, pp. 15–16. Pac. Northwest For. Range Expt. Sta., Portland, OR.

Keen, B. A. 1928. The limited role of capillarity in supplying water to plants roots. *Proc. Papers, 1st Internatl. Cong. Soil Sci.* 1:504–511.

Kelliher, F. M., T. A. Black, and D. T. Price. 1986. Estimating the effects of understory removal from a Douglas fir (*sic*) forest using a two layer canopy evapotranspiration model. *Water Resour. Res.* 22:1891–1899.

Kerfoot, O. 1968. Mist precipitation on vegetation. *For. Abstr.* 29:8–20.

Kittredge, J. 1948. *Forest influences.* McGraw-Hill, New York.

Kittredge, J. 1953. Influences of forests on snow in the ponderosa-sugar pine-fir zone of the central Sierra Nevada. *Hilgardia* 22:1–96.

Kramer, P. J. 1942. Species difference with respect to water absorption at low temperatures. *Am. J. Bot.* 29:828–832.

Krumbach, A. R. Jr., and D. P. White. 1964. Moisture, pore space, and bulk density changes in frozen soil. *Soil Sci. Soc. Am. Proc.* 28:422–425.

Leonard, R. E. 1961. Interception of precipitation by northern hardwoods. Station Pap. No. 159, Northeast. For. Expt. Sta., Upper Darby, PA.

Leonard, R. E. and A. R. Eschner. 1968. Albedo of intercepted snow. *Water Resour. Res.* 4:931–936.

Lewis, D. C. and R. H. Burghy. 1964. The relationship between oak tree roots and ground water in fractured rock as determined by tritium tracing. *J. Geophys. Res.* 69:2579–2588.

Lopushinsky, W. and G. O. Klock. 1974. Transpiration of conifer seedlings in relation to soil water potential. *For. Sci.* 20:181–186.

Lovett, G. M., W. A. Reiners, and R. K. Olson. 1982. Cloud droplet deposition in subalpine balsam fir forests: hydrological and chemical inputs. *Science* 218:1303–1304.

Lowe, D. S., F. C. Polcyn, and R. Shay. 1965. Multispectral data collection program. pp. 667–680. In: *Proc. 3rd Symposium on Remote Sensing of the Environment.* 14–16 Oct., 1964. Inst. of Sci. and Tech., University of Michigan, Ann Arbor, MI.

Lull, H. W. and F. M. Rushmore. 1961. Further observations on snow and frost in the Adirondacks. USDA Forest Service Res. Note 116. Northeast For. Expt. Sta., Upper Darby, PA.

Marshall, J. D. and R. H. Waring. 1986. Comparison of methods of estimating leaf-area index in old-growth Douglas-fir. *Ecology* 67:975–979.

Massman, W. J. 1980. Water storage on forest foliage: a general model. *Water Resour. Res.* 16:210–216.

Massman, W. J. 1983. The derivation and validation of a new model for the interception of rainfall by forests. *Agric. Meteorol.* 28:261–286.

Merriam, R. A. 1973. Fog drip from artificial leaves in a fog wind tunnel. *Water Resour. Res.* 9:1591–1598.

Miller, P. C. 1982. Nutrients and water relations in Mediterranean-type ecosystems. pp. 325–332. In: *Dynamics and management of Mediterranean-type ecosystems.* USDA. Forest Service. Gen. Tech. Rept. PSW-58. Pac. Southwest For. Range Expt. Sta., Berkeley, CA.

Miller, P. C. and D. K. Poole. 1980. Partitioning of solar and net irradiance in mixed and chamise chaparral in southern California. *Oecologia* 47:328–332.

Molchanov, A. A. 1960. *The hydrological role of forests.* Academy of Sciences of the U.S.S.R. Institute of Forestry, Moscow. (Translated from the Russian by I.P.S.T., Cat. No. 870.) Available from Office of Technical Services, U.S. Dept. Commerce, Washington, DC. OTS63-11089.

Molz, F. J. 1981. Models of water transport in the soil-plant system: a review. *Water Resour. Res.* 19:1245–1260.

Monteith, J. L. 1965. Evaporation and environment. *Symp. Soc. Expl. Biol.* 19:205–234.

Moore, C. J. 1976. A comparative study of radiation balance above forest and grassland. *Quart. J. R. Met. Soc.* 102:889–899.

Munn, R. E. 1961. Energy budget and mass transfer theories of evaporation. Nat. Res. Council, Assoc. Comm. on Geodesy and Geophysics. subcomm. on Hydrology. Second Canadian Hydrology Symposium, Toronto.

Novak, M. D. and T. A. Black. 1982. Test of an equation for evaporation from bare soil. *Water Resour. Res.* 18:1735–1737.

Oberlander, G. T. 1956. Summer fog precipitation on the San Francisco Peninsula. *Ecology* 37:851–852.

Pearce, A. J., L. K. Rowe, and J. B. Stewart. 1980. Nighttime wet canopy evaporation rates and the water balance of an evergreen mixed forest. *Water Resour. Res.* 16:955–959.

Penman, H. L. 1956. Estimating evaporation. *Trans. Am. Geophys. Union* 37:43–50.

Penman, H. L. 1963. *Vegetation and hydrology.* Commonwealth Bureau of Soils, Harpendon, Tech. Comm. No. 53.

Phillips, W. D. 1963. Depth of roots in soil. *Ecology* 44(2):424.

Ritchie, J. T. 1972. Model for predicting evaporation from a row crop with incomplete cover. *Water Resour. Res.* 8:1204–1213.

Roberts, J. 1983. Forest transpiration: a conservative hydrological process? *J. Hydrol.* 66:133–141.

Roberts, J., C. F. Pymar, J. S. Wallace, and R. M. Pitman. 1980. Seasonal changes in leaf area, stomatal and canopy conductances and transpiration from bracken below a forest canopy. *J. Appl. Ecol.* 17:409–422.

Roberts, S. W., P. C. Miller, and A. Valamanesh. 1981. Comparative field water relations of four co-occurring chaparral shrub species. *Oecologia* 48:360–363.

Roberts, S. W., B. R. Strain, and K. R. Knoerr. 1980. Seasonal patterns of leaf water relations of four co-occurring forest tree species: parameters from pressure volume curves. *Oecologia* 46:330–337.

Rohwer, C. 1931. Evaporation from free water surfaces. U.S. Dept. Agric. Tech. Bull. No. 271.

Rothacher, J. 1963. Net precipitation under a Douglas-fir forest. *For. Sci.* 9:423–429.

Running, S. W., R. H. Waring and R. A. Rydell. 1975. Physiological control of water flux in conifers. *Oecologia* 18:1–16.

Rutter, A. J. 1975. The hydrological cycle in vegetation. pp. 111–154. In: J. L. Monteith, Ed. *Vegetation and the Atmosphere. Vol. 1. Principles.* Academic, London.

Rutter, A. J. and A. J. Morton. 1977. A predictive model of rainfall interception in forests. III. Sensitivity of the model to stand parameters and meteorological variables. *J. Appl. Ecol.* 14:567–588.

Rutter, A. J., A. J. Morton, and P. C. Robins. 1975. A predictive model of rainfall interception in forests. II. Generalization of the model and comparison with observations in some coniferous and hardwood stands. *J. Appl. Ecol.* 12:367–380.

Rutter, A. J., K. A. Kershaw, P. C. Robins, and A. J. Morton. 1971. A predictive model of rainfall interception in forests, I. Derivation of the model from observations in a plantation of Corsican pine. *Agric. Meteorol.* 9:367–384.

Sammis, T. W. and L. W. Gay. 1979. Evapotranspiration from an arid zone plant community. *J. Arid Environ.* 2:313–321.

Satterlund, D. R. 1959. Vegetation storage on the watershed. *J. For.* 57:12–14.

Satterlund, D. R. and A. R. Eschner. 1965. The surface geometry of a closed conifer forest in relation to losses of intercepted snow. USDA Forest Service Res. Pap. NE-34. Northeastern Forest Expt. Sta., Upper Darby, PA.

Satterlund, D. R. and H. F. Haupt. 1967. Snow catch by conifer crowns. *Water Resour. Res.* 3:1035–1039.

Satterlund, D. R. and H. F. Haupt. 1970. Disposition of snow caught by conifer crowns. *Water Resour. Res.* 6:649–652.

Sharma, M. L. 1984. Evapotranspiration from a eucalyptus community. *Agric. Water Manage.* 8:41–56.

Shepard, W. 1972. Some evidence of stomatal restriction of evaporation from well-watered plant canopies. *Water Resour. Res.* 8:1092–1095.

Shevelev, N. N. 1977. Interception of vertical and horizontal precipitation in the forests of the central Ural. *Lesovedeniye*, No. 6., pp. 38–46. (seen in: *Sov. Hydrol.: Selected Papers* 16:313–318).

Shuttleworth, W. J. 1977. The exchange of wind-driven fog and mist between vegetation and the atmosphere. *Boundary-Layer Meteorol.* 12:463–489.

Singh, B. and G. Szeicz. 1979. The effect of intercepted rainfall on the water balance of a hardwood forest. *Water Resour. Res.* 15:131–138.

Stewart, J. B. 1977. Evaporation from the wet canopy of a pine forest. *Water Resour. Res.* 13:915–921.

Stewart, J. B. 1983. A discussion of the relationships between the different forms of the combination equation for estimating crop evaporation. *Agric. Meteorol.* 30:111–127.

Stewart, J. B. 1984. Measurement and prediction of evaporation from forested and agricultural catchments. *Agric. Water Manage.* 8:1–28.

Swanson, R. H. 1980. Surface wind structure in forest clearings during a chinook. pp. 26–30. In: *Proc. 48th Ann. Meeting, Western Snow Conf.*

Tajchman, S. J. 1971. Evapotranspiration and energy balances of forest and field. *Water Resour. Res.* 7:511–523.

Tan, C. S. and T. A. Black. 1976. Factors affecting canopy resistance of a Douglas fir (sic) forest. *Boundary Layer Meteorol.* 10:475–488.

Thornthwaite, C. W. 1948. An approach toward a rational classification of climate. *Geog. Rev.* 38:85–94.

Thornthwaite, C. W., and B. Holzman. 1942. Measurement of evaporation from land and water surfaces. US Dept. Agric. Tech. Bull. No. 817.

U. S. Geological Survey. 1954. Water-loss investigations: Lake Hefner studies; Tech. Report, US Geol. Surv. Prof. Paper 269.

Van Hylckama, T. E. A. 1969. Photosynthesis and water use of saltcedar. *Bull. Int. Assoc. Sci. Hydrology* 14:71–83.

Voigt, G. K. 1960. Distribution of rainfall under forest stands. *For. Sci.* 6:2–9.

Wronski, E. 1984. A model of canopy drying. *Agric. Water Manage.* 8:243–262.

Zahner, R. 1967. Refinement in empirical functions for realistic soil moisture regimes under forest cover. pp. 261–274. In: W. E. Sopper and H. W. Lull, Eds. *Proc. Internatl. Symposium on Forest Hydrology.* Pergamon Press, Oxford.

Zinke, P. J. 1967. Forest interception studies in the United States. pp. 137–161. In: W. E. Sopper and H. W. Lull, Eds. *Proc. Internatl. Symposium on Forest Hydrology.* Pergamon Press, Oxford.

9 Snow Accumulation, Melt, and Vaporization

Snow occupies a unique position in the water cycle. Its position derives from its distinctive physical characteristics, its build up near, or as, the major active surface over extensive areas and the climatic conditions under which it occurs. These factors combine so that snow may contribute to streamflow in amounts and at times that greatly differ from its occurrence as precipitation. Snow may constitute well under one-half of the annual precipitation in an area, yet be the source of most of the annual streamflow. In the United States, it is the principal source of water yield throughout the mountainous West inland from the coastal ranges, across the northern Lake States into parts of the Northeast. Mountain snowpacks may release their water slowly, sustaining streamflow well into summer, but sometimes they are rapidly translated into runoff, as during the Columbia River flood of 1948. Water users and flood plain dwellers watch snow surveys anxiously for indications of spring flood and summer water supply potential.

Though a general seasonal and spatial pattern of snow accumulation seems regularly evident, close examination reveals amazing variation. During autumn and spring, the climate fluctuates across the threshold between rain and snow, freeze and thaw. In many places, this irregular fluctuation continues throughout the winter. Thus, the margin of the snowpack shifts, and its water equivalent changes as snow accumulates, melts, or is lost by evaporation or *sublimation*. (If vaporization occurs directly from a solid, the process is sublimation, but if a water film covers the snow crystals, as is frequently the case, then vaporization is a two step process of melt, then evaporation. In this chapter, no distinction will be made between the two, as the exact process in any given circumstance may not be known.) Only rarely does the snow on the ground represent all the antecedent snowfall. The water equivalent of the snowpack at the time of maximum accumulation is commonly only 40–80% of winter snowfall, with the remainder being lost by interception, melt, and direct evaporative losses during the period since the first snowfall.

The snowpack water balance may best be understood if the distinctive physical characteristics are considered the woof, and energy exchange processes the warp, from which any of the varying patterns may be woven. The complex variations among physical characteristics and energy exchange may be clarified by assuming the physical conditions of snow to be constant and subjecting it to prevailing energy conditions over a short time interval. At the end of the period, not only is the snowpack water balance changed, but the physical properties of the snow may also change significantly. These new characteristics are used as the starting point for the next interval and the process is reiterated.

9.1 PHYSICAL CHARACTERISTICS OF SNOW

The most obvious physical characteristic of snow is its occurrence as a porous solid, usually of low density. This characteristic is responsible for highly irregular deposition of snowfall on windy days. It settles out in areas protected from the wind, whereas exposed areas may remain bare. Snow deposited in exposed areas during calm weather may later be redistributed by drifting. It is very difficult to obtain an accurate measure of watershed snowfall during windy weather. In fact, a truly accurate snow gage for windy conditions has yet to be constructed.

Unlike a liquid, all snowfall reaching the ground is stored until it melts or is lost by vaporization. In exposed areas, it may be suspended in the air by drifting winds, where the process of sublimation may be sharply intensified. Snow in the trees or on the ground greatly modifies the active surface. It also has the capacity to store liquid water from rain or melt, in the same manner as a soil, though usually in a much lesser degree.

Snow on the ground may undergo many changes from deposition to final disappearance. Among the important characteristics that change are crystal form, snow density, temperature, thermal conductivity, albedo, liquid water content, and permeability. These changes may strongly affect and are affected by energy exchange and influence the amount and timing of delivery of liquid water to the ground.

The density of newly fallen snow ranges mostly between 0.05 and 0.15 g cm^{-3}, although some forms such as *graupel* (a soft snow pellet) may be considerably more dense. The U.S. Weather Service considers that on the average, 10 units depth of fresh snow is equivalent to 1 unit depth of water (or a density of 10%). As snow accumulates, metamorphism begins. Compression, abrasion by wind, melt or rain and refreezing, and vapor transfer within the pack change crystalline structure and usually increase overall density. However, snowpack characteristics frequently vary with depth and some layers, such as depth hoar may become less dense during temperature gradient metamorphism (La Chapelle, 1969; Schaerer, 1981). At the time of maximum accumulation, the average density of deep snowpacks often reaches 40–60% in the West although it seldom exceeds 35% in the eastern United States where snowpacks tend to be shallower.

9.1.1 Water in the Snowpack

The snowpack seldom consists of solid particles alone. Thin films of liquid water may cover ice crystals at temperatures well below freezing, held in this state by the extreme curvature of the surface, which depresses the freezing point (Chalmers, 1959). At temperatures near freezing and at the freezing point, liquid water is almost always present in small quantities. The *water content* of snow refers to the portion of the pack that consists of liquid water. It may reach as much as 5–10% of the mass of solid particles. The term water content is frequently misused to express the *water equivalent* (we) of the snowpack, which is the depth of water that would be released if all the snow were melted. The water equivalent includes both the solid and liquid phases of the snowpack, and is a function of its depth times its density. It may be determined by weighing a column of snow with a given cross-sectional area cut from the pack with a snow sampling tube; each gram per square centimeter is equivalent to 1 cm depth, we, as water has a density of 1 g cm^{-3}.

The temperature of the snowpack may vary in time and with depth, but it can never exceed 0 °C under normal circumstances. In temperate climates, the lowest layers of deep snow usually approach 0 °C and show little fluctuation, but where low winter temperatures persist for long periods, as in much of Canada and the northern Lake States where air temperatures may remain at −10 °C or lower for several weeks at a time, soils freeze deeply and the lowest layers of the snowpack become and may remain well below freezing until the melt season begins. Snow temperatures fluctuate most widely and rapidly at the surface and the fluctuations decrease rapidly with depth because of the poor heat conductivity of loose snow and its *low volumetric heat capacity* (the amount of heat required to change the temperature of a unit volume of a substance by 1 °C; snow with a density of 0.1 g cm^{-3} and a specific heat of 0.5 has a volumetric heat capacity of only 0.21 J cm^{-3}).

The amount of energy needed to convert a given mass of snow to liquid water depends on the temperature and water content of the snowpack. For example, a cold snowpack of −8 °C and no liquid water in it would require 351.5 J g^{-1} to convert the snow to meltwater; 16.7 J to raise the temperature to the melting point and 334.8 J to satisfy latent heat of fusion requirements. (The amount of energy required to bring a snowpack to a temperature of 0 °C throughout is referred to as its *cold content*. It represents the water equivalent times specific heat times the difference between 0 °C and its mean temperature.) On the other hand, a snowpack at 0 °C consisting of 5.3% liquid water would require only 318.1 J g^{-1} to melt since only 0.95 g of ice needs to be melted to release 1 g of water $[0.95 + (0.053 \times 0.95) = 1]$. The amount of energy required to release 1 g of water by melting, expressed as a percentage of that needed to melt 1 g of pure ice at 0 °C is the *thermal quality* of the snowpack. In the examples above, thermal quality is 105 and 95%, respectively.

Water is not necessarily released from the snowpack whenever melting occurs. In a cold snowpack, much of the water from melt at the surface percolates into colder portions of the pack, where refreezing takes place until release of the latent heat of fusion brings the temperature up to 0 °C. If the snow layer is stratified, a common condition, the movement of melt water is complicated. Water may be ponded at the interface between strata until flow fingers develop that penetrate the lower stratum. Some drainage may occur before percolating water wets other parts of the pack. Elsewhere, ice layers may form above impeding horizons and frozen ground, significantly delaying penetration of melt (Marsh and Woo, 1984).

Excess water continues to move within the snowpack, refreezing, until the whole snowpack becomes *isothermal* (iso = same, thermal = heat) throughout at 0 °C. Part of the water is retained within the pack as a liquid before free drainage begins. Unless impermeable ice lenses exist, the water content seldom exceeds 5% by weight before drainage begins throughout the pack. Snow in forests is less likely to exhibit ice layers than that in the open. When a snowpack becomes isothermal throughout at 0 °C and its water holding capacity is reached, it is considered *ripe*, and any further melt is accompanied by release of water from the pack.

In early spring, a diurnal cycle of surface melt in daylight hours, percolation and refreezing at night may occur many times before the snowpack becomes ripe. On the other hand, where a deep snowpack overlies unfrozen ground, heat conducted upward from the ground may keep the bottom of the snowpack melting slowly and releasing water during long periods of subfreezing weather. Where soils are frozen, no such release of water occurs. In either case, substantial quantities of water are seldom released from the snowpack until it is ripe throughout and melting occurs at the surface.

Ripe snowpacks respond with release of water whenever melt occurs, at rates closely related to surface melt rates, whereas cold, dry snowpacks may show no release with surface melt. For this reason, snow hydrologists may divide the snow zone into two subgroups (Smith, 1974). First is the *warm* (or *wet*) *snow zone*, where the snowpack remains in ripe or near-ripe condition throughout most of the snow season, melting conditions occur frequently, and high winter streamflows are not unusual. Second is the *cold* (or *dry*) *snow zone*, where considerable melt must take place in most circumstances before the pack becomes ripe, winter melt is infrequent, and high winter flows are rare. There is no sharp dividing line between these two zones, and the boundaries shift from year to year.

In general, the warm snow zone occurs in regions with a maritime winter climate. In North America, it extends southward along the Pacific Coast along the western mountains from southeastern Alaska to the Sierra Nevada Mountains of California. A wedge penetrates inland to nearly across the northern Rocky Mountains through Oregon, Washington, northern Idaho and northwestern Montana, and across southern British Columbia. Only a narrow zone extends inland from the Atlantic into northeastern United States and the Maritime Provinces of Canada.

The cold snow zone occupies the North American Interior and extends nearly to the eastern coast, in the region of continental winter climate. The southern border is a transition zone. Where winter melt is frequent a snowpack seldom persists for more than a few weeks except in high elevations of the West. The rest of the region is usually considered to lie outside the snow zone, although transient snows may occur several times each winter.

9.1.2 Snowpack Albedo

Before snow melts, it must show an energy gain and the energy balance of any active surface is greatly conditioned by its albedo. Among all the substances on the watershed, few can compare with snow in this respect. The albedo of snow is highly variable. The high albedo of fresh snow is well known to skiers and mountain climbers, for whom protective eye covering is a necessity to prevent snow-blindness on sunny or even bright overcast days. As snow ages, accumulates surface dust, becomes wet and dense, characteristics of snow change, and the albedo drops sharply. The relation of albedo to these characteristics is shown in Table 9.1.

The albedo of snow is also influenced by its depth and the albedo of underlying material, for snow reflects not only from its surface, but from underlying layers as well. As much as 50% of the solar radiation that reaches the surface may penetrate to a depth of 10 cm in fresh, loose snow (Geiger, 1965). Thus, a thin layer of fresh, dry snow lying upon old dirty snow would exhibit an albedo intermediate between the two types. Wet snow has a lower albedo than dry snow because water films are highly absorptive of solar radiation (Kuzmin, 1956).

9.1.3 Longwave Radiation Exchange Characteristics

The whiteness of snow is totally misleading as an indication of its longwave radiation characteristics. Snow is one of the blackest bodies found in nature with respect to longwave radiation. This characteristic, combined with the transmissivity of snow to solar radiation explains why depth is important in determining reflection of solar radiation. Solar radiation that penetrates a thin snow layer and is absorbed by an underlying layer, such as dark coniferous foliage, may be emitted as longwave radiation and be totally absorbed by the overlying snow. Similarly, snow near the base of a tree melts away from the trunk because it absorbs so much of the longwave radiation from the stem, which is warmed by the sun or warm air.

TABLE 9.1. The Albedo for Solar Radiation in Relation to Characteristics of Snow

Snow Character	Albedo
New, dry	0.75–0.95
New, wet	0.70–0.75
Old, dry	0.60–0.70
Old, wet	0.50–0.60
Old, wet, dirty	0.35–0.45

Source: Data from Kuzmin (1956) and others.

The unique physical characteristics of snow: its maximum temperature of 0 °C and maximum vapor pressure of 611 Pa, its low volumetric heat capacity and poor heat conductivity, its highly variable albedo and transmissivity for solar radiation, and its capacity to store and transmit liquid water, must all be kept in mind if processes of snow accumulation and *ablation* (wasting away) are to be understood and modified. Failure to consider any of the above factors will almost invariably lead to erroneous conclusions and unsatisfactory snowpack management.

9.2 SNOW DEPOSITION

The accumulation of snow on the ground at any given time is a function of the deposition of precipitation as snow and subsequent gains by hoar frost deposition, condensation, and retention of rain, either as a liquid or frozen in the pack, minus losses by the release of melt water and evaporation. Losses due to interception have been previously considered. Drifting may redistribute snow reaching the ground and be accompanied by substantial losses in open, windswept terrain (Tabler, 1973).

Snow falling from the clouds does not necessarily reach the ground as snow, as those who live in hilly or mountainous country are well aware. Hilltops may be covered with a crest of white, while only rain reaches the valley floor below. Sometimes the difference in the form of precipitation is clearly evident over a change in elevation of only a few tens of meters and in temperature of less than a full degree.

Falling snow, particularly if it is in the form of stellar crystals or plates rather than graupel or pellets (La Chapelle, 1969), is admirably constructed for high energy exchange in turbulent air. It has a very large surface in relation to its mass, and a low settling speed even in calm air, which is considerably reduced by turbulence as wind speeds increase (Businger, 1965). It may be suspended through an exceptionally long trajectory before reaching the ground. If the air is above freezing or has a vapor pressure of less than 611 Pa (the saturated vapor pressure of ice at 0 °C), melting and/or sublimation may proceed rapidly. An analytical evaluation of sublimation of wind blown snow indicates the potential for large losses (Schmidt, 1972). Empirical studies in Russia indicated that the rate of sublimation of blowing snow may be 20 times that of stationary snow (Dyunin, 1956). Drifting snow losses will be considered in more detail later.

Snow deposition increases with elevation, as a consequence of the cooler air temperatures that result in a greater proportion of precipitation reaching the ground as snow and because of the normal increase in total precipitation with elevation. Vapor losses from falling or blowing snow may or may not be greater at one elevation than another, depending on the energy status of the atmosphere and the distance over which snow travels before reaching the ground.

In northern Idaho, the average increase in snowpack water equivalent at the time of maximum accumulation was about 8 cm per 100 m of increased elevation between 800 and 1700 m. Melting was common at lower and intermediate elevations, which suggests deposition differences during the period of accumulation were somewhat less (Packer, 1962). In northern New York, the increase of snowfall with elevation is extremely variable, depending on the location of topographic features with respect to winter storm paths and type of storm, ranging from extremely small amounts to as much as 21 cm per 100 m on Tug Hill, in the lee of Lake Ontario (Muller, 1960).

Other sources of deposition of moisture in the snowpack include frost and condensation, intercepted rime that falls to the surface, and rain that may be retained as liquid or that freezes within the pack. Most rain and condensation occur during periods when water is being released from a melting pack, however, and the water equivalent is usually decreasing during such periods. Most of the ice layers formed in the snowpack are due to melting and refreezing of water previously present, but are sometimes due to rain or condensation that later freezes. Frost may occur frequently, adding small amounts of moisture, but the source of vapor may be from warmer lower layers of the snowpack or soil beneath as well as from the atmosphere.

9.3 MELT AND VAPORIZATION

Just as snow accumulation represents the balance between additions to and losses from the snowpack and cannot easily be considered independently of each other, so too it is most fruitful to consider melt and sublimation together. Both represent latent energy exchange and their relative occurrence depends on the state of the atmosphere that controls how energy absorbed by the snowpack is partitioned.

Two related physical characteristics dominate energy exchange with snow; the maximum temperature of 0 °C that snow can attain and the corresponding maximum vapor pressure of 611 Pa for ice at 0 °C. Except for these important limitations, heat exchange with the atmosphere is the same over snow as over water. The temperature limitation, however, limits the storage of sensible heat within the snowpack. Therefore, energy absorbed that is in excess of that needed to warm the snowpack to 0 °C throughout and satisfy evaporation requirements must be used to melt the snowpack.

9.3.1 Sources of Heat

There are six sources of energy that can supply heat to the snowpack.

1. Solar radiation is both direct beam and diffuse. In early winter at high latitudes, direct beam solar radiation may be weak, and exceeded in amount by the diffuse component, but the total flux may be too low to

offset losses from reflection and longwave outgoing radiation except for short periods of the day around noon. In spring, the direct beam component dominates and total solar radiation increases rapidly (Table 9.2). Furthermore, the decreasing albedo of the melting snowpack increases the proportion that is absorbed.

2. Longwave radiation from sky, clouds, and vegetation is usually insufficient to offset losses of radiation emitted by the snowpack (Table 9.3). This is because the snowpack radiates as a blackbody, and unless the sky is completely obscured by vegetation or clouds, it does not. Even when the sky is completely overcast and radiating at near blackbody efficiency, the temperature of the cloud base is usually lower than that of the air near the ground The only time a net longwave radiation surplus by a snowpack is likely is when air temperatures exceed 0 °C with clouds in the sky or a forest canopy over the snowpack. Average snow surface temperatures approximate average air temperatures (though the range is larger) until air temperatures exceed 0 °C.
3. Convection of sensible heat may contribute substantial energy to a snowpack, particularly during chinooks and in the lee of large water bodies (Table 9.4). It is unlikely that temperatures as high as 10 °C with

TABLE 9.2. Average Daily Solar Radiation and Extremes, Spokane, Washington[a]

	10-Year Record		
Annual Period	Mean	Highest	Lowest
January 1–7	356	779	59
February 5–11	699	1703	142
March 5–11	1147	1925	373
April 2–8	1787	2880	419
April 30–May 6	1950	3411	486

Source: Phillips (1965).

[a](47° 37′ N Latitude, 117° 31′ W Long.), 10-Year Record (1955–1964) (J cm^{-2} day^{-1})

TABLE 9.3. Longwave Radiation Exchange between Snow and Sky Under Various Sky Conditions (J cm^{-2} day^{-1})

Snow and Air Temperature (°C)	Snow Emitted	Clear Sky		50% Clouds		Overcast	
		In	Net	In	Net	In	Net
−20	2009	1323	−686	1649	−360	1946	−63
−10	2344	1661	−683	2067	−277	2281	−63
0	2721	2047	−674	2453	−286	2661	−60
5[a]	2721	2260	−461	2578	−143	2867	146
10[a]	2721	2490	−231	2888	167	3085	364

[a]Maximum snow surface temperature is 0 °C.

TABLE 9.4. Approximate Sensible Heat Transfer from Air of Different Temperatures and Wind Speeds to Snow

Air Temperature (°C)	Snow Temperature (°C)	Wind Velocity ($m\ s^{-1}$)	Heat Transfer ($J\ cm^{-2}\ day^{-1}$)
5	0	1	188
5	0	5	942
5	0	10	1883
5	0	20	3767
10	0	1	377
10	0	5	1883
10	0	10	3767

Source: Diamond (1953) and the authors.

wind velocities as great as 20 $m\ s^{-1}$ exist for any great distance over snow except during extreme chinooks, but temperatures of 5 °C with wind speeds up to 10 $m\ s^{-1}$ are common in coastal western North American and parts of northeastern United States in winter. Treidl (1970) estimated a flux of 250 $J\ cm^{-2}\ 9\ h^{-1}$ to snow between Kentucky and northern Michigan during movement of a warm air mass across the region.

4. Convective transfer of latent heat of condensation can be an especially important source of energy when warm, moist air masses travel over snow. Precipitation occurring as rain often accompanies such air masses, but much of the snow melt popularly attributed to warm rain is due to condensation. Condensation melt accounted for as much as 38% of hourly melt when the dewpoint temperature was between 3 and 4 °C during a January storm in eastern Oregon (Zuzel et al., 1982). One gram of condensate will melt 7.5 g of snow, which is the ratio of the latent heat of fusion to the latent heat of condensation. For condensation to occur, the vapor pressure of the air must exceed 611 Pa when the snow surface temperature is 0 °C. The amount of condensation melt from saturated air over a snowpack is shown in Table 9.5.

   It should be obvious that no vaporization from snow takes place when condensation occurs. It should also be clear that for every 7.5 g of snow melt, 8.5 g of liquid water are produced, for the condensate is in addition to the melted snow.

5. Sensible heat from warm rain is usually overrated as an important source of heat for melting snow, but some heat is available from rain. The temperature of falling rain is approximately that of the wet bulb temperature and rarely exceeds 10 °C over snow. As the latent heat of fusion of snow is 335 $J\ g^{-1}$, this means 1 cm of warm rain (10 °C) would melt only 1.25 mm (we) of snow. At lower wet bulb temperatures, snow melt would be even less.

6. Conduction of heat from unfrozen soil to an overlying snowpack may provide enough heat to sustain winter streamflow from melt at the bottom of the pack (Federer, 1966). The amount of energy from this source is very small, being greatest in autumn and decreasing into spring. It rarely accounts for as much as 60 J cm^{-2} day^{-1} in autumn and may be negligible by spring. Smaller quantities of heat may be conducted upward into the snowpack even from frozen soils, but is not used in melt. When the snow is at less than 0 °C, heat transferred upward may be lost as radiation, sensible heat or sublimation.

9.3.2 Energy Partitioning

The first demand that must be met in partitioning the thermal energy absorbed by a snowpack is that of emitted radiation. Under clear or partly cloudy skies in early winter, daily net radiation deficits may occur even when air temperatures are above freezing, cooling both the snowpack and the overlying air. With the approach of spring, daily net radiation surpluses become more common, for incoming allwave radiation increases, snow albedo decreases and longwave outgoing radiation is limited by the maximum snow temperature of 0 °C.

If a surplus of net radiation occurs, it may be stored in the snowpack as sensible heat until the snowpack temperature reaches 0 °C, used in sensible or latent heat exchange with the air, to melt the snow, or any combination of the above. Only small quantities of heat are needed to raise the temperature of a cold snow surface due to snow's low specific heat and density that results in a low volumetric heat capacity, and its poor heat conductivity. For example, the surface temperature of fresh, dry snow rose from −13 °C in the shade to −5 °C

TABLE 9.5. Condensation Melt of an Open Snowpack from Saturated Air Masses at Various Temperatures and Wind Speeds

Air Temperature (°C)	Saturated Vapor Pressure (mb)	Vapor Pressure Difference, Snow–Air (mb)	Wind Velocity (m s^{-1})	Snow Melt Rate: Liquid Mass (g cm^{-2} day^{-1})	Snow Melt Rate: Thermal Equivalent (J cm^{-2} day^{-1})
0	6.11	0	Any	0	0
			1	0.43	142
			5	2.15	720
5	8.72	−2.61	10	4.31	1444
			20	8.61	2884
			1	1.10	339
10	12.28	−6.17	5	5.09	1703
			10	10.17	3407

after only 15 min exposure to the winter sun on a clear December morning in Pullman, Washington (measured with a Barnes PRT-5 radiation thermometer). The air was calm, with a temperature of −12.8 °C.

If the air temperature is lower than the snow surface temperature, both sensible heat and latent heat of sublimation will be transferred to the air, partitioned according to the temperature and vapor pressure gradients. If the air temperature exceeds the snow surface temperature, sensible heat will always be transferred to the snow. However, transfer of latent heat may be in either direction at warmer air temperatures, depending on the humidity of the air. Figure 9.1 may be used to determine the direction of vapor transfer over a wide range of snow surface temperatures and air temperature–humidity combinations.

The amount of vaporization that can occur from a snowpack under a wide range of conditions, provided energy is available to sustain it, is shown in Table 9.6. Vaporization at very low temperatures is always slight, regardless of the humidity of the air. The most rapid rates usually occur on bright days with air temperatures from −10 °C to slightly above freezing, with decreases at both lower and higher temperatures. During subfreezing temperatures, relative humidities seldom persist for long below 60%, and are often higher, so rates of sublimation are usually less than those shown in Table 9.6.

TABLE 9.6. Evaporation from an Open Snowpack to the Atmosphere Under Favorable Conditions[a]

Snow Temperature (°C)	Air Temperature (°C)	Relative Humidity (%)	Vapor Pressure Difference, Snow–Air (mb)	Wind Speed ($m\ s^{-1}$) 1	5	10	20
				Evaporation ($J\ cm^{-2}\ h^{-1}$)			
0	Any	0	6.11[b]	14[b]	71[b]	141[b]	283[b]
0	7	60	0.14	1	2	3	6
0	5	60	0.88	2	10	21	41
0	0	60	2.45	6	24	57	115
0	−5	60	3.58	8	41	83	165
0	−10	60	4.39	10	51	102	203
−5	Any	0	4.02[b]	9[b]	46[b]	93[b]	186[b]
−5	0	60	0.35	1	4	8	16
−5	−5	60	1.61	4	19	37	75
−5	−10	60	2.30	5	27	53	106
−10	Any	0	2.60[b]	6[b]	30[b]	60[b]	120[b]
−10	−10	60	1.03	3	10	24	48
−20	Any	0	1.03[b]	3[b]	10[b]	24[b]	48[b]
−30	Any	0	0.38[b]	1[b]	4[b]	9[b]	18[b]

[a] 500 $J\ cm^{-2}$ is the thermal equivalent of 1.12-mm depth, water equivalent, of evaporation.
[b] Designates maximum possible for given snow surface temperature and wind speed.

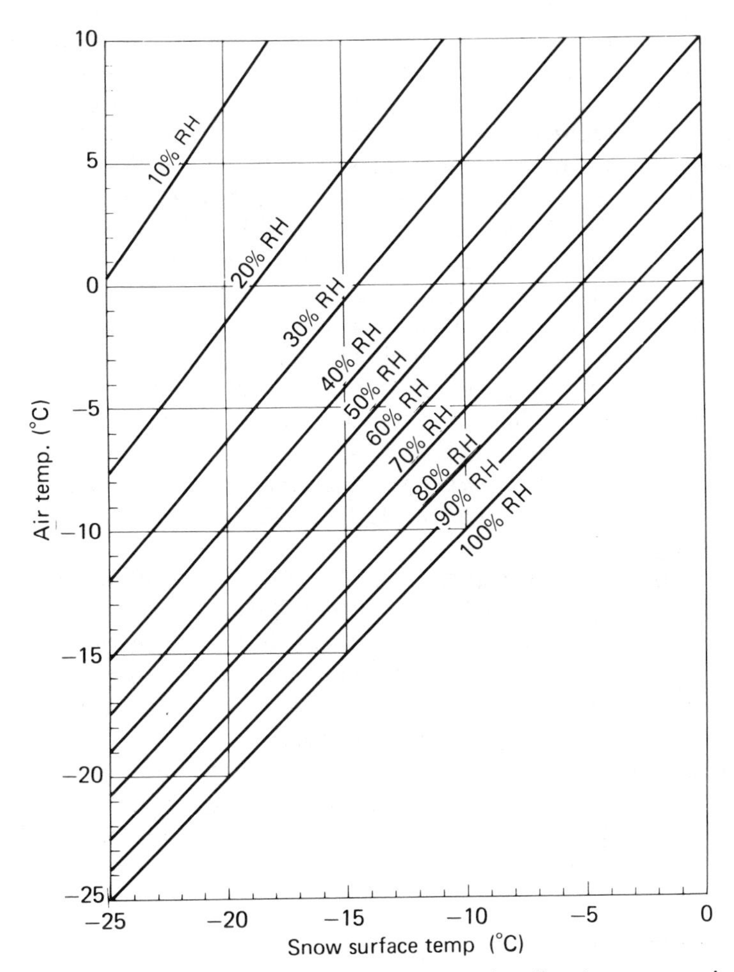

FIG. 9.1 Chart to determine occurrence of condensation (frost) or evaporation (sublimation) from a snow surface. *Directions*: Extend a line vertically from the snow surface temperature. Evaporation will occur under all air temperature and relative humidity (RH) combinations to the left of the line and whenever air temperatures are lower than snow temperatures (blank area below 100% RH line). Condensation will occur under all temperature and RH combinations to the right of the line and above the 100% RH line.

Sensible and latent heat transfer to the atmosphere is always severely limited by snow's maximum temperature of 0 °C and vapor pressure of 611 Pa, regardless of the net energy absorbed by the snowpack. Any surplus of energy after these needs are met remains in the snowpack. Since it is physically impossible to store this energy as sensible heat after the snow reaches 0 °C, the remaining energy causes the snow to melt. Melt water drained from the snowpack represents latent energy lost, albeit in another direction from that lost to the atmosphere as vapor. Melt water retained in the snowpack, however, represents stored latent heat and may be released upon refreezing. It may then supply energy needed to sustain outgoing radiation or be transferred as sensible heat or vapor to a colder atmosphere. Melt water retained in the snowpack is the principal means by which a surplus of energy absorbed during the day may be carried over to night when a deficit exists. Miller (1955) estimated that as much as 420 J cm^{-2} day^{-1} can be so carried over.

The conditions under which snow will warm or cool, vaporize, melt, or freeze depends on the temperature of the snow surface and atmospheric conditions. Many combinations of change can occur simultaneously: vaporization and melt, vaporization and cooling or freezing, condensation and melt, frost and cooling. However, melting has the last claim upon the net energy supply. The snowpack will tend to cool and freeze whenever net radiation is negative, and even when net radiation is positive and sensible heat flow is to the snowpack, the energy is first used to sustain vaporization demands according to vapor pressure deficits and air resistance. Only after these needs are met is any surplus used in melt.

It is useful to combine snowpack energy exchange relations presented in this section into an energy balance to determine the relative amount of snow vaporization or melt under any given set of conditions. Such an accounting is shown in Table 9.7 for one set of conditions. Energy coming to the snowpack from any source is entered as a gain and energy removed is entered as a loss. If gains exceed losses (as when meltwater remains in the snowpack and may later refreeze) the remainder may be carried forward to the next day's account. Net energy surpluses are largely used to melt snow whenever air temperatures are at freezing or above, for the low vapor pressure gradients that usually exist limit latent heat transfer to the air.

9.3.3 Drifting Snow

Drifting snow represents a unique situation that may result in sublimation loss many times greater than that from a snowpack on the ground. It also results in the redistribution of snow on the ground, concentrating it in some places protected from the wind to depths many times the accumulation due to snow fall alone, while diminishing it in others.

Kind (1981) describes the drifting process in detail. Essentially, wind moving over a snow surface exerts a drag force (shear stress), causing snow particles to move when sufficient force occurs to overcome the cohesion and inertial forces that oppose motion. Where streamlines of wind flow converge, acceleration occurs; where they separate (diverge), flow decelerates. Wind striking solid objects that are not streamlined breaks away (diverges) behind, and sometimes to a lesser degree in front of them, creating a region of low velocity and hence, low shear stress. Elsewhere, vortices of excess velocity may occur. Snow is picked up in regions of excess velocity and dropped in regions of low velocity. Ultimately, regions of low velocity may become filled with

TABLE 9.7. An Example of a Daily Snowpack Energy Balance[a]

Items	Gains	Losses	Water Equivalent
	($J\ cm^{-2}\ day^{-1}$)		(cm depth)
Radiation:			
1. Solar incoming	1046		
2. Reflected (albedo 0.75)		785	
3. Net solar (1–2)	262		
4. Longwave incoming	2616		
5. Longwave outgoing (snow temp. 0 °C)		2773	
6. Net radiation (3 + 4 − 5)	105		
Conduction:			
7. From soil	8		
Convection:			
8. Sensible heat (from Table 9.4)	565		
9. Latent heat (evaporation Table 9.6)		142	0.058
10. Surplus used in melt (6 + 7 + 8 − 9)		536	1.600
Total	4236	4236	

[a]Given: snow surface temperature 0 °C, albedo 0.75. Air temperature 5 °C, relative humidity 60%, wind speed 3 $m\ s^{-1}$. Sky partly cloudy, incoming solar radiation 1046 $J\ cm^{-2}$, incoming longwave radiation, 2616 $J\ cm^{-2}$.

snow, creating smoother flow lines that reduce and ultimately stop divergence and hence, deposition.

Once a threshold wind speed sufficient to initiate drifting begins, the capacity of wind to transport snow varies as the cube of velocity (v^3). Threshold speeds vary, depending on snow cohesion and density, but usually exceed 3 m s^{-1} at 1 m above the snow surface (Schmidt, 1982). Snow drifting is not significant at above-freezing temperatures, because water films under tension bind snow crystals together.

Once snow particles are lifted from the surface, their entire surface is exposed to energy exchange with the atmosphere. Provided the air is not saturated with vapor, sublimation will occur, depressing the particle temperature somewhat and forcing sensible heat flow to it. The rate of sublimation varies, depending on several factors, including: (a) vapor pressure deficit of the air, (b) air temperature, (c) solar radiation, (d) air pressure (and hence elevation), (e) particle size, (f) density (concentration) distribution of particles in the air, and (g) the degree of air turbulence. Sublimation varies directly with vapor pressure deficit, approximately doubles with each 10 °C increase in temperature between −20 °C and 0 °C, can double due to heat transfer from solar radiation, increases by 30–50% as elevation increases to 4000 m from sea level (other factors equal), increases with particle size and concentration in the air stream, and apparently increases with increasing turbulence (Schmidt, 1972).

A basic concept useful to estimate sublimation losses from windblown snow is *transport distance*, defined as the average distance a snow particle must travel in the air before completely evaporating (Tabler, 1971). The transport distance estimated during five winters at various sites in southeastern Wyoming ranging from open, rolling foothills at 2300-m elevation to large forest openings in mountains at 2900 m was a remarkably consistent 3050 m. Although this value is probably not universally appropriate, it suggests that a given transport distance may be widely applicable within any given region (Tabler, 1975).

In an infinitely large, open windswept area all the snow relocated by the wind would be lost to sublimation. However, natural barriers exist on all terrain. They may be spaced at distances less than the transport distance. Then, the contributing distance or "fetch" between barriers is less than the distance between barriers, because wind is reduced (and drifts will form) for distances up to 30 times beyond the height of the barrier (Tabler, 1980). In most terrain, not all the snow that falls in open areas is relocated. Some is retained in surface irregularities and behind low vegetation even in windy weather, some forms surface strength sufficient to resist abrasion by the wind and some may melt in place after snowfall before drifting winds occur.

If conditions are reasonably uniform between barriers, the evaporation loss is related to transport distance, fetch and the proportion of snow that is relocated by wind. It may be estimated as (Tabler, 1975):

$$Q_L = \frac{P_r R_c - \frac{1}{2} P_r R_u \left[1 - (0.14)^{R_c/R_u}\right]}{P_r R_c} \tag{9.1}$$

where Q_L is the loss as a fraction of P_r, P_r is the amount of snow that is relocated (m depth), R_c is the fetch or contributing distance between barriers (m) and R_u is transport distance (m). Figure 9.2 shows the losses of relocated snow as a percentage of that relocated for various contributing distances in southeastern Wyoming. Losses are probably similar in areas of similar climate and elevation, the amount decreasing with decreases in elevation, windiness, and temperature. The proportion of snow that is relocated by the wind varies widely with climate (windiness, frequency of melting, frequency, and amount of snowfall) and vegetation and terrain roughness. Tabler (1973) estimated that as much as 52% of the total snowfall may be lost from high elevation sagebrush lands in southeastern Wyoming.

9.4 THE INFLUENCE OF VEGETATION

Vegetation influences snow accumulation and dissipation in several ways. It influences deposition through interception processes and its effects on wind

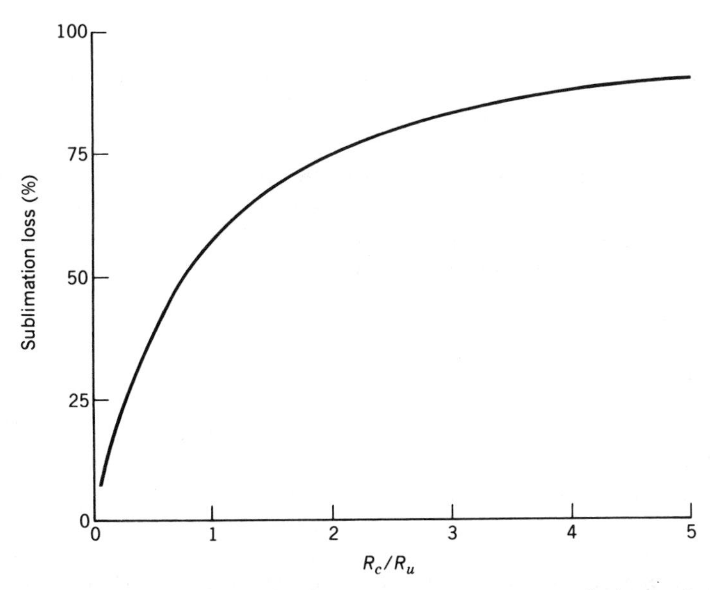

FIG. 9.2 Sublimation loss from blowing snow, as a percentage of relocated snow is a function of contributing distance (R_c) in relation to transport distance (R_u) in southeastern Wyoming (Tabler, 1975).

transport of falling or drifting snow. It influences melt and vaporization by modifying radiative and advective energy exchange. Interception of snow and rime has already been discussed, and will not be considered further except as it influences melt and vaporization of snow on the ground.

9.4.1 Effects on Wind

Where vegetation protrudes above the snowpack, it influences wind patterns, and hence deposition, redistribution, and advective energy exchange of snow. Scattered, open vegetation provides a rough surface causing irregular spatial variation in wind movement, which is accelerated in some places and slowed in others. As the vegetation becomes more dense, the active surface is displaced upward and air motion in the stem space is greatly reduced. Snow that enters regions of reduced velocity tends to be deposited, whereas that in regions of high velocity is carried beyond, or if on the ground, may be scoured away. Thus, falling snow is focused into areas of calm; behind barriers or snow fences and into small openings in forests (Hoover, 1971). Within small forest openings, deposition tends to be greatest in the center, with the least snow deposited in the lee (windward-facing) forest border zone and intermediate amounts in the windward forest (Gary, 1974, 1975; Kind, 1981).

Vegetation also influences energy exchange with the snowpack. Considering first advective energy exchange, snow tends to be deposited precisely in those areas with the lowest rates of exchange. The lack of wind that induces deposition also restricts sensible and latent heat transfer to drift sites. Snow on the ground under a forest canopy, even of leafless hardwoods, is subject to much less wind than a snowpack in the open. In general, one would expect that convective heat exchange with the snowpack is greatest where the vegetation is very open and protrudes only moderately above the snowpack so that roughness is increased, but the primary active surface remains that of the snow on the ground. It is minimum under a dense canopy that displaces the primary active surface upward into the crown layer.

9.4.2 Effects on Radiation Exchange

Radiative energy exchange is also greatly modified by vegetation. Even low vegetation that does not protrude above the surface may have a noticeable effect. Solar radiation that penetrates the snow surface may be absorbed by underlying vegetation, heating and melting the surrounding snow, first by conduction, then as a cavity forms as the snow recedes, by emission of longwave radiation from the heated vegetation. During cold, clear weather, the snow may melt away from the plant at depth, but the surface loses heat fast enough to remain frozen. Then, the snow surface may metamorphose into a clear ice lens and a covered air pocket is created around the plant that acts as a greenhouse. The same process often creates a "well" in the snow surrounding either buried or protruding plants under warmer conditions (Fig. 9.3). For example, the surface temperature of a 45-cm diameter Douglas-fir trunk near the snow line and facing toward the sun was found to be as much as 26 °C warmer than the shady side of the trunk in May 1970, when ambient air temperature was 15 °C. An irregular snow well had melted away all around the trunk where longwave radiation from the warm trunk was absorbed by the snow (Unpublished data of D. R. Satterlund and H. F. Haupt, taken 5 May, 1970 at Priest River Experimental Forest, Priest River, ID).

The greatest effect of vegetation is when a tall forest canopy shelters the snowpack below. Radiation reaching the snowpack may be expressed by the following equation modified from Leaf and Brink (1975):

$$R_n = R_s\,[(1-\alpha)\tau] + [V(\varepsilon\sigma T_a^4 - \sigma T_s^4)] + [(1-V)(\sigma T_a^4 - \sigma T_s^4)] \qquad (9.2)$$

where R_n is the net radiation, R_s is the incoming solar radiation, α is the albedo, τ is the transmissivity of the canopy for solar radiation, V is the average view factor of the snowpack, ε is the emissivity of the atmosphere, σ is the Stefan-Boltzmann constant, T_a is the air temperature (K), and T_s is the snow surface temperature. Both the transmissivity of the canopy and average view factor are functions of the degree of canopy closure, crown density, and height. For the same degree of canopy closure, transmissivity and view factor decrease with increasing crown density and height. Solar altitude also influences the transmissivity of the canopy for solar radiation. Solutions to Eq. 9.2 over a complete

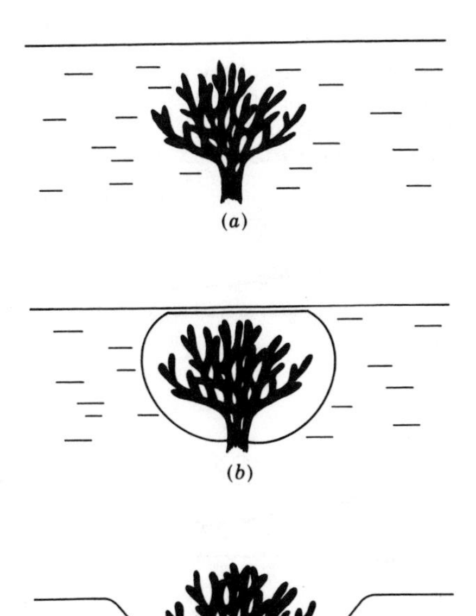

FIG. 9.3 Development of snow melt pocket (*b*) or well (*c*) when solar radiation that penetrates the snowpack (*a*) is absorbed by vegetation, causing melting of the surrounding snow.

range of crown cover from open to completely closed and for two different air temperatures are shown in Fig. 9.4. Petzold (1981) found that net allwave radiation to snow increased up to a forest density of 50% in an open, boreal forest, whereas Bohren and Thorud (1973) found from model studies that maximum net radiation could occur at any density, depending on albedos and radiant fluxes. Snow in the canopy further reduces solar radiation reaching the snowpack below and keeps the canopy from warming above 0 °C until the crowns become bare of snow.

Radiation balances for points within and adjacent to openings of any size and shape may be estimated by combining shadow boundary and view factor estimates with solar and longwave radiation estimates, taking into account the different temperatures of border trees on the sunny and shaded sides of the opening (Tuller, 1973). The trunk space below conifer tree crowns may be an "energy trap" beneath the sunny border where low angle winter direct beam solar radiation penetrates, and with reflected radiation, is trapped beneath the crown canopy. On the other hand, the effect of the forest shade extends well into the opening on the shady border (Fig. 9.5).

9.4.3 Combined Effects

The effects of vegetation on snow deposition and energy exchange tend to accentuate each other. When advective heat exchange is coupled with radia-

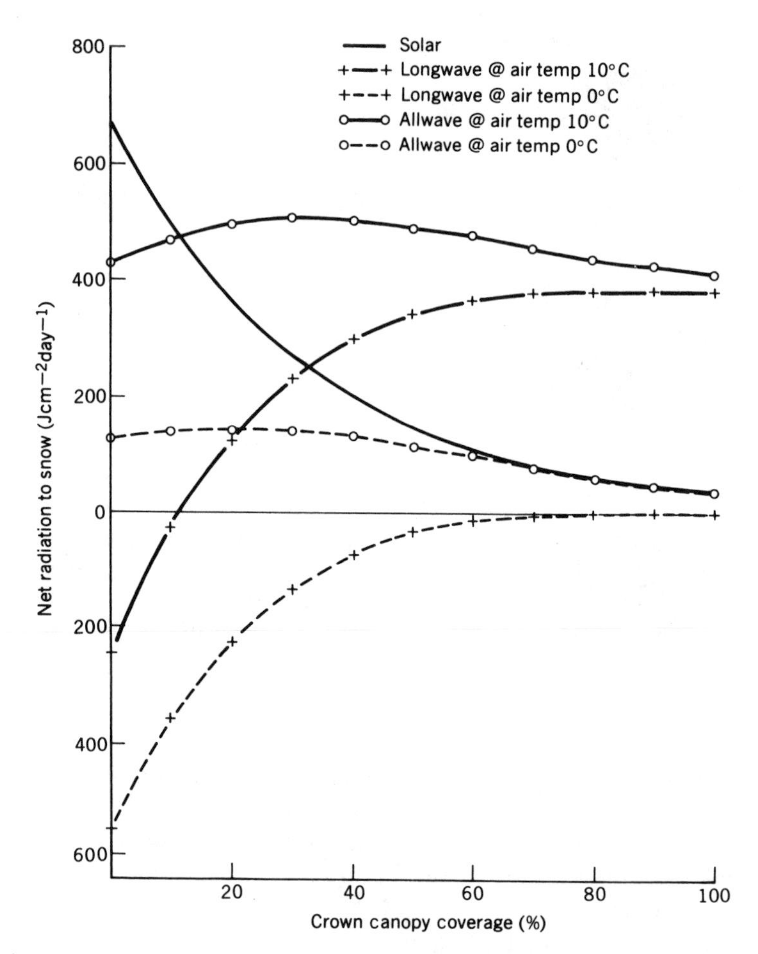

FIG. 9.4 Net solar, longwave and all wave radiation to a snowpack as related to crown cover on a clear day at two air temperatures. Albedo of the snowpack is 60%.

tion exchange in forests of varying canopy coverage, peak energy availability is shifted toward very open stands. There, enough vegetation protrudes above the surface to create roughness without unduly slowing wind. Such stands also absorb solar radiation and reduce the view factor without unduly shading the snow, as would be the case under dense canopies. Measured snowmelt was most rapid under just such an open brushy stand in central New York (Eschner and Satterlund, 1963).

Similarly, the redistribution of snow into and within forest openings combined with an asymmetrical radiation balance distribution results in earliest disappearance of snow along and under sunny forest borders with the last remnants of snow persisting along the shaded boundary (Gary, 1974, 1975; Golding and Swanson, 1986).

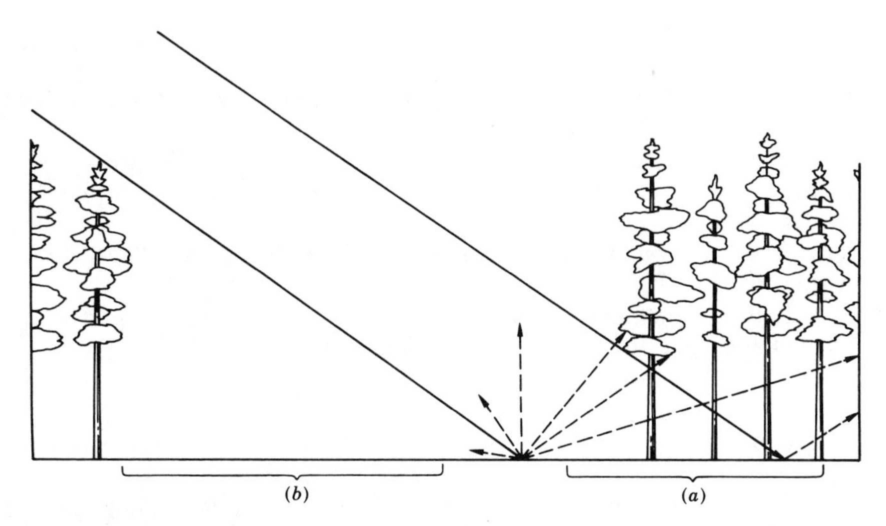

FIG. 9.5 The radiation balance of a snowpack varies widely within and adjacent to a forest opening. The sunny border is a trap for solar radiation (area *a*), which heats the nearby trees that occupy most of the view, emitting strong longwave radiation to the nearby snow surface. Snow on the shaded boundary receives less solar radiation as the solar path length through the adjacent canopy increases (area *b*). Snow melts most rapidly near the sunny boundary and least rapidly near the shaded one.

These relationships offer promise for management of vegetation for modification of snow deposition, melt, and vaporization. Nevertheless, much remains to be learned, for as stated by Colbeck et al. (1979) "In forested areas or other settings where vegetation protrudes above the snow, the energy exchange is very complicated. . . . Few quantitative measurements have been made of the energy exchange at the forest canopy or between the canopy and the snow surface."

LITERATURE CITED

Bohren, C. F. and D. B. Thorud. 1973. Two theoretical models of radiation heat transfer between forest trees and snowpacks. *Agric. Meteor.* 11:3–16.

Businger, J. A. 1965. Eddy diffusion and settling speed in blown snow. *J. Geophys. Res.* 70:3307–3313.

Chalmers, B. 1959. How water freezes. *Sci. Am.* 200:114–122.

Colbeck, S. C., E. A. Anderson, V. C. Bissell, A. G. Crook, D. H. Male, C. W. Slaughter, and D. R. Wiesnet. 1979. Snow accumulation, distribution, melt and runoff. *EOS, Trans. Am. Geophys. Union* 60:465–468.

Diamond, M. 1953. Evaporation or melt of a snow cover. *US Army Corps of Engineers Res. Pap.* 6.

Dyunin, A. K. 1956. The structure of storm snow and the laws of snow transport. Institute Geografii, Akad. Nauk SSSR. Voprosy ispol' zovaniya snega. pp. 106–119. (Seen in: Selected articles on snow and snow evaporation, OTS 63-11005, Office Tech. Services, US Dept. Commerce, pp. 1–13.)

Eschner, A. R. and D. R. Satterlund. 1963. Snow deposition and melt under different vegetative covers in central New York. USDA Forest Service Res. Note NE-13. Northeastern For. Expt. Sta., Upper Darby, PA.

Federer, C. A. 1966. Sustained winter streamflow from groundmelt. USDA Forest Service Res. Note NE-41. Northeastern For. Expt. Sta., Upper Darby, PA.

Gary, H. L. 1974. Snow accumulation and snowmelt as influenced by a small clearing in a lodgepole pine forest. *Water Resour. Res.* 10:348–353.

Gary, H. L. 1975. Airflow patterns and snow accumulation in a forest clearing. *Proc. West. Snow Conf.* 43:106–113.

Geiger, R. 1965. *The climate near the ground.* (Translated from the German by Scripta Technica, Inc.) Harvard University Press, Cambridge, MA.

Golding, D. L. and R. H. Swanson. 1986. Snow distribution patterns in clearings and adjacent forest. *Water Resour. Res.* 22:1931–1940.

Hoover, M. D. 1971. Snow interception and redistribution in the forest. *Proc. Int. Semin. Hydrol. Prof.* (W. Lafayette, IN, July, 1971) 3:114–122.

Kind, R. J. 1981. Snow drifting. pp. 338–359. In: Gray, D. M. and D. H. Male, Eds. *Handbook of snow.* Pergamon Press Canada Ltd., Willowdale, Ontario.

Kuzmin, P. P. 1956. Determination of the coefficient of reflection of snow cover by the method of successive approximation. In: *Snow and Snowmelt, their study and use.* Akad. Nauk SSSR. Institute Geografii:30–43 (Translated by Mrs. A. Dodge, in US Army Corps of Engineers, N. Pac. Div., Tech, Bull. 19, Portland, OR, 1959).

La Chapelle, E. R. 1969. *Field guide to snow crystals.* University of Washington Press, Seattle, WA.

Leaf, C. F. and G. E. Brink. 1973. Computer simulation of snowmelt within a Colorado subalpine watershed. USDA Forest Service Res. Pap. RM-99, Rocky Mt. For. Range Expt. Sta., Ft. Collins, CO.

Marsh, P. and M. K. Woo. 1984. Wetting front advance and freezing of meltwater within a snow cover. 1. Observations in the Canadian arctic. *Water Resour. Res.* 20:1853–1864.

Miller, D. H. 1955. Snow cover and climate in the Sierra Nevada, California. *Univ. Calif. Pub. Geog.* 11.

Muller, R. A. 1960. The complex snowfall distribution in New York State. Paper presented at New York Meeting, AAAS. Dec. 27, 1960.

Packer, P. E. 1962. Elevation, aspect, and cover effects on maximum snowpack water content in a western white pine forest. *For. Sci.* 8:225–235.

Petzold, D. E. 1981. The radiation balance of melting snow in open boreal forest. *Arctic Alpine Res.* 13:287–293.

Phillips, E. L. 1965. *Washington climate for these counties: Adams, Lincoln, Spokane. Whitman.* Ag. Ext. Service, Washington State University, Pullman.

Schaerer, P. A. 1981. Avalanches. pp. 475–518. In: D. M. Gray, and D. H. Male, Eds. *Handbook of snow.* Pergamon Press Canada, Ltd., Willowdale, Ontario.

Schmidt, R. A., Jr. 1972. Sublimation of wind-transported snow—*a model.* USDA Forest Service Res. Pap. RM - 90. Rocky Mt. For. Range Expt. Sta. Ft. Collins, CO.

Schmidt, R. A. 1982. Properties of blowing snow. *Rev. Geophys. Space Phys.* 20:39–44.

Smith, J. L. 1974. Hydrology of warm snowpacks and their effects upon water delivery . . . some new concepts. pp. 76–89. In: *Advanced concepts and techniques in the study of snow and ice resources.* Natl. Acad. Sciences, Washington, DC.

Tabler, R. D. 1971. Design of a watershed snowfence system, and first-year snow accumulation. *Proc. Western Snow Conf.* 39:50–55.

Tabler, R. D. 1973. Evaporation losses from windblown snow and the potential for recovery. *Proc. Western Snow Conf.* 41:75–79.

Tabler, R. D. 1975. Estimating the transport and evaporation of blowing snow. pp. 85–104. In: *Proc. Snow Mgt. on the Great Plains Symposium.* Great Plains Agr. Council. Pub. 73. Agric. Expt. Sta., University of Nebraska, Lincoln, NE.

Tabler, R. D. 1980. Geometry and density of drifts formed by snow fences. *J. Glaciol.* 26:405–419.

Treidl, R. A. 1970. A case study of warm air advection over a melting snow surface. *Bound. Layer Meteor.* 1:155–168.

Tuller, S. E. 1973. Effects of vertical vegetation surfaces on the adjacent microclimate: the role of aspect. *Agric. Meteorol.* 12:407–422.

Zuzel, J. F., R. R. Allmaras, and R. Greenwalt. 1982. Runoff and soil erosion on frozen soils in northeastern Oregon. *J. Soil Water Conserv.* 37:351–354.

10 Erosion and Sediment

In many locations, the major control over water quality in watershed management rests upon control of erosion and sedimentation. To be sure, other factors may also be important, such as pollution by pesticides and nutrients, changes in stream temperature, and so on. Though the latter are not to be minimized, their importance in wildland management often is dwarfed by comparison with effects of erosion and sediment. Throughout much of the world, wildland watershed management consists primarily of protecting land and water from erosion that can result from timber harvesting, grazing, fire, construction, mining, and other activities on the land.

Erosion and sedimentation refer to two phases in the process of detaching material in one place, transporting and depositing it in another. *Erosion* refers to the detachment and transport of the material and *sedimentation* to its deposition. Particulate material is called *sediment* once transport has begun. Erosion can also include dissolution and transport of dissolved substances that may considerably exceed in mass the particulate substances removed in some areas of heavy precipitation. Dissolved loads are generally low, though concentrations may be high in dry climates (Clayton, 1981; Swanson et al., 1982 a). Only particulate material will be treated in this section. It may be organic or inorganic in nature, and ranges in size from colloidal to more or less coherent masses covering more than a square kilometer in area.

Erosion and sediment may have many adverse effects. They can occur at the source, in transit and or upon deposition.

At the source, both biotic productivity and hydrologic functions of the land may be impaired by erosion of porous, fertile topsoil. The infiltration capacity of the site can be reduced (though reduced infiltration is usually an original cause of erosion, adverse feedback effects exacerbate the situation), as can nutrient and moisture storage. Erosion can damage transportation systems and other facilities. The costs of land management ultimately may increase while the value of the goods and services produced by the land decrease.

Nor might it end there. Sediment in transport can render water unsuitable for many uses and reduce its value for others. Fish and wildlife may suffer. The necessity for, and costs, of water treatment increase. Dirty water is aesthetically repugnant. The list could go on.

The deposition of sediment may also be costly. It may clog stream channels, increasing floods and requiring dredging to maintain navigation. It can fill reservoirs, reducing their capacity to store water. In irrigation water, fine sediment particles may clog soil pores, reducing infiltration by sealing the surface against the entry of water.

Sediment often contributes to other water quality problems as well. For example, many pesticides and nutrients are carried into the aquatic system adsorbed on sediment. Sediment clogging stream channels can contribute to increased stream temperatures as water flow is slowed and spread over a greater surface area. Stream productivity is decreased as sediment reduces light penetration necessary for photosynthesis of aquatic plants. Biochemical oxygen demand may be increased, and reaeration may be reduced. Sometimes the effects are synergistic.

Considering all the foregoing, it is little wonder that erosion and sediment control have been a primary focus of watershed management. At the same time, we must recognize that some amount of erosion and sedimentation can be expected to occur naturally. Indeed, natural erosion and sedimentation processes have helped shape our landscape and provide spawning gravels for fish, fertile floodplains, and productive estuaries. The challenge to the watershed manager is to characterize local natural erosion and sedimentation processes, and then identify and avoid practices that are expected to augment these processes to levels yielding unacceptable resource impacts.

10.1 BASIC PRINCIPLES

Erosion occurs whenever sufficient force is applied to overcome the resistance and initiate movement of material on the earth's surface. The forces acting to move sediment are wind, water, and gravity, but wind will not be considered here as it is seldom important on most wildland watersheds and wind transported sediment has little effect on water quality.

Sediment is derived primarily from erosion of the watershed and channel cutting. All watersheds produce sediment, for erosion is a natural geologic phenomenon. Sediment can occur without erosion, however, on fully vegetated areas. Some of the litter produced by vegetation growing in the *riparian zone* (along the land–water interface) invariably reaches the water where it may disintegrate and decompose and be carried downstream. In an old-growth forest in western Oregon, 0.30 t of organic material and 0.53 t of inorganic material entered the stream channel from a 10.2 ha watershed annually (Swanson et al., 1982 a). Undisturbed areas in the Allegheny Mountains of West Virginia may yield water with up to 15 parts per million (ppm) of sediment, which is primarily organic (Reinhart et al., 1963). This amount of sediment exceeds US Public Health Service standards for drinking water. Such concentrations occur infrequently and are often more beneficial than detrimental, for they may represent the primary energy input into aquatic food chains of shaded headwater streams. Therefore, the sediment problem in watershed management is almost entirely related to soil erosion.

Though erosion is a natural geologic phenomenon, erosion rates are highly variable. They tend to be greatest in semiarid climates and in regions of recent geologic uplift. Importantly, almost any use of the land by people can accelerate the erosion process. It can be unrealistic to expect to hold erosion rates to their geologic norm and still use the land. Therefore, erosion control often implies minimizing accelerated erosion under the impacts of land use.

The sediment output from any watershed is ordinarily much less than the erosion occurring in the watershed, for large quantities of sediment tend to be deposited near their source. Of the sediment produced by erosion (as estimated by the truncation of soil profiles) for 10 large river basins in the southeastern United States since settlement began in the seventeenth century, only about 5% has yet reached the sea. The rest has been deposited on the watersheds and in reservoirs in the basins where it was produced (Trimble, 1975). Sediment control for water quality protection therefore requires an understanding of the entire erosion–sedimentation system; a step-by-step routing from the point of production, including all intermediate stages, until it leaves the watershed. Sediment routing is the weakest link in our understanding of the system. Sediment may be stored and remobilized many times before it leaves the system, and each time it is deposited represents another opportunity for stabilizing it.

The first step in understanding the system is an examination of the processes of erosion. There are three main classes of erosion on wildland watersheds: surface erosion, mass wasting, and channel cutting.

Surface erosion is the detachment and transport of individual soil particles or small aggregates from the land surface. It is caused by the action of raindrops and surface runoff, either in thin films or concentrated in rivulets. It may remove soil in more or less thin layers (*sheet erosion*), in rills or in gullies. On steep, dry slopes gravity alone may be sufficient to cause movement (*dry ravel*). Surface erosion is seldom serious in forest areas except when they are disturbed; where bare soil is exposed by roads, fire, timber harvesting (Brown, 1985; Patric, 1976). It is an important source of sediment from many range lands and is the dominant type of erosion from most cultivated lands.

Mass wasting includes all forms of erosion in which more or less cohesive masses of soil are displaced. It includes creep, landslides, slumps, earthflows and debris avalanches and torrents, and is frequently the dominant form of erosion in steep country. The rate of movement may be rapid, as in a debris avalanche or rockfall, or it may be almost imperceptibly slow, as in soil creep down a hillside. It is a major geologic process in mountainous areas and occurs under the best of conditions, but human activities may greatly accelerate the process (Sidle et al., 1985).

Channel cutting, or the detachment and movement of material from a stream channel, is the third major type of erosion. It may result from the movement of individual particles, as in shifting grains of sand in bars, or from mass movement, as when a large part of an undercut bank falls and is swept downstream (Heede, 1980).

There are a few minor forms of erosion that do not fit neatly into the above classes, such as *piping*, the erosion of an underground cavity by water flowing through cracks in the soil (Heede, 1971), or *solifluction*, a form of erosion intermediate between creep and surface erosion. It takes place under conditions of alternate freezing and thawing, whereby freezing water lifts a thin mass of soil particles, which upon thawing are displaced downward on slopes as the saturated, viscous mass settles under the force of gravity.

Except for dry ravel, which would be a self-limiting form of erosion in the absence of water (Anderson et al., 1959), most forms of erosion are associated with excess water. Sometimes it is difficult to separate between types, for rills may become gullies that in turn become stream channels. And a debris torrent may be considered a catastrophic form of channel erosion. Any classification is arbitrary, for in reality there is a continuum of erosion forms. Nevertheless, each class will be discussed separately.

10.2 SURFACE EROSION

At any time there are always two opposing forces acting on the material of any sloping surface; those acting to move material and those acting to resist movement. The force of gravity is the ultimate downward acting force, but except in mass wasting and dry ravel, it is usually expressed through the force of moving water. The primary resisting forces are the inertia and forces of cohesion and friction that must be overcome to initiate movement. When moving water is a direct force acting to cause erosion, control may be exerted by reducing the forces tending to cause erosion, increasing those that resist them, or both.

10.2.1 The Universal Soil Loss Equation

Agricultural scientists in the United States have sought to learn how to control surface erosion since Thomas Jefferson (author of the Declaration of Independence and third president of the United States) wrote of the ravages of erosion in the eighteenth century (Ackermann, 1976). Quantitative data from thousands of plot-years and watershed-years of precipitation, soil loss and site data were assembled in the late 1950s to see if it was possible to estimate soil loss from field size upland agricultural areas. These studies led to the development of the universal soil loss equation (USLE), presented in 1965 (Wischmeier and Smith, 1965) and revised and expanded in 1978 (Wischmeier and Smith, 1978; Meyer, 1984). The USLE has also been adapted for use in estimating surface erosion from wildlands (Wischmeier, 1975; Curtis et al., 1977; Dissmeyer and Foster, 1980) with varying degrees of success.

The USLE is useful in summarizing the primary factors controlling surface erosion:

$$A = RKLSCP \tag{10.1}$$

where A is estimated soil loss per unit area and time, usually ton acre^{-1} year^{-1}; R is a rainfall and runoff factor, equal to the number of rainfall erosion index (EI) units plus a factor for runoff from snowmelt or thaw; K is the soil erodibility factor, equal to the soil loss rate per EI unit for a clean-tilled fallow soil on a 72.6-ft (22.1 m) length of uniform 9% slope; L is a slope length factor, which is the ratio of soil loss from a given slope length to that from a 72.6-ft long slope under identical conditions; S is a slope steepness factor, the ratio of soil loss from a given slope to that from a 9% slope under identical conditions; C is a cover and management factor, the ratio of soil loss from a given cover and management condition to that from an identical area in tilled, continuous fallow; P is the support factor or ratio of soil loss with a practice like contouring or terracing as compared to tillage up and down the slope.

Of the factors, R, L, and S are primarily associated with the availability of energy to cause erosion. The term K is primarily associated with resistance to erosion, and C and P influence both erosive energy and resistance, but will be considered here as factors of resistance for the purpose of convenience.

10.2.2 Erosive Energy

The primary source of erosive energy is rainfall and runoff. Dry ravel moves under the force of gravity, but initial dislodgement may be from heating or cooling, surface drying, or biotic action. However, dry ravel would move only a few meters at most if it were not moved farther by moving water. Meltwater from snow and surface ice in soils may be important in moving large quantities of soil under some circumstances. All these factors, except dry ravel, are taken into account in the R factor in the USLE, which includes two parts: the storm erosion index (EI) and the snowmelt and thaw factor (R_s), or

$$R = EI + R_s \tag{10.2}$$

where R is annual erosivity, EI is the accumulated storm erosivity, and R_s is erosivity from snowmelt and thaw.

The EI index represents the statistical interaction between the total energy in falling rain and rainfall intensity. It equals the product of total storm energy in hundreds of foot tons per acre times the maximum 30-min intensity in inches per hour. [Conversion of the Universal Soil Loss Equation to SI units has been done by Foster et al. (1981). Applications in the United States are all based on English units, so they will be used herein.] Together, these two factors indicate (a) how much energy is available, (b) how concentrated the energy is, and (c) the amount of surface runoff available to transport sediment. Surface runoff is not determined explicitly; rather, the EI term indicates how particle detachment is combined with transport capacity. The EI calculations do not include rain showers of less than 0.5 in. (12.7 mm) separated from other rain periods by more than 6 h unless at least 0.25 in. (6.4 mm) of rain fell in 15 min or less. The average annual sum of storm EI values for the continuous United States is shown in Fig. 10.1.

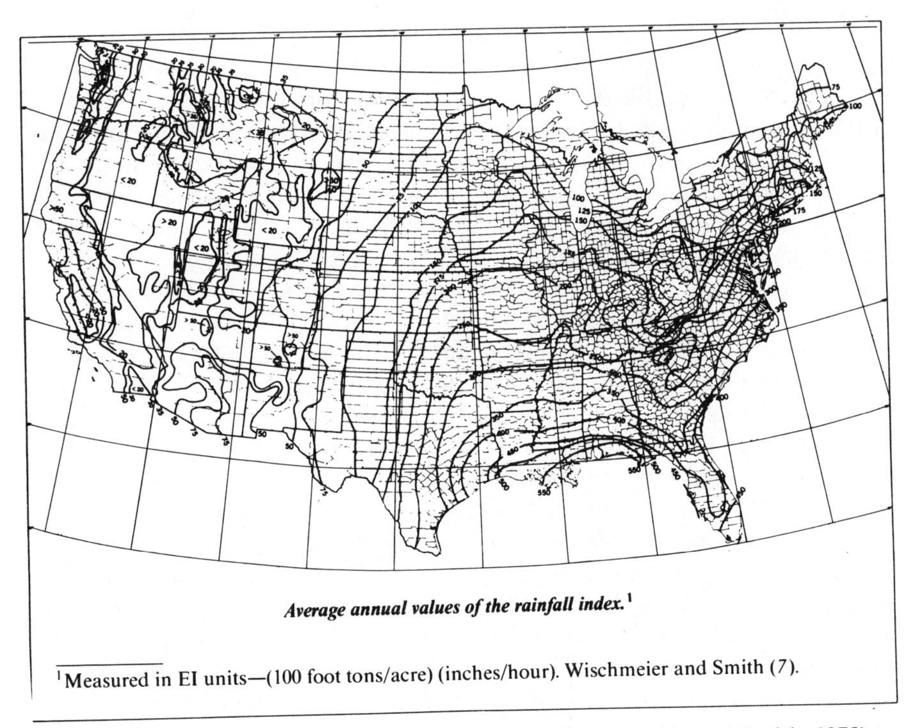

Average annual values of the rainfall index.[1]

[1] Measured in EI units—(100 foot tons/acre) (inches/hour). Wischmeier and Smith (7).

FIG. 10.1 Iso-erodent map of annual *EI* values (Wischmeier and Smith, 1978).

All other factors being equal, annual surface erosion would, on the average, vary in direct proportion to annual *EI* values. Annual *EI* values are greatest in the southeastern United States where annual precipitation is high in amount, is nearly all in the form of rain, and where a high proportion of storms yield $\frac{1}{2}$ in. or more of precipitation at high intensity. Values decrease toward the north where a greater proportion of annual precipitation falls as snow and inland in the Great Plains where total annual precipitation decreases. In the West, *EI* values are low or because of lack of precipitation as rain or the generally low intensity storms where rainfall is high.

10.2.2.1 Raindrops as Erosive Forces. The greatest force of moving water on most watersheds is represented by the impact of falling rain. More than 900 ft tons (2.44 MJ) of force are expended on 1 acre (0.4 ha) when 1 in. (25.4 mm) of rain falls in 1 h. The amount varies with the amount and rate of rainfall. The basic kinetic energy equation is applicable to falling rain:

$$K = \frac{1}{2}mv^2 \tag{10.3}$$

where K is the kinetic energy, m is the mass, and v is the velocity.

In natural rainfall, there is a great increase in kinetic energy as the rate of rainfall increases, for not only does the mass of water striking the surface in any time period increase, but the velocity of fall increases as well. The increase in velocity with rainfall intensity derives from two factors: (a) as rainfall intensity increases, the median size of individual raindrops increases; and (b) large drops have a higher *terminal velocity*, or rate of fall when air resistance prevents them from accelerating further, than small drops. Most raindrops approach their terminal velocity closely after a free fall of only about 7 or 8 m. Almost all rainfall, and drip from that height, has reached its terminal velocity by the time it strikes the ground (Table 10.1). Thus, it should be clear that the protective function of a tall forest does not derive from breaking the force of falling rain by the canopy. Drip from the canopy is usually in larger drops than natural rain. Mosley (1982) observed that rainfall energy under a beech forest was always greater than that of rainfall energy out in the open, interception losses notwithstanding. Only material on or near the ground surface is effective in breaking the force of rain, as was reported long ago by Gleissner (1914) in Bavaria, where erosion ravaged forest lands from which the litter was gathered by peasants for animal bedding.

Large raindrops striking bare soil often have the capacity to detach and move soil particles. At times, the drops may be likened to tiny bombs that blast small craters in the soil, throwing loosened particles beyond the crater. Sometimes soil particles are moved a meter or more by the force of raindrop splash. This type of surface erosion is called *splash erosion* and can sometimes be detected by the presence of *soil pedestals*, small columns of soil extending above the surface like small tree stumps, usually capped by a stone or piece of litter that protected them from the falling rain. Splash erosion can be seen readily around new building construction, where drip from the eaves causes the foundation to be stained by displaced soil particles. On sloping areas splash erosion systematically displaces soil downhill, but on level soils there is no net soil movement, and thus no net erosion, unless surface runoff is available to transport the detached particles. Particles dispersed by splash erosion tend to clog surface pores; hence, where splash erosion is common, surface runoff usually follows.

Despite the great force of falling rain, not all of the energy is available for erosion. A portion is expended in breaking the cohesive forces bonding the drop together, as it usually disintegrates upon striking the ground. Another part may merely pack surface particles closer together. Furthermore, though a great deal of force may be expended on a single square meter during a single rainstorm, it is expended drop by drop, spread out in small quantities over both space and time. Unless the individual drops possess enough force to overcome the inertia necessary to move a particle, or overcome the forces by which it is bound to others, no splash erosion will result. It is for this reason that soils tend to be less erodible with a decrease in the silt fraction, regardless of whether the corresponding increase is of sand or clay (Wischmeier and Smith, 1978). Larger sand particles have greater inertia, and clays tend to increase soil cohesion.

TABLE 10.1. Energy Characteristics of Rainfall[a]

Precipitation Type	Intensity		Median Drop Diameter		Fall Velocity		Kinetic Energy			
	(in. h^{-1})	(mm h^{-1})	(in.)	(mm)	(ft s^{-1})	(m s^{-1})	(10^{-6} ft lb $drop^{-1}$)	(Kgfm)	(ft lb ft^{-2} h^{-1})	(Kgfm)(m^2 h^{-1})
Mist	0.002	0.05	0.004	0.10	0.7	0.2	*b*	*b*	*b*	*b*
Drizzle	0.01	0.25	0.038	0.97	13.5	4.1	9.5	0.04	0.5	0.02
Light rain	0.04	1.02	0.049	1.24	15.7	4.8	27.4	0.17	2.6	0.12
Moderate rain	0.15	3.81	0.063	1.60	18.7	5.7	82.7	0.36	13.6	0.63
Heavy rain	0.60	15.24	0.081	2.06	22.0	6.7	243.2	1.05	75.5	3.49
Cloudburst	4.00	101.60	0.112	2.84	25.9	7.9	891.0	3.83	697.6	32.25
Drip from trees			0.20±	5.08±	30.0±	9.1±	6807.1	29.25		

[a] Derived from data in Kittredge (1948).
[b] Less than 0.1.

It also explains why litter is effective in preventing splash erosion—the single raindrop seldom possesses enough energy to overcome the litter's inertia. Thus, splash erosion, despite the very great kinetic energy of heavy storms, is seldom a problem unless the surface is bare.

10.2.2.2 ***Flowing Water as an Erosive Force.*** Though the same basic kinetic energy equation is applicable to falling rain or flowing water, the force distribution in flowing water is very complex. Water may flow in thin films, in smooth laminar fashion, or it may be wildly turbulent in torrents. The simple kinetic energy equation is no longer sufficient to express the erosive energy or sediment carrying capacity of widely differing flows.

The key element in erosion is detachment because it takes more energy than is needed to transport soil particles. Furthermore, if initial detachment is avoided, control of erosion is achieved. On bare soils, most of the energy for detachment is supplied by falling rain. *Sheet erosion*, the removal of a more or less continuous layer of soil, is usually the combined result of raindrop splash detaching the particles and surface runoff transporting them away. Surface flow has enough force for detachment only as sheets of water deepen beyond a thin film or where it becomes concentrated in rills or other channels.

The erosive force of flowing water is primarily a function of the drag forces exerted upon the surface. Determination of these forces is highly complex and is related to the velocity distribution in the flow (Fig. 10.2). The erosive power of running water has been expressed empirically as a function of the mean velocity of flow.

There is a direct relation between the quantity of runoff and its velocity, and any measure that reduces the amount of runoff will usually reduce erosion. The ultimate case is when all surface runoff is prevented and surface erosion is restricted to splash erosion or dry ravel.

The velocity of runoff may also be reduced without reducing the amount of runoff. For flow in a channel, Robert Manning, an Irish engineer of the nineteenth century, developed one of the most commonly used equations in hydraulics to express velocity (Searcy, 1965). Manning's equation is

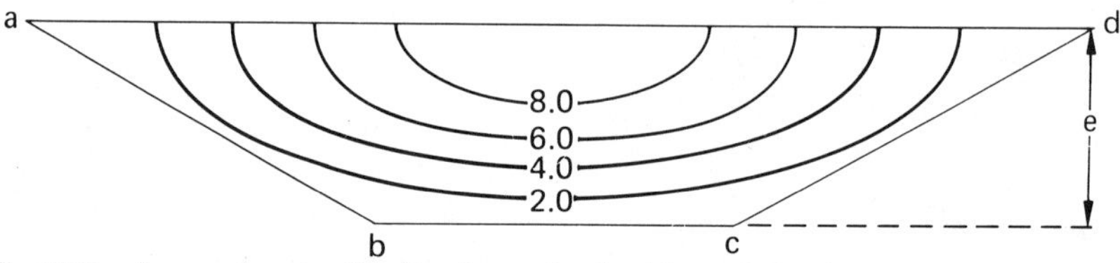

FIG. 10.2 Approximate distribution of velocities of flowing water in a straight trapezoidal channel. The mean velocity is determined by channel gradient, shape, roughness and depth of flow (Searcy, 1965). Shape is expressed as cross-sectional area divided by channel perimeter (line abcd), both measured perpendicular to direction of flow.

$$V = \frac{R^{2/3} S^{1/2}}{n} \tag{10.4}$$

where V is the mean velocity in meters per second (m s^{-1}), n is the Manning coefficient of channel roughness, R is the hydraulic radius in meters (m), and S is the slope in meter per meter (m m^{-1}). Some n values useful in land management are given in Table 10.2. Good vegetation or litter on the soil surface can increase roughness by up to a factor of 10 over bare surfaces.

The *hydraulic radius*, R, is a channel shape factor that depends on channel cross-section dimensions and depth of flow so that:

$$R = \frac{A}{WP} \tag{10.5}$$

where A is the cross-sectional area of flowing water in square meters and WP is the length in meters of the *wetted perimeter*, or contact length between the water and its containing channel, both measured at right angles to the direction of flow. In Fig. 10.2, A is $[(ad + bc)/2]e$ and WP is $(ab + bc + cd)$. Outside of a channel, as in surface runoff that flows as a film or sheet over a surface, the factor R reduces to approximately the depth of flow in meters. Its value is always very small for surface runoff, usually 0.001 or less.

TABLE 10.2. Manning Coefficients for Land and Channel Surfaces

Surface		n
Unlined open channels, fairly uniform		
Earth, no vegetation		0.020–0.025
Grass, some weeds		0.025–0.030
Sides clear earth, gravel bottom		0.025–0.030
Grassy swales, smooth		
Good stand:	ht. 4–6 in. (10–15 cm)	0.050–0.090
	ht. about 12 in. (30 cm)	0.090–0.18
	ht. about 24 in. (61 cm)	0.150–0.30
Fair stand:	ht. about 12 in. (30 cm)	0.080–0.140
	ht. about 24 in. (61 cm)	0.130–0.150
Natural stream channels, irregular		
Some grass and weeds, no brush		0.040–0.055
Steep gradient and sides, bottom gravel, cobbles, few boulders		0.040–0.050
Steep gradient and sides, bottom cobbles, with large boulders		0.050–0.070
Overland flow		
Packed clay		0.030
Light turf		0.20
Dense sod		0.35
Dense shrubs and forest litter		0.4

Source: Adapted from Searcy (1965) and Crawford and Linsley (1966).

It is clear that the depth of water is of critical importance in erosion, and depth of water is related to all the elements in the USLE, including, though perhaps only indirectly, *K* or soil erodibility. The *EI* factor has already been discussed, but the amount of water is also influenced by R_s, or the water available for erosion from melting snow or thawing soil.

Erosion from snowpack melt is probably unimportant where soils are unfrozen (Hart and Loomis, 1982), but may be very important where snowpacks are shallow and frozen soils prevent infiltration of the meltwater. The recommended value for R_s is 1.5 times the cumulative precipitation during December–March in the snow zone (Zuzel et al., 1982). Erosive runoff from snowmelt may also occur where ice layers formed in the snowpack concentrate runoff in channels (Sturgis, 1975). Both frozen soils and ice layers in the snowpack are most likely to occur on crop and rangelands where the snowpack is shallow, but frozen soils and surface runoff from snowmelt also occur in the cold snow zone of forests in the northern Lake States and interior Canada.

The amount and velocity of surface runoff increases with increases in slope length and steepness. Slope position is also important, as water tends to concentrate on lower slopes. For example, on a uniform slope exceeding a 5% gradient divided into three equal segments, the upper, middle, and lower thirds would be expected to contribute 19, 35, and 46%, respectively, of the total soil loss (Wischmeier and Smith, 1978).

Cover influences the erosivity of flowing water as it influences the amount of surface runoff through its effect on infiltration and the velocity of a given depth of surface runoff through its effect on surface roughness. Similarly, the erodibility of soil is determined partly by its infiltration capacity.

The practice factor, *P*, of the USLE applies mostly to croplands, but terracing, contour ditches, and contour furrows are sometimes applied on rangelands and heavily prepared forest plantations, where they reduce effective slope length and steepness and the amount of runoff. Shaping the cross section of a ditch to minimize hydraulic radius for a given flow may also control erosive energy. On disturbed areas such as strip mines, shaping of the surface is a means of controlling slope length and steepness to minimize erosion.

10.2.3 Resistance to Erosive Forces

The forces that permit a soil to resist surface erosion are much more complex and difficult to quantify than are the forces of moving water. Our information is largely empirical and descriptive. However, among these forces are the cohesion of soil particles to one another, the mass, volume and surface area of materials on the eroding surface, and the amount of plant material, living or dead, that may anchor the eroding material and protect it from receiving the direct force of the eroding agent.

Furthermore, some of the same characteristics that cause material to resist the erosive force of water also act, by influencing infiltration and surface roughness, to decrease the amount and velocity of surface runoff. All these factors interact. For example, living and dead organic material protects the soil surface directly from the impact of rain, provides the cementing substances that bind aggregates together as it decomposes, and helps to maintain maximum porosity and infiltration capacity. Gravel and stone on eroding surfaces resist erosion by virtue of their large mass and inertia and also increase roughness. Large debris such as logging slash also exhibits the twofold action of resisting and reducing erosive forces simultaneously. Thus, measures of resistance to erosive forces are much less precise than measures of the erosive force.

In channels, resistance is often expressed in terms of *maximum permissible velocities*, or the highest velocity of flow that the channel material can withstand without displacement. Fine, loose soils can safely withstand only low velocities of flow, whereas hard bedrock is unmoved by even swift currents. Vegetation established on almost any soil increases its resistance, hence the importance of grassed waterways to carry runoff from cultivated fields. Most of the information on permissible velocities has been derived from studies in ditches and channels (Table 10.3). Determining safe velocities for overland flow is complicated by the impact of raindrops on thin films, which often accounts for much of the detaching force available.

Most work on the erodibility of surface soils has been concentrated on measuring the resistance of soil aggregates to breakdown, the size of water-stable aggregates on the surface and correlations with soil parent material,

TABLE 10.3. Maximum Permissible Velocities of Clear Water in Erodible, Smooth Channels[a, b]

Material and Cover	Maximum Permissible Velocity ($m\ s^{-1}$)
Fine sand (noncolloidal)	0.46
Sandy loam (noncolloidal)	0.52
Silt loam (noncolloidal)	0.61
Ordinary firm loam	0.96
Above, with good grass cover	0.76–1.83
Fine gravel	0.76
Stiff clay (very colloidal)	1.13
Graded, loam to cobbles (noncolloidal)	1.13
Graded, silt to cobbles (colloidal)	1.22
Above, with good grass cover	1.07–2.44
Coarse gravel	1.22
Cobbles and shingles	1.52
Shales and hard pans	1.83

[a] Adapted from Tables 3 and 4, Searcy (1965).

[b] Maximum permissible velocity is lower for clear water than water carrying sediment, as clear water retains an unsatisfied sediment carrying capacity.

vegetation cover, organic matter content, and other factors (Dyrness, 1967 a; Trott and Singer, 1983). In the USLE, the erodibility factor, K, is estimated on the basis of soil texture, percent organic matter, soil structure, and soil permeability. Soils high in silt are most erodible, decreasing in erodibility with increases in either the sand or clay fraction. Erodibility decreases as organic matter content increases, as the size of structural aggregates increase, and as granular structure changes to blocky, platy, or massive. Erodibility is least in rapidly permeable soils and greatest in those with very low permeability (Wischmeier and Smith, 1978).

The USLE should not be uncritically applied to wildland conditions. Limited precipitation data for many areas reduces the accuracy of R factors for local use. Slope data in the USLE were derived from lengths mostly less than 100 m, and steepnesses of less than 20%. The soil erodibility (K factor) nomograph does not include soils with organic matter contents greater than 4%, a level not unusual for forest soils. Wischmeier (1975) recommends that K be multiplied by 0.7 when organic matter exceeds 4%, in lieu of better information. Erosion dynamics are not addressed directly by the USLE. For example, Megahan (1974) found that the erodibility of bare disturbed granitic soils decreased rapidly with time, apparently as a result of armoring of the surface as fines were removed, leaving larger particles behind. Hart (1984) suggests that the USLE might be used on wildlands if modified for steep slopes, residual roots, and organic matter in exposed soils and antecedent moisture. Without considerable modification and validation for wildlands, however, the USLE is likely to remain primarily useful in reminding resource managers of the key factors that often control surface erosion.

Where bare slopes are at or exceed the *natural angle of repose* (the maximum slope angle at which loose soil can be piled), any force initiating detachment causes erosion, for the force of gravity alone will transport the material downslope. Frost, raindrop impact, biotic activity, and even light winds may be sufficient to initiate movement. Many bare surfaces, such as road cuts and compacted fills are constructed to steeper angles. So long as surface particles remain bound to the underlying soil by forces of cohesion, roots, or even water under tension, surface erosion does not occur. As soon as these forces are overcome, however, the soil particles are carried away by gravity.

10.2.4 Soil Frost

Soil frost is often a major factor in surface erosion from bare or poorly vegetated soils. Water expands by about 9% upon freezing and exerts pressures great enough to fracture solid rock, crush the hulls of large ships, and create potholes in our roads. Freezing also withdraws water from lower soil depths to the surface, causing formation of ice lenses that grow to exceed normal pore volume, leaving the surface supersaturated with water upon melting.

Frost may have a threefold effect on surface erosion: (a) it may overcome cohesive forces, causing detachment of individual particles from the soil surface; (b) it often prevents water reaching the surface from infiltrating; and (c) it can be a source of water for overland flow when it melts, even in the absence of rain or melting snow. Needle ice or stalactite frost forms near the surface and commonly lifts and detaches soil particles and creates supersaturated conditions. Concrete frost, or solid ice lenses, may greatly inhibit infiltration.

Needle frost may form to a height of nearly 1-cm overnight, melt the next day, and reform again, repeatedly. It can be a major factor in surface erosion from bare or poorly vegetated slopes where the frequency of freeze–thaw cycles of moist soils may be closely related to soil loss. One freeze–thaw cycle occurs when the soil surface temperature drops below freezing and subsequently rises back above it. Both the *amplitude* (the range in temperature) and the *period* (the length of time the soil remains below freezing) of the cycle varies. Usually no more than one occurs daily, but a single cycle can last for months in northern regions.

Hershfield (1974, 1979) provided maps of the number of annual and winter freeze–thaw cycles in the conterminous United States, based on the analysis of air temperature data. They show that the number of cycles is greatest in the mountains of the West, but the data are somewhat misleading as an indicator of the erosive effects of soil frost. Snow, particularly where it exceeds 30 cm or so in depth, greatly dampens the fluctuation in temperature at the soil surface, and reduces frost action. Also, soil freezing may have little erosive effect when the surface moisture content is well below field capacity. Soil frost is most important as an erosive factor on bare soils where freezing temperatures are common but snow seldom covers the ground, as in much of central and southeastern United States. Swift (1984) attributed heavy erosion of cutslopes primarily to frost action, and frost action was a primary factor in soil losses of unvegetated strip mine dumps in Illinois (Haigh and Wallace, 1982).

In addition to the direct effects on surface erosion, soil frost indirectly contributes to erosion by hindering the establishment of vegetation on bare soils. Seedling roots are easily broken by frost heaving. This is often a problem on bare soils at high elevations where frost occurs frequently throughout the growing season, or where winds sweep the surface free of insulating snow.

10.3 MASS WASTING

Mass wasting occurs when the cohesive and frictional strength of earth or rock is overcome across an area of weakness and the mass is displaced downward. The failure may be in the form of a sudden rupture along a distinct surface or it may take place as slow, plastic deformation throughout a large mass. Both types of failure may be involved at different times in the displacement process. Slow creep may cause an over steepening of a slope, which is accompanied by increasing downward stress and weakening cohesive and frictional resistance until complete rupture occurs. Mass wasting can reduce the productivity of the land, damage transportation systems, and end up in stream channels where it may continue to provide sediment downstream over long periods of time.

Mass wasting is the primary form of natural erosion in most mountainous areas of the world. The rate may be so high as to be a major factor in tropical forest disturbance, causing large areas to be retained in a primary successional state (Garwood et al., 1979). Unusually large landslides also have resulted in notable losses of life and property over the past century (Sidle et al., 1985). Slope stability is a major concern wherever human activity occurs near steep slopes, but some areas are much more hazardous than others. Radbruch-Hall et al. (1981) presented a general map and description of landslide incidence and susceptibility for the major physical subdivisions of the conterminous United States. Burroughs (1985) compiled maps based on the interaction of landslide hazards, aquatic resource concerns, and land management activities for much of the western United States.

Among the major factors influencing landslide incidence and susceptibility are slope, precipitation, rock types, faults, joints and bedding planes, soil type, and degree of weathering. The most susceptible sites are on steep slopes in wet climates where recent uplift has taken place, as in the coast ranges of western United States. The problem is compounded where soft, easily weathered, fine-grained rocks consisting of silt and clay sized particles predominate, and in areas of frequent tectonic activity, such as earthquakes. Landslides are few in flat areas and in those with dry climates, such as the High Plains and Basin and Range provinces.

10.3.1 Types of Mass Wasting

There are many definitions and classifications of mass wasting (Sharp, 1938; Varnes, 1958; Terzaghi, 1960; Sheng, 1966; Hutchinson, 1968; Swanston, 1974; Dunne and Leopold, 1978). *Landslide* is often used as a generic term that includes all types of mass wasting that exhibit perceptible motion. Large amounts of mass movement also may take place as relatively imperceptible creep or solifluction. Each type is described in more detail below.

10.3.1.1 Creep. *Creep* is slow, largely irreversible downslope movement in which gravitational stresses cause deformation, but not complete failure, of plastic or highly viscous material. Movement may be confined to highly weathered zones and occurs predominately during wet seasons, or it may be deep seated and continuous. It is sometimes caused by volume changes in clays caused by alternate wetting and drying. The high water retention capacity of clayey materials contributes to their susceptibility to creep.

Slopes on which creep occurs may become progressively more prone to failure. Tension cracks may appear and spread, reducing remaining shear strength until a slide takes place. If soils become saturated, the rate of creep may increase until the soil begins to flow. Some formations such as marine "*quick clays*," unconsolidated cohesionless clays, exhibit creep on very slight slopes when at or near saturation, but they may suddenly liquify when stressed and turn into a clayey paste that forms a mudflow. Such marine quick clays are found in coastal areas of Washington State on the Olympic Peninsula and in the Puget Sound region, the Champlain–St. Lawrence Valley of eastern United States and Canada, Scandinavia, and elsewhere.

The contribution of creep in supplying sediment to streams is often underappreciated. For example, if a soil layer 1 m deep encroaches on both sides of a stream channel at the rate of only 1 mm $year^{-1}$, and it has a density of 1.5 g cm^{-3}, then it will supply 3 tons of sediment per kilometer of stream per year. These values may be easily exceeded on unstable terrain under natural conditions. Creep is often a major supplier of sediment to road ditches and other channels as well.

Slopes subject to creep can often be recognized by the tilting and displacement of objects on the surface. Fenceposts set vertically become noticeably tilted after a few years, and where creep is nonuniform across a slope a straight line of posts becomes curved and wires may be broken by the strain. Tree trunks form broad arcs as the tips grow vertically upward while the lower stems become more tilted. Tension cracks may be evident, but are often hidden by vegetation. Bulges may appear in road cuts and over land where creep is active, giving a hummocky appearance to the area (Burroughs et al., 1976). Creep may also be monitored by noting the displacement of holes drilled into moving material (DeGraff and Olson, 1980).

10.3.1.2 Solifluction and Other Freeze–Thaw Movement. Downhill soil mass movement from natural processes of freezing and thawing of soil moisture is widespread in temperate and cold regions. It often takes place on slopes with only slight gradients. Expansion and contraction of freezing and thawing water may provide the direct force for movement or the freezing of wet soils may thrust the overlying material outward, and upon melting it may become saturated and creep or flow downward under the force of gravity (Lambert, 1972). Higashi and Corte (1971) show in detail the process of solifluction.

10.3.1.3 Landslides. Landslides have many subforms, but all are distinguished by shearing failure and a perceptible rate of movement. The initial failure may take place across a more or less straight surface or by backward rotation through a more or less circular arc. Once failure occurs, the mass may move by falling, sliding, or flowing downward.

Most *slides* are characterized by planar failure and may disintegrate as they move downhill. They are often narrow, long, and shallow. Rapidly moving slides are often classed as *debris avalanches*, *debris slides* or, if viscous, *debris flows*. Sometimes a debris slide reaches a stream where it may mix with water and entrain bank and bed materials, sluicing them downstream. Such highly fluid *debris torrents* may leave tributary channels scoured to bedrock all the way to their junction with the main stream, which may be blocked by the resulting debris jam.

Rotational failures are called *slumps* or *landslips* if the movement is short and the material settles as a cohesive mass. If the material exhibits cohesive characteristics as it moves downhill, it may be classed as an *earthflow*. Slumps and earthflows may be either shallow or deep seated, but they tend to be deeper seated than most debris slides.

Falls are rapid movement of material freely down steep gradients. Failure may be either planar or rotational.

Other types of mass movement occur, but they are usually of little importance in land management.

10.3.2 Mechanics of Slope Failure

Regardless of the type of mass wasting, slope failure takes place only when shear stress exceeds shear strength or resistance. *Shear stress* includes all of the forces acting to displace material. The principal component is the weight of the earth material on the slope, but it may include water under positive pressure, biomass (a single large tree may weigh several tons), earthquakes, passing traffic, explosive, or other vibratory forces that impart momentum to the mass, and even wind that sways tall trees (Bell, 1963; Sheng, 1966; Burroughs et al., 1976).

Shear strength or *resistance* includes all the forces that act to resist movement of the mass. It is composed primarily of frictional and cohesive properties of the soil and rock along its weakest plane and the binding strength of plant roots. The relations among the different forces maybe clarified by examining the mechanics of planar and rotational failure. Analysis depends on idealized and simplified models of the complex and highly variable forces acting on real slopes, but proves useful in identifying the relationships of the more important elements involved (Swanston,1970).

10.3.2.1 Planar Failure. Figure 10.3 illustrates the forces contributing to shear stress and shear resistance on a sloping plane surface. The primary driving forces contributing to shear stress are the weight of the soil column, the weight of the biomass or other loads on the slope and wind pressure that causes the tree to sway so that the forces are transmitted into the soil. Earthquakes frequently exert stress by imparting vibratory momentum to the slope. Water under positive pressure may exert force against the soil or rock that acts to displace it. This *pore water pressure* may be considerable when a large head of water is confined by an impermeable layer. The total shear stress is

$$S_s = (W_s + W_b)\sin\alpha + P_w + M + P_p \tag{10.6}$$

where S_s is shear stress, W_s and W_b are weights of soil and biomass (or other load) in a soil column, α is slope inclination in degrees, P_w is the pressure transmitted by wind on trees, M is momentum, and P_p is water under positive pressure. In the absence of wind, momentum, and pore water pressure, stress

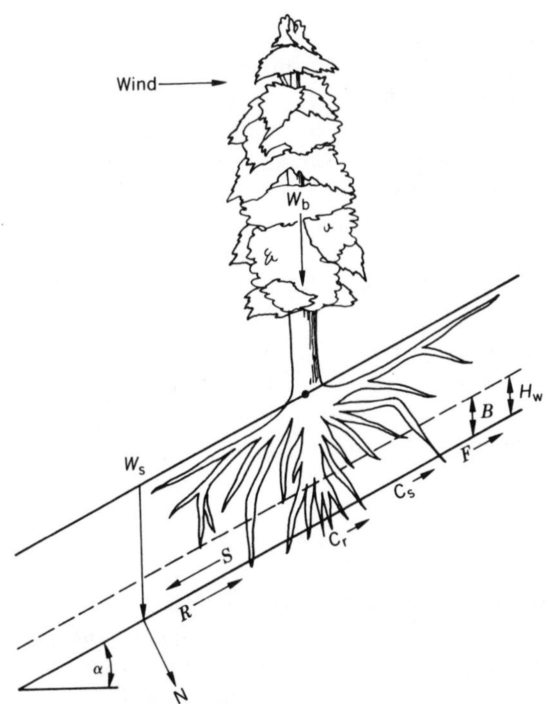

FIG. 10.3 Forces contributing to shear stress and resistance on a sloping plane surface. Stress (S) is caused by the weight of soil (W_s) and biomass (W_b) acting down the slope [$(W_s + W_b)(\sin\alpha)$], plus the pressure of the wind transmitted to the soil through the tree and other forces such as earthquakes that impart momentum to the soil. Resistance (R) is primarily due to the weight of the soil and biomass acting normal to the slope [$(W_s + W_b)(\cos\alpha)$], reduced by the lifting force of bouyancy (B) in any saturated layer (H_w = height of the water table), plus the forces of friction (F), soil cohesion (C_s) and root cohesion (C_r) along the potential failure surface.

may be calculated readily provided slope angle, soil density, and weight of biomass are known. However, the contribution of wind, momentum, and pore water pressure is seldom known.

The forces acting to resist stress are frictional forces along the plane, cohesive forces of the soil material, and the binding forces of roots or effective cohesion due to roots. The frictional resistance is a function of the force that acts normal to the surface and the coefficient of friction, which ranges from about 0.5 to nearly 1.0, averaging about 0.7 for many soils (Burroughs et al., 1976). The force normal to the surface is a function of the weight of the soil and biomass for unsaturated soils:

$$N = (W_s + W_b)\cos\alpha \tag{10.7}$$

where N is the normal force, W_s and W_b are the soil and biomass weights, and α is the slope inclination in degrees. The normal force is reduced if some layer of the soil is saturated. *Buoyant forces* are exerted upward on the soil mass in the layer saturated by water. The bouyant force is equal to the density of water times its depth in a column of unit surface area. For soils that contain a saturated layer, the normal force becomes:

$$N = (W_s + W_b - B)\cos\alpha \tag{10.8}$$

where B is the buoyant force.

Frictional resistance is equal to the normal force times the coefficient of friction. However, in soil mechanics, frictional resistance to sliding is often expressed as a function of the tangent of the *angle of internal friction*, which is the slope angle at which shear stress is equal to shear resistance in cohesionless material, or:

$$F = N(\tan\phi) \tag{10.9}$$

where F is the frictional resistance and ϕ is the angle of internal friction. The angle of internal friction is also known as the *natural angle of repose*.

In addition to frictional forces, cohesive forces that bind soil particles together provide additional resistance to sliding, as does the strength of root reinforcement. Therefore, shear resistance may be expressed as:

$$S_r = N(\tan\phi) + C_s + C_r \tag{10.10}$$

where S_r is the shear resistance, ϕ is the angle of internal friction, and C_s and C_r are the cohesion provided by the soil and roots, respectively.

The relationship between shear stress and shear resistance determines the stability of the slope, often expressed in terms of a *factor of safety*, the ratio between shear resistance and shear stress:

$$FS = S_r/S_s \tag{10.11}$$

where FS is the factor of safety and S_r and S_s are the shear resistance and stress, respectively. Whenever the factor of safety exceeds one ($FS > 1.0$), the slope is stable. Failure occurs as soon as the FS drops below that value.

10.3.2.2 Rotational Failure. Stress and resistance forces in rotational slope failure are much more complex than in planar failure. In order to reduce this complexity, the mass is often divided into vertical sections (slices) and the stability of each section is analyzed separately and then summed to determine the stability of the whole (Fig. 10.4). All sections to the right of the axis of rotation tend to cause downward movement along the failure arc, but any section lying to the left resists movement.

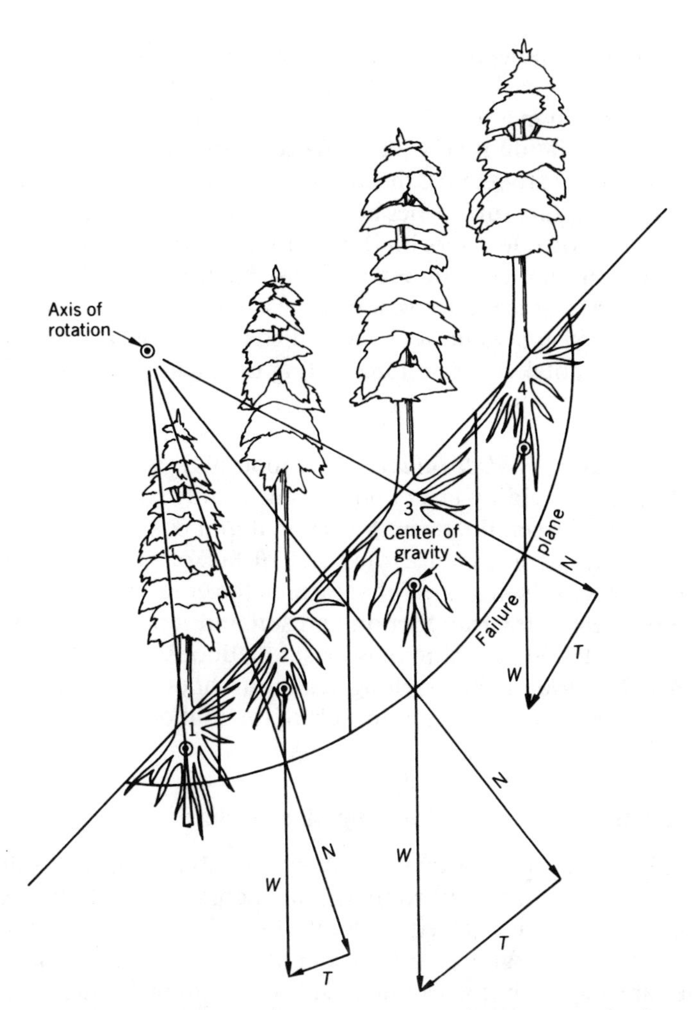

FIG. 10.4 Forces contributing to shear stress and resistance on a sloping rotational surface. The mass is divided into vertical sections for analysis. Stress and resistance components are comprised of the same elements as on a plane surface, but the slope used is that of the tangent (*T*) to the failure arc at a point below the center of gravity of each section. *W* is the vector of weight and *N* is the normal to the tangent. Total stress and resistance is the sum of section values.

The downward force components are comprised of the same elements as in planar failure, but the slope used is that of the tangent to the failure arc at a point directly below the center of gravity of each section. Forces can be determined graphically by scaling the weight vector, setting the normal through the tangent point, and the axis of rotation and extending the tangent at a right angle from the normal vector to the end of the weight vector.

Frictional forces along the failure surface vary only with the difference in the normal force for each vector if the angle of internal friction is assumed constant. Soil cohesion is usually not considered to vary by direction of shear, although some variation is very likely. As root strength contributes to shear resistance only where roots cross the failure arc, it may vary greatly among sections. Where the failure surface lies below the root zone, as in Sections 2 and 3 of Fig. 10.4, cohesion due to root binding is absent. Similarly, buoyancy forces due to saturation may vary among sections. Finally, the failure area along each section varies. Since cohesive forces are usually expressed on a unit area basis, cohesive resistances must be multiplied by the area across which they act to obtain the total cohesive force acting on each section and the mass as a whole.

10.3.2.3 Soil Arching in Slope Failure. If the roots of trees (or piling) are firmly anchored in an unyielding layer and are sufficiently strong, soil movement can only occur between the trees. A frictional and cohesive resistance may then act along the lateral, as well as the basal, surface of the slide, with trees acting as abutments for "soil arches" that form in the ground. Depending on the spacing, the angle of internal friction and soil and root cohesion within the soil mass, considerable restraint may exist to reduce slope failure. Gray and Megahan (1981) found that arching restraint should be common at tree spacings of 6.3 m or less on granitic soils of the Idaho Batholith.

10.3.3 Variation of Factors Influencing Slope Failure

In wildland management, evaluations of slope stability through solution of soil mechanics equations are uncommon, but increasing in number. Analyses of past slope failures have helped to reveal which site characteristics are most important in causing failure, and in identifying the circumstances under which they are likely to become critical. This information is often sufficient to enable the land manager to recognize and avoid many hazardous situations without the necessity of detailed analysis.

The factors most often associated with slope failure are (a) slope steepness that equals or exceeds the angle of internal friction; (b) wet soils; (c) geology and soil types susceptible to failure; and (d) removal of vegetation. These factors can be changed by management activities, particularly timber harvesting and construction. Other factors, such as wind causing trees to sway, earthquakes or others are so complex that they are seldom analyzed. Instead, a higher factor of safety in seismically active regions, exposed windy sites, or other situations where such impacts can be expected is usually required to reduce the risk of slope failure.

10.3.3.1. Slope Steepness and Angle of Internal Friction. If loose soil particles are poured into a pile, they will form a slope of such steepness that shear stresses are in balance with shear resistance. The resulting slope is the *natural* *angle of repose* or at the angle of internal friction. If cohesion of the soil mass is slight, any slight disturbance of the balance can result in slope failure.

The angle of internal friction of most soil materials has been shown repeatedly to lie mostly in the range of about 30°-45°, with a mean of about 35°, as shown in Table 10.4. In some loose, clayey soils the angle of internal friction when saturated is much less, declining to only a few degrees. The angle of internal friction varies with (a) particle shape—angular particles have a higher friction angle than round, (b) mineralogy—mica-rich and certain non-cohesive clays slide very easily, and (c) density—firm or compact materials have greater strength than loose material.

Although it is possible for soils having a gradient less than their saturated angle of internal friction to fail, most failures occur on slopes that equal or exceed those angles. Thus, most studies have shown increasing occurrence of slope failure as steepness increases (Table 10.5). The close correspondence between Tables 10.4 and 10.5 suggest that slope gradient alone may serve as a strong indicator of landslide hazard. The critical slope threshold under a wide variety of circumstances lies near 30° or 55%. Anything that steepens a slope, whether natural or through land use or construction activities, brings it closer to failure. Landsliding tends to be common where slopes are steepened by downcutting of streams in regions of recent uplift, by creep or by undercutting by streams or human activity such as road construction.

10.3.3.2 Soil Moisture. Landslides are generally few in arid regions (Radbruch et al., 1981) and even in humid regions most occur during heavy storms or snowmelt. The hazard increases greatly as storm rainfall exceeds about 125 mm, but lesser storms may trigger slides if antecedent conditions have been wet and intensity is high (Swanston, 1967b, 1969,1974; Sheng, 1966; Endo, 1969; Rice et al., 1969; Dyrness, 1967b; Rice and Foggin, 1971; Klock, 1972; Swanston and Dyrness, 1973; Campbell, 1975; Day et al., 1977; Wu and Swanston, 1980; Gray and Megahan, 1981; Carter and Galloway, 1981; Sidle and Swanston, 1982; Eschner and Patric, 1982). Adding water to dry soil always increases shear stress by its additional weight, but its effect on shear resistance varies with moisture content and soil plasticity.

The effect of moisture content on shear strength of nonplastic soils may readily be observed by walking a short distance across a sandy beach from dry upland into the water. Where the sand is dry, there is no cohesion. The stresses of each foot striking readily displace the sand. Nearer the waterline, where moisture from lapping waves is draining, the sand becomes firm, the grains held together by water under tension. As one approaches the water, the sand again becomes easier to displace until beneath the water it is only slightly more firm than where dry. The greater firmness under water results from tighter packing of the grains by wave action, for the matric potential becomes positive at saturation. The moisture content at which moisture tension contributes most to soil cohesion varies with pore size distribution and the amount of pore space, but it usually lies somewhere near the permanent

TABLE 10.4. Angle of Internal Friction (ϕ) of Soil Masses in Western United States

Location	ϕ (degrees)	Soil Conditions	Source
Oregon Coast Range	28–34	Saturated, sandy	Yee and Harr, 1977
Oregon Coast Range	39–42	Dry, sandy	Yee and Harr, 1977
Idaho, southeastern	31–40	Mine overburden	Jeppson et al., 1974
Oregon, Washington Coast	24–49	Wide soil range	Schroeder and Alto, 1983
Alaska, southeastern	37–44	Shallow soils over bedrock	Wu et al., 1979
Alaska, southeastern	37	Glacial till	Swanston, 1969
Idaho, Batholith	29–37	Coarse texture	Gray and Megahan, 1981

TABLE 10.5. Slope Gradients Associated with Mass Failure

Location	Gradients (degrees)	Comments[a]	Source
Idaho Batholith	31–60	Mean 39, $n = 16$	Gray and Megahan, 1981
Southeast Alaska	30+	Most about 37	Swanston, 1970
San Gabriel Mts., California	40–48	Mean 42, $n = 17$	Rice et al., 1969
Ashio Mt., Japan	35–65	Most from 35 to 50	Ichikawa, 1952
Shimane Prefecture, Japan	20–41	89% over 31	Kohno et al., 1968
Clearwater NF, Idaho	Up to 42	Mean 33, $n = 629$	Day et al., 1977
Southeast Alaska	32–60	Widespread survey	Swanston, 1967a
Santa Monica Mts., California	27–56	Deposition <12	Campbell, 1975
Cascade Mts., Oregon	0–42	78% from 24 to 37	Dyrness, 1967b
Taiwan	25–65	$n = 631$	Sheng, 1966

[a] n = number of samples.

wilting point, except in uniform coarse material that has more cohesion when moist.

At saturation, cohesive forces due to moisture tension disappear and are replaced by buoyancy forces. Under static saturation, only buoyancy acts to reduce the normal force. The greater the thickness of the saturation zone, the greater the reduction. When soil water is moving under positive pressure, it also exerts a drag force that not only tends to lift, but also to separate individual particles, keeping them in suspension if the movement is upward. Quicksand occurs where soil water moves upward under positive pressure. The phenomenon can be observed in clear springs where "sand boils" form in regions of moderate upward movement. Even slow seepage can exert important destabilizing forces (Burroughs et al., 1976).

Soil water also influences the plasticity of soils. Soil clods high in clay may require great force to crush when they are dry, but as they are wetted shear strength decreases. Water films tend to separate the clay particles and thus reduce its cohesive strength (Burroughs et al., 1976). As the moisture content increases, the soil may remain cohesive, but is readily deformed under pressure. With additional water it may even become liquid. The *plastic limit* is the moisture content at which a soil becomes plastic. The *liquid limit* is the moisture content at which the soil passes from a plastic to a liquid state. Sloping plastic soils usually exhibit creep and may show other forms of mass instability as well. Not all soils exhibit plasticity, especially those low in clays. Most modern soil surveys include plasticity and moisture characteristics in their descriptions of soils (e.g., Barker, 1981).

The shrink–swell characteristics of many soils upon alternate wetting and drying also causes weakening. Shrinkage cracks developed during drying offer a ready path of water entry and create planes of weakness that may persist during wet periods. The expansion of clays as they are rewetted during each wet season also contributes to shear stresses, causing downhill movement by creep and slides.

Finally, soil moisture influences the kind and intensity of weathering that soils undergo. For example, iron and humus compounds leached downward in certain upland soils may act as strong cementing agents under oxidizing conditions, creating high soil strength (Paeth et al., 1971). Under saturated conditions, reduction may occur and they then contribute little to soil strength.

All in all, it becomes easy to see why mass wasting is a dominant form of erosion wherever steep slopes occur with plentiful moisture. These two factors alone are sufficient to discriminate 85% of areas exhibiting mass failures in an eastern Ohio watershed (Lanyon and Hall, 1983). They are almost always included in any slope stability classification.

10.3.3.3 Geology and Soils. The differences in slope stability within a given region may frequently be related to differences in bedrock or soil material. For example, volcaniclastic terrain and lava flows occupy similar portions of the H. J. Andrews Experimental Forest of western Oregon, but more than 98% of the slides took place in the volcaniclastic area (Swanson and Dyrness, 1975). Similar examples may be cited elsewhere throughout the world (Oyagi, 1968; Laffan, 1979; Radbruch-Hall et al., 1981; among many others).

Major factors contributing to soil stability related to geology and soils include: (a) structure of the parent material and (b) ease and nature of the weathering products. In any given region, either may be of prime importance.

The structure of parent material may control stability in several ways. Sedimentary rocks with bedding planes parallel to the slope may provide little frictional resistance to sliding by overlying materials. Fractures between them may create zones of weakness. On the other hand, vertical fractures and joints or bedding planes perpendicular to the slope can increase stability if roots can penetrate into them, and by absorbing water. Impermeable bedrock can cause saturated zones to develop during wet weather (Swanston and Dyrness, 1973; Burroughs et. al., 1976; Day et al., 1977; Swanston and Swanston, 1977; Durgin, 1977). Contact zones between easily weathered and resistant rocks or between rocks of high and low strength may develop steep slopes that are susceptible to sliding (Swanson and James, 1975; Swanston, 1978, 1981).

The type and intensity of weathering reflects the interaction among climate, vegetation, and parent material and determines the degree of weathering and the nature of the weathered products. Certain rock types, such as siltstones, mudstones, shales, pyroclastics (sometimes called volcaniclastics), serpentine rocks, and many others are easily weathered, particularly in warm, humid climates where chemical weathering rapidly transforms primary minerals into clays (Swanson and Swanston, 1977; Swanston and Dyrness, 1973; Swanston, 1978). Others, including granitics and sandstones may form shallow, coarse, cohesionless soils where debris slides are common (Clayton and Arnold, 1972; Durgin, 1977; Clayton et al., 1979). Materials containing large amounts of mica are often landslide-prone (Day et al., 1977).

Soil cohesion and frictional resistance are related to clay mineralogy and soil moisture. When below the plastic limit, clays may exhibit great cohesive strength, but as moisture content increases, water may be absorbed into clay lattices, reducing granular friction in extreme cases to near zero. Water between loose clay grains may permit sudden liquification, as occurs in quick clays. Swelling clays of the smectite group (montmorillonite) tend to be particularly unstable (Paeth et al., 1971; Burroughs et al., 1976; Swanston, 1978), as are the amorphous clays (Taskey et al., 1978; Istok and Harward, 1982).

10.3.3.4 Removal of Vegetation. Three characteristics contributing to slope stability are strongly influenced by vegetation: (a) the weight of biomass on the slope, (b) moisture relations of slope materials, and (c) cohesion of the soil mass due to binding by roots (Gray and Megahan, 1981). The effect of timber harvesting on the weight of biomass on the slope is straightforward; weight is directly reduced, although the effect on stability is usually negligible.

The effect of timber harvesting on soil moisture relations is also straightforward in principle: Soils will remain wetter longer than under uncut stands.

Whether the wetter soils cause failure, however, depends on whether reductions in cohesion and increases in possible saturation combine sufficiently to reduce the factor of safety below unity. This, in turn, depends on soil cohesive properties, permeability, and the probability of heavy storms or rapid snowmelt. Provided moisture content-cohesion data and water table response to precipitation data are available, they may be combined with antecedent moisture condition estimates and rainfall or snowmelt data to predict the probability that the critical factor of safety will be exceeded. For example, Wu and Swanston (1980) showed how piezometric levels (elevation of groundwater) were likely to change, and thus change the probability of slope failure after clearcutting in southeastern Alaska, based on climatic data and previously established groundwater–precipitation relations. Regardless of whether it is possible to estimate site-specific probabilities of failure or not (e.g., naturally large variations in key controlling factors presents sampling challenges), it should be understood that increases in soil water beyond a certain point always tend to increase failures on unstable slopes. It bears repeating that nearly all slope failures occur during wet conditions, and soils remain wetter longer when vegetation is removed.

On some sites, plant roots may provide the margin of safety to keep soil from sliding. For example, on a granitic soil with a safety factor of unity ($FS = 1.0$) under dry soil conditions, root cohesion of only 3.2 kPa was sufficient to keep a fully saturated soil from sliding (Gray and Megahan, 1981).

Effective root cohesion is a function of the tensile strength of roots times the fraction of the soil cross-section occupied by roots. Great variation exists in both strength and root occupancy; for example, live Douglas-fir roots varied in tensile strength from 27.6 to 61.6 MPa in several different studies (Burroughs and Thomas, 1977; O'Loughlin and Watson, 1979; Ziemer, 1981), and Ziemer (1981) found extreme variation of root biomass within and among different areas. Root concentrations consistently decline with increasing soil depth, however, so the deeper the potential failure surface, the less likely roots will add significant shear strength to unstable soils. Despite these variations, it is generally agreed that live tree roots add to soil strength.

Numerous studies have shown that root strength decreases rapidly after roots die and begin to decay (Kitamura and Namba, 1966, 1968; Burroughs and Thomas, 1977; Ziemer and Swanston, 1977; O'Loughlin and Watson, 1979; Wu and Swanston, 1980; Ziemer, 1981). On the other hand, the growth of new or released vegetation adds strength as the root zone is reoccupied. Grass, forbs, and shrubs may contribute significantly while trees are becoming reestablished (Endo and Tsuruta, 1969; Ziemer and Swanston, 1977; Ziemer, 1981; Waldron and Dakessian, 1982; Waldron et al., 1983). As a result of this interaction of decay and growth, root shear strengh may follow a pattern of decline after cutting coniferous forests, reaching a minimum some 4–15 years later, and then increasing again, depending on species composition and other factors (Fig. 10.5).

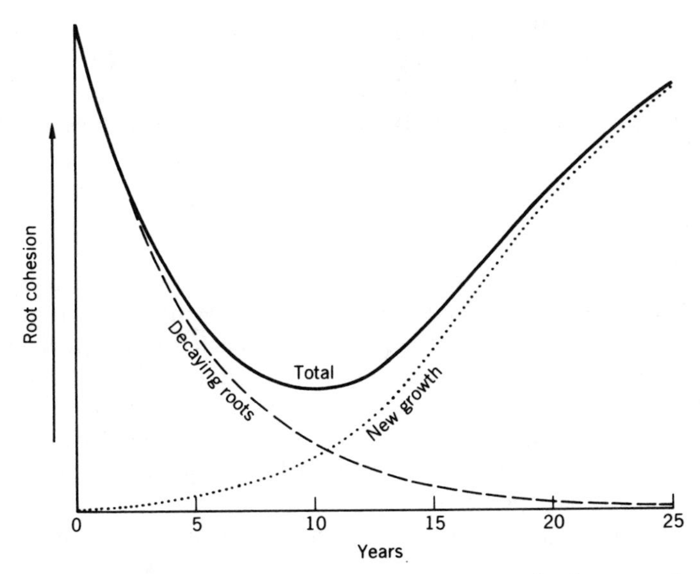

FIG. 10.5 General relationship between decay and regrowth of roots and soil reinforcement that may occur following clearcutting of conifer forests at year zero.

Few data are available to reveal the shear strength pattern that develops when sprouting species are cut or top-killed. Some root death follows top death, but the amount of death, rate of recovery, and effects on soil shear strength are largely unknown. Nevertheless, studies such as those by Endo and Tsuruta (1969) and Ziemer (1981) suggest that surviving root systems of understory species may contribute significantly to soil strength during the period when the contribution of tree roots is at its minimum. The strength retention from sprouting tree species should be at least as great.

10.4 CHANNEL EROSION

Channels range from upland gullies formed by ephemeral runoff to those containing large perennial rivers. They are the primary transport route of sediment derived from the watershed, as well as possible sediment sources. Stable channels can deliver sediment from a watershed without actively eroding themselves. A channel system is said to be at equilibrium when, over the long run, the stream power is sufficient to transport the sediment delivered to the channel and neither deposition or erosion takes place. Equilibrium is dynamic, not static, for sediment may be delivered continuously from the mouth of a stream system although it enters only sporadically.

Anything that upsets the equilibrium of the channel system can cause deposition or erosion in the channel. Chief among such features are (a) changes in stream power, (b) changes in sediment supply to the stream, and (c) changes in resistance of bed and bank materials to cutting. Stream dynamics are highly complex and are explained well for nonspecialists by Dunne and Leopold (1978) and Heede (1980); only a few basic principles will be considered here.

10.4.1 Stream Power

The power of a stream comes from the movement of water from a high elevation to a lower one. Not all flows are equally effective in determining channel equilibrium. Low flows are too tranquil to move much material (their efficiency per unit of energy is low) and extreme flows, although capable of moving large amounts when they do occur, are usually rare and unpredictable. In most streams, nearly all sediment is moved by flows that are bankfull or greater, for the cumulative product of their power times their frequency accounts for most of the effective stream power.

Stream power may be expressed for any unit width of stream as:

$$P = \gamma Q s \tag{10.12}$$

where P is the stream power per unit width (kg m^{-1} s^{-1}), γ is the unit weight of water (kg m^{-3}), Q is the discharge (m^3 s^{-1}), and s is the slope of the water surface. From this equation it is clear that there are two major controls of stream power: discharge and stream slope. Anything that changes these parameters in a stream at equilibrium will cause degradation or aggradation of the channel provided the change is sufficient to overcome resistance of the channel materials to movement (for erosion to occur) or a supply of sediment is available (for deposition to take place).

The major factor that causes gullies to develop or become reactivated is usually an increase in runoff, with development following fairly consistent and identifiable stages (Harvey et al., 1985). Gullies develop frequently in cultivated and mined areas and may form in natural vegetation zones such as desert, grasslands, shrub, and open woodlands or forests. They rarely occur on sites supporting a healthy, dense vegetative cover unless they invade from the outside (Heede, 1975 a). In areas of natural vegetation, gullies can be caused by disturbances such as overgrazing, fire, drainage from roads, or others that provide a source of concentrated runoff. Climatic change cannot be ruled out as a cause in some areas.

Not all increases in runoff will initiate stream channel degradation. Peak discharges from many storms are below the critical threshold of stream power needed to overcome channel resistance. It seems obvious that the stream power needed to initiate degradation would considerably exceed that necessary to transport sediment at the equilibrium rate, for most stable stream systems handle infrequent very high flows without more than temporary disruption. Indeed, in a study of the effects of strip mining on channel stability in 176 forested first-order basins in Pennsylvania, channel degradation was slight below a threshold of disturbance of either 0.45 km^2 of mined area or 50% of total basin area mined. Thereafter, a sharp increase in channel cutting developed. Apparently a natural safety factor exists, enabling most stable channels to resist occasional strong impulses of stream power (Touysinhthiphonexy and Gardner, 1984). Thus, the assumption may not be warranted that temporary increases in streamflow, such as might result from partial cutting or burning of small forested watersheds, are likely to lead to channel degradation. Such an assumption underlies timber harvest guidelines used in the Northern Region, United States Forest Service (Benoit, 1973).

Stream power is a function of slope or stream gradient, as well as discharge. Stream gradient is a major control of flow velocity. Where gradients are slight in small stream systems, velocities are low; where steep, they increase. Slopes can be increased by lowering of a base level, removal of an erosion-resistant control point, or shortening of a stream reach, as when a channel is straightened to accommodate road construction in a narrow valley (Galay, 1983). Many small mountain streams have steep average gradients that would imply high erosion rates unless something existed to prevent them. In most cases, erosion is prevented by: (a) bed armoring with gravel and boulders; (b) bar formation perpendicular to flow; and, most important in forest headwater streams, (c) steps formed by logs and other large forest debris (Heede, 1977, 1981).

Armoring is caused by the removal of fines, leaving a layer of larger gravel or stones that create a stable bed. The armor size is determined by the largest flows because lighter particles available for transport are moved out by such flows. Thereafter, degradation is halted as the armor prevents deep cutting and increases channel roughness (Heede, 1980,1981).

Bars form by the accumulation of gravel or rock caused by the irregular distribution of forces of flowing water. They also increase bed roughness and may cause steps to form (Heede, 1980, 1981).

Steps from large forest debris and bars that impede flow dissipate energy by creating a flatter channel gradient upstream, and the overflow energy is dissipated in the plunge pool below. Heede (1980) found average velocities at the outlet of such pools to be 0.12 m s^{-1} or less in a stream with discharges up to 0.49 m^3 s^{-1}. The reduction in stream power by steps can be considered to be in direct proportion to the total stream relief that is comprised of steps. Many studies have indicated up to 50% or more reduction (Heede, 1972, 1975 b, 1976; Keller and Swanson, 1979; Swanson et al., 1976). However, Marston (1982) found that neither log steps nor falls caused a significant difference in the equilibrium condition of stream networks in the Oregon Coast Range. He found that log steps were most frequent in third-order streams because although tree fall occurred most often in first- and second-order streams, steps did not develop in the steep V-shaped topography that prevented their positioning within and across the channels. In fourth- and fifth-order streams, storm flows tended to remove large debris before steps could develop.

10.4.2 Sediment Supply

Most streams at equilibrium exhibit a balance between stream power and sediment supply, and over the long run neither degrade nor aggrade. However, a change in the sediment input to the stream may change its stability. Clear water in motion always has an unsatisfied sediment carrying capacity, and any bed or bank materials that are of appropriate size and not protected from the water's tractive force or exhibiting sufficient mechanical strength to resist it will be removed. Thus, erosion control on uplands may initiate a cycle of stream cutting, especially in streams flowing in alluvial channels.

On the other hand, increased sediment inputs may overtax transport capacity and the channel will aggrade. The streambed rises and tends to overspill its banks, creating a smaller hydraulic radius for a given discharge and thus reducing the velocity. The velocity is reduced most as the water leaves the channel, so much of the sediment is dropped on the banks, creating a natural levee.

Every stream has its equilibrium sediment carrying capacity, depending on its channel dimensions and velocity. However, channel dimensions and velocity change with rate of discharge and, since stream channels are seldom uniform, from place to place in the stream. Thus, streams are characterized by a constantly shifting equilibrium, picking up sediment as they rise and depositing it as they fall. At any given time, sediment may be picked up in one part of the channel while it is being deposited in another (Leopold and Maddock, 1953; Heede, 1980). The net result is that most sediment is moved downstream in a sort of "leap frog" fashion, being moved at times of high flow and in places of higher velocity and deposited during falling flows and in places where velocity slows. When such channel storage and supply are accounted for, sediment concentrations during storm flows can be predicted with greater accuracy (Van Sickle and Beschta, 1983).

Obviously, protection against channel cutting must be directed toward protection from the infrequent high flows and at locations of higher velocities. In irregular channels, the higher velocities are located in deep water on the outside of curves (Fig. 10.6) and in areas of high channel slope such as falls, rapids, and riffles. Unless they are formed in erosion-resistant portions of the channel, falls, and rapids may retreat upstream, supplying large amounts of sediment. Where such retreat is evident, the stream channel has not reached equilibrium.

10.4.3 Resistance of Bed and Bank Materials

Mountain streams of high average gradient are usually prevented from rapid erosion because they are divided into steps of high erosion resistance, as by logs, transverse bars, rock outcrops, and others, separated by sections of low gradient. Should the steps lack erosion resistance or be removed, their loss would increase the stream power available for removal and transport of channel bed materials. Beschta (1979a) showed that removal of large organic debris

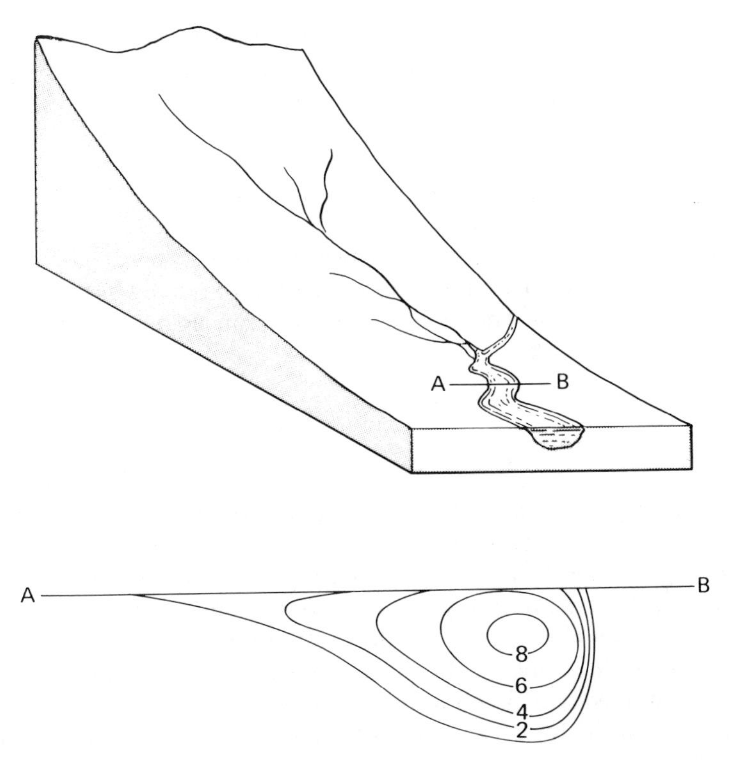

FIG. 10.6 Channel cross-section and distribution of relative velocities of flow at a curve in an irregular channel.

in a Coast Range stream in Oregon resulted in downcutting and removal of more than 5000 m^3 of sediment in a 250 m reach the first winter following removal. Bilby (1984) indicated that clearing logging debris from a fourth-order channel caused movement of older debris that remained and bed scouring up to 80 cm deep. However, older debris exceeding 10 m in length or anchored in the bed or banks generally resisted movement. Downcutting may be self-limiting where the channel bed contains materials suitable for armoring, such as large gravel and rocks, or where resistant bedrock is exposed.

Where the base level of a channel is fixed by erosion-resistant materials, downcutting is largely prevented and any stream cutting tends to be lateral. Streams at equilibrium grade may still do considerable bank cutting and change their course, cutting here and depositing there, especially if the channel is cut in weak alluvial materials. Debris in streams may reduce stability by deflecting flow so that banks are undercut or new channels are cut (Swanson et al., 1982 b). If vegetation can be maintained on the banks, it may add enough strength to resist cutting.

Where downcutting takes place, no lateral bank of soil material is likely to remain stable. Banks that are undermined or steepened beyond their limit of shear resistance will fail no matter how well their upper portions are protected. Where tree roots lie largely above the level of cutting, the presence of the trees may decrease rather than enhance bank stability.

10.5 SEDIMENT ROUTING AND DELIVERY

There may be a very poor correlation between erosion on a watershed and the discharge of sediment out of it. The difference is represented by the change in storage within the basin. The sediment delivery concept is simple in essence and may be expressed as:

$$I \pm \Delta S = O \qquad (10.13)$$

where I is the sediment input or production, ΔS is the change in storage within the system, and O is the output from the system. However, there are many different sources, types, and processes of erosion; transport methods and routes of sediment movement; and locations, sizes, and types of storage reservoirs. Furthermore, all elements of Eq. 10.13 are erratic over time and most responsive to extreme events. Yet, the better the details of the system are understood, the better the productivity of both land and water may be sustained under the effects of human activity. Each transfer, storage, and linkage element represents an opportunity for control, or threat of damage, depending on the way they are modified when using the land.

Unfortunately, even such a basic element as sediment transport past a given point in a channel is difficult to determine accurately (Glymph, 1975). Sediment is carried in suspension or as *bedload* (by rolling, sliding, or bouncing along the streambed) and often only suspended sediment is measured. Both the amount of sediment and the proportion carried in suspension may vary markedly in time and in space. Suspended loads vary with discharge, usually increasing with increasing discharge, but even at the same discharge rate, the amounts differ according to whether stage is rising or falling and through a sequence of successive storm events (Beschta, 1979b).

Despite these variations, one of the most commonly used methods to estimate suspended sediment yields is the use of a sediment rating table whereby sediment concentration is statistically related to streamflow rate, and then integrated over the period of interest according to the appropriate flow duration curve. Other methods include reservoir sediment deposition surveys whereby repeated measurements of the bed of reservoirs reveal deposition in the still waters. Sometimes estimates of the *sediment delivery ratio* (the ratio of sediment leaving a basin to that produced in it) are developed based on watershed characteristics. Reservoir deposition surveys and sediment delivery ratios include both suspended and bedloads (Holeman, 1975; Livesey, 1975). Sometimes bedload is measured separately with special samplers (Beschta, 1981a), but problems of integrating over space and time remain (O'Leary and

Beschta, 1981; Mosley, 1981). Bedload may vary markedly over short distances within a channel, and in time during the same runoff event (Mosley, 1981; Beschta, 1983). Such variations suggest that monitoring programs to determine sediment discharge will be difficult and costly at best and regulatory programs based on detecting moderate changes are likely to be unrealistic.

Nevertheless, sediment routing and delivery processes are becoming better understood (Swanson et al., 1982 b; Moore, 1984). Already such models are helping to identify critical portions in the delivery system (Schumm et al., 1976; Trimble, 1976, 1981; Wilkin and Hebel, 1982; Meade, 1982). Sediment budgets constructed from measures of most of the elements of a real system (in contrast with gross estimates) are rare, but provide confirmation of the essential validity of the budgeting concept (Lahre, 1982).

Many factors influence sediment routing and delivery from a basin. Most obvious is simply the amount of sediment produced. While there may be a poor correlation between production and delivery, production still sets the upper limit to delivery. If erosion is low, there can be little sediment removed from any basin. However, even if production is high, most sediment is stored before it moves very far. The factors influencing sediment delivery include the nature of the sources, the sinks, and the delivery system of the basin.

10.5.1 Sediment Sources

Not all parts of a watershed or sources of sediment are equal when it comes to delivery from the basin. Sediment from different places, of different kinds, and derived by different processes contributes unequally.

The size of sediment that can be transported by water varies as a small fractional exponent of velocity, so fines are moved much more readily than coarse materials. A watershed that produces fine sediment tends to yield a greater proportion at its outlet than one that produces coarse sediment for the same regime. In a similar manner, fines would be more easily moved if the texture is uniform than if it is well graded (i.e., showing a wide range of particle sizes), for the larger particles in a graded mixture tend to protect the remaining fines by building a protective armor as the fines in the surface layers are selectively removed. Fines are also more readily moved by nearly all storms yielding runoff, for the power threshold needed to initiate and sustain motion is low. Fine sediment delivered from a channel is often considered to be "source limited," in that it is moved so readily that virtually all that enters the channel system is carried through it. The delivery of coarse bedload is much more a function of stream energy than supply, so it is sometimes considered to be "energy limited" (Renfro, 1975; Robinson, 1977; Beschta, 1981 b; Walling, 1983).

This type of erosion has strong bearings on sediment delivery. Surface erosion may be well dispersed with many opportunities to enter storage before reaching a channel. On the other hand, the very nature of the gully and channel erosion process implies efficient transport simultaneous with removal.

Similarly, debris torrents are characterized by efficient transport, whereas rotational slumps may exhibit little displacement and retain a degree of cohesion that inhibits further transport (Renfro, 1975; Anderson, 1975; Walling, 1983).

The location of erosion with respect to the channel system is an important determinant of sediment delivery. Other factors being equal, the closer a source is to a stream channel, the greater the probability of rapid transport from the watershed. The nature of the terrain between the source of sediment and the channel may be as important as distance. Where slopes are steep, convex, and bare, little opportunity may exist to trap sediment once it is en route to the channel. Accumulated evidence suggests that most of the sediment load carried by streams originates in or near the channel system most of the time (Renfro, 1975; Anderson, 1975; Patton and Schumm, 1981; Mosley, 1981; Lisle, 1981; Beschta, 1981c; Meade, 1982; Wilken and Hebel, 1982; Pearce and Watson, 1983; Burns and Hewlett, 1983; Walling, 1983).

10.5.2 Sediment Sinks and Storage

Most sediment is deposited and remains in storage in the watershed where it was produced. Upland areas provide many storage sites. Potential depositional areas occur wherever slope gradients decrease, where cover and other barriers may reduce energy or disperse flow, where surface flows infiltrate or in surface irregularities in which water is trapped (Trimble, 1981). Most valleys other than youthful V-shaped ones undergoing active downcutting are major sediment storage reservoirs. Valley terraces, floodplains, and natural levees are all *alluvial* (water-laid) deposits. The entire Mississippi River Delta below Cairo, Illinois, is a massive deposit of stored sediment (Ritter et al., 1973; Trimble, 1981).

Sediment can also be stored in river channels, lakes, ponds, and reservoirs wherever space and streamflow energy permit. Such storage may be virtually permanent, as in lakes and reservoirs; quasipermanent, as that stored behind a log barrier that may take 100 years or more to decay and fail; or it may be temporary as in shifting bars. The amount of storage in many streams may be many times the annual sediment discharge. Sometimes, excessive inputs of sediment into a system, as by landslides, will cause channel aggradation that can persist for many years (Lyons and Beschta, 1983). The same log and rock steps that reduce stream power often create sediment basins for short- to long-term storage (Beschta, 1979; Megahan and Platts, 1980; Lisle, 1981; Heede, 1981; Pearce and Watson, 1983).

10.5.3 Transport System

The limiting factor in the sediment yield from most watersheds is the capacity of the transport system, for more sediment is produced in nearly every basin than is delivered through its outlet; few watersheds of any size exhibit a sediment delivery ratio of unity. Whether sediment reaches a channel depends partly on channel density and distribution and source areas that vary with season and storm size. It also depends on the amount of streamflow and the streamflow regimen. Other factors equal, the greater the annual discharge, the greater the sediment carrying capacity. But most sediment is carried by infrequent high flows. Of two streams with equal annual flow, the one with widely varying flow, where most of the annual yield occurs in a few large pulses, has a far greater sediment transport capacity than one with uniform flow. Williams (1975) related transport capacity to storm flow volume times peak flow rate, and by substitution for the rainfall energy factor (R) in the USLE, developed a sediment yield model.

Stream gradient and stream power also influence sediment transport capacity. These factors have already been discussed and will not be repeated here.

10.5.4 Relations to Watershed Characteristics

Many attempts have been made to relate sediment yield to various watershed characteristics. Some models are process oriented and others are statistical, and still others are mixtures of the two. Many of the process models still retain a strong "black box" component. A good discussion of the several kinds are found in Swanson et al. (1982 b) and USDA Agricultural Research Service publication ARS-S-40 (1975). Among the major factors related to sediment yield are watershed size, land use, and watershed morphometric characteristics such as shape, steepness, drainage patterns, and others.

Nearly all studies have shown that the sediment delivery ratio decreases with increasing watershed size in the manner:

$$DR = kA^{-x} \tag{10.14}$$

where DR is the delivery ratio, k is a watershed coefficient related to watershed characteristics, A is the area, and x is an exponent that varies from 0.125 to 0.20 (Renfro, 1975; Walling, 1983). The reasons for such a decrease in a delivery ratio with increasing watershed area are many. Local storms in large basins may generate much erosion and sharp peaks, but the effects are largely lost by dilution and damping as they are routed downstream, even if the sediment is not stored en route. The average slope of a large basin is always less than in a small one in the same physiographic region (Bogolyuba and Karanshev, 1974; Renfro, 1975; Walling, 1983). The average slope of a large river basin such as the Mississippi River only modestly departs from the horizontal, despite mountains of over 4000 m in tributary headwaters. Opportunities for storage are numerous in most large basins because of the numerous reservoirs that have been constructed throughout the world. All in all, it is not unexpected that relative sediment yield decreases with watershed area.

Land use in a watershed is also highly correlated with sediment delivery. Both production and trapping are involved. Activities on wildlands will be discussed in detail in Chapter 12.

Many attempts have been made to relate sediment delivery to morphological characteristics of drainage basins. Such characteristics as channel density, mainstem channel length, relief-length ratios, average slope, and others have yielded good correlations, accounting for 90% or more of the variation in sediment yield in some basins (Renfro, 1975). However, extrapolations to other watersheds that differ from those used to establish the correlation may be risky. Furthermore, few of the correlation models include consideration of bedload transport. Most have been developed in largely agricultural regions, although Anderson (1975) and his co-workers have done many studies in wildlands of the Pacific Coast states. Unfortunately, many correlation models fail to provide information identifying the key processes or locations where control efforts would be effective in reducing erosion or stabilizing sediment.

LITERATURE CITED

Ackermann, W. C. 1976. Soil and water conservation. *EOS Trans. Am. Geophys. Union* 57:708–711.

Anderson, H. W. 1975. Relative contribution of sediment from source areas and transport processes. pp. 66–82. In: *Present and prospective technology for predicting sediment yields and sources*. ARS-S-40, USDA Agric. Res. Service, Sedimentation Lab, Oxford, MS.

Anderson, H. W., G. B. Coleman, and P. J. Zinke. 1959. Summer slides and winter scour . . . dry-wet erosion in southern California mountains. USDA Forest Service Tech. Pap. 36. Pac. Southwest For. Range Expt. Sta., Berkeley, CA.

Barker, R. J. 1981. Soil survey of Latah County area, Idaho. USDA Soil Cons. Service in coop. with University of Idaho, College of Agr. and Idaho Soil Conserv. Comm., G.P.O., Wash., DC.

Bell, J. R. 1963. Mechanics of mass soil movement. pp. 141–162. In: *Symposium of forest watershed management*. Oregon State University, Corvallis, OR.

Benoit, C. 1973. *Forest hydrology. Pt. II. Hydrologic effects of vegetation manipulation*. USDA Forest Service, Northern Region, Missoula, MT.

Beschta, R. L. 1979a. Debris removal and its effects on sedimentation in an Oregon Coast Range stream. *Northwest Sci.* 53:71–77.

Beschta, R. L. 1979b. The suspended sediment regime of an Oregon Coast Range stream. *Water Resour. Bull.* 15:144–154.

Beschta, R. L. 1981a. Increased bag size improves Helley-Smith bedload sampler for use in streams with high sand and organic matter transport. pp. 17–25. In: *Proc. of the erosion and sediment transport measurement symposium* I.A.H.S. Pub. No. 133.

Beschta, R. L. 1981b. Management implications of sediment routing research. pp. 18–23. In: *National Council of the Paper Industry for Air and Stream Improvement Tech. Bull.* 353, Portland, OR.

Beschta, R. L. 1981c. Patterns of sediment and organic matter transport in Oregon Coast Range streams. pp. 179–188. In: *Erosion and sediment transport in Pacific Rim steeplands.* I.A.H.S. Pub. No. 132.

Beschta, R. L. 1983. Sediment and organic matter transport in mountain streams of the Pacific Northwest. pp. 1–69 to 1–89. In: *Proc. of the D. B. Simons symposium on erosion and sedimentation.* Simons, Li & Assoc., Ft. Collins, CO.

Bilby, R. E. 1984. Removal of woody debris may affect stream channel stability. *J. For.* 82:609–612.

Bogolyuba, I. V. and A. V. Karanshev. 1974. Water erosion and sediment discharge. *Trans. State Hydrologic Inst. (Trudy GGI)* 210:5–21 (seen in: *Sov. Hydrol.: Selected Papers* No. 3:143–154).

Brown, G. W. 1985. Forestry and water quality. 2nd ed. OSU Bookstore, Corvallis, Or.

Burns, R. G. and J. D. Hewlett. 1983. A decision model to predict sediment yield from forest practices. *Water Resour. Bull.* 19:9–14.

Burroughs, E. R., Jr. 1985. Survey of slope stability problems on forest lands in the west. pp. 5–16. In D. Swanston, Ed., *Proc. Workshop on Slope Stability. Problems and Solutions in Forest Management*. USDA For. Serv. Gen. Tech. Rep. PNW-180. Pac. Northwest Forest and Range Exp. Sta., Portland, OR.

Burroughs, E. R., Jr., G. R. Chalfant, and M. A. Townsend. 1976. *Slope stability in road construction.* Bureau of Land Management, Oregon State Office, Portland, OR.

Burroughs, E. R., Jr., and B. R. Thomas. 1977. Declining root strength in Douglas-fir after felling as a factor in slope stability. USDA Forest Service Res. Pap. INT-190. Intermountain For. Range Expt. Sta., Ogden, UT.

Campbell, R. H. 1975. Soil slips, debris flows, and rainstorms in the Santa Monica Mountains and vicinity, southern California. Geol. Surv. Prof Pap. 851, GPO, Washington, DC.

Carter, L. D. and J. P. Galloway. 1981. Earth flows along Henry Creek, northern Alaska. *Arctic* 34:325–328.

Clayton, J. L. 1981. Magnitude and implications of dissolved and sediment elemental transport from small watersheds. pp. 83–98. In: D. M. Baumgartner, Ed. *Interior west watershed management.* Coop. Ext., Washington State University, Pullman, WA.

Clayton, J. L. and J. F. Arnold. 1972. Practical grain size fracturing density and weathering classification of intrusive rocks of the Idaho Batholith. USDA Forest Service Gen. Tech. Rept. INT-2. Intermountain For. Range Expt. Sta., Ogden, UT.

Clayton, J. L., W. F. Megahan, and D. Hampton. 1979. Soil and bedrock properties: weathering and alteration products in the Idaho Batholith. USDA Forest Service Res. Pap. INT-237. Intermountain For. Range Expt. Sta., Ogden, UT.

Crawford, N. H., and R. K. Linsley. 1966. *Digital Simulation in Hydrology. Stanford Watershed Model IV.* Dept of Civil Engineering, Stanford, Univ. Tech. Rept. No. 39.

Curtis, N. M., Jr., A. G. Darrach, and W. J. Sauerwein. 1977. Estimating sheet-rill erosion and sediment yield on disturbed western forest and woodlands. Tech. Notes, Woodland, No. 10. USDA Soil Cons. Service, West Tech. Service Ctr., Portland, OR.

Day, N., W. Megahan, and D. Hattersley. 1977. Landslide occurrence in the Clearwater National forest 1974–1976. USDA Forest Service, Clearwater National Forest, Orofino, ID (mimeo).

DeGraff, J. V. and E. P. Olson. 1980. Landslide monitoring techniques for wildland management. *J. Soil Water Conserv.* 35:241–242.

Dissmeyer, G. E. and G. R. Foster. 1980. A guide for predicting sheet and rill erosion on forest land. Tech. Pub. SA-TP 11. USDA Forest Service, State and Pvt. Forestry, Southeastern Area, Atlanta, GA.

Dunne, T. and L. B. Leopold. 1978. *Water in environmental planning.* W. H. Freeman, San Francisco, CA.

Durgin, P. B. 1977. Landslides and the weathering of granitic rock. pp. 127–131. In: *Reviews in Engineering Geology,* Vol. III, Geological Soc. Am.

Dyrness, C. T. 1967a. Erodibility and erosion potential of forest watersheds. pp. 599–611. In: W. Sopper and H. Lull, Eds. *International Symposium on Forest Hydrology,* Pergamon Press, Oxford.

Dyrness, C. T. 1967b. Mass soil movements in the H. J. Andrews Experimental Forest. USDA Forest Service Res. Pap. PNW-42. Pac. Northwest For. Range Expt. Sta., Portland, OR.

Endo, T. 1969. Probable distribution of the amount of rainfall causing landslides. pp. 122–136. *1968 Annual Rept., Hokkaido Branch, Tokyo Forest Expt. Sta., Tokyo* (translated by J. M. Arata and R. R. Zeimer, Pac. Southwest For. Range Expt. Sta., Arcata, CA).

Endo, T. and T. Tsuruta. 1969. A report in regard to the reinforcement action of vegetational roots upon the tensile strength of the natural soil. pp. 183–189. *1968 Annual Rept., Hokkaido Branch, Forest Expt. Sta.,* Sapporo (translated by J. M. Arata and R. R. Ziemer, Pac. Southwest For. Range Expt. Sta., Arcata, CA).

Eschner, A. R. and J. H. Patric. 1982. Debris avalanches in eastern upland forests. *J. For.* 80:343–346.

Foster, G. R., D. K. McCool, K. G. Renard, and W. C. Moldenhauer. 1981. Conversion of the universal soil loss equation to SI metric units. *J. Soil Water Conserv.* 36:355–359.

Galay, V. J. 1983. Causes of river bed degradation. *Water Resour. Res.* 19:1057–1090.

Garwood, N. C., d. P. Janos, and N. Brokaw. 1979. Earthquake-caused landslides: a major disturbance to tropical forests. *Science* 205:997–998.

Gleissner, M. J. 1914. The relation of the surface cover and ground litter in a forest to erosion. *For. Quart.* 12:37–40.

Glymph, L. M. 1975. Evolving emphases in sediment yield predictions. pp. 1–4. In: *Present and prospective technology for predicting sediment yields and sources.* ARS-S-40, USDA Agric. Res. Service Sedimentation Lab, Oxford, MS.

Gray, D. H. and W. F. Megahan. 1981. Forest vegetation removal and slope stability in the Idaho Batholith. USDA Forest Service Res. Pap. INT-271. Intermountain For. Range Expt. Sta., Ogden, UT.

Haigh, M. I. and W. L. Wallace. 1982. Erosion of strip-mine dumps in LaSalle County, Illinois: preliminary results. *Earth Surface Processes and Landforms* 7:79–84.

Hart, G. E. 1984. Erosion from simulated rainfall on mountain rangeland in Utah. *J. Soil Water Conserv.* 39:330–334.

Hart, G. E. and S. A. Loomis. 1982. Erosion from mountain snowpack melt. *J. Soil Water Conserv.* 37:55–57.

Harvey, M. D., C. G. Watson, and S. A. Schumm. 1985. Gully erosion. USDI Bureau of Land Mgmt. Tech. Note 366. US Govt. Printing Office, Washington, DC.

Heede, B. H. 1971. Characteristics and processes of soil piping in gullies. USDA Forest Service Res. Pap. RM-68. Rocky Mt. For. Range Expt. Sta., Ft. Collins, CO.

Heede, B. H. 1972. Flow and channel characteristics of two high mountain streams. USDA Forest Service Res. Pap. RM-96. Rocky Mt. For. Range Expt. Sta., Ft. Collins, CO.

Heede, B. H. 1975a. Stages of development of gullies in the West. pp. 155–161. In: *Present and prospective technology for predicting sediment yields and sources.* ARS-S-40. USDA Agric. Res. Service Sedimentation Lab, Oxford, MS.

Heede, B. H. 1975b. Mountain watersheds and dynamic equilibrium. pp. 407–420. In: *Proc. of the watershed management symposium, Logan, UT,* J. Irrig. and Drain. Div., ASCE.

Heede, B. H. 1976. Equilibrium condition and sediment transport in an ephemeral mountain stream. *Hydrology and water resources in Arizona and the Southwest* 6:97–102.

Heede, B. H. 1977. Influence of forest density on bedload movement in a small mountain stream. *Hydrology and water resources in Arizona and the Southwest* 7:103–107.

Heede, B. H. 1980. Stream dynamics: an overview for land managers. USDA Forest Service Gen. Tech. Rept. RM-72. Rocky Mt. For. Range Expt. Sta., Ft. Collins, CO.

Heede, B. H. 1981. Dynamics of selected mountain streams in the western United States of America. *Z. Geomorph.* 25:17–32.

Hershfield, D. M. 1974. The frequency of freeze–thaw cycles. *J. App. Meteorol.* 13:348–354.

Hershfield, D. M. 1979. Freeze–thaw cycles, potholes, and the winter of 1977–78. *J. App. Meteorol.* 18:1003–1007.

Higashi, A. and A. E. Corte. 1971. Solifluction: a model experiment. *Science* 171:480–482.

Holeman, J. N. 1975. Procedures used in the Soil Conservation Service to estimate sediment yield. pp. 5–9. In: *Present and prospective technology for predicting sediment yield and sources.* ARS-S-40. USDA Agric. Res. Service Sedimentation Lab, Oxford, MS.

Hutchinson, J. H. 1968. Mass movement. pp. 688–695. In: R. W. Fairbridge, Ed. *The encyclopedia of geomorphology.* Dowden Hutchinson & Ross, Stroudsburg, PA.

Ichikawa, M. 1952. A study of the landslides and their factors in the upper drainage area of the Watarase River. *Geographical Critique of Geophysical Institute,* Tokyo Burrika University 25:495–504 (translated by J. M. Arata and R. R. Ziemer, Pac. Southwest For. Range Expt. Sta., Arcata, CA).

Istok, J. D. and M. E. Harward. 1982. Clay mineralogy in relation to landscape in the Coast Range of Oregon. *Soil Sci. Soc. Am. J.* 46(6):1326–1331.

Jeppson, R. W., R. W. Hill, and C. E. Israelsen. 1974. Slope stability of overburden spoil dumps from surface phosphate mines in southeastern Idaho. PRWG 140-1. Utah Water Res. Lab., Utah State Univ. in coop. with USDA Forest Service, Intermountain For. Range Expt Sta., Ogden, UT.

Keller, E. A. and F. J. Swanson. 1979. Effects of large organic material on channel form and fluvial processes. *Earth Surface Processes and Landforms* 4:361–380.

Kitamura, Y. and S. Namba. 1966. A field experiment on the uprooting resistance of tree roots. *Proc. 77th Meeting of the Japanese Forestry Society.* pp. 568–570 (translated by J. M. Arcata and R. R. Ziemer, Pac. Southwest For. Range Expt. Sta., Arcata, CA).

Kitamura, Y. and S. Namba. 1968. A field experiment on the uprooting resistance of tree roots. *Proc. 79th Meeting of the Japanese Forestry Society.* pp. 360–361 (translated by J. M. Arcata and R. R. Ziemer, Pac. Southwest For. Range Expt. Sta., Arcata, CA).

Kittredge, J. 1948. *Forest influences.* McGraw-Hill Book Co., Inc., New York.

Klock, G. O. 1972. Snowmelt temperature influence on infiltration and soil water retention. *J. Soil Water Conserv.* 27:12–14.

Kohno, Y., S. Namba, K. Takiguchi, Y. Kitamura, T. Kurobori, K. Arimitsu, K. Miyagawa, and C. Kobayashi. 1968. Roles of topography, soil and forest in the landslides of a weathered granite area. *Rept. Coop. Res. on Disaster Prevention.* No. 14. Natl. Res. Ctr. for Disaster Prevention, Science and Tech. Agency, Tokyo (translated by J. M. Arcata and R. R. Ziemer, Pac. Southwest For. Range Expt. Sta., Arcata, CA).

Laffan, M. D. 1979. Slope stability in the Charleston-Punakaiki region, New Zealand. *New Zealand J. Sci.* 22:183–192.

Lambert, J. D. H. 1972. Plant succession on tundra mudflows: preliminary observations. *Arctic* 25:99–106.

Lanyon, L. E. and G. F. Hall. 1983. Land-surface morphology: 2. predicting potential landscape instability in eastern Ohio. *Soil Sci.* 136:382–386.

Lehre, A. K. 1982. Sediment budget of a small Coast Range drainage basin in north-central California. pp. 67–77. In: Swanson et al., Eds. *Sediment budgets and routing in forest drainage basins.* USDA Forest Service Gen. Tech. Rept. PNW-141. Pac. Northwest For. Range Expt. Sta., Portland, OR.

Leopold, L. B. and T. Maddock, Jr. 1953. The hydraulic geometry of stream channels and some physiographic implications. *Geol. Survey Prof. Paper* No. 252.

Lisle, T. E. 1981. The recovery of aggraded stream channels at gauging stations in northern California and southern Oregon. pp. 189–211. In: *Erosion and sediment transport in Pacific Rim steeplands.* I.A.H.S. Pub. 132.

Livesey, R. H. 1975. Corps of Engineers methods for predicting sediment yields. pp. 16–32. In: *Present and prospective technology for predicting sediment yield and sources.* ARS-S-40. USDA Agr. Res. Service Sedimentation Lab, Oxford, MS.

Lyons, J. K. and R. L. Beschta. 1983. Land use, floods, and channel changes: Upper Middle Fork Willamette River, Oregon (1936–1980). *Water Resour. Res.* 19(2):463–471.

Marston, R. A. 1982. The geomorphic significance of log steps in forest streams. *Ann. Assoc. Am. Geograph.* 72:99–108.

Meade, R. H. 1982. Sources, sinks, and storage of river sediment in the Atlantic Drainage of the United States. *J. Geol.* 90:235–252.

Megahan, W. F. 1974. Erosion over time on severely disturbed granitic soils: a model. USDA Forest Service Res. Pap. INT-156. Intermountain For. Range Expt. Sta., Ogden, UT.

Megahan, W. F. and W. S. Platts. 1980. Riverbed improves over time: South Fork Salmon. pp. 380–395. In: *Symposium on watershed management,* Vol. I, ASCE, New York.

Meyer, L. D. 1984. Evolution of the universal soil loss equation. *J. Soil Water Conserv.* 39:99–104.

Moore, R. J. 1984. A dynamic model of basin sediment yield. *Water Resour. Res.* 20:89–103.

Mosley, M. P. 1981. The influence of organic debris on channel morphology and bedload transport in a New Zealand forest stream. *Earth Surface Processes Landforms* 6:571–579.

Mosley, M. P. 1982. The effect of a New Zealand beech forest canopy on the kinetic energy of water drops and on surface erosion. *Earth Surface Processes Landforms* 7:103–107.

O'Leary, S. J. and R. L. Beschta. 1981. Bedload transport in an Oregon Coast Range stream. *Water Resour. Bull.* 17:886–894.

O'Loughlin, C. and A. Watson. 1979. Root-wood strength deterioration in radiata pine after clearfelling. *New Zealand J. For. Sci.* 9:284–293.

Oyagi, N. 1968. Weathering-zone structure and landslides of the area of granitic rocks in Karno-Daito, Shimane Prefercture. *Rept. Coop. Res. on Disaster Prevention* No. 14. Natl. Res. Ctr. for Disaster Prevention, Science and Tech. Agency, Tokyo (translated by J. M. Arata and P. B. Durgin, Pac. Southwest For. Range Expt. Sta., Arcata, CA).

Paeth, R. C., M. E. Howard, E. G. Knox, and C. T. Dyrness. 1971. Factors affecting mass movement of four soils in the western Cascades of Oregon. *Soil Sci. Soc. Am. Proc.* 35:943–947.

Patric, J. H. 1976. Soil erosion in the eastern forest. *J. For.* 74(10):671–677.

Patton, P. C. and S. A. Schumm. 1981. Ephemeral-stream processes: implications for studies of quaternary valley fills. *Quat. Res.* 15:24–43.

Pearce, A. J. and A. Watson. 1983. Medium-term effects of two landsliding episodes on channel storage of sediment. *Earth Surface Processes Landforms* 8:29–39.

Radbruch-Hall, D. H., R. B. Colton, W. E. Davies, B. A. Skipp, I. Lucchitta, and D. J. Varnes. 1981. *Landslide overview map of the conterminous United States.* Geol. Survey Prof. Pap. 1183. GPO, Washington, DC.

Reinhart, K. G., A. R. Eschner, and G. R. Trimble, Jr. 1963. Effect on stream-flow of four forest practices in the mountains of West Virginia. USDA Forest Service Res. Pap. NE-1. Northeastern For. Expt. Sta., Upper Darby, PA.

Renfro, G. W. 1975. Use of erosion equations and sediment-delivery ratios for predicting sediment yield. pp. 33–45. In: *Present and prospective technology for predicting sediment yields and sources.* ARS-S-40. USDA Agric. Res. Service Sedimentation Lab, Oxford, MS.

Rice, R. M., E. S. Corbett, and R. G. Bailey. 1969. Soil slips related to vegetation, topography and soil in southern California. *Water Resour. Res.* 5:647–659.

Rice, R. M. and G. T. Foggin, III. 1971. Effect of high intensity storms on soil slippage on mountainous watersheds in southern California. *Water Resour. Res.* 7:1485–1496.

Ritter, D. F., W. F. Kinsey, III, and M. E. Kauffman. 1973. Overbank sedimentation in the Delaware River Valley during the last 6,000 years. *Science* 179:374–375.

Robinson, A. R. 1977. Relationship between soil erosion and sediment delivery. pp. 159–167. In: *Erosion and solid matter transport in inland waters symposium.* I.A.H.S. Pub. No. 122.

Schroeder, W. L. and J. V. Alto. 1983. Soil properties for slope stability analysis: Oregon and Washington coastal mountains. *For. Sci.* 29:823–833.

Schumm, S. A., M. P. Mosely, and G. L. Zimpfer. 1976. Unsteady state denudation. *Science* 191:871.

Searcy, J. K. 1965. Design of roadside drainage channels. Bur. Public Roads, US Dept. Commerce. Hydraulic Design Series No. 4, Washington, DC.

Sharpe, C. F. S. 1938. *Landslides and related phenomena: a study of mass-movements of soil and rock.* Columbia University Press, New York.

Sheng, T. C. 1966. *Landslide classification and studies in Taiwan.* Chinese-American Joint Commission on Reconstruction, Taipei, Forestry Series, No. 10.

Sidle, R. C., A. J. Pearce, and C. L. O'Loughlin. 1985. *Hillslope stability and land use.* Water Res. Monograph II. Am. Geophys. Union, Wash. DC.

Sidle, R. C. and D. N. Swanston. 1982. Analysis of a small debris slide in coastal Alaska. *Can. Geotech. J.* 19:167–174.

Sturgis, D. L. 1975. Oversnow runoff events affect streamflow and water quality. pp. 105–117. In: *Proc. Snow Management on the Great Plains Symposium.* Great Plains Agr. Council. Pub. 73. Agric. Expt. Sta., University of Nebraska, Lincoln, NE.

Swanson, F. J. and C. T. Dyrness. 1975. Impact of clear-cutting and road construction on soil erosion by landslides in the western Cascade Range, Oregon. *Geology* 3:393–396.

Swanson, F. J., R. L. Fredricksen, and F. M. McCorison. 1982a. Material transfer in a western Oregon forested watershed. pp. 233–266. In: R. L. Edmonds, Ed. *Analysis of coniferous forest ecosystems in the western United States.* Hutchinson Ross Pub. Co., Stroudsburg, PA.

Swanson, F. J. and M. E. James. 1975. Geology and geomorphology of the H. J. Andrews Experimental Forest, western Cascades, Oregon. USDA Forest Service Res. Pap. PNW-188. Pac. Northwest For. Range Expt. Sta., Portland, OR.

Swanson, F. J., G. W. Lienkaemper, and J. R. Sedell. 1976. History, physical effects, and management implications of large organic debris in western Oregon streams. USDA Forest Service Gen. Tech. Rept. PNW-56. Pac. Northwest For. Range Expt. Sta., Portland, OR.

Swanson, F. J. and D. N. Swanston. 1977. Complex mass-movement terrains in the western Cascade Range, Oregon. pp. 113–124. In: *Reviews in Engineering Geology,* Vol. III. Geological Soc. Am.

Swanson, F. R., R. J. Janda, T. Dunne, and D. N. Swanston, tech Eds. 1982b. *Sediment budgets and routing in forested drainage basins.* USDA Forest Service Gen. Tech. Rept. PNW-141. Pac. Northwest For. Range Expt. Sta., Portland, OR.

Swanston, D. N. 1967a. Debris avalanching in thin soils derived from bedrock. USDA Forest Service Res. Note PNW-64. Pac. Northwest For. Range Expt. Sta., Portland, OR.

Swanston, D. N. 1967b. Soil-water piezometry in a southeast Alaska landslide area. USDA Forest Service Res. Note PNW-68. Pac. Northwest For. Range Expt. Sta., Portland, OR.

Swanston, D. N. 1969. Mass wasting in coastal Alaska. USDA Forest Service Res. Pap. PNW-83. Pac. Northwest For. Range Expt. Sta., Institute of Northern Forestry, Juneau, AK.

Swanston, D. N. 1970. Mechanisms of debris avalanching in shallow till soils of southeast Alaska. USDA Forest Service Res. Note PNW-103. Pac. Northwest For. Range Expt. Sta., Portland, OR.

Swanston, D. N. 1974. *The forest ecosystem of southeast Alaska. 5. Soil mass movement.* USDA Forest Service Gen. Tech. Rept. PNW-17. Pac. Northwest For. Range Expt. Sta., Portland, OR.

Swanston, D. N. 1978. Effect of geology on soil mass movement in the Pacific Northwest. pp. 89–115. In: C. T. Youngberg, Ed. *Forest soils and land use.* Proc. Fifth N. Am. For. Soils Conf. Colorado State University, Fort Collins, CO.

Swanston, D. N. 1981. Watershed stability based on soil stability criteria. pp. 43–58. In: D. M. Baumgartner, Ed. *Interior west watershed management.* Coop. Ext., Washington State University, Pullman, WA.

Swanston, D. N. and C. T. Dyrness. 1973. Stability of steep land. *J. For.* 71:264–273.

Swift, L. W., Jr. 1984. Soil losses from roadbeds and cut and fill slopes in the southern Appalachian Mountains. *South. J. Appl. For.* 8:209–215.

Taskey, R. D., M. E. Harward, and C. T. Youngberg. 1978. Relationship of clay mineralogy to landscape stability. pp. 140–164. In: C. T. Youngberg, Ed. *Forest soils and land use.* Proc. Fifth N. Am. For. Soils Conf. Colorado State University, Fort Collins, CO.

Terzaghi, K. 1960. Mechanism of landslides. pp. 83–123. In: *Application of Geology to Engineering Practice.* Geol. Soc. Am. New York.

Touysinhthiphonexay, K. C. and T. W. Gardner. 1984. Threshold response of small streams to surface coal mining, bituminous coal fields, central Pennsylvania. *Earth Surface Processes Landforms* 9:43–58.

Trimble, S. W. 1975. Denudation studies: can we assume a steady state? *Science* 188:1207–1208.

Trimble, S. W. 1976. Unsteady state denudation. *Science* 191:871.

Trimble, S. W. 1981. Changes in sediment storage in the Coon Creek Basin, Driftless Area, Wisconsin, 1853–1975. *Science* 214:181–183.

Trott, K. E. and M. J. Singer. 1983. Relative erodibility of 20 California range and forest soils. *Soil Sci. Soc. Am. J.* 47:753–759.

USDA Agric. Res. Service. 1975. Present and prospective technology for predicting sediment yield. ARS-S-40. USDA Sedimentation Lab, Oxford, MS.

VanSickle, J. and R. L. Beschta. 1983. Supply-based models of suspended sediment transport in streams. *Water Resour. Res.* 19(3):768–778.

Varnes, D. J. 1958. Landslide types and processes. pp. 20–47. In: E. Eckel, Ed. *Landslides in engineering practice.* Highway Research Board. Spec. Rept. 29. Natl. Acad. Sci. Natl. Research Council. Pub. 544.

Waldron, L. J. and S. Dakessian. 1982. Effect of grass, legume, and tree roots on soil shearing resistance. *Soil Sci. Soc. Am. J.* 46:894–899.

Waldron, L. J., S. Dakessian, and J. A. Nemson. 1983. Shear resistance enhancement of 1.22-meter diameter soil cross sections by pine and alfalfa roots. *Soil Sci. Soc. Am. J.* 47:9–14.

Walling, D. E. 1983. The sediment delivery problem. *J. Hydrol.* 65:209–237.

Wilkin, D. C. and S. J. Hebel. 1982. Erosion, redeposition and delivery of sediment to Midwestern streams. *Water Resour. Res.* 18:1278–1282.

Williams, J. R. 1975. Sediment-yield prediction with universal equation using runoff energy factor. pp. 244–252. In: *Present and prospective technology for predicting sediment yield and sources.* ARS-S-40, USDA Agr. Res. Service Sedimentation Lab, Oxford, MS.

Wischmeier, W. H. 1975. Estimating the soil loss equation's cover and management factor for undisturbed areas. pp. 118–124. In: *Present and prospective technology for predicting sediment yield and sources.* ARS-S-40, USDA Agr. Res. Service Sedimentation Lab, Oxford, MS.

Wischmeier, W. H. and D. D. Smith. 1965. *Predicting rainfall-erosion losses from cropland east of the Rocky Mountains.* Agr. Handbook 287. USDA, Washington, DC.

Wischmeier, W. H. and D. D. Smith. 1978. *Predicting rainfall-erosion losses—a guide to conservation planning.* Agr. Handbook 537. USDA, Washington, DC.

Wu, T. H., W. P. McKinnell, III, and D. N. Swanston. 1979. Landslides on Prince of Wales Island, Alaska. *Can. Geotech. J.* 16:19–33.

Wu, T. H. and D. N. Swanston. 1980. Risk of landslides in shallow soils and its relations to clearcutting in southeastern Alaska. *For. Sci.* 26:495–510.

Yee, C. S. and R. D. Harr. 1977. Influence on soil aggregation on slope stability in the Oregon Coast Ranges. *Environ. Geol.* 1:367–377.

Ziemer, R. R. 1981. Roots and the stability of forested slopes. pp. 343–361. In: *Erosion and sediment transport in Pacific Rim steeplands.* I.A.H.S. Pub. No. 132.

Ziemer, R. R. and D. N. Swanston. 1977. Root strength changes after logging in southeast Alaska. USDA Forest Service Res. Note PNW-306. Pac. Northwest For. Range Expt. Sta., Portland, OR.

Zuzel, J. F., R. R. Allmaras, and R. Greenwalt. 1982. Runoff and erosion on frozen soils in northeastern Oregon. *J. Soil Water Conserv.* 37:351–354.

PART III
Managing Wildland Watersheds

INTRODUCTION

Watershed management is only one factor among many that determine the nature of water and aquatic resources on any watershed. Thus, difficulties in answering simple questions such as, "What happens to streamflow when a forest is cut?" are much greater than most people imagine. Forests themselves are highly variable, as are the environments in which they grow. Even in any given forest, no two years are likely to have the same weather; and extremes of any one year may not be repeated for decades or more. In view of all the uncontrolled and often unpredictable factors that influence streamflow, such a simple question can only be answered in terms of probability of response.

To reduce the range of variation and thus increase the predictability of response, two basic approaches are used: (a) researchers attempt to control variation by several means; and (b) they seek to understand each hydrologic process in detail and, if possible, derive the underlying laws that control them.

To study the effects of management practices, the paired (control) watershed technique has been widely used to reduce variation. In this approach, two or more similar watersheds in proximity are selected for study. Streamflow characteristics from each are then measured until a statistical relationship is established whereby the flow of the one(s) to be treated can be predicted within suitable limits from measurements of flow from the control watershed. Then, the control watershed is left undisturbed and treatments representing one or more management practices are applied to the others. If measured streamflow from the treated watersheds changes in amounts exceeding previously established limits from that predicted based on flow from the control watershed, then the difference is assumed to be a result of the treatment applied. This general approach also has been used to evaluate the effects of management on water quality and aquatic resources. Variation is minimized by selecting study watersheds as similar as possible in every respect except treatment. The proximity of treated and control watersheds helps insure that weather conditions during the study will be similar.

Control watershed studies are extremely expensive and require many years to generate useful information. Fewer than 150 have been reported as this is written. The cost and duration of watershed studies may limit their use in the future, and even data collection from existing study areas may be reduced.

Paired watersheds provide excellent information on stream response to treatment, but they have limitations. By themselves, such studies seldom provide a detailed understanding of the mechanism of response to treatment. Furthermore, the results cannot be extrapolated with confidence to watersheds that differ in substantial degree from the study watersheds. Therefore, studies to elucidate the underlying hydrologic processes are usually conducted separately or in conjunction with control watershed studies.

Hydrologic processes are studied in the laboratory, as well as the field. Ultimately, however, all must be validated by field studies. No attempt will be made to describe the various means of study, as they are too numerous to cover here. However, a major problem that arises from studying separate processes is that of integrating each into a holistic hydrologic system and correctly assessing interactions that arise among individual elements of the system. Hydrologists have made considerable progress through the use of systems analysis and many hydrologic models have been developed. Few have been designed specifically to take wildland management effects into account, and fewer still have been validated over any wide range of conditions.

Nevertheless, research has provided a sound basis for developing watershed management techniques to maintain or improve water resources on wildlands. In the following chapters we shall examine that potential.

11 Control of Amount and Timing of Streamflow

The water yield characteristics of quantity and regimen are inextricably interrelated. Both are expressed completely by the hydrograph of any stream. Any management activity that modifies either modifies the other. Therefore, they will be considered together in this treatment of management to control water yields.

Problems of water quantity are chronic throughout many subhumid to arid regions of the world and occur irregularly during times of extended drought even in humid regions. Such problems are often compounded by irregular streamflow regimen that results in waste, damage, and destruction. Floods may be followed by periods of low flow that are incapable of meeting withdrawal demands or instream needs such as providing suitable habitat for fish (Wesche, 1974; Reiser and Bjornn, 1979).

In the long run, there are only two major environmental constraints determining the total quantity of runoff: (a) the amount of precipitation and (b) the amount of evaporative water losses. Therefore, to increase total yields, precipitation must be increased, evapotranspiration reduced, or both. Most wildland watershed management practices to increase water yields have been based on measures designed to reduce evapotranspiration losses.

Where streamflow regimen is the problem, several watershed management alternatives exist: (a) modifying the route of water followed from its place of impact on the watershed until it reaches the stream channel, (b) modifying the distribution and timing of melt of snow, and (c) modifying the distribution and amount of evapotranspiration losses. Together, these modifications may change both the amount and timing of water yields.

Land management activities that modify a watershed offer incomplete control of streamflow at best, and all are accompanied by at least some risk of damage, as well as potential for improvement. Damage may result from improper application due to ignorance or lack of care, but risk of damage is not eliminated even by the best of application. The adverse effects of severe storms may be increased if they occur at the wrong time; for example, while a normally beneficial treatment, such as a reseeding operation, is underway on a depleted range.

The potential for modifying the amount and timing of water yields through modification of the watershed is highly variable from one watershed to another and even within different parts of the same watershed. Not all watersheds or portions of watersheds are equal; rather, they frequently respond differently to similar practices, depending on their climatic, topographic, soil, and other characteristics. Yield changes are usually greatest the first few years following treatment and decline more or less rapidly as the watershed reverts to its previous condition.

In this chapter, we shall examine the results of watershed management practices that have been applied on various sites to determine management possibilities for controlling the amount and timing of streamflow. Each of the major water yield zones will be examined for their potential, from high elevation alpine regions down to desert phreatophytes.

11.1 THE ALPINE ZONE

Most alpine regions are areas of relatively high precipitation in the form of snow that often persists well into the summer. The growing season is short and cool, and soils are frequently thin and poorly developed. The zone supports a fragile, low-growing vegetation cover consisting mostly of sedges, grasses, and low herbs with some low shrubs and dwarfed trees (*krummholz*) on protected sites. Land is used primarily for grazing and recreation. Individual areas may be small and widely scattered, and many are incorporated into parks, wilderness, and primitive areas.

Alpine areas are important water source areas in the western United States and Canada, particularly in the drier regions. For example, in Utah 10% of the area at the highest elevations yields 60% of the state's runoff (Bagley et al., 1964). Furthermore, much of the streamflow from snowmelt is generated during the dry summer months. Martinelli (1965) showed that alpine snowfields in several watersheds along the Front Range of the Rocky Mountains of Colorado contributed from 59 to 95% of total runoff during July and August, 1956. In the north Cascades of Washington, a minimum discharge hydrograph of the Nooksack River for the years 1947–1959 revealed that the lowest rates of flow from June through September were greater than the minimum flows of any other month (Fig. 11.1). The river heads at East Nooksack Glacier near the base of Mt. Shuksan and is fed by tributaries that originate in snowfields and small glaciers. The snowfields normally provide a major part of streamflow through July, while glacier melt occurs somewhat later after most of the snow cover has disappeared, and continues into the fall (Garling, 1960).

Alpine areas seldom contribute importantly to snowmelt floods (Anderson, 1966). The high elevation and cold climate causes melt to lag behind that of the more extensive lower elevations. However, exceptions do occur, as the spring floods following the exceptionally heavy snowfall of 1983 in Utah clearly demonstrated.

Alpine lands have also been the source of summer flash floods. Many of the small communities located at the mouths of steep canyons along the Wasatch

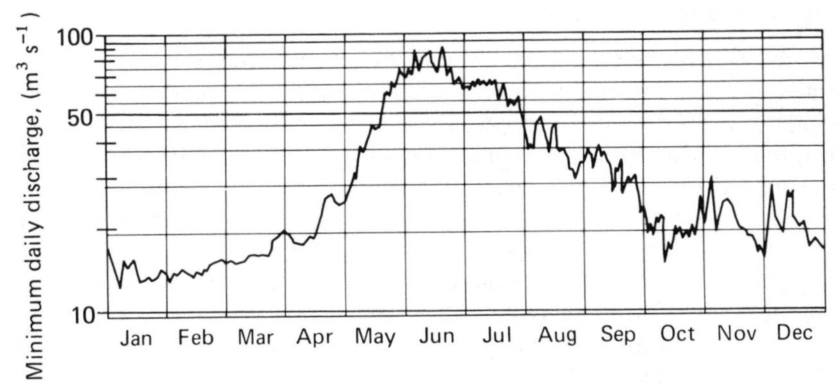

FIG. 11.1 Minimum discharge hydrograph for the years 1947–1959. Nooksack River below Cascade Creek, Washington (adapted from Garling, 1960).

Front in Utah suffered from disastrous flooding following intense, brief thunderstorms on steep alpine slopes. Fragile vegetation depleted by fires and overgrazing may have compounded the problem. Farmington, Utah, was subjected to several such floods between 1923 and 1936, and one flood resulted when less than 13 mm of rain fell on the upper watershed. Nevertheless, six lives were lost and property damage was extensive (Bailey et al., 1947). The same communities depended on irrigation agriculture to support their economies, and water was in short supply nearly every summer.

The nature of alpine regions and their problems suggest that management opportunities to control streamflow quantity and regimen are limited. The areas are mostly inaccessible, they are often small and scattered, and many are incorporated into parks and wilderness areas. The harsh environment and fragile vegetation restrict management options and make any intensive use risky. However, critical needs for summer water supplies and protection from floods make it necessary to consider management of some of these lands. Precipitation enhancement, snow management, and surface runoff control offer the primary potential for water yield improvement.

11.1.1 Precipitation Enhancement

Cloud seeding, discussed in Chapter 2, is most effective in winter orographic storms. It should be most effective in alpine regions where most of the precipitation occurs as snow. Furthermore, natural losses of vapor are low and should be satisfied by natural precipitation, so any increase should be almost completely converted to streamflow. Nevertheless, cloud seeding is limited by many constraints, including restrictions on use in wilderness areas.

11.1.2 Alpine Snow Management

Summer snowpacks are almost immune to net evaporation. They frequently gain as much moisture during periods of condensation as they lose during periods of evaporation (Kittredge, 1953; Martinelli, 1959). In areas bare of snow, the growing season is short and cool. Though moisture losses in bare areas are greater than from snow (Hutchison, 1966), they are limited in comparison to warmer regions by the short season and low vapor pressure gradients prevailing under cool conditions. However, strong winds are normal in the alpine environment (Judson, 1965). The distribution of snow is more the result of wind deposition than direct snowfall. During subfreezing weather, snow is blown from exposed sites and deposited in protected spots behind topographic barriers and vegetation, or lost (Lull and Orr, 1950; Berndt, 1964; Martinelli, 1975).

The potential for increasing water yield from alpine snows is strongly related to the reduction of sublimation losses of blowing snow (Tabler, 1973) and the reduction of losses from the snowpack on the ground. Increasing the amount of snow retained in alpine areas also tends to modify streamflow regimen, for large meltwater flows tend to come later in the season than smaller ones (Court, 1961).

There are basically three methods to capture blowing snow in drifts: (a) build snow fences, (b) establish shelter belts of vegetation, or (c) reshape the terrain over which snow blows. Only fencing offers practical potential for most alpine areas, but any method may be appropriate in lower elevation open snow zone lands where climate and land use permit their use. The same principles of design apply everywhere. Therefore, all will be considered together in this section.

The first consideration in capturing blowing snow in drifts is to determine the direction of winter winds. Storm winds tend to be relatively consistent in direction, but records are seldom available for any specific wildland site. Under most circumstances, this means the manager must determine wind direction by observation. A late winter or spring survey of drift orientation behind natural features will reveal the significant wind direction of the previous season (Lull and Orr, 1950). Aerial photographs provide an inexpensive means of surveying large areas. An examination of vegetation near drifts may tell whether a given drift varies from the long-term norm. Vegetation developing on sites covered by late-lying drifts commonly differs in density and species composition from that on sites where the snow disappears earlier (Ellison, 1954; Smith, 1969; Wight et al., 1975). The deformation pattern of krummholz is also a good indicator of the direction of drifting winds. Barriers to trap snow should be placed perpendicular to prevailing winds wherever possible.

Designing a system to trap drifting snow requires more than determining its orientation. Among the factors that must be considered are (a) annual snowfall, (b) the proportion of snow that is relocated, (c) the contributing distance (fetch) between natural or human-made storage features, and (d) the storage capacity of each barrier. Further considerations may include such things as specific design features of fences, plant barriers, or reshaped terrain, alone or in combination, and the economic feasibility. All of these factors interact. For example, one would expect a greater proportion of snow to be relocated in years of above normal snowfall, for once the storage capacity of natural vegetation and terrain features is reached any additional snow is subject to movement.

It is very difficult to determine snowfall in the windy, open alpine zone. Estimates based on snow accumulation may be grossly in error, as sublimation losses from blowing snow may be heavy and the trap efficiency of barriers declines drastically as they approach their maximum storage capacity (Tabler, 1975). Goodison et al. (1981) discuss snowfall measurement in detail and indicate that dual fences raised well above the surface and surrounding shielded gages will yield catches similar to those located in protected locations.

The amount of snow reaching any barrier may be estimated by a modification of Eq. 9.1 (Sturges and Tabler, 1981) as follows:

$$Q = \tfrac{1}{2}P_r R_u \,[1 - (0.14)^{R_c/R_u}] \qquad (11.1)$$

where Q is the quantity of snow reaching the barrier ($m^3\ m^{-1}$), P_r is the depth of snow relocated (m), R_u is the transport distance (m), and R_c is the fetch or contributing distance between barriers (m). The term P_r can be considered to be the annual snowfall less melt and the storage capacity of natural vegetation and terrain features in the fetch between barriers. If the natural storage can be considered a constant, then

$$P_r = Q_s - C \qquad (11.2)$$

where Q_s is the annual snowfall and C is the appropriate constant.

Once the quantity of blowing snow reaching any point is known, the capacity of the barrier needed at that point to capture the drifting snow can be estimated from barrier dimensions, porosity, topography behind the barrier and drift form. Barrier capacities can be determined for topographic catchments and fences from the work of Tabler (1974, 1975, 1980). Similar information can be derived for other features such as shelter belts or shrubs or trees of various types by scale modeling (Tabler and Jairell, 1980; Peterson and Schmidt, 1984).

Though a system of fences spaced within drift length of each other would almost prevent sublimation losses due to blowing snow, such a system normally would not be economically justified. However, if cost functions for barriers of various capacities are known, standard economic analysis can be used to determine optimum fence spacing for snow water having any given value (Tabler, 1968).

Savings of snow water from blowing snow can be estimated by solving Eq. 9.1 for natural systems and from a system with barriers. Additional savings of snow will result from concentrating snow into drifts through reduction of advective heat supplies, for drifts form precisely in those areas where air movement is reduced most. Furthermore, concentrating snow in drifts reduces the amount of surface per unit of mass. Energy exchange between the atmosphere and the snowpack is a surface phenomenon, so the smaller the surface per unit mass of snow, the less energy absorbed under any given conditions.

A number of less common techniques can be used in snow management. For example, evaporation losses from snowpacks may be further reduced by applying monolayers of long-chain fatty alcohols to the snow surface (Meiman and Slaughter, 1967; Smith and Halverson, 1971). Other substances, such as carbon black or sawdust, can be applied to snow surfaces to accelerate or retard the rate of melt (Regelin and Wallmo, 1975; Martinelli, 1975; Tarum and Meiman, 1979; Drake, 1981). Martinelli (1975) suggested intentional avalanching of snow might carry a good deal of snow over into the summer, but analysis by Martinec and de Quervain (1975) indicates that snowmelt from avalanches moving 760-m downslope would tend to be accelerated by the higher air temperatures at the lower elevation. None of these techniques are likely to find widespread application.

11.1.3 Runoff Control

Control of runoff timing in the alpine depends primarily on snowpack management, just discussed, or on controlling water movement to stream channels. During snowmelt, source areas include most of the area under and adjacent to the melting snowpack and near stream channels. Little can be done to prevent the rapid appearance of meltwater or rain as streamflow, except to reduce snowmelt rates by concentrating snow in drifts. However, as soil moisture is depleted by drainage and evapotranspiration after most of the snow melts, enough storage space may be available to hold the water needed to prevent flash floods if it can be made to enter the soil.

If flash floods are to be prevented from fragile alpine areas in regions of intense summer storms, the delicate ecological balance characteristic of the zone must not be allowed to deteriorate. In some instances, complete exclusion of certain uses, such as grazing, may be necessary to maintain adequate infiltration capacities. Elsewhere, the use must be carefully controlled. Assessment of use potential of alpine regions will require a thorough knowledge of ecology, soils, range, and recreation management, which cannot be covered in detail in this book. However, it should be emphasized that no alpine watershed manager is likely to be successful lacking training and knowledge in these fields.

A more difficult problem exists where past use has destroyed cover and led to the development of flood source lands that must be rehabilitated. Brown et al. (1978) estimated that 12% of the zone in the western United States, exclusive of Alaska, was in need of rehabilitation. Often revegetation, the ultimate hope to restore infiltration capacities and prevent surface runoff, requires supporting measures to be achieved.

In view of the typical severe growing conditions in the alpine, any measures of rehabilitation short of complete control of surface runoff are likely to be unsuccessful. Revegetation is slow and often uncertain, even with proper site preparation, and control by this means is gradual. Therefore, additional measures may be needed to insure success.

Some of the best examples of control over surface runoff are the contour trenches installed on flood source watersheds in Davis County, Utah. *Contour trenches* are ditchlike structures aligned along the contour and partitioned by cross dams at short intervals. In effect, they are a series of small reservoirs that intercept surface runoff before it has a chance to become concentrated and scour the surface to form a gully system. The water slowly seeps from the trench into the soil in the poststorm period. The size and spacing of the trenches depend on soil depth, slope, and the amount and intensity of rainfall. They must have sufficient capacity to hold all the surface runoff to be expected from the design storm.

The cross dam partitions of the trench are designed to avoid excessive concentration of water if a portion of the trench system should fail by collapse or cutting by excess water. If the trench were not partitioned, the entire catch of water would be lost when failure occurred. Worse, the trench would then serve as a collecting ditch that concentrated flow so that lower lying trenches would likely be successively overwhelmed. With partitions, failure of one part of the system does not damage the function of the rest of it.

Where trenches are built, they are usually supplemented by revegetation by seeding or planting. As the vegetation develops, it controls sediment movement and enhances infiltration capacity in the area between trenches. Less sediment and runoff are generated and by the time the trenches have lost important amounts of storage capacity, it is hoped that they will no longer be needed. The development of structures and reestablishment of vegetation must be supported by a properly designed management program for big game and livestock if the watershed is to retain its effectiveness (Croft and Bailey, 1964).

Where gullies are too deep to permit the building of contour trenches, ditches may be constructed to divert water away from the gully head. The gully itself is often reshaped and seeded, with protection by mulches and check dams where necessary.

It should be emphasized that intensive runoff control measures such as those mentioned above are very expensive. They are justified only by heavy flood damages and extreme environmental degradation that may take place in their absence. The alpine zone is one area where prevention is worth far more than most cures.

Arctic tundra is similar in many respects to alpine regions. Snow is the major type of precipitation, low growing vegetation is characteristic, ecosystems tend to be fragile, and extensive land management for water quantity or regimen control is unlikely. Techniques of snow management should be the same as elsewhere. Where rehabilitation of disturbed surfaces is necessary, certain differences must be taken into account. For example, frost heaving frequently occurs during midsummer nights in mid-latitude alpine regions, but is not likely at high latitudes. On the other hand, permafrost conditions tend to be more widespread and important in the arctic. The fragility of arctic ecosystems is proportional to the ice content of the permafrost and where ice content is low, as in large areas underlain by coarse materials, ecosystems may be surprisingly robust (Webber and Ives, 1978).

11.2 THE FOREST SNOWPACK ZONE

The forests below the alpine are the single most important water source area in western North America. Though they yield less runoff per unit area than the alpine zone, their total aggregate area is so much greater and their unit water yields sufficiently high that they provide most of the annual yield. Most of the western forest region inland from the coastal forests and above 300-m elevation in Canada to 1500-m southward into the Sierra Nevada Mountains of California receives a major portion of its annual precipitation as snow, which accumulates to build an extensive snowpack each year. In eastern North America, a persistent snowpack develops every winter northward from the Lake States, the Allegheny Plateau and Mountains of Pennsylvania, and inland from the coastal regions of New York and New England.

Areas characterized by a maritime winter climate and along the southern border of the forest snowpack zone have frequent above-freezing winter temperatures and comprise the *warm snow zone*. There, snowmelt is generated frequently throughout the winter. This zone includes most of the forests of southeastern Alaska, southern British Columbia inland almost to Alberta, Washington, northern Idaho and northwestern Montana, Oregon, California, and Arizona in the west. It narrows to the eastern parts of the continent across the southern Lake States, Pennsylvania, and New England.

The *cold forest snow zone* occupies the continental interior eastward almost to the Atlantic and northward from the warm snow zone into the arctic. Winter snowmelt is infrequent except where winter chinooks occur along the Rocky Mountain eastern front from Alberta southward. Even in the chinook region, the amount of winter runoff is seldom very great.

There is a transition region on the warm border between the forest snow zone and forests outside the snow zone where winter precipitation alternates between rain and snow. A shallow snowpack may build up, only to melt shortly thereafter, frequently during the next storm that comes as rain. Rain on snow floods may then result, particularly in winter in and below the warm snow zone in northwestern United States and western British Columbia.

The forest snow zone is more moderate in climate than the alpine or arctic regions. Growing seasons are as long as 7 months at lower elevations and latitudes. Moisture ranges from plentiful to sufficient to support tall closed forests in most of the zone, but extended summer water deficiencies are common in many western forest types and low annual or summer streamflows are a problem in much of the zone. Heavy economic development is concentrated along many rivers draining the forest snow zone, requiring protection from floods and dependent on continuous supplies of water for cities, farms, or industry. Intensity of land use within the zone is highly variable, ranging from urban areas to essentially undisturbed wilderness. All of these characteristics combine to create a variable need for maintenance or improvement of streamflow quantity and regimen and a variable potential for watershed management to meet the needs.

11.2.1 Increasing Water Yields

Increased water yields from the forest snow zone may result from precipitation enhancement, from reductions in evapotranspiration losses, or both. Precipitation enhancement in the snow zone forests seems most likely in western mountain regions and, where successful, should be only slightly less effective in generating runoff than in the alpine zone (Leaf, 1975). Most watershed management practices designed to increase water yields seek to modify snow distribution and decrease evapotranspiration losses by means of timber harvesting, thinning, or the conversion of forests to grasslands. Natural factors such as wildfire or insect attack may also influence streamflow.

On the whole, most studies in the forest snow zone, as in other forests, have indicated an increase in water yield following forest removal or a decrease following forest regrowth (Bosch and Hewlett, 1982), but the response is highly variable (Table 11.1). Responses range from a decrease in average annual flow of 138 mm over 9 years after patch logging in small drainages within the Bull Run municipal watershed, Oregon (Harr, 1980), to increases as great as 472 mm annually following a devastating wildfire in Washington (Helvey, 1980), and decreases up to 493 mm annually per acre planted to conifers in central New York (Schneider and Ayer, 1961).

Many factors combine to cause the highly variable streamflow response to forest removal or regrowth; among them site, climate (and the particular weather subsequent to treatment), the kind and degree of treatment, forest type and density, and natural variation in streamflow that may mask or exaggerate response as estimated by statistical tests. Though the relations are never simple, much of the variation can be accounted for on the basis of water availability and energy supply on the site and the amount by which they are modified by treatment.

For example, the larger responses listed in Table 11.1 tend to result from heavier vegetation change in moist climates or during exceptionally wet years.

TABLE 11.1 Water Yield Responses to Forest Treatment in the Forest Snowpack Zone

Decrease In Forest Cover (%)	Treatment	Forest Type	Location	Annual Water Yield Increase (Decrease) (mm year^{-1})	References
40	Strip cuts	Conifer	CO	109, 147, 89, 76, 91	Troendle and Leaf, 1981
36	Patch cuts	Conifer	CO	36, 58, 66, 20	Troendle, 1982; Alexander et al., 1985
100	Clearcut, herbicides 3 years	Hardwoods	NH	343, 274, 240, 191, 93	Hornbeck, 1975
30	Strip cuts	Hardwoods	NH	20, 50	Hornbeck, 1975
(40)	Increase in density	Mixed	NY	(196) after 39 years	Eschner and Satterlund, 1966
100	Clearcut	Aspen, conifer	CO	34, 47, 25, 22, 13	Reinhart et al., 1963
25	Insect kill	Conifer	CO	Avg. 58	Love, 1955
35	Insect kill	Conifer	MT	Avg. 45	Potts, 1984
100	Wildfire	Conifer	WA	91	Helvey, 1980
100	Wildfire	Conifer	WA	112	Helvey, 1980
100	Wildfire	Conifer	WA	74, 472[a], 178	Helvey, 1980
100	Clearcut, burned	Conifer	OR	425, 390, 325, 350, 300	Harr, 1983
60	Shelterwood	Conifer	OR	200, 240, 190, 200, 100	Harr, 1983
25	Patch cuts	Conifer	OR	Avg. (138), not significant	Harr, 1980
25	Patch cuts	Conifer	OR	Avg. (58), not significant	Harr, 1980
(47)	Reforested	Conifer	NY	(106) after 26 years	Schneider and Ayer, 1961
(35)	Reforested	Conifer	NY	(172) after 24 years	Schneider and Ayer, 1961
(58)	Reforested	Conifer	NY	(154) after 24 years	Schneider and Ayer, 1961
31	Patch cut	Conifer	Quebec	Not significant	Plammondon and Ouellet, 1980
80–90	Typhoon, salvage	Mixed	Japan	99, 268, 175, 225, 103	Nakano, 1971
100	Clearcut, no regrowth	Hardwood	Japan	95, 36, 31, 158, 212	Nakano, 1971
50	Cut merch. groups	Mixed	Japan	120, 76, 152, 106, 161	Nakano, 1971
100	Clearcut	Conifer	Japan	43, (47), 45, 38, (1)	Nakano, 1971
100	Fire	Conifer	Japan	140, 167, 197, 217, 62	Nakano, 1971
100	Clearcut overstory	Mixed	Japan	102, 137, 92, 143, 107	Nakano, 1971
25	Moderate fire	Conifer	AZ	Avg. 14	Campbell et al., 1977
70	Severe fire	Conifer	AZ	Avg. 21	Campbell et al., 1977

25	Insect defoliation	Conifer	OR	132[a]	Helvey and Tiedemann, 1978
16	Insect defoliation	Conifer	OR	Not significant	Helvey and Tiedemann, 1978
13	Insect defoliation	Conifer	OR	Not significant	Helvey and Tiedemann, 1978
100	Clearcut	Conifer	AZ	96, 23, 46, 37, 35	Brown et al., 1974
32	Strip cut	Conifer	AZ	50, 15, 10, 21, 44	Brown et al., 1974
75	Thinned	Conifer	AZ	22, 37, 38, 74	Brown et al., 1974
50	Strip cut, thinned	Conifer	AZ	18, 18, 41	Brown et al., 1974
65	Strip cut, thinned	Conifer	AZ	142	Brown et al., 1974
70	Clearcut	Conifer	Norway	200–250	Haveraaen, 1981
53	16 Clearcut, rest partial	Conifer	AZ	10, 36, 10, 10	Rich, 1972
20	Clearcut	Aspen	UT	Not significant	Johnston, 1984
20	Clearcut	Hardwood	PA	70, 32, 63, 73, 70	Lynch et al., 1975
40	Above + midslope cut	Hardwood	PA	96	Lynch et al., 1975
0.6	Riparian cut	Hardwood	AZ	Not significant	Rich and Gottfried, 1976
32	Above + clearcut on moist site	Conifer	AZ	21–37 range, 8 years	Rich and Thompson, 1974
75	Above + clearcut on dry site	Conifer	AZ	30–194 range, 7 years	Rich and Thompson, 1974
46	Selection, roads, fire	Conifer	AZ	1 of 13 years significant	Rich and Thompson, 1974
80	Above + thinning	Conifer	AZ	21–325 range, 7 years	Rich and Thompson, 1974
100	Clearcut	Aspen or pine	MN	90–200, first year	Verry, 1986
16	Road, patch cut	Conifer	ID	57, 65	King, 1984
27	Road, patch cut	Conifer	ID	85, 108	King, 1984
29	Road, patch cut	Conifer	ID	80, 151	King, 1984
33	Road, patch cut	Conifer	ID	122, 115	King, 1984
70	Clearcut	Conifer	Sweden	233, 270	Rosen, 1984
100	Clearcut	Conifer	Sweden	163, 238	Rosen, 1984

[a]Estimate based on extrapolation beyond limits of calibration data.

Lesser or nonsignificant responses tend to occur where 30% or less of the watershed is treated or in dry climates, especially where single tree selection, thinning, shelterwood, or small group cuttings are used. The decreases following patch cutting in Oregon, though statistically not significant, are probably real, and result from a reduction of fog drip and rime input into the watershed that more than offset transpiration savings from cutting (Harr, 1982). In one respect, snow zone forests differ from others in that precipitation tends to be strongly redistributed into small openings. The snow is most concentrated in openings that range in diameter from 2 to 8 times the height of the trees in the surrounding forest. As openings reach or exceed 15 heights, wind scour and sublimation of blowing snow increases and net losses may take place. Concentrated snow is associated with increased runoff efficiency when it melts that may more than offset the increased transpiration of the trees on the edge of small openings (Golding and Swanson, 1978; Troendle, 1983).

11.2.2 Control of Snowmelt Regimen

Streamflow regimen in the forest snowpack zone may respond to changes in routing precipitation from the watershed surface to the stream channel, changes in the amount and distribution of evapotranspiration losses, or modifications of snow accumulation and melt. Only the latter is unique to the forest snowpack zone and will be discussed in this section.

Annual peak flows from forests of the snowpack zone almost invariably occur during the melt season. If these high flows are to be reduced, it requires that melt be either enhanced earlier or delayed (or both). The melt season is different in different places. In subalpine or subarctic forests, melt may not begin until late spring and sometimes extends into early summer. Melt occurs intermittently throughout the winter at low elevations and latitudes in the warm snow zone.

The classification of snowpacks into warm and cold reflect differences in the melt process and release of snowmelt more than any differences in amount of snow. The warm snow zone, as in the maritime winter climatic region of western North America, may include some of the heaviest snowpacks on the continent as well as shallow ones, just as cold snowpacks may be deep or shallow. In mountainous terrain, both warm and cold snowpacks may be found at different altitudes in the same watershed.

Snow also occurs in forests outside the snowpack zone near the border of the warm snow zone where winter storms commonly occur as snow that melts within a few days to a week or more after falling. These transient snows also influence winter streamflow regimen and their presence may augment the effects of rain and melt in the warm snowpack zone.

Changes in forest cover may modify snow accumulation and melt in any of the situations under which snow occurs. Snow tends to accumulate more rapidly in openings because of redistribution of snow by the wind, lower interception losses, and the greater melt of intercepted snow that may move like rain through the warm snowpack beneath the trees. Snowmelt rates tend to be higher in openings, but the effect varies with weather conditions. In warm, cloudy weather when convective transport of sensible and latent heat are the major sources of energy, the greater wind in openings almost always causes faster melt. When skies are clear, winds calm or light, and snow is clean, snowmelt in openings may be less than in the forest. On the average, however, melt in openings is accelerated. Thousands of measurements over many years by Kuusisto (1980) showed that average melt rates in Finland increased by 0.16 mm $°C^{-1}$ day^{-1} (we) for each 10% decrease in forest cover.

Where snow accumulation is increased by forest cutting the additional water usually appears during the melt season, generating a greater volume of runoff. In parts of the cold snowpack zone, if soil moisture recharge is not completed before the snowpack forms, the snowmelt runoff volume may be further augmented by evapotranspiration savings of the previous summer. In many parts of the interior western United States, nearly all the increase from timber harvesting appears during the spring snowmelt season. In the warm snowpack zone and summer-moist portions of the cold snowpack zone, recharge is completed earlier and most savings appear in summer, autumn, or early winter (Satterlund and Eschner, 1964; Troendle and Leaf, 1981). Several studies may be used to illustrate the different runoff responses to accelerated melt induced by forest harvesting or reforestation of open fields.

In northwestern United States and adjacent Canada, most flooding occurs in the winter season when heavy rains may be accompanied by melt from the warm snow zone and transient snows below the snowpack zone. One study in the transient snow zone in the Cascade Mountains of Oregon revealed that average peak streamflows were 36% lower and delayed in time from a logged watershed as compared to one that was uncut. The earlier runoff from the forested watershed was attributed to rapid melting of intercepted snow that drained directly through the snowpack under the trees, whereas melt of the snow on the ground in the opening was delayed (Harr and McCorison, 1979). However, a comprehensive study of six larger watersheds extending into the warm snowpack zone in the same area showed that winter peak flows from rain on snow events increased over the past 30–40 years as the cumulative area of timber harvested increased (Christner and Harr, 1982). Updates of earlier studies also suggest a potential relation between clearcutting and changes in peak rain-on-snow flow events, but the degree and specific nature of this relation are poorly understood. (Harr, 1986).

In central New York, winters are characterized by frequent periods of melt, with temperatures only slightly above freezing. The snowpack in the open alternately builds and declines, sometimes with rain, throughout the winter. The snowpack in the forest accumulates more slowly due to interception losses and melt in the canopy, but also has fewer periods and lower rates of melt. Hence, when spring arrives, more snow remains in the forest than on open areas. Snow in the open disappears gradually under the cool conditions of winter and early spring and may be entirely gone before the advent of warm weather. Then, snow retained in the forest may melt more rapidly than rates observed earlier in the open under cooler conditions.

In addition, the longer snow remains in the forest, ripe and draining so as to keep the soil charged to capacity with meltwater, the greater the probability of rain on the snow. Consequently, in spring more rapid rates of melt and greater amounts of runoff may be yielded from forested than open areas. Such conditions are illustrated by two nearby watersheds, one mostly open (Albright Creek) and the other mostly forest (Shackham Brook) for the winter of 1961–1962 (Figs. 11.2 and 11.3). Though winter peak flows from open lands were greater than from forested lands, rates tended to be low in comparison with those of spring. Then, high flows were much greater from the forested watershed and concentrated in a short period. On Shackham Brook, one-quarter of the annual streamflow was discharged in just 10 days of late March and early April, whereas it took 18 days to discharge the same proportion of annual flow from Albright Creek (Satterlund and Eschner, 1964, 1965).

In the cold snowpack zone, winter melt is slight or none and any increase in snowmelt runoff from timber harvesting is added to the high flows of spring, so volume is increased. In any given year, peak flow rates may be increased or decreased as illustrated for Fool Creek, Colorado, where 40% of the subalpine forest was removed in strips (Fig. 11.4). The peak rate of discharge increased by 50% in 1956, but in 1957 it decreased by 23%. In 1956, the weather warmed steadily throughout the melt season. Snow melted faster in the treated

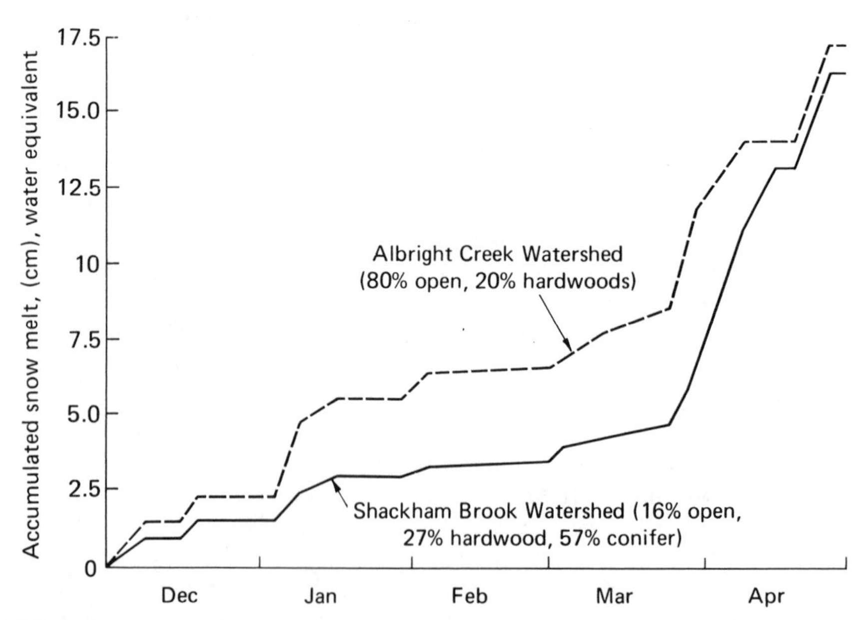

FIG. 11.2 Accumulated snowmelt from an open and reforested watershed in central New York, winter 1961–1962 (from Satterlund and Eschner, 1964).

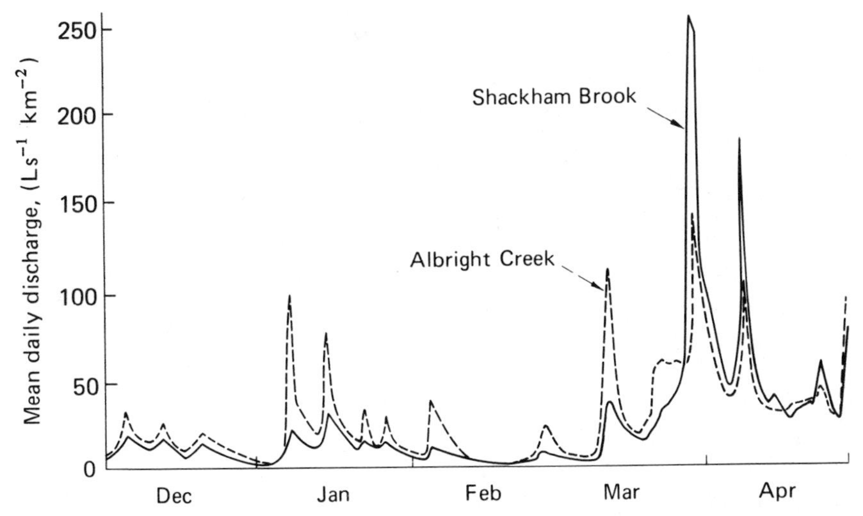

FIG. 11.3 The hydrographs for Albright Creek and Shackham Brook reflect the different snowmelt timing of the two watersheds, winter, 1961–1962 (from Satterlund and Eschner, 1964).

watershed than its untreated control, East St. Louis Creek, until after runoff peaked. In 1957, a warm spell initiated rapid runoff in the open, but that in the forest lagged. Then a cool spell slowed melt in both areas and allowed drainage of soil water. The return of warm weather again caused rapid melt, but most of the snow at low elevations in the cut area was gone and little remained to contribute to the delayed main peak (Goodell, 1958).

The study continued and hydrographs for 28 years after cutting showed an average 23% increase in peak flows in comparison with the 12 years prior to treatment. Flow analysis also indicated that advanced snowmelt due to cutting resulted in peak discharges occurring about 7.5 days earlier in the year, and that peak flows and the date of their occurrence were returning to preharvest levels at a very slow rate (Troendle and King, 1985).

The consensus of available evidence shows that runoff increases induced by timber harvest appear early in the melt period. Annual peak runoff rates appear little changed unless they take place early, when increases are common. This suggests that it may be possible to alter the snowmelt hydrograph by inducing earlier melt, but it may be difficult to delay melt by harvesting in the forest snow zone.

In planning cutting operations to influence the timing of snowmelt runoff, several important factors should be noted, including (a) the normal snowmelt patterns (i.e., are snows transient, warm, cold, or some combination?); (b) physiography (topographic position, slope direction and steepness, and elevation); (c) climate; (d) timber types, density and distribution; and, most importantly, (e) the desired runoff regime.

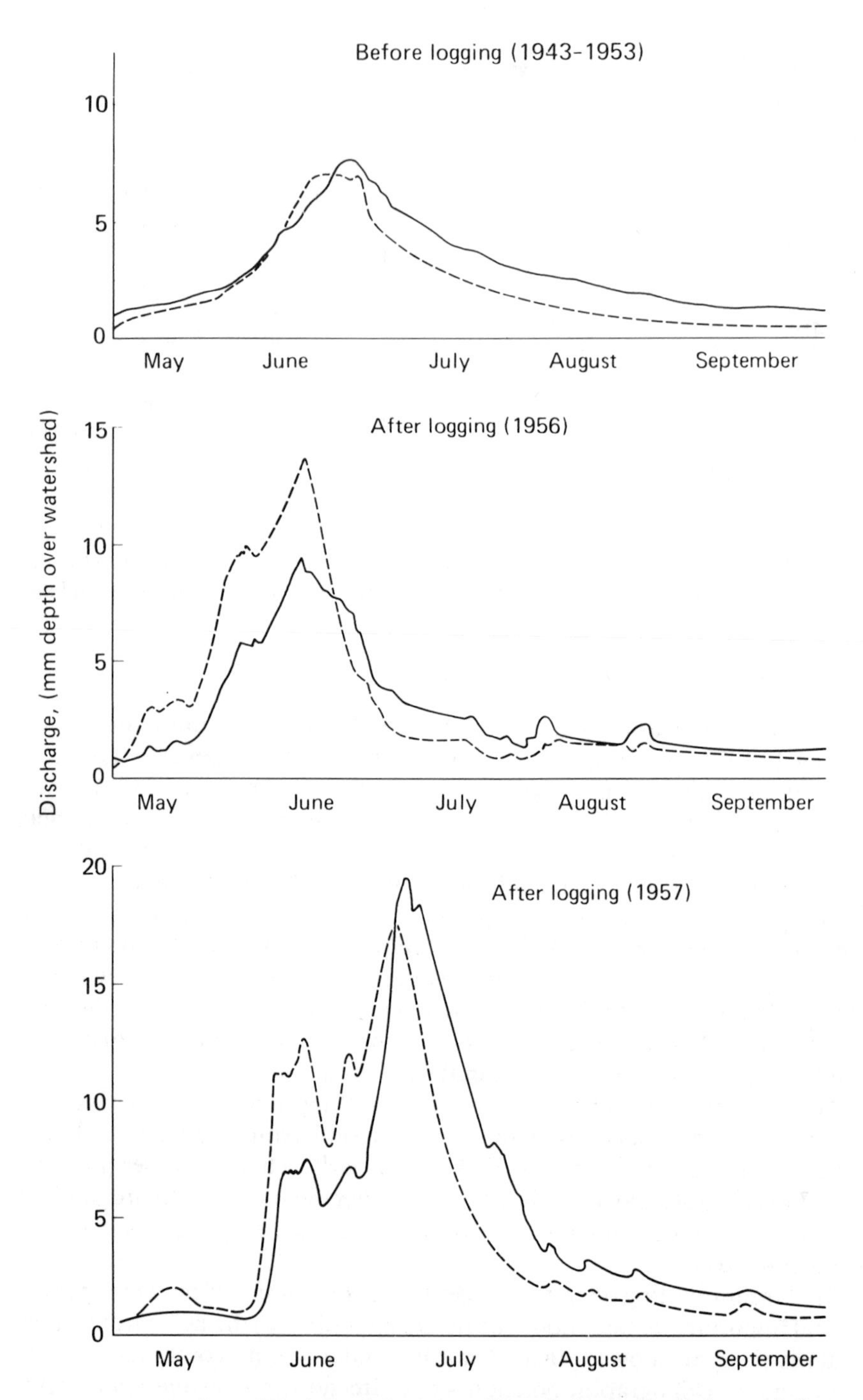

FIG. 11.4 Streamflow of Fool Creek (----) and East St. Louis Creek (———), Fraser, Colorado before and 2 years after timber was harvested from 40% of Fool Creek in 1954–1956, but not cut on East St. Louis Creek (from Goodell, 1958).

Most of the time the objective is to reduce the highest flows and spread the period of concentrated runoff over a longer period of time. In the transient and warm snowpack zones, highest flows tend to occur in winter as a result of rain on the melting snowpack. In the cold snowpack zone they are delayed until spring. Snowmelt is initiated later, but average melt rates are as fast or faster with increases in elevation. Similarly, melt is initiated earlier on high energy slopes facing southward in the northern hemisphere than on northerly slopes, but melt rates may be faster or slower, depending on persistence of the snowpack, exposure to winds, and cloudiness. At similar elevations, snow on ridges disappears faster than in valley bottoms. In many areas high energy slopes have less vegetation cover, so cutting results in less change than on low energy slopes. Timber type, evergreen or deciduous, is especially important during snowmelt, which usually is completed before deciduous species are leafed out. All in all, it appears that management systems that are designed to increase the natural heterogeneity of a watershed will flatten and broaden the snowmelt hydrograph. Cutting systems that increase homogeneity will sharpen it.

On a watershed managed for timber as well as water, cutting normally will not be restricted to a few favorable locations. Ultimately, all except streamside buffer strips are likely to be temporarily removed. Nevertheless, the design of the cutting system may critically influence snowmelt runoff. Both accumulation and melt may be modified to accomplish a degree of control.

In some places cutting should be designed to delay melt, or at least prevent it from coming earlier. In others, the effort might be directed toward acceleration, but under most circumstances a combination of both will be most advantageous. Ordinarily, if a range of natural heterogeneity exists, one would not try to accelerate melt where it normally occurs late or delay melt where it occurs early, as this would tend to concentrate runoff. Exceptions occur, however. Where winter flooding from rain on snow may be a problem, as in parts of northwestern United States, one might seek to delay melt and reduce winter melt rates in the transient and warm snow zones, as the heaviest storms tend to occur early in the winter and the odds of heavy rain decrease steadily into late winter and early spring. Snowmelt in the absence of rain is less likely to cause serious flooding.

To delay melt, two major principles apply. Concentrate the snow, reduce the rate of energy exchange, or both. Reducing energy exchange to snow is difficult, but possible, in certain situations. For example, where prevailing winds are from south or west, snow can be moved from high energy to low energy slopes across mountain ridges. Openings should be created from just to the lee of the ridge, across and down the windward side. They should be large, usually more than 8 H, and oriented with wind direction to permit wind movement to reach the snow. Snow will be moved from high to low energy slopes and deposited in concentrations where melt is delayed (Haupt, 1972).

Whenever snow is concentrated, the period of melt may be extended even if melt rates are unchanged and whenever the increase in rate of melt is insufficient to offset the increase in amount of snow. Snow tends to be concentrated most in openings from 2 to 8 H in diameter or narrow strips oriented perpendicular to prevailing winds. Melt in forest openings tends to decrease with decreasing size of the opening. Wind movement that transports latent and sensible heat to the melting pack is reduced and shade extends over a greater portion of the opening. Therefore, optimum opening size for extended snow melt may be smaller than desired for maximum accumulation alone.

Sometimes it is possible to orient openings to enhance cold air drainage to delay melt. Bulldozing slash to the downhill boundary helps trap cold air (West, 1961). However, frost effects on regeneration during the growing season may make such practices undesirable.

Where solar radiation is a major source of energy, unmelted snow persists longest not in the forest, but in openings next to the shaded forest border. Rapid melt occurs along the sunny border from direct solar radiation and strong longwave radiation from the heated trees. This phenomenon has led to a design of a cutting system called the "wall and step" for increased snow accumulation and extended melt. Anderson (1963) described the procedure as follows:

> . . . If maximum accumulation of the snowpack is the management objective and strip cutting is silviculturally desirable and economic, cut the forest in strips oriented across slopes perpendicular to the direction of maximum solar radiation. Generally the strips would be east–west on north and south slopes, northeast–southwest on east slopes, and northwest–southeast on west slopes. Successive strip cutting (Fig. 11.5) would proceed generally southward; that is, toward the maximum solar radiation. Once through a cutting rotation, we would have established a wall-and-step forest, with the wall to the south providing shade, and the steps to the north giving the least back radiation (Fig. 11.6).
>
> The width of the cut strip would depend on the slope of the terrain, on the height of the trees expected at the next rotation, and on our objectives. For maximum delayed snowmelt, we suggest strips one-half times the tree height on steep south and west slopes, one to two times the tree height on level areas and one to four times the tree height on steep north slopes.

Tests of this cutting pattern have indicated that much of the increase in the snowpack persists after peak snowmelt.

Where accelerated snowmelt is desired, openings should be larger to reduce snow accumulation and increase exposure to wind and sun. They should not permit snow to be removed to low energy slopes in the lee of a ridge. The wall and step can be reversed to attain increased exposure of the opening to solar radiation and maximum back radiation from residual trees.

It seems clear that forest management in the snowpack zone can be designed to improve or maintain desirable snowmelt hydrographs. Many snowmelt models are available to help select an optimum management system (US Army, 1956, 1960; Anderson and Crawford, 1964; Anderson, 1968, 1973; Leaf and Brink, 1973; Holtan et al., 1975; Solomon et al., 1976; Troendle and Leaf, 1980; among others). They range from simple to complex, and evaluations suggest that one or more will fit most situations (Baker and Car-

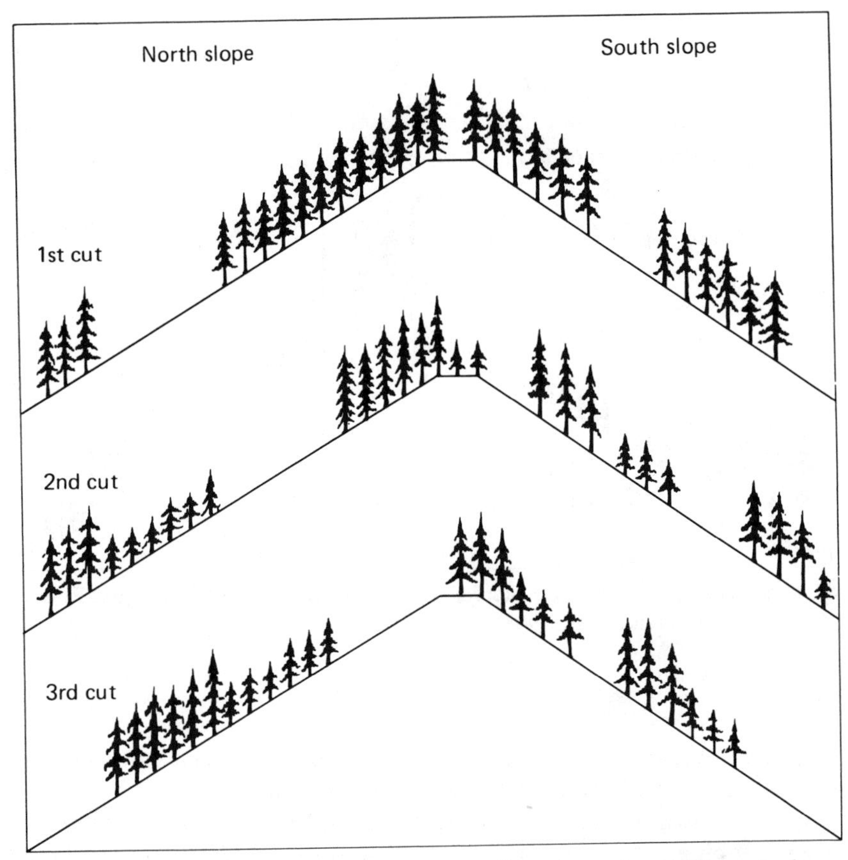

FIG. 11.5 Successive strip cutting creates a wall-and-step forest (from Anderson, 1963).

der, 1977; Haverly et al., 1978; Harr, 1981; Huber, 1983). By applying the proper cutting system to each appropriate site, managers may harvest timber without damaging snowmelt regimen and sometimes even improve it.

11.3 FORESTS OUTSIDE THE SNOW ZONE

Forests that receive little or no snow, or where a snowpack rarely persists through the winter, occur in areas of moderate to heavy rainfall through the warmer portion of the temperate zones into the tropics. They include a wide variety of forests. Evergreen conifers occur in western and southeastern United States, western and southern Europe, southern and eastern Asia, and are widely planted in parts of the southern hemisphere. Deciduous hard-

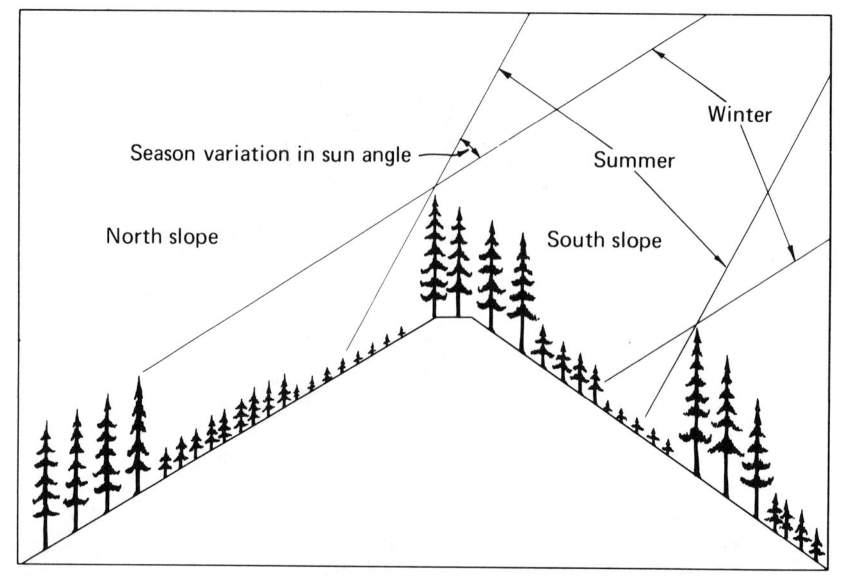

FIG. 11.6 Wall-and-step forest increases snow accumulation without greatly accelerating melt (from Anderson, 1963).

woods are also common in much of the world, and evergreen hardwoods are common in parts of the temperate zone and the tropics.

Some of the world's most productive forests are found outside the snow zone where a combination of ample rainfall and long growing seasons exist. Though seasonal water deficiencies may develop in some temperate and tropical forests, forest lands comprise the major water source areas wherever they are found. The combination of ample rainfall and plentiful energy suggests a high potential for modification of streamflow through forest management.

Land use intensity where forests are the natural vegetation varies widely. Many forests have been permanently cleared for agriculture throughout the world, and shifting cultivation is common in much of the wet tropics. Though nearly uninhabited jungle can still be found, population pressures upon many forest areas are growing. Demands for well-regulated water supplies are growing both within forests and downstream of them in adjacent drier or heavily populated regions where pressures may be even greater.

Most of the gaged watershed studies and much other forest watershed management research has taken place in the United States, but scattered studies have been conducted throughout the world. Unfortunately, though much is written about the problems in the wet tropics, limited data are available from the region. Nevertheless, principles derived elsewhere should be applicable to the wet tropics if properly adapted to their often extreme conditions and used conservatively.

11.3.1 Increasing Water Yields

Increased water yields from forests outside the snow zone may be attained primarily by reducing transpiration and interception losses by timber harvesting, thinning or burning, and by conversion to grass or cropland.

Consideration of the moderate to high amounts of precipitation and energy available in these forests suggests that annual responses soon after treatment may exceed those found in the snow zone. The data in Table 11.2 support such reasoning. Though many studies show no or limited increases in streamflow following forest removal (or reductions in yield due to forest growth), responses of 300 mm or more annually have been widely observed; in North Carolina and Oregon in the United States, and in Kenya, South Africa, and New Zealand. Figure 11.7 summarizes observations from 65 widely located study basins of changes in forest cover versus annual streamflow.

Several factors appear to account for the variation in observed responses, many of which are the same as those in other forests. The larger responses tend to occur where changes are most complete, as with clearcutting or complete revegetation of bare or open sites, on productive sites in areas of high water availability. Lesser or nonsignificant responses occur on drier sites and where partial cutting systems remove only a small portion of the cover at any one time. Removal of understory vegetation may be largely offset by increases in overstory transpiration (Kelliher et al., 1986).

Other factors equal, the heavier responses are associated with evergreen rather than deciduous vegetation. This results from the greater interception by

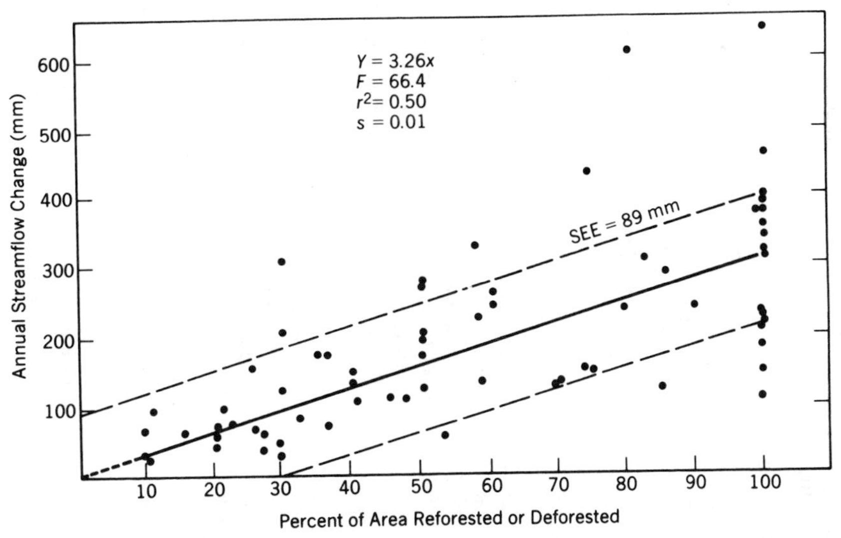

FIG. 11.7 Percent increase or decrease in forest cover versus change in annual streamflow (adapted from Trimble and Weirich, 1987). Typically, reforestation decreases streamflow, whereas deforestation increases it.

TABLE 11.2 Water Yield Response to Treatments of Forests Outside the Snowpack Zone

Decrease In Forest Cover (%)	Treatment	Forest Type	Location	Annual Water Yield Increase (Decrease) (mm year^{-1})	References
?	Pine overstory cut	Mixed	Japan	205, 98, 36, 102, 138	Nakano, 1971
82	Clearcut, burned	Mixed	OR	370, 520, 615, 465, 615	Harr, 1983
25	Patch cut, part burned	Mixed	OR	Not significant	Harr, 1983
100	Clearcut, burned	Conifer	OR	520, 457, 450, 430, 390	Harr, 1983
30	Patch cut, burned	Conifer	OR	150, 160, 260, 300, 230	Harr, 1983
100	Clearcut	Conifer	OR	120, 350, 400, 80, 50	Harr, 1983
100	Clearcut	Conifer	OR	350, 280, 320, 230, 250	Harr, 1983
50	Road, shelterwood	Conifer	OR	30, 80, 70, 100, 90 (1st year not significant)	Harr, 1983
30	Patch cut, road	Conifer	OR	80, 90, 110, 90, 90	Harr, 1983
100	Clearcut, regrowth	Hardwoods	NC	362, 275, 281, 255, 198	Bosch and Hewlett, 1982
100	Clearcut, regrowth	Hardwoods	NC	375, 218, 130, 100, 70	Bosch and Hewlett, 1982
65	51% Clearcut, thinned	Hardwoods	NC	220, 98, 108, 41, 40	Bosch and Hewlett, 1982
100	Clearcut, regrowth	Hardwoods	NC	255, 100, 85, 0, 0	Bosch and Hewlett, 1982
100	Clearcut, no regrowth	Hardwoods	NC	414, 337, 231, 160, 228	Bosch and Hewlett, 1982
(100)	Pine planted	Conifers	NC	(662) relative to bare (248) relative to hardwood	Bosch and Hewlett, 1982
50	Strips poisoned	Hardwoods	NC	189, 155, 130, 112, 100	Bosch and Hewlett, 1982
22	Understory cut	Hardwoods	NC	71, 64, 55, 47, 39	Bosch and Hewlett, 1982
100	Clearcut	Hardwoods	NC	150, 51, 60, 32	Bosch and Hewlett, 1982
(100)	Pine planted	Conifers	NC	(400) relative to bare (250) relative to hardwood	Bosch and Hewlett, 1982
100	Cleared, no regrowth	Hardwoods	NC	127, 95, 59, 113, 80	Bosch and Hewlett, 1982
30	Logger selection	Hardwoods	NC	25 average	Bosch and Hewlett, 1982
53	Selection	Hardwoods	NC	55 average	Bosch and Hewlett, 1982
27	Selection	Hardwoods	NC	Not significant	Bosch and Hewlett, 1982
80	Converted to grass	Hardwoods	NC	Not significant	Bosch and Hewlett, 1982
100	Cleared, regrowth	Hardwoods	NC	260, 200, 160, 120	Swank et al., 1982

34	Cleared for tea	Hardwoods, bamboo	Kenya	103	Bosch and Hewlett, 1982
34	Cleared, pine planted	Hardwoods, bamboo	Kenya	457, 229, 178	Bosch and Hewlett, 1982
85	Commercial clearcut, regrowth	Hardwoods	WV	130, 86, 89	Reinhart et al., 1963
36	Diameter limit, regrowth	Hardwoods	WV	64, 36	Reinhart et al., 1963
20	Selection, regrowth	Hardwoods	WV	Not significant	Reinhart et al., 1963
13	Selection, regrowth	Hardwoods	WV	Not significant	Reinhart et al., 1963
14	Selection, regrowth	Hardwoods	WV	Not significant	Bosch and Hewlett, 1982
8	Selection, regrowth	Hardwoods	WV	Not significant	Bosch and Hewlett, 1982
6	Selection, regrowth	Hardwoods	WV	Not significant	Bosch and Hewlett, 1982
91	Clearcut	Hardwoods	WV	253, 85, 60, 80	Bosch and Hewlett, 1982
50	Clearcut upper half, no regrowth	Hardwoods	WV	155, 145	Bosch and Hewlett, 1982
100	Above, plus lower half	Hardwoods	WV	251, 261	Bosch and Hewlett, 1982
50	Clearcut lower half, no regrowth	Hardwoods	WV	165, 142	Bosch and Hewlett, 1982
100	Above, plus upper half	Hardwoods	WV	259	Bosch and Hewlett, 1982
(74)	Pine planted	Conifers	S. Africa	(440) after 22 years	Bosch and Hewlett, 1982
(84)	Pine planted	Conifers	S. Africa	(13)	Bosch and Hewlett, 1982
(57)	Pine planted	Conifers	S. Africa	(325) after 23 years	Bosch and Hewlett, 1982
(98)	Pine planted	Conifers	S. Africa	(400) after 15 years	Bosch and Hewlett, 1982
(36)	Pine planted	Conifers	S. Africa	(170) after 8 years	Bosch and Hewlett, 1982
(84)	Pine planted	Conifers	S. Africa	Not significant, 8 years	Bosch and Hewlett, 1982
(100)	Eucalyptus planted	Hardwoods	S. Africa	(403) after 5 years	Bosch and Hewlett, 1982
52	30 Cleared, rest selection	Mixed	AL	102, 36	Bosch and Hewlett, 1982
86	Cut, burned, replanted	Mixed	AL	297, 244, 91	Bosch and Hewlett, 1982
45	Thinned, herbicide	Mixed	AR	107, 58, 89, 58	Bosch and Hewlett, 1982
100	Clearcut, herbicide	Mixed	AR	226, 142, 114, 145	Bosch and Hewlett, 1982
(75)	Pine planted	Mixed	TN	(152) after 16 years	TVA, 1962
(34)	Pine planted	Mixed	TN	Not significant	TVA, 1961
100	Cleared, planted	Conifer	GA	254	Hewlett, 1979
(70)	Pine planted	Mixed	OH	(144) after 18 years	Hill, 1960

(*Continued*)

TABLE 11.2 *Continued*

Decrease In Forest Cover (%)	Treatment	Forest Type	Location	Annual Water Yield Increase (decrease) (mm $year^{-1}$)	References
100	Cleared, planted	Conifer	FL	50	Swindel et al., 1983a
100	Cleared, planted	Conifer	FL	152	Swindel et al., 1983a
100	Cleared, pine planted	Evergreen hardwood	New Zealand	650	Pearce et al., 1980
75	Cleared, pine planted	Evergreen hardwood	New Zealand	540	Pearce et al., 1980
64	Logging, 2% per year	Mixed	WA	Not significant	Bosch and Hewlett, 1982
90+	Wildfire, reburns	Mixed	OR	Avg. 229 over 15 years	Anderson, 1976
(90+)	Pines planted	Mixed	MD	(138–304) over 15 years	Corbett and Spencer, 1975
50+	Riparian cut, thinned	Mixed	MD	46–122 over 4 years	Corbett and Spencer, 1975
100	Herbicide injection	Hardwoods	NJ	119	Corbett and Heilman, 1975
100	Herbicide spray	Hardwoods	NJ	198, 129, 140, 90	Corbett and Heilman, 1975
75	Insect defoliation, recovery	Hardwoods	NJ	148	Corbett and Heilman, 1975
100	Clearcut, regrowth	Mixed	Japan	Avg. 90 over 5 years	Fujieda and Abe, 1982
100	Fire, planted	Conifer	Japan	Avg. 60 over 5 years	Fujieda and Abe, 1982
[a]	Fire, regrowth	Eucalyptus	Australia	(240) over prefire, 21 years	Langford, 1976
[a]	Fire, regrowth	Eucalyptus	Australia	(220) over prefire, 21 years	Langford, 1976
[a]	Fire, regrowth	Eucalyptus	Australia	(145) over prefire, 21 years	Langford, 1976
[a]	Fire, regrowth	Eucalyptus	Australia	(155) over prefire, 21 years	Langford, 1976
90	Clearcut	Conifer	BC	Avg. 360 over 2 years	Hetherington, 1983
1–40	Progressive clearcut, 6 years	Conifer	BC	No significant change	Hetherington, 1983
(10–28)	Reforested	Conifer	GA, AL, SC	(25–99) after many years	Trimble and Weirich, 1987

[a] Postfire stand shorter and more dense than prefire stand.

evergreens and their ability to transpire when deciduous trees may be leafless. Another factor influencing water yield response to treatment is not evident from Table 11.2, but is related to slope direction. In the temperate zones, slopes receiving the greater solar radiation exhibit the lesser response (Douglass, 1983). Apparently, on many sites exposed to high insolation, internal water deficits develop in the trees increasing stomatal resistance and thus limiting water loss. Frequent internal water stress also tends to reduce vegetation growth, so forests are shorter and less dense. Thus, removing vegetation saves less water from transpiration and interception.

Another factor causing a lesser response from equatorward facing slopes in the temperate zone may be that these slopes receive a greater proportion of their total energy supply from solar radiation and tend to be convective heat sources. Poleward slopes are often heat sinks. Cutting equatorward slopes does little to reduce the efficiency of convective heat transfer, as increased thermal mixing from higher surface temperatures offsets the reduction of roughness that accompanies forest removal and is important in forced convection. Poleward forests acting as heat sinks may be quite dependent on forced convection for a large part of their total energy supply. Reducing the roughness of these sites by harvesting forests decreases advective energy exchange that is not otherwise offset, so the total energy supply is reduced to a greater extent by cutting poleward than equatorward slopes. There are few data supporting this, but it is consistent with the principles of energy exchange. If true, the effect could be expected to diminish and disappear into the tropics as the differences between north and south slopes as heat sources and sinks decline. The same effect may exist to a lesser degree on easterly and westerly slopes. Westerly slopes are more frequently strong heat sources than sinks. There is no reason that any differential east–west slope response should disappear in the tropics.

Interception studies (Isaac, 1946; Oberlander, 1956; Ekern, 1964; Kerfoot, 1968; Azevedo and Morgan, 1974; and Lovett et al., 1982) suggest that streamflow increases from tree removal in fog belts may be limited and flows may even decrease. This suggestion is supported by the studies of Harr (1980, 1982) in the forest snow zone. He found growing season streamflow was reduced by timber harvest in a watershed where fog drip was an important input into the hydrologic cycle during summer.

11.3.2 Control of Streamflow Regimen

Changes in streamflow regimen from vegetation change in any forest may be due to changes in the route of inflow to the stream channel or changes in evapotranspiration losses, or both. Important streamflow responses include changes in peak flow rates, changes in stormflow volume, and changes in low flow regimen.

11.3.2.1 Controlling the Route of Movement to Channels. The soil surface is a critical interface that largely determines the route that rain falling on a watershed will follow to the stream channel. Land use effects are most pronounced at the surface, but extend to some degree throughout the root zone. Sometimes they extend even farther, as when deep cuts during road construction may intercept subsurface flow and convert it to channel flow through the ditch system.

On most undisturbed forest lands, surface runoff is uncommon except from rock outcrops, areas of shallow soils or source areas that generate saturated overland flow. Soil frost is occasionally a cause of surface runoff in undisturbed forests in the cool temperate zone.

Forests outside the snow zone are usually accessible for timber harvest throughout the year, and operations may continue even when soils are wet and especially susceptible to damage by heavy machinery. Soil impacts may include mineral soil exposure or displacement, compaction, puddling, and rutting. To provide continuity in operations, harvesting should be concentrated on more resistant soils during wetter periods, reserving easily damaged sites for dry period operations. However, it may be necessary to suspend operations even on relatively resistant sites during poor weather to avoid soil and water impacts. Proper selection of harvesting techniques may permit operations on some hazardous sites if disturbance to the forest floor and soil can be avoided. Harvesting techniques are discussed more fully in Chapter 12.

Soil frost is not often a serious problem where deep snowpacks occur because deep snow serves as excellent insulation to keep the soil from freezing (Hart and Lull, 1963). However, it may become an important factor in the generation of overland flow when cold weather causes freezing of moist, bare soils. Ordinarily, only concrete frost importantly reduces infiltration. It tends to develop in dense, fine textured soils low in organic matter and under thin litter, and where soils are very wet. Compaction favors development of concrete frost and is usually associated with soil exposure that increases the frequency and severity of soil freezing.

On many forest lands, practices designed to maximize infiltration may be offset by road systems or skid trails that require deep cuts into the subsoil. Subsurface flow intercepted by roads is concentrated in ditches and quickly routed to stream channels. This alteration may cause either increases or decreases in peak flow rates, depending on whether the surface flow is synchronized or desynchronized with natural peaks (King and Tennyson, 1984). Road surfaces are also deliberately designed to shed rain. Unless this surface water is well dispersed where it can infiltrate, it, too, will quickly appear in the channels. Roads and landings may be the primary source areas for surface runoff on many managed forest watersheds. A dense road network may contribute an important increment to peak flow rates. When compacted areas exceeded 10% of several small watersheds, peaks increased by 33–48%; but when 5% or less of the watersheds was in roads, no such effect was observed (Harr et al., 1975, 1979; Ziemer, 1981). Harr (1987) emphasized, however, that the curvilinear relationship between compacted area and increased peak flows that he earlier reported does not indicate any specific compaction threshold above which detrimental streamflow changes can be expected.

Forest fires also may result in conditions that cause overland flow by inducing water repellancy in the surface soil or from exposing bare surfaces to rain (Tiedemann et al., 1979). Severe wildfires, such as the Tillamook burns in Oregon and the Entiat burn in Washington, may cause substantial increases in overland flow and peak flow rates (Anderson, 1976; Helvey et al., 1976). Campbell et al. (1977) observed peak discharges to be 58 times that of an unburned control following a severe fire in a ponderosa pine watershed in Arizona.

Damage to watersheds by fire is not restricted to wildfire. Small amounts of overland flow were evident from summer rainfall on two logged watersheds where slash was disposed of by broadcast burning in western Montana (DeByle and Packer, 1972) and has been widely observed under similar circumstances elsewhere. Prescribed burning may also be used in vegetation conversion and in site preparation for planting trees. In Mississippi, a prescribed burn consumed the L (litter) layer, but removed less than 1% of the F (fermentation) layer of a cull hardwood stand that was killed by injection of 2,4-D (2,4-dichlorophenoxyacetic acid). Stormflows, overland flows, and peak discharges all increased during the first 3 years after treatment by as much as 50% (Ursic, 1970).

Studies in southeastern United States show that mechanical site preparation may be associated with increased stormflow volume and peak flow rates. Overland flow is involved; it was directly observed on one study (Hewlett and Doss, 1984) and left sediment deposits in another (Swindel et al., 1983b). However, the stormflow volume and peak flow rate increases may have been partly caused by a reduction in transpiration as well as overland flow.

A major storm provided an opportunity to observe the effects of complete clearing and a major reduction in timber and associated litter cover of a ponderosa pine forest in Arizona. Peak flows were increased by 167% on a cleared watershed where slash was bulldozed into windrows 3 years earlier; by 89% on a cleared watershed planted to grass and grazed to utilize 60% of the grass; and by 88% where 76% of the basal area was removed by a group selection cut with windrowed slash a year earlier. Two treatments of lesser degree, approximately 32 and 20% removal, exhibited smaller increases that were "only a marginal response, if indeed significant." The high peak flows from the heavy treatments could be expected to occur only once every 100–200 years under normal conditions (Brown et al., 1974).

Some forests outside the snow zone have been severely depleted in the past by overcutting, clearing, and grazing. Such deteriorated watersheds can be rehabilitated, reducing overland flow and local flooding and erosion. The rehabilitation of the White Hollow watershed in eastern Tennessee by the Tennessee Valley Authority (TVA) illustrates the potential improvement (TVA, 1961).

Before rehabilitation, the White Hollow watershed of 694 ha consisted of 66% depleted woodland, 4% cultivated land, 4% grassland, and 26% abandoned cropland with a thin cover of broom sedge (*Andropogon* sp.). After 6 years, practically all the cleared land was converted to forest and numerous check dams and diversion ditches were built. Despite protection, 61 ha burned, destroying several pine plantations. An insect outbreak later killed 10 ha and required the removal of 300 M board feet of additional timber to control it. Another 90 M board feet were removed a few years later. Despite these setbacks, the area of forest floor covered by litter increased from 48 to 96%, and its average depth increased by 41% in a 20-year period. The greatest increases were on abandoned fields where they were most needed.

The effects of runoff appeared primarily as a reduction in peak flows, particularly in summer. Summer peaks were reduced by 90–92%, and winter peaks were reduced by up to 28%, although runoff volumes did not change appreciably (Fig. 11.8). Similar results are evident from other studies in eastern United States (Dils, 1953; Hill, 1960; TVA, 1962; Ursic, 1966). When grazing of a small, partially wooded watershed was stopped, mean peak flow rates were reduced by 96% after 3 years (Sartz and Tolsted, 1974).

Overland flow may occur frequently in undisturbed tropical rain forest catchments, even from soils with high saturated hydraulic conductivity characteristics, as a consequence of the high intensity rains that are common (Bonell and Gilmour, 1978). When tropical or semitropical forests are subjected to shifting cultivation, surface runoff is increased, but the site tends to recover during subsequent fallow intervals. Increasing pressure on the land has shortened the cycle between successive clearing and burning in India to as little as 5 years from the 30 years in forest cover common in the past. As the interval was shortened from 30 to 10 years on a porous soil, surface runoff during the cultivation period increased by 16%; when shortened from 30 to 5 years, it increased by 25%. On the latter site, 5 years after cultivation ceased runoff was still 92% of that observed during cultivation on a site subjected to a 30-year fallow cycle. By the 10th year of recovery, it was still 63% (Toky and Ramakrishnan, 1981).

11.3.2.2 Modification of Evapotranspiration Losses. The generation of runoff from any wildland watershed is likely to be modified by changes in evapotranspiration resulting from vegetation management. The effects may appear as changes in stormflow volume and peak flow rate, changes in baseflow, or all three. Reductions in evapotranspiration as by cutting, fire, or other treatment will increase runoff, whereas regrowth of vegetation will decrease it. Changes in streamflow are affected by many interacting factors, including the amount, location and pattern of vegetation treatment, soil moisture storage and hydraulic conductivity characteristics, climate and the particular weather subsequent to treatment.

11.3.2.2.1 Changes in Stormflow. Stormflow will be increased in volume whenever (a) runoff is generated from rainfall that would otherwise be par-

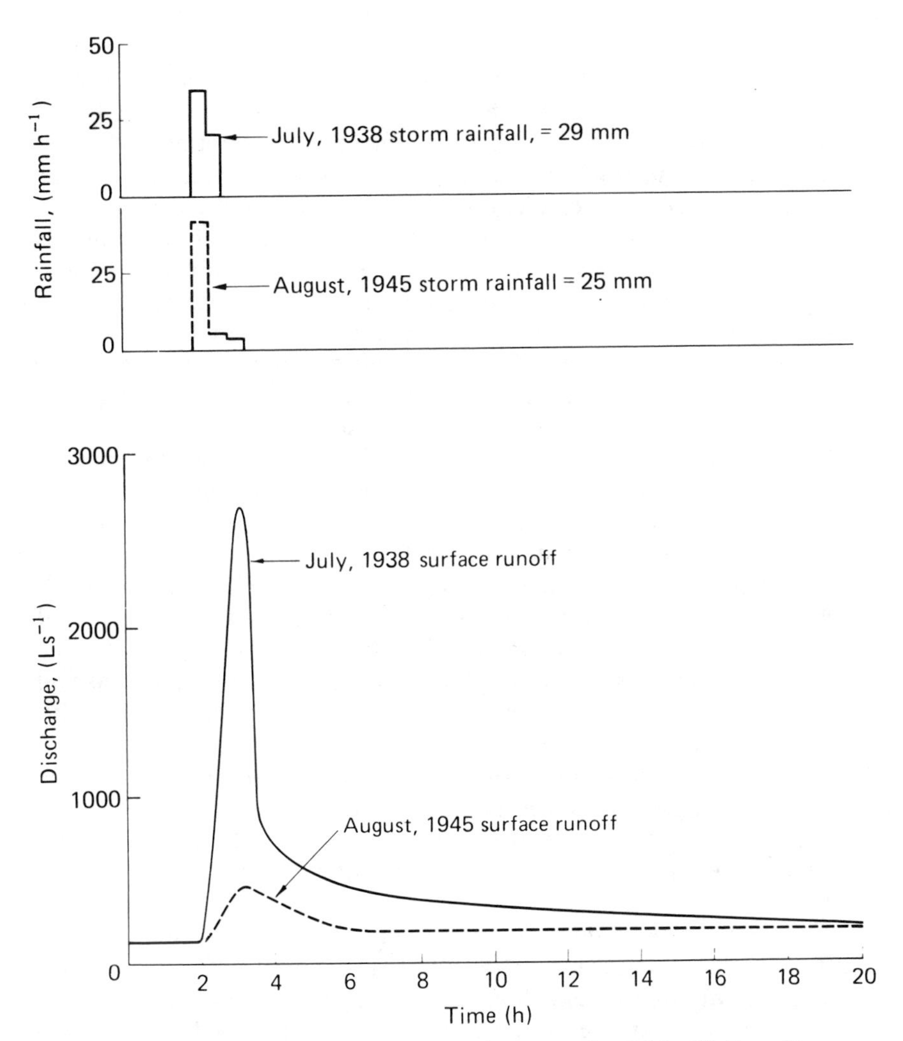

FIG. 11.8 Comparative rainfall and storm hydrographs, White Hollow, Tennessee, before and after rehabilitation (adapted from TVA, 1961).

tially or fully stored in the absence of vegetation treatment, and (b) the additional runoff is transmitted into the channel system quickly enough to appear on the storm hydrograph. Storage may be prevented when infiltration capacity is reduced sufficiently to generate surface runoff, as previously discussed. It may also be prevented when reduced evapotranspiration from vegetation treatment maintains soil moisture at higher levels. Then less infiltration is needed to satisfy retention storage requirements, and any excess moves through or over the watershed mantle to the channel system. Depending on how rapidly it moves, some or all of it may appear on the storm hydrograph.

The most pronounced stormflow volume increases would be expected with the largest expansion of source areas in response to reduction in evapotranspiration. This means that treatment on the moist sites around the edges of existing source areas is likely to be most efficient in generating additional storm runoff. The rate at which source areas shrink during rainless periods will be reduced. Furthermore, soil moisture will be sustained at a higher level beyond source area boundaries so that a lesser increment of infiltration is needed to bring the moisture content above the threshold for runoff generation. Thus, the rate of expansion during storms will be increased.

Treatment on both active source areas and drier sites well beyond their margins will tend to have lesser effects on stormflow volume. On active source areas, runoff will result from all rainfall, regardless of vegetation status. About the only increase from removing vegetation would come from reduced interception losses during the storm. On the dry sites, most storms are too small to provide even the reduced increment needed to bring up the soil moisture content to the threshold of runoff generation. Dry sites are typically farther from the channel system than moist sites, so even if runoff is generated, it is less likely to reach the channel quickly enough to appear on the storm hydrograph.

Stormflow resulting from evapotranspiration reduction would be expected to show a variable response in temperate climates and in tropical areas having a distinct dry season. The greatest relative increases should appear as soil moisture recharge is being completed at the end of the dry season. Then, differences in moisture depletion the previous season would result in maximal differences in moisture storage between treated and untreated sites. The treated site would generate stormflow from a large area before recharge is completed on most of the untreated site. Later, when recharge approaches completion everywhere, little difference in stormflow between different sites should be evident.

Two characteristics of stormflow from small watersheds are important: stormflow volume and peak rate of flow. Of the two, stormflow volume is likely to be more important, for peak flow rates are rapidly damped as they move downstream and peaks from individual tributaries are likely to be desynchronized. Despite the greater importance of stormflow volume from small watersheds, only a few studies have considered volume as well as peak rates of stormflow in relation to forest treatment.

One study on very small (0.61–1.25 ha) watersheds in the South Carolina Piedmont compared prescribed burning and clearcutting of three pine plantations established on eroded old fields. No changes were evident after three annual prescribed burns; after the third, more than 16 tons ha^{-1} of unburned litter remained on the site. However, both peak flows and stormflow volume increased following cutting, peak flows by 55–150% and stormflow volumes by 33–100% (Douglass et al., 1983).

Another study was at the Coweeta Hydrologic Laboratory in western North Carolina where summer storms commonly recharge soil moisture deficits. In much of the eastern United States, tropical storms frequently generate record flooding in late summer and early autumn when they move inland. Two such storms struck in quick succession on September 28 and 29 and October 3, 4, and 5, 1964, generating the largest regional flood on record on the larger rivers in the area. The first storm dropped in excess of 26.9 cm of rain in each of two small basins, one clearcut and the other untreated. The second dropped in excess of 23.8 cm on each. During the first storm, stormflow from the treated watershed was increased in volume by 33% (3 cm) with respect to the control, and peak flows were increased by about 35%. Recharge from the first storm reduced the difference in moisture storage between the clearcut and uncut control, and the volume of stormflow from the second, smaller storm was greater from both watersheds. However, the increase in stormflow volume was only 13% from the treated watershed, and peak flow was reduced by 19% from that predicted on the basis of stormflow from the control (Hewlett and Helvey, 1970). The lesser increase in stormflow volume from the first to second storm was as expected, but the decrease in peak flow below that predicted for undisturbed conditions was not, and its cause remains uncertain.

Most of the time, even where tropical storms do occur, they are not frequent and major flooding during the recharge season is unusual. Prestorm streamflow is usually low until soil moisture recharge nears completion. Under these more normal circumstances, the effect of reducing evapotranspiration is relatively large on the minor stormflows that occur when antecedent flow is low, but absolute increases tend to be unimportant except as they may contribute to improvement of the low flow regime. Rothacher (1973) found that clearcut logging of Douglas-fir in western Oregon caused increased rates of stormflow that were concentrated among the smaller peaks, but the higher peaks from major storms (10 L s^{-1} ha^{-1} or more) were not greater after logging than before (Figure 11.9). Similar results have been found by Reinhart et al. (1963) and Patric (1973) in West Virginia; Harr (1976), Harr and McCorison (1979) and Harr et al. (1975) in Oregon, Duncan (1986) in Washington, Hetherington (1983) in Canada, Hewlett and Doss (1984) in Georgia, and Swindel and Douglass (1984) in Florida.

In summation, reductions in evapotranspiration from forest clearing outside the snow zone tend to increase both stormflow volume and peak flow rates, but most increases occur during periods of lower streamflow. Increased flows at these times pose little danger and may actually improve the low flow regime. Stormflow volumes and peak flow rates during periods of high runoff are little changed, except in the unusual circumstance of large tropical storms occurring in rapid succession.

11.3.2.2.2 Changes in Low Flows. Modifying evapotranspiration losses tends to have a relatively large effect on low flows for the same reason that the largest relative increases in stormflow are concentrated in the low flow range; it does not take much additional water to double the size of a small trickle. Nevertheless, though absolute increases may be modest, they may be very important

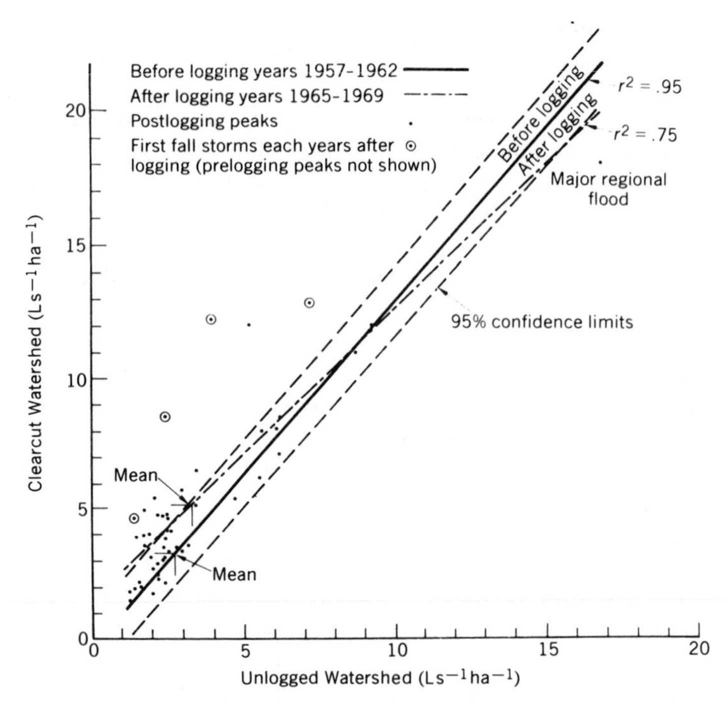

FIG. 11.9 Peak streamflow relations of two experimental watersheds on the H. J. Andrews Experimental Forest before logging and after clearcutting. Before-logging data points not shown (from Rothacher, 1973).

during these periods when lack of water is a limiting factor in aquatic productivity and water supply for people.

Reductions in evapotranspiration increase the amount of water during low flow in three ways: (a) by increasing the amount of stormflow during low flow periods, as already discussed; (b) by reducing the extraction of soil water that is moving into the channel system; and (c) by increasing the amount of water available for deep percolation to recharge soil moisture and groundwater that moves through the mantle to provide baseflow.

In summer-dry regions the greatest potential contribution to improvement of low flows probably comes from reducing the extraction of water that is in motion to the stream channel; from source areas and that moving into source areas that is within reach of the roots of transpiring plants.

By their very nature, source areas are the part of the watershed that is most tightly linked to the flowing stream. While they offer great potential for improvement of low flows if evapotranspiration is reduced, they also offer high risk for degradation of water quality if they are abused.

Where source areas occupy riparian zones, disturbance of the ground surface must be avoided. If trees are to be removed, felling should be directed away from the stream and heavy equipment for skidding or yarding kept off the moist source area. Logs should be winched to the vehicle, or, at least partially suspended if a cable system is used. Even when logged with care, it may be necessary to leave undisturbed such sensitive areas as seeps and buffer strips along the channel if water quality standards are to be met. More will be said about these concerns when control of water quality is discussed.

Low flows are also improved when evapotranspiration is reduced because soil moisture recharge may occur more often and to a greater degree than when the draft upon soil moisture supplies is unabated. This process is most important in summer-moist climates. Timber harvest causes the season of soil moisture depletion to begin later and that of soil moisture recharge to begin earlier. Therefore, flows are not only higher during the low flow period, but the period does not last as long. Stream segments that are intermittent may be converted to perennial flow following timber harvest.

When all the above factors are combined, improvement in low flows following timber harvest may be considerable. In the summer-moist regions of eastern United States, most of the annual increases in streamflow following timber harvest appeared during the growing season for nearly every paper cited in Table 11.2. In the summer-dry West, flow increases during the dry season tend to be small in absolute quantity, but relative increases reach as much as 100%. These increases may be relatively short-lived, however, in small watersheds where riparian regrowth is rapid (Harr, 1983).

One important exception to increased flow during the dry season took place on Bull Run municipal watershed in Oregon where timber harvesting reduced fog drip input into the watershed and low flows were further diminished as a result (Harr, 1980). Reduced fog drip also may have been responsible for lack of a summer streamflow increase following cutting on Vancouver Island (Hetherington, 1983).

11.4 THE WOODLAND AND BRUSHY ZONE

Woodlands and brush include such vegetation types as oak woodlands and savannahs, low conifers such as junipers and pinyon pines, and tall shrub types variously known as scrub, chaparral, maquis, matorral, fynbos, and others. They are widespread throughout the world and grow in semiarid or Mediterranean-type climates. Tree communities are never both tall and dense: if tall, as in some digger pine and oak woodlands of California, the crown canopies are open and trees widely spaced; if canopies are closed, as in some pinyon–juniper stands, the trees are seldom more than 10 m in height. Shrubs range from impenetrably dense to widely scattered.

A key characteristic of all the types is the existence of a moisture deficiency for much of the year. Annual precipitation is widely variable. In most types, snow is unimportant as a source of moisture. They tend to occupy the warmer parts of the temperate zones, extending into the tropics.

Lands in these types are used mostly for grazing. Some are highly urbanized, as the southern California region. Irrigated agriculture is common in and around these types. Water demands are heavy and often greatly exceed supplies.

Water yields in these types tend to be low to moderate in amount and irregular in nature. Few perennial streams originate in them; most are intermittent or ephemeral. Despite the relatively low flows, floods can be a major problem.

Nearly all of the gaged watershed studies in these types have been conducted in the southwestern United States and in South Africa. Similarities of climate and vegetation throughout many parts of the world suggest that results should be widely applicable elsewhere.

11.4.1 Increasing Water Yields

Streamflow increases following removal or killing of woody vegetation in this zone have been much less consistent than in high, closed forests (Table 11.3). Nevertheless, on the better sites in the higher rainfall areas and in wet years, important increases in water yield have been obtained. They exceeded 400 mm annually twice on one small chaparral watershed in Arizona when precipitation exceeded 1200 mm (Hibbert et al., 1982). About one-half of all the treatments in Table 11.3 resulted in increases that were less than 25 mm annually or were not significantly changed.

Probably the most important factors influencing potential yield increases are the productivity of the site and rainfall. The productivity of the site is a function of soil depth, exposure, and climate. It is best expressed in the field by vegetation type, height, and density. For a given type, the taller and more dense it is, the greater and more consistent treatment response is likely to be. However, different types are not directly comparable. For example, pinyon–juniper woodlands grow taller and more dense in a drier climate and on shallower soils than chaparral. Removal or killing pinyon–juniper woodland seldom yielded more than 13 mm of additional runoff even in wet years (Table 11.3).

In the climates characteristic of woodlands and brush, rainfall is the most important single factor determining response to treatment. Figure 11.10 indicates that mean response in seven chaparral drainages in Arizona more than quadrupled as annual precipitation increased from 500 to 1000 mm. Data from two of the studies suggest further the importance of moisture availability as a factor determining treatment response. When only moist site vegetation on 40% of Three Bar B and 15% of Whitespar B watersheds was killed, streamflow increases did not vary significantly from those when treatment was expanded to include drier uplands on 100% and 35% of the respective watersheds, given similar annual precipitation for each pair (Hibbert et al., 1982).

The type and degree of treatment also influences response. Unless partial treatments are restricted to moist sites, they are always less effective in generating additional runoff. There are many types of treatment used for killing

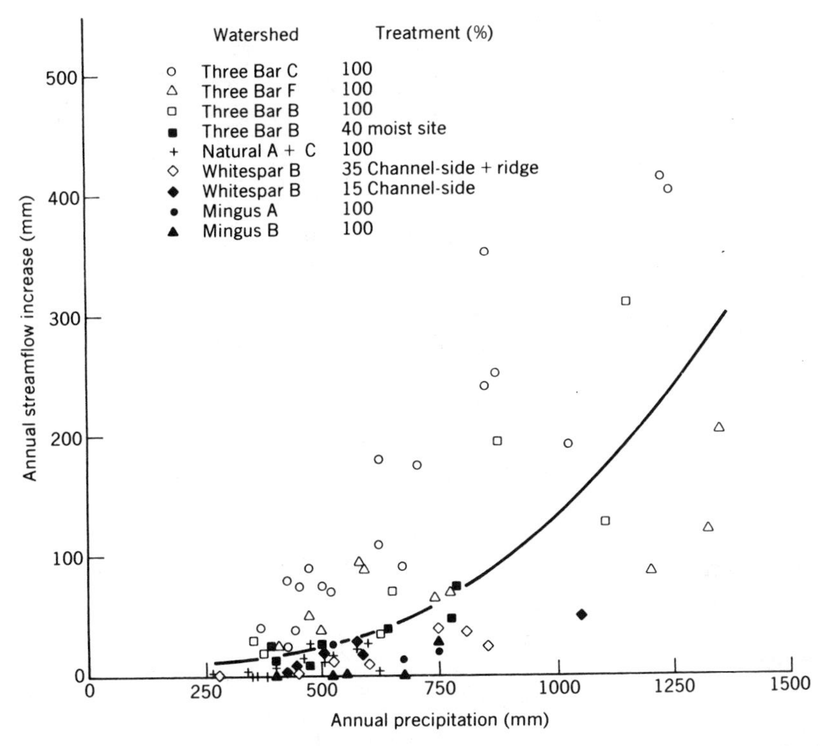

FIG. 11.10 Annual streamflow response increased with increasing precipitation from nine treatments that removed or killed all or part of the brush cover on seven small watersheds in Arizona (data from Hibbert et al., 1982).

or removal of woodlands or brush. They include mechanical methods such as cutting with saws or shears; use of bulldozers equipped with root plows, brush blades, rollers that cut and crush brush, or anchor chains or cables strung between them to uproot pinyon and juniper trees. Burning is used, alone or in conjunction with other methods (Hibbert et al., 1974; Ffolliott and Thorud, 1975).

Fire tends to reduce vegetation most completely and may render soils nonwettable. However, most woodland and brush types other than pinyon-juniper are well adapted to fire and quickly regenerate. Water yields may be maximized following fire, but often consist largely of surface runoff that scours the land and cause heavy mudflow floods.

Where type conversion is desired, mechanical or chemical methods are most commonly used, either alone or in combination. Most sites are seeded to grass. It is usually necessary to repeat treatment at intervals, as the successional forces leading to reestablishment of woody vegetation are very strong in most of the brush types. Many mechanical treatments such as chaining, cabling, root plowing, and windrowing with brush blades leave a rough

TABLE 11.3 Water Yield Response to Treatment of Woodland and Brush Vegetation

Decrease In Forest Cover (%)	Treatment	Vegetation Type	Location	Annual Water Yield Increase (Decrease) (mm year^{-1})	References
100	Convert to grass, 2-4,D and 2,4,5-T	Mixed oak-digger pine woodland	CA	57[a], 90[a], 172[a], 94[a]	Lewis, 1968[a]
100	2,4-D and picloram spray dead trees remained	Pinyon–juniper woodland	AZ	Avg. 5 over 8 years	Baker, 1984
100	As above, trees removed	As above		Avg. 1 (not significant), 6 years	Baker, 1984
100	Convert to grass, karbutilate granules	Chaparral	AZ	74, 63, 72, 215, 41	Davis, 1984
15	Fenuron and picloram on shrubs near channels	Chaparral	AZ	16, 25; avg. 16, 5 years	Ingebo, 1971; Hibbert et al., 1982
35	Above plus ridges treated	Chaparral	AZ	Avg. 13 over 5 years	Hibbert et al., 1982
100	Convert to grass with herbicides	Oak woodland	CA	111, 154, 75	Bosch and Hewlett, 1982
2	Cleared along channels, sprayed 2-4,D and 2-4-5T	75% Oak woodland 25% Phreatophytes	CA	11	Rowe, 1963
4	Repeat above over more area	90% Oak woodland 10% Phreatophytes	CA	5 (dry season only)	Rowe, 1963

(20)	Scrub invasion of grass	Grass, fynbos	S. Africa	Not significant	Bosch and Hewlett, 1982
(100)	Fire, scrub regrowth	Fynbos scrub	S. Africa	Max. (211) 20 years	Bosch and Hewlett, 1982
100	Converted to grass	Chaparral	AZ	Avg. 148 for 18 years	Hibbert et al., 1982
40	Individual shrubs killed	Chaparral	AZ	Avg. 30 for 7 years	Hibbert et al., 1982
100	Individual shrubs killed same watershed as above	Chaparral	AZ	Avg. 87 for 7 years	Hibbert et al., 1982
100	Individual shrubs killed	Open chaparral	AZ	Avg. 9 for 17 years	Hibbert et al., 1982
100	Uprooted by cables, grass seeded	Utah juniper	AZ	(9), (1), 7, (2), 5 Avg. 1, not significant	Clary et al., 1974
100	Killed by 2–4,D and picloram, trees remained	Utah juniper	AZ	10, 30, 1, 4, avg. 11	Clary et al., 1974
100	Felled, herbicide followup	Alligator juniper	AZ	14, 23, (3), 4, 19 Avg. 10, not significant	Clary et al., 1974
100	Burned, regrowth	Chaparral	AZ	Avg. 10, 5 years	Hibbert et al., 1982
100	Killed with karbutilate	Chaparral	AZ	Avg. 5 for 5 years not significant	Hibbert et al., 1982

[a] Results recalculated from original.

[b] 2,4,5-T is 2,4,5-trichlorophenoxyacetic acid.

surface that enhances surface water storage and infiltration, so water yields may be less than when chemicals are used.

11.4.2 Control of Streamflow Regimen

11.4.2.1 Floods. Most flood flows in the woodland and brushy zone are derived from surface runoff. In much of the southwestern United States, the chaparral type is managed almost exclusively for its watershed protection function. Flash floods also develop from the pinyon–juniper type where some stands so dominate the site that the soil surface is largely barren of understory grasses and forbs (Arnold et al., 1965). In such cases, neither the watershed protective function nor forage production needs are met.

The major control over surface runoff rests upon maintaining or establishing a ground cover of litter or living vegetation sufficient to protect the soil from the impact of rain, and to maintain infiltration capacities sufficient to dispose of the intense rates of rain that are characteristic of the region.

Fire is a major hazard in the chaparral type. In southern California, a combination of long dry spells, highly flammable fuels, steep slopes, nonwettable soils, increasing human use, and heavy fall and winter rains make the fire-flood hazard particularly great (Anderson et al., 1976). In most of the world, such brush types are associated with repeated fire and offer similar hazards. In many of the brush types, watershed management consists primarily of protection from fire with only minor emphasis on increasing water yields. A study of the effects of fire on three small watersheds on the San Dimas Experimental Forest (Dunn et al., 1988) near Los Angeles, California, shows why.

The three watersheds, I-Wolfeskill Canyon, II-Fern Canyon, and III-Upper East Fork of the San Dimas Canyon, lie adjacent to one another in the headwaters of San Dimas Creek. In 1938, a wildfire burned 26% of watershed II and 3% of watershed III. In 1953, another wildfire burned 32% of watershed I and an additional 3% of watershed II. The results, before and after each of the fires, are shown in Table 11.4. After the 1938 fire, peak flows from watershed II increased by 4.4–13.0 times and from watershed III (with only 3% burned) by 2.5–4 times. After the 1953 fire, peak flows from watershed I increased 15.6–67.7 times, and watershed II (again, with only 3% burned) by 1.2–3.7 times normal peak flows (San Dimas Experimental Forest, 1954).

In some areas, chaparral has been converted to grass to provide forage and reduce fire hazards (Bentley, 1967; Hibbert et al., 1974). In burned areas, rehabilitation includes seeding to grasses, both annual and perennial, and contour trenching (Rice et al., 1965; Krammes and Hill, 1963). However, fire is a part of the natural system in these types and in the long run is unlikely to be eliminated. Fire suppression must be accompanied by fuel management strategies that will reduce the size, intensity, and frequency of wildfire. Major elements of these strategies are fuel breaks and prescribed burning (Pase and Lindenmuth, 1971; Pase and Knipe, 1977; Rice et al., 1982). Despite all efforts,

TABLE 11.4 Peak Flows from Three Chaparral Watersheds as Affected by Fire[a]

	Rainfall (mm)		Peak Flow Rates (L s^{-1} km^{-2})					
			Soils Dry			Soils Wet		
Date of Storm	During Storm	Before Storm	I	II	III	I	II	III
Before 1938 fire								
1/31–2/12/36	110	56	22.4	15.5	10.7			
12/14–16/36	103	87	11.4	10.2	3.4			
12/9–11/37	80	1	15.5	6.3	7.9			
Average	98	48	16.4	10.7	7.3			
1/31–2/1/38	43	228				16.4	12.5	13.0
2/2–4/38	95	271				81.5	64.8	41.1
Average	69	249				49.0	38.7	27.1
After 1938 fire[b]								
12/17–21/38	185	56	71.4	601.4	129.4			
1/5/39	60	241				40.0	137.9	55.6
1/31–2/4/40	91	226				16.8	95.6	33.5
Before 1953 fire								
Average dry	201	61	255.1	220.0	216.6			
Average wet	124	320				81.1	74.8	60.9
After 1953 fire[c]								
1/17–20/54	167	66	4687.7	220.9	58.7			
1/23–25/54	121	235				3247.0	233.2	156.2

[a] Adapted from Tables 5 and 6 (San Dimas Experimental Forest, 1954).
[b] Fire burned 26% of Watershed II and 3% of Watershed III. Watershed I unburned.
[c] Fire burned 32% of Watershed I and 3% of Watershed II. Watershed III unburned.

conflagrations will continue in these types to be followed by floods. If damage is to be minimized, people should avoid the canyons and alluvial cones that are expected to be claimed by the resulting torrents.

Floods are less of a problem in the pinyon–juniper type where the absence of fire on grazed ranges permits invasion by these species with consequent deterioration of a ground cover. Much of the type now has insufficient litter and ground cover to carry a fire. Many means have been used to convert the site to grass. Prescribed fire may be used where surface fuels are adequate to carry a fire or individual trees are burned with a flamethrower; trees may be cut by hand or machine; they are grubbed by hand or with bulldozers equipped with clearing blades or root plows. Chemical control and cabling or chaining are the most popular methods. In the latter method, a cable or long, heavy anchor chain is attached to two crawler tractors. The tractors move on a parallel path between the trees, dragging the chain in an arc behind them. The trees are knocked over and uprooted where they are just left, bulldozed into windrows, or burned. More than 400,000 ha of pinyon–juniper had been treated in Arizona alone by 1961 (Arnold et al., 1965). Generally, water yield responses have been minor, but there is some evidence of peak flood flow increases as a result of chaining in one study (Ffolliott and Thorud, 1975).

11.4.2.2 ***Low Flows.*** An important benefit of treatment to increase water yields in the woodland and brushy zone has been the conversion of several small streams from intermittent to perennial flow (Hibbert et al., 1982). For example, on Whitespar B watershed that received about 587 mm of precipitation annually, treatment of the channel-side brush over just 15% of the watershed not only increased annual water yields by 16 mm, but created continuous flow in the main channel for 5 years after treatment. Prior to treatment, the stream dried up for 8 or 9 months each year (Ingebo, 1971). However, there is evidence that riparian vegetation downstream of treated watersheds may change in composition and increase in density, which may limit the downstream extent of perennial flow (DeBano et al., 1984). Concerns about protecting or enhancing streamside vegetation for fish and wildlife benefits are another important consideration in the use of source area vegetation treatments to increase water yields.

Colman (1953) pointed out how gully cutting in brush and grassland watersheds lowered the water table and dried out streams in late summer and fall in southern California. Heede (1981) showed that check dams prevented downcutting and trapped sediment sufficient to restore perennial flow in a formerly ephemeral gully system in southern Colorado.

11.5 THE PHREATOPHYTE ZONE

Phreatophytes (*phreat* = well + *phyte* = plant) are plants that obtain their supply of water directly from groundwater or the capillary fringe just above it. They occur in all parts of the world, growing near stream banks, on flood plains, and wherever groundwater is within reach of their roots. Some plants are facultative phreatophytes, that is, they are capable of using groundwater, but they commonly grow on well-drained uplands. Many forest tree species, both conifer and hardwood, are facultative phreatophytes.

Phreatophytes use tremendous quantities of water. Davenport et al. (1979) reported water use of 243 mm by saltcedar in southwestern United States in just the month of June. Robinson (1958) estimated that nearly 6.5 million ha are covered by phreatophytes in the 17 western states of conterminous United States, and that they consume over 28 billion m^3 of water annually. Croft (1948) estimated that evapotranspiration from Farmington Creek in the Wasatch Mountains of Utah consumed one-third the total flow in late summer.

By definition, water seldom limits evapotranspiration from phreatophyte communities. Furthermore, since many are surrounded by dry, hot country that is a source of large amounts of advected heat, they typically exhibit an "oasis effect," whereby they absorb sensible heat from their surroundings by convection. Oasis effects are not limited to deserts; they are common nearly everywhere. Phreatophyte communities within closed forests exhibit the effect whenever an advective heat source exists, whether it is a nearby slope warmed by more direct solar radiation or surrounding forests that have dried enough so that soil moisture limits evapotranspiration. Thus, water losses per unit of phreatophyte cover are probably greater than in any other vegetation type. Correspondingly, the potential for reducing water losses should also be greater. In addition, the location of phreatophytes in seeps, around springs, and on source areas immediately adjacent to stream channels indicates that any transmission lag in routing evapotranspiration savings to the stream channel are likely to be slight.

Treatment of phreatophyte communities has been by cutting, killing by herbicides, and application of antitranspirants. However, the total streamflow response has been very difficult to detect consistently, as the area treated is often so small that the effects on annual and seasonal yield are masked by natural variation of channel inflow from the rest of the watershed or treated areas extending beyond the phreatophyte zone (Rowe, 1963; Rich, 1965; Hibbert, 1967). Where vegetation was eradicated along one segment of a stream and streamflow entering and leaving the segment was measured, an increase of 518 mm per unit area treated was indicated. Other methods, such as observation of ground water wells, evaporation tanks, evaporation tents, and others have been used to derive estimates of water use and potential savings, but all such estimates contain considerable uncertainty (Horton and Campbell, 1974; Seyhan et al., 1983). Based on energy budget measurements, Davenport et al. (1979) estimated transpiration reduction of 20–25% when an antitranspirant was sprayed by a back-pack mist blower, but the reduction declined to only 10% after 1 month. Helicopter application was unsuccessful in reducing transpiration.

Though there may be some doubt as to how much water yields may be increased by treating phreatophytes, logic as well as some data suggest that increases can be substantial. Phreatophyte control should be more effective in increasing flow in effluent than in influent streams. Along effluent streams, the plants intercept water moving toward the stream, sometimes over large distances (Mader et al., 1972), so savings may be great and all appear as increased water in the channel. Along influent streams, the plants remove water moving out from the channel. By increasing the hydraulic gradient, they may increase outflow from the channel slightly, but their greatest effect is to reduce the amount of ground water recharge. Only part of their effect would be reflected in channel flow if they were removed.

Regardless of which is the case, there is little doubt of the fact of an increase and its almost immediate impact on streamflow (Fig. 11.11). Reducing or eliminating transpiration by phreatophytes causes the diurnal fluctuation in streamflow to be reduced or disappear. Furthermore, the rate of streamflow recession is reduced when transpiration is reduced (Tschinkel, 1963; Federer, 1973; Weisman, 1977). Thus, treating phreatophytes is particularly effective in increasing streamflow during periods of maximum water use.

Despite the potential water yield improvement possible from treating phreatophytes, any proposed treatment should be approached with caution. Removal may prove to be extremely difficult. Many species sprout, and growing conditions on these moist sites are so favorable it is nearly impossible to eliminate the problem with one treatment. The probability of reinvasion is high. Though herbicides may be effective, there are many limitations to their use, especially near water bodies and shallow groundwater. Many phreatophyte sites are on soft, wet stream banks or in canyon bottoms that are not suited to machine treatment, as with bulldozers using brush blades, chains, or root plows.

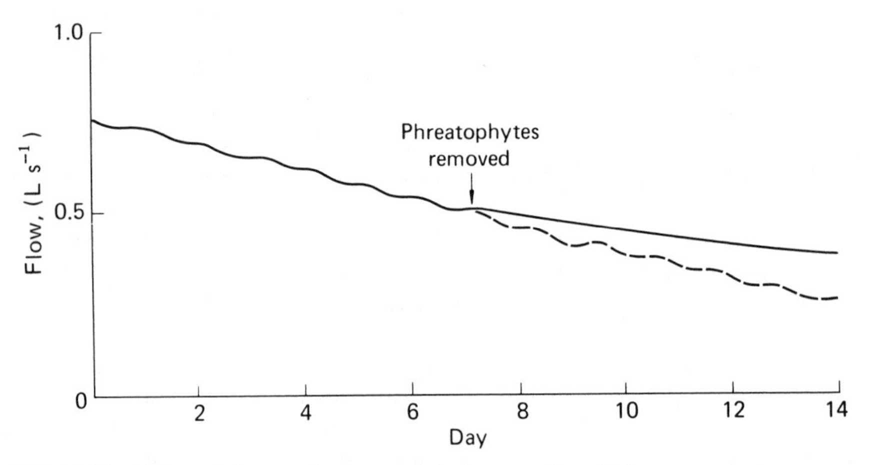

FIG. 11.11 Effect of phreatophyte removal on streamflow. Prior to removal, transpiration withdraws moisture during daylight hours, depressing discharge, which partially recovers at night. After removal, only minor streamflow fluctuations are evident. Dashed line represents projected streamflow under continued influence of phreatophytes.

There are further limitations to the desirability of phreatophyte control (Johnson and McCormick, 1978). Removal of streamside vegetation may increase stream temperatures and sediment. Treatment also seriously deteriorate visual, recreational, and wildlife and fish habitat values associated with the important streamside environment. Riparian areas should be evaluated with respect to all values, not just streamflow improvement. Solving the problem of conflicting benefits and costs will require consideration of more than just technical feasibility.

11.6 THE LOW SHRUB AND GRASSLAND ZONE

Low shrubs and grasslands are widely distributed throughout the world, but only the arid and semiarid types such as steppes, desert shrubs, and dry grasslands of the temperate and tropical parts of the world will be discussed here. The moister, high elevation, high latitude types were included in discussion of the alpine zone.

The dominant hydrologic characteristic of this zone is limited precipitation that may occur as either rain or snow. Rain often falls in highly localized, intense storms. Frozen soils are common wherever subfreezing temperatures occur. Streams are mostly intermittent or ephemeral and the low annual water yields frequently consist mostly of surface runoff. The limited vegetation is used mostly for grazing.

11.6.1 Increasing Water Yields

Opportunities for substantially increasing water yields become increasingly rare as sites become too dry to support woodland or tall brush vegetation. Efforts to provide increased flow into stock ponds and other local storage facilities have been successful on carefully selected sites, but the small increases are rapidly diminished or lost if they must be transported very far downstream. Most streams in this zone are influent through much of their length, and water moving down dry channels is lost by seepage and absorption by phreatophytes. The rate of downstream transmission loss varies with antecedent moisture in channels, channel size and shape, nature of bed and bank materials, and depth and duration of flow. Losses as high as 23,600 m^3 km^{-1} of channel have been measured from Walnut Gulch, Arizona (Hickok, 1967). Therefore, increased yields are likely to provide only local benefits, as for livestock or wildlife watering and irrigation.

Nevertheless, though benefits are local and small in absolute quantity, they may be quite valuable. The availability of water is frequently the limiting factor in the use of many lands in arid and semiarid regions.

There are basically two ways to increase water yields in the low shrub and grassland zones: (a) control blowing and drifting snow and (b) harvest water by rendering a portion of the watershed surface impermeable or nearly so.

Limited possibilities may exist by conversion of big sagebrush to grass on deep soils where precipitation exceeds 375 mm annually (Sturges, 1973, 1975, 1977).

The principles of controlling blowing and drifting snow were discussed earlier. In one study, snow fencing increased the water available to fill an irrigation reservoir by 54 m^3 m^{-1} of fence; and in another, removal of sagebrush from the windward size of a ridge successfully increased a drift on the leeward side intended to recharge a livestock pond. However, even though the drift was within a few hundred meters of the pond, little additional water reached the pond because of seepage losses beneath the drift and in the conveyance ditch leading to it (Sturgis and Tabler, 1981). In another study where soil and geological conditions were more favorable, fencing increased snow accumulation by 58% and streamflow by 137% at an estimated cost of just \$0.067 m^{-3} of additional water (Tabler and Sturgis, 1986). Snowdrift management can increase water yields effectively when combined with water harvesting techniques (Saulmon, 1973).

Water harvesting is the intentional management of a catchment to prevent infiltration and generate surface runoff (Frasier and Myers, 1983). Surface runoff may be an unintended consequence of grazing, especially where utilization is heavy and on sites subject to compaction (Sartz and Tolsted, 1974; Gifford and Hawkins, 1978). Ironically, rehabilitation of depleted rangeland has been known to reduce runoff so much that in order to graze it, it became necessary to haul water when the stock pond failed for lack of runoff.

In water harvesting the runoff is usually stored in tanks or ponds, but is sometimes spread over fields with a water spreading system for irrigation. Many different techniques are used to treat catchments to induce runoff. They include coating the surface with asphalt or cement, asphalt–fiberglass membranes, standard roofing felts, rubber and plastic sheeting, sodium salts to disperse clays, silicons and waxes to repel water, and others. Runoff efficiencies vary from a few to nearly 100%. Stored water may be protected from evaporation by storing in tanks or rubber bags, use of monomolecular films, floating covers, and others. Costs of water harvesting and storage vary with precipitation, site and method of catchment treatment and storage, but usually compare favorably with alternative systems such as wells and water hauling. In some situations, no other alternative exists (Fairbourn et al., 1972; Fink et al., 1979; Frasier, 1980; Oron et al., 1983).

On the whole, except for minor local water supplies, management of most low shrub and grassland should seek to maximize infiltration and route precipitation through the vegetation on the site rather than try to induce additional runoff.

11.6.2 Control of Streamflow Regimen

There are few opportunities to improve the irregular flow of the intermittent and ephemeral streams typical of the low shrub and grassland zone. Possibilities for controlling snow have already been mentioned. If snow with a depth of 30 cm or more can be retained over a watershed, soil freezing may be reduced, enhancing infiltration. Most soil moisture is lost by evapotranspiration nearly every year, regardless of cover type and density, so control of evapotranspiration offers little hope of success. The only means of regimen control over large areas is to control surface runoff.

It should be made clear that most streamflow in this zone is derived from surface runoff. Therefore, improvement of infiltration to control surface runoff is likely to reduce or eliminate water yields, for most water that enters the soil is ultimately lost as evapotranspiration.

Three general means can be used to control surface runoff from the low shrub and grassland zone. Depending on the situation, any one or a combination of methods may be necessary. They are

1. A well-designed program for management of livestock and game that maintains a suitable cover of native vegetation to retain snow and enhance infiltration of rain.
2. Improvement of vegetation, either wholly or in part, by fertilizing, reseeding with adapted grasses, forbs or shrubs, or removal and conversion of unsatisfactory native vegetation.
3. The use of structural measures such as contour trenches or plowing, gully control, range pitting, or others.

The above measures are listed in order of simplicity and ease of controlling surface runoff. Each should be combined with the preceding one(s) if success is to be insured. All of them help to insure maximum infiltration, reduce the extent of impermeable frost, and prevent surface runoff.

In deeply gullied ephemeral channels, check dams may trap sediments that serve as aquifers and extend the period of flow, as explained in a previous section (Heede, 1981). Systems can be developed to store and extract water from sandy sediments behind check dams for livestock use (Sivils and Brock, 1981). Controlled grazing to stabilize riparian vegetation may similarly extend seasonal flows from trapped fines, although annual flows may well decrease as evapotranspiration is enhanced.

11.7 PATTERN AS A FACTOR IN STREAMFLOW RESPONSE[1]

Pattern refers to location, form, size, and concentration of treatments on a watershed. For example, timber harvest or other treatments may be concentrated in a small area or dispersed widely over a watershed. They may be con-

[1]The general topic of pattern in forest management was discussed by J. D. Hewlett in an unpublished paper presented to a seminar at the School of Forest Resources, University of Georgia, Athens, GA.

fined to persistent source areas or areas that generate runoff only infrequently. They may exhibit different forms, as clearcuts that are round, long and narrow, or irregular in shape. There may be one large treated area or many small ones.

Pattern has a strong influence on the nature and degree of response to any treatment that modifies a watershed, whether it is the killing or removal of vegetation, compaction or exposure of soil, or rehabilitation of disturbed lands. Pattern, along with type and amount of treatment, is one of the few factors subject to a high degree of management control. It also should be a key consideration in analyses of the watershed effects of cumulative land management activities, although too often such analyses are oversimplified (Harr, 1987).

Generally speaking, the farther a treatment occurs from a stream channel, the smaller the streamflow response. Whenever changes must be transmitted a long distance over or through the watershed mantle, there is a good opportunity to buffer them. Surface runoff may infiltrate. Excess water in one area may be stored in another. Concentrations of flow, either subsurface or surface, may be dispersed and waves may be damped.

Treatments that are imposed on areas of topographic divergence, such as broad ridges, would be expected to have a lesser effect on the hydrograph, particularly on storm runoff, than the same treatment located on areas of topographic convergence, such as swales (O'Loughlin, 1986).

By definition, responses to modification of source areas are more pronounced than the same modifications of nonsource areas, for source areas are the portion of any watershed that is tightly linked to the stream system. Any treatment effect is readily transmitted to the channel with little delay or buffering.

Similarly, concentrated treatments tend to have greater effects than the same amount of treatment dispersed widely over a watershed. There is greater opportunity to buffer many small dispersed treatments than one large one. In addition, small units may show less response per unit area treated than large ones because of edge effects.

The relative amount of edge is important because the effects of plant communities extend beyond their above-ground boundaries. For example, the lateral root systems of many plant species extend beyond their crowns. Energy exchange is modified wherever dissimilar surfaces lie adjacent to one another, as pointed out in Chapter 7. The edge effect is more than just a function of the relative length of the perimeter of an opening; it is also a function of the height of the perimeter elements. The taller the vegetation, the greater the edge effect, as shadows are longer, view factors lower and roughness is likely to be increased. Furthermore, openings tend to be most rapidly reoccupied along their edges, so the greater the relative amount of edge, the more quickly the treatment effect is likely to disappear.

Three major factors determine the relative amount of edge, or edge per unit area of treatment: (a) the size of the treated unit, (b) the shape of the treated unit, and (c) the height of the perimeter elements surrounding the treated units. If height remains constant for any circle, regular polygon, or rectangle, quadrupling the area decreases the relative edge by one-half. However, for a given area of opening, different shapes have different relative edges. A circle has the smallest perimeter of any geometric form. The *shape factor* of a unit of any shape is its relative edge expressed relative to that circle.

$$S = \frac{P}{2\pi(A/\pi)^{\frac{1}{2}}} \tag{11.3}$$

where S is the shape factor, P is the perimeter of the unit, and A is the area of the unit. The divisor of the equation is the perimeter of a circle of area A. The relative edge for regular geometric forms of different types and sizes with a height of 1 unit, and the shape factor for each form is given in Table 11.5. The relative edge would vary directly with the height of the perimeter elements where heights differ.

Though relative edge is readily quantified for many forms and sizes of openings, interpretation of its effect is not so straightforward. What little evidence is available suggests an inverse, curvilinear relationship, that is, response tends to be highest with low relative edge, but very quickly diminishes and then levels out as relative edge increases. For example, if 25% of a forest in several different watersheds was removed by different methods ranging from a single clearcut to smaller, more dispersed patch cuts or group selection or a uniformly dispersed single-tree selection, the water yield response might range from substantial to negligible, respectively (Fig. 11.12).

11.8 PERSISTENCE OF TREATMENT EFFECTS

When watershed treatments stop, the watershed tends to return to its original condition unless some critical threshold is exceeded and the change becomes irreversible. Thus, when vegetation is cut, streamflow increases and as it regrows the increase is diminished and finally disappears. Should treatment be severe, as occurred when fire on a mountaintop near Cranberry Lake, New York, completely consumed a thick organic soil allowing the underlying mineral soil to erode to bedrock, runoff may be permanently altered.

The persistence of the treatment effect is directly related to the recovery of the factors that caused it. For example, at the Coweeta Hydrologic Laboratory in North Carolina, Watershed 13 was cut in 1940 and allowed to regrow. Regrowth from hardwood sprouts was rapid; by the 12th year, basal area reached about one-half of the original stand and crown coverage and leaf mass were equal to the original cover. The initial streamflow increase of 370 mm had declined to about 130 mm annually (Fig. 11.13). Extrapolation of the trend indicated that increases would be negligible after 35 years of recovery.

TABLE 11.5 Shape Factor and Relative Edge of Openings of Various Shapes and Sizes

Shape	Shape Factor		Size of Opening (m^2)						
			1	8	64	512	4096	32,768	262,144
Circle	1.00	radius (m)	0.56	1.60	4.51	12.77	36.11	102.13	288.86
		edge ($m\ m^{-2}$)	3.545	1.253	0.443	0.157	0.055	0.020	0.007
Equilateral triangle	1.29	side (m)	1.52	4.30	12.16	34.39	97.26	275.09	2334.2
		edge ($m\ m^{-2}$)	4.559	1.612	0.570	0.201	0.021	0.025	0.009
Square	1.13	side (m)	1.00	2.83	8.00	22.63	64.00	181.02	512.0
		edge ($m\ m^{-2}$)	4.000	1.414	0.500	0.177	0.063	0.022	0.008
Hexagon	1.05	side (m)	0.62	1.76	4.96	14.04	39.71	112.31	317.65
		edge ($m\ m^{-2}$)	3.722	1.316	0.465	0.165	0.058	0.021	0.007
Rectangle									
1 × 2	1.20	side[a] (m)	0.71	2.00	5.66	16.00	42.26	128.00	362.04
		edge ($m\ m^{-2}$)	4.243	1.500	0.530	0.188	0.066	0.023	0.008
1 × 4	1.41	side[a] (m)	0.50	1.414	4.00	11.31	32.00	90.51	256.00
		edge ($m\ m^{-2}$)	5.00	1.768	0.625	0.221	0.078	0.028	0.010
1 × 8	1.80	side[a] (m)	0.354	1.00	2.83	8.00	22.63	64.00	181.02
		edge ($m\ m^{-2}$)	6.364	2.250	0.795	0.281	0.099	0.035	0.012
1 × 16	2.40	side[a] (m)	0.25	0.71	2.00	5.66	16.00	45.26	128.00
		edge ($m\ m^{-2}$)	8.500	3.005	1.063	0.376	0.133	0.047	0.017
1 × 32	3.29	side[a] (m)	0.18	0.50	1.41	4.00	11.31	32.00	90.51
		edge ($m\ m^{-2}$)	11.667	4.125	1.458	0.516	0.182	0.064	0.023

[a]Shortest side.

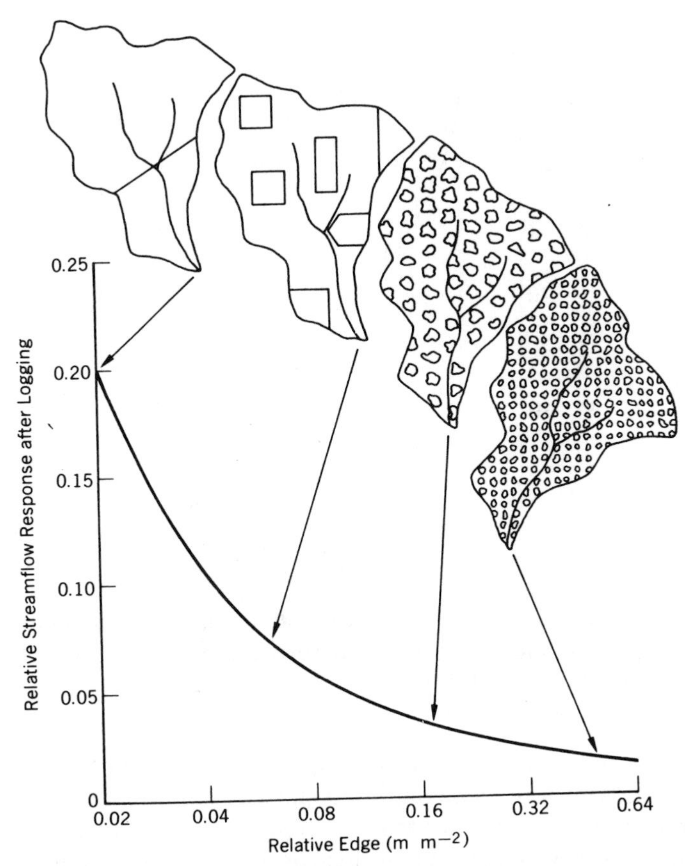

FIG. 11.12 Removal of 25% of the timber on several watersheds yields a variable streamflow response, depending on the shape and size of the opening(s) that determine relative edge. Note log scale for relative edge.

The watershed was again cleared 23 years after the first cutting, and the declining trend of increase established earlier was still valid (Hibbert, 1967).

Many factors influence the rate of watershed recovery to original conditions. Among them are (a) the characteristics of the original ecosystem, (b) the kind, (c) intensity, (d) amount, (e) pattern of disturbance, and (f) the particular hydrologic and growing conditions subsequent to the initial treatment. The nature of the watershed ecosystem is of dominant importance. Where ecosystems are robust and highly productive, recovery tends to be rapid and complete; where they are fragile, recovery is likely to be slow and tenuous. Where growing conditions are optimal, ecological pressures for site reoccupancy are almost overwhelming; where severe, they inhibit recovery. Some vegetation types show more consistent resilience than others. Those that reproduce by sprouting or suckering from surviving roots or stumps exhibit

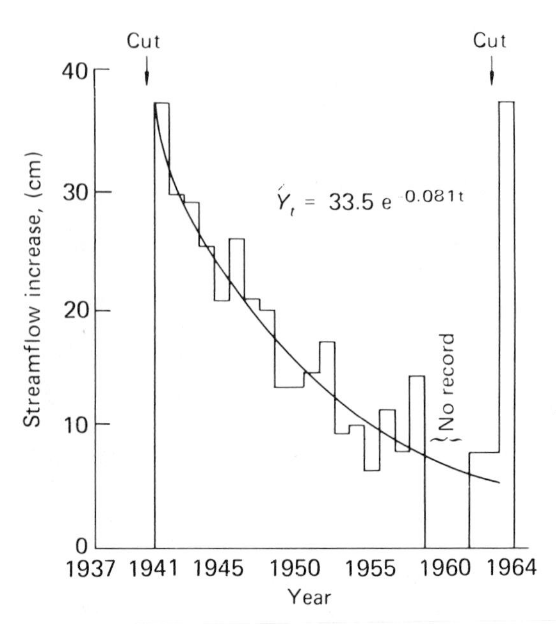

FIG. 11.13 Declining trend in annual streamflow following clearcutting and forest regrowth. Equation shows mean streamflow increase ($\hat{Y}$) by number of years (t) since cutting (data from Hibbert, 1967).

more certain and rapid recovery from topkill than those types that reproduce by seed alone. The particular conditions subsequent to treatment may modify the normal rate of recovery as when exceptionally good or poor seed years occur. Unusual moisture or drought, heat or cold may not only modify the rate of watershed recovery, but directly influence the degree of streamflow response itself.

The kind, intensity, amount, and pattern of treatment also influences recovery. Fire may have a different effect than timber harvest. Whether a site is heavily compacted or not may greatly influence the persistence of any treatment effect. And some disturbances are essentially permanent, as when a deep roadcut converts subsurface to surface flow and reroutes it through a ditch system to a channel. Generally, the greater the intensity and amount of treatment, the slower the recovery. The effect of pattern has been elaborated previously with respect to degree of response and the effects on persistence are similar.

11.9 INCREASING WATER YIELD RESPONSE TO WATERSHED TREATMENT

Nearly all studies designed to increase water yields have been based on reduction of evapotranspiration losses alone. However, it also appears feasible to increase the initial effect of treatment by combining such treatments with attempts to increase precipitation. In nearly all studies, the gain in water yield has been greater in wet years than in dry ones, in humid climates than in subhumid and semiarid ones, and on moist sites rather than on dry uplands. Therefore, if precipitation could be increased over a treated watershed, the streamflow response might be enhanced.

Satterlund (1969) has tested the effect of combining weather modification to increase precipitation with watershed treatment to reduce losses by means of a water balance model. The results indicated that a synergistic interaction takes place when both practices are successfully combined. Such an interaction can result in a total increase in streamflow greater than the sum of the increases that could be expected from separate practices to augment precipitation or reduce evapotranspiration losses.

The reasons for the more effective combined response are quite simple. When water is limiting, actual evapotranspiration becomes less than the potential evapotranspiration. At such times additional moisture may simply increase water losses by removing the moisture limitation. On the other hand, cutting the forest has the primary effect of reducing energy available on the watershed to evaporate water. If water, not energy, is already limiting losses, reducing energy supplies is unlikely to be very effective in reducing losses. However, if precipitation is increased and available energy reduced simultaneously, then a surplus of water may be made available for streamflow. A simple example illustrates the point.

Assume precipitation is 100 units, actual evapotranspiration is 100 units, and potential evapotranspiration is 110 units. (Actual evapotranspiration is limited by lack of water.) If precipitation is increased 10 units by cloud seeding, then enough water is available to satisfy potential evapotranspiration demand, is lost, and none is left over for streamflow. Similarly, decreasing potential evapotranspiration by 10 units has no effect since lack of water already limits actual losses to 100 units. But, if precipitation is increased by 10 units (to 110 units) at the same time potential evapotranspiration is reduced by 10 units (to 100 units), then 10 units of water (110 units of precipitation less 100 units of evapotranspiration) is potentially available for runoff.

What has been shown above is that *under some circumstances* two practices applied simultaneously can be effective when both applied separately are not. How frequently the critical circumstances exist in nature can be determined by taking actual climatic records from various stations and computing their water balances, first modifying the natural precipitation data, then the potential evapotranspiration, then both together. When this was done, it was revealed that the combined practices promise a synergistic gain of varying amount almost everywhere in the United States. The model indicated synergistic gains that ranged up to 43 mm of runoff annually in addition to that which would be obtained from both practices applied separately. Greatest gains were indicated in the wooded areas of the Southwest and in the Black Hills regions where summer precipitation followed the same trend as potential evapotranspiration (Fig. 11.14). Lesser synergistic gains were evident

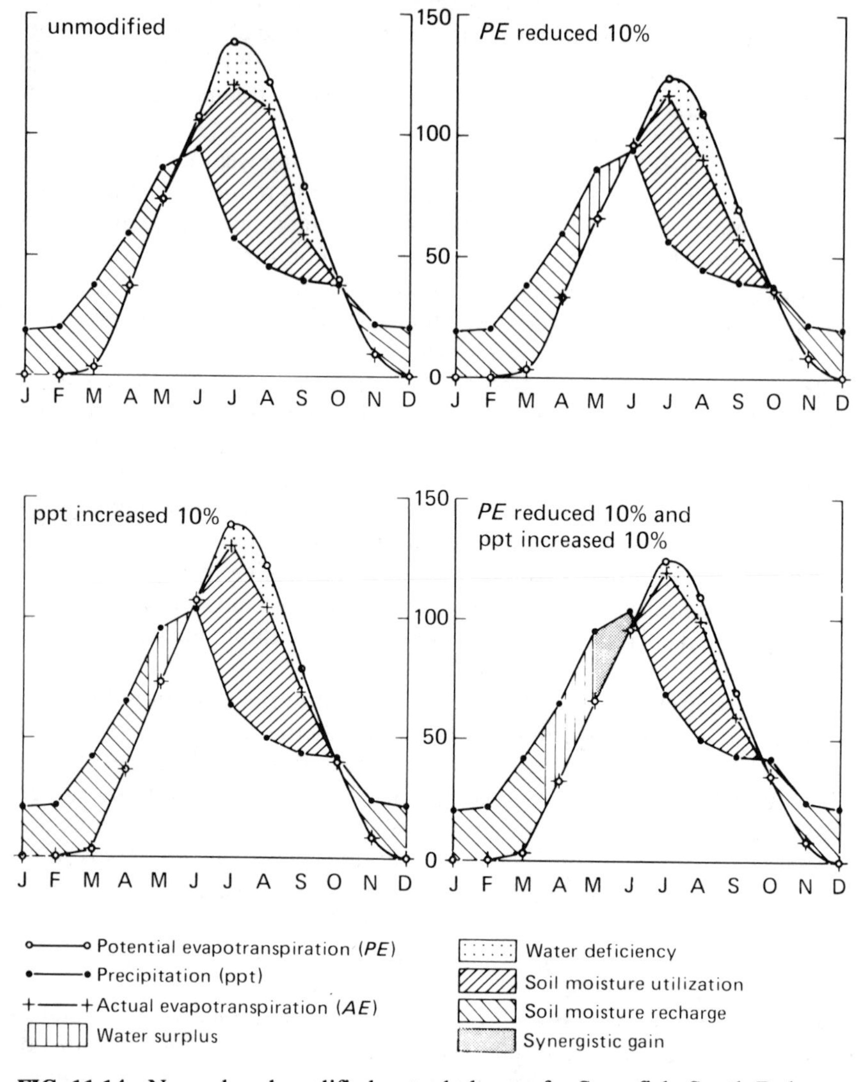

FIG. 11.14 Normal and modified water balances for Spearfish, South Dakota.

where precipitation was concentrated in winter, or in humid regions where precipitation often exceeded potential evapotranspiration (Fig. 11.15).

Only one analysis of the effects of combining weather modification with watershed treatment on streamflow regimen from the forest snow zone has been conducted. Where the increased precipitation falls as snow, spring melt rates rather than the amount of snowpack usually determine the runoff pat-

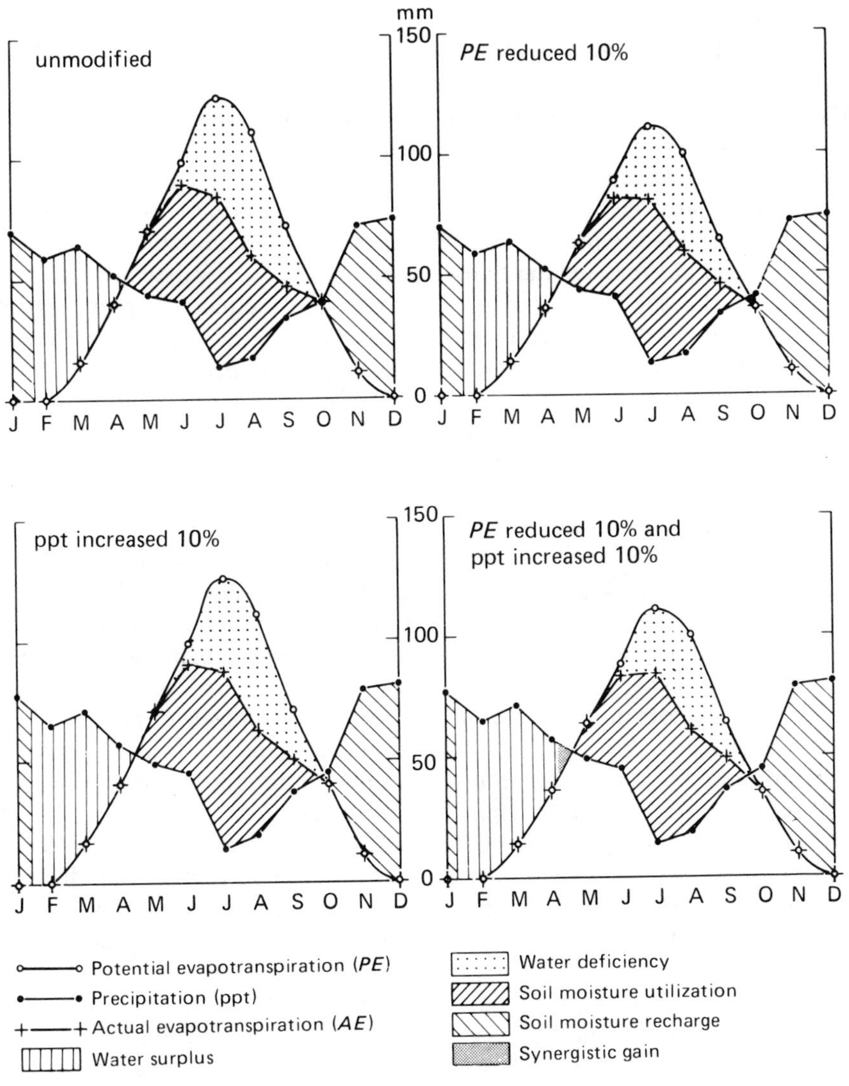

FIG. 11.15 Normal and modified water balances for Potlatch, Idaho.

tern, but the greater amount of snow extends the melt period (Leaf 1975). Elsewhere, under most circumstances, it can be safely surmised that responses would be greatest during low to moderate streamflow conditions as there would be no reason to increase precipitation when water supplies are ample. Small to moderate flood peaks might be increased in the pattern shown by the early fall peaks in Fig. 11.9, but increases in major peaks are unlikely.

11.10 RELATION BETWEEN CLEARING OF FORESTS AND THE DISAPPEARANCE OF SPRINGS

It is appropriate in this chapter to comment upon the frequently cited phenomenon of the disappearance of springs following the removal of forests in many parts of the world. Such observations seem to be as old as writing about forests, as noted in Chapter 3. Though many statements to this effect derive from the faulty memory of the observer, the phenomenon is too well accepted, and at times substantiated by records, to be dismissed lightly.

It is clear that the removal of forests on deep soils reduces evapotranspiration, and that nearly every recent study has shown that base flow is thereby increased. Since sustained springflow represents the appearance of base flow, this would tend to indicate that the removal of forests should increase springflow rather than cause its decrease. However, in none of the experimental studies was there a serious disturbance of the forest floor or any substantial reduction in infiltration.

On the other hand, when forests are removed and the land converted to cultivated fields or pastures, the subsequent land use may result in a considerable reduction in infiltration, especially in humid climates where heavy grazing on moist soils can cause serious compaction. Similar effects may occur locally when a new road diverts subsurface water away from its natural course toward a downhill spring (Sartz, 1969).

Thus, two factors may operate simultaneously in opposite directions when forests are removed: a reduction in removal of moisture from the soil by transpiration, and a possible reduction in the amount of water entering the soil (infiltration). If evapotranspiration is reduced to a greater degree than infiltration, the effect of forest clearing will be to increase springflow. But if infiltration is reduced more than evapotranspiration, springflow may decline. Therefore, it is not the forest clearing that causes springs to dry up, but rather the reduction of infiltration due to the clearing operation or subsequent land use.

There is no reason to expect that the clearing of forests will cause springs to decline or dry up so long as favorable conditions for infiltration are maintained. A possible exception exists where forests in fog zones capture large quantities of drip or rime that would be lost if the forests were cut.

LITERATURE CITED

Alexander, R. R., C. A. Troendle, M. R. Kaufmann, W. D. Shepperd, G. L. Crouch, and R. K. Watkins. 1985. The Fraser Experimental Forest, Colorado: Research program and published research 1937–1985. USDA For. Serv. Gen. Tech. Rep. RM-118, Rocky Mt. For. Range Expt. Sta., Ft. Collins, CO.

Anderson, E. A. 1968. Development and testing of snowpack energy balance equations. *Water Resour. Res.* 4:19–37.

Anderson, E. A. 1973. National Weather Service river forecast system—snow accumulation and ablation model. NOAA Tech. Memo. NWS HYDRO-17, US Dept. Commerce, Silver Spring, MD.

Anderson, E. A. and N. H. Crawford. 1964. The synthesis of continuous snowmelt runoff hydrographs on a digital computer. Tech. Rept. No. 36, Dept. Civ. Eng., Stanford University, Palo Alto, CA.

Anderson, H. W. 1963. Managing California's snow zone lands for water. USDA Forest Service Res. Pap. PSW-6, Pacific Southwest For. Range Expt. Sta., Berkeley, CA.

Anderson, H. W. 1966. Integrating snow zone management with basin management. pp. 355–373. In: S. Kneese, Ed. *Water Research.* Johns Hopkins Press, Baltimore, MD.

Anderson, H. W. 1976. Fire effects on water supply, floods and sedimentation. *Proc. Tall Timbers Fire Ecol. Conf.* 15:249–260.

Anderson, H. W., M. D. Hoover, and K. G. Reinhart. 1976. Forests and water: effects of forest management on floods, sedimentation, and water supply. USDA For. Serv. Gen. Tech. Rep. PSW-18. Pac. Southwest For. Range Expt. Sta., Berkeley, CA.

Arnold, J. F., D. A. Jameson, and E. H. Reid. 1965. The pinyon–juniper type in Arizona: effects of grazing, fire and tree control. USDA Forest Service Prod. Res. Rept. No. 84, Washington, DC.

Azevedo, J. and D. L. Morgan. 1974. Fog precipitation in coastal California forests. *Ecology* 55:1135–1141.

Bagley, J. M., R. W. Jeppson, and C. H. Milligan. 1964. Water yields in Utah. Spec. Rept. 18, Agric. Expt. Sta., Utah State University, Logan, UT.

Bailey, R. W., G. W. Craddock, and A. R. Croft. 1947. Watershed management for summer flood control in Utah. USDA Misc. Pub. No. 639, Washington, DC.

Baker, M. B., Jr. 1984. Changes in streamflow in an herbicide-treated pinyon–juniper watershed in Arizona. *Water Resourc. Res.* 20:1639–1642.

Baker, M. B., Jr. and D. R. Carder. 1977. Comparative evaluations of four snow models. *Proc. Western Snow Conf.* 45:58–62.

Bentley, J. R. 1967. *Conversion of chaparral grassland: techniques used in California.* Ag. Handbook No. 328, USDA, Washington, DC.

Berndt, H. W. 1964. Inducing snow accumulation on mountain grassland watersheds. *J. Soil Water Conserv.* 19:195–198.

Bonell, M. and D. A. Gilmour. 1978. The development of overland flow in a tropical rain-forest catchment. *J. Hydrol.* 39:365–382.

Bosch, J. M. and J. D. Hewlett. 1982. A review of catchment experiments to determine the effect of vegetation changes on water yield and evapotranspiration. *J. Hydrol.* 55:3–23.

Brown, H. E., M. B. Baker, Jr., J. J. Rogers, W. P. Clary, J. L. Kovner, F. R. Larson, C. C. Avery, and R. E. Campbell. 1974. Opportunities for increasing water yields and other multiple use values on ponderosa pine forest lands. USDA Forest Service Res. Pap. RM-129, Rocky Mt. For. Range Expt. Sta., Ft. Collins, CO.

Brown, R. W., R. S. Johnson, and K. Van Cleve. 1978. Rehabilitation problems of arctic and alpine regions. pp. 23–44. In: *Reclamation of drastically disturbed lands.* Am. Soc. Agron., Madison, WI.

Campbell, R. E., M. B. Baker, Jr., P. F. Ffolliott, F. R. Larson, and C. C. Avery. 1977. Wildfire effects on a ponderosa pine ecosystem: an Arizona case study. USDA Forest Service Res. Pap. RM-191, Rocky Mt. For. Range Expt. Sta., Ft. Collins, CO.

Christner, J. and R. D. Harr. 1982. Peak streamflow in the transient snow zone, western Cascades, Oregon. *Proc. Western Snow Conf.* 50:127–138.

Clary, W. P., M. B. Baker, Jr., P. F. O'Connell, T. N. Johnsen, Jr., and R. E. Campbell. 1974. Effects of pinyon–juniper removal on natural resource products and uses in Arizona. USDA Forest Service Res. Pap. RM-128, Rocky Mt. For. Range Expt. Sta., Ft. Collins, CO.

Colman, E. A. 1953. *Vegetation and watershed management.* The Ronald Press, New York.

Corbett, E. S. and J. M. Heilman. 1975. Effects of management practices on water quality and quantity: the Newark, New Jersey, municipal watersheds. pp. 47–57. In: *Proc. Municipal Watershed Management Symp.* USDA Forest Service Gen. Tech. Rept. NE-13, Northeastern For. Expt. Sta., Upper Darby, PA.

Corbett, E. S. and W. Spencer. 1975. Effects of management practices on water quality and quantity: Baltimore, Maryland, municipal watersheds. pp. 25–31. In: *Proc. Municipal Watershed Management Symp.* USDA Forest Service Gen. Tech. Rept. NE-13, Northeastern For. Expt. Sta., Upper Darby, PA.

Court, A. 1961. Large meltwater flows come later. USDA Forest Service Res. Note 189, Pac. Southwest For. Range Expt. Sta., Berkeley, CA.

Croft, A. R. 1948. Water loss by stream surface evaporation and transpiration by riparian vegetation. *Am. Geophys. Union Trans.* 29:235–239.

Croft, A. R. and R. W. Bailey. 1964. Mountain water. USDA Forest Service, Intermountain Region, Ogden, UT.

Davenport, D. C., J. E. Anderson, L. W. Gay, B. E. Kynard, E. K. Bonde, and R. M. Hagan. 1979. Phreatophyte evapotranspiration and its potential reduction without eradication. *Water Resour. Bull.* 15:1293–1300.

Davis, E. A. 1984. Conversion of Arizona chaparral to grass increases water yield and nitrate loss. *Water Resour. Res.* 20:1643–1649.

DeBano, L. F., J. J. Brejda, and J. H. Brock. 1984. Enhancement of riparian vegetation following shrub control in Arizona chaparral. *J. Soil Water Conserv.* 39:317–320.

DeByle, N. V. and P. E. Packer. 1972. Plant nutrient and soil losses in overland flow from burned forest clearcuts. pp. 296–307. In: S. C. Csallany, T. G. McLaughlin, and W. D. Striffler, Eds. *Watersheds in Transition.* Am. Water Resour. Assoc., Urbana, IL.

Dils, R. E. 1953. Influence of forest cutting and mountain farming on some vegetation, surface soil and surface runoff characteristics. USDA Forest Service Sta. Pap. No. 24, Southeastern For. Expt. Sta., Asheville, NC.

Douglass, J. E. 1983. The potential for water yield augmentation from forest management in eastern United States. *Water Resour. Bull.* 19:351–358.

Douglass, J. E., D. H. VanLear, and C. Valverde. 1983. Stormflow changes after prescribed burning and clearcutting pine stands in the South Carolina Piedmont. pp. 454–460. In: E. P. Jones, Jr., Ed. *Proc. second biennial Southern silvicultural res. conf.* USDA Forest Service Gen. Tech. Rept. SE-24, Southeastern For. Expt. Sta., Asheville, NC.

Drake, J. J. 1981. The effects of surface dust on snowmelt rates. *Arctic and Alpine Res.* 13:219–223.

Duncan, S. H. 1986. Peak stream discharge during thirty years of sustained timber management in two fifth order watersheds in Washington state. *Northwest Sci.* 60(4):258–264.

Dunn, P. H., S. C. Barro, W. G. Wells II, M. A. Poth, P. M. Wohlgemuth, and C. G. Colver. 1988. The San Dimas Experimental Forest: 50 years of research. USDA For. Serv. Gen. Tech. Rep. PSW-104. Pac. Southwest For. Range Expt. Sta., Berkeley, CA.

Ekern, P. C. 1964. Direct interception of cloud water on Lanoihale, Hawaii. *Soil Sci. Soc. Am. Proc.* 28:419–421.

Ellison, L. 1954. Subalpine vegetation of the Wasatch Plateau, Utah. *Ecol. Mono.* 24:89–184.

Eschner, A. R. and D. R. Satterlund. 1966. Forest protection and streamflow from an Adirondack watershed. *Water Resour. Res.* 2:765–783.

Fairbourn, M. L., F. Rauzi, and H. R. Gardner. 1972. Harvesting precipitation for a dependable, economical water supply. *J. Soil Water Conserv.* 27:23–26.

Federer, C. A. 1973. Forest transpiration greatly speeds streamflow recession. *Water Resour. Res.* 9:1599–1604.

Ffolliott, P. F. and D. B. Thorud. 1975. Water yield improvement by vegetation management. Focus on Arizona. PB-246 055. Nat. Tech. Inf. Serv., US Dept. Commerce, Springfield, VA.

Fink, D. H., G. W. Frasier, and L. E. Myers. 1979. Water harvesting treatment evaluation at Granite Reef. *Water Resour. Bull.* 15:861–873.

Frasier, G. W. 1980. Harvesting water for agricultural, wildlife, and domestic uses. *J. Soil Water Conserv.* 35:125–128.

Frasier, G. W. and L. E. Myers. 1983. *Handbook of water harvesting.* USDA Agr. Res. Serv. Ag. Handbook No. 600, Washington, DC.

Fujieda, M. and T. Abe. 1982. Effects of regrowth and afforestation on streamflow on Tatsunokuchiyama Experimental Watershed. *Bull. For. For. Prod. Res. Inst.* (*Japan*) 317:113–138.

Garling, M. E. 1960. Surface water and hydrology. pp. 34–111. In: *Water resources of the Nooksack River Basin and certain adjacent streams.* Water Supply Bull. No. 12, Div. Water Resour., Dept. Cons., Olympia, WA.

Gifford, G. F. and R. H. Hawkins. 1978. Hydrologic impact of grazing on infiltration: a critical review. *Water Resour. Res.* 14:305–313.

Golding, D. L. and R. H. Swanson. 1978. Snow accumulation and melt in small forest openings in Alberta. *Can. J. For. Res.* 8:380–388.

Goodell, B. C. 1958. A primary report on the first year's effect of timber harvesting on water yield from a Colorado watershed. Station Pap. No. 36. Rocky Mt. Forest and Rge. Expt. Sta., Ft. Collins, CO.

Goodison, B. E., H. L. Ferguson, and G. A. McKay. 1981. Measurement and data analysis. pp. 191–274. In: D. M. Gray and D. H. Male, Eds. *Handbook of snow.* Pergamon Press Canada, Ltd., Willowdale, Ontario.

Harr, R. D. 1976. Forest practices and streamflow in western Oregon. USDA Forest

Service Gen. Tech. Rept. PNW-49, Pac. Northwest For. Range Expt. Sta., Portland, OR.

Harr, R. D. 1980. Streamflow after patch logging in small drainages within the Bull Run municipal watershed, Oregon. USDA Forest Service Res. Pap. PNW-286, Pac. Northwest For. Range Expt. Sta., Portland, OR.

Harr, R. D. 1981. Some characteristics and consequences of snowmelt during rainfall in western Oregon. *J. Hydrol.* 53: 277–304.

Harr, R. D. 1982. Fog drip in the Bull Run municipal watershed, Oregon. *Water Resour. Bull.* 18:785–789.

Harr, R. D. 1983. Potential for augmenting water yield through forest practices in western Washington and western Oregon. *Water Resour. Bull.* 19:383–393.

Harr, R. D. 1986. Effects of clearcutting on rain-on-snow runoff in western Oregon: a new look at old studies. *Water Resour. Res.* 22:1095–1100.

Harr, R. D. 1987. Myths and misconceptions about forest hydrologic systems and cumulative effects. pp. 137–141. In Proc. Calif. Watershed Mgmt. Conf., Nov. 1986, Sacramento. Report No. 11, Wildland Resources Center, University of California, Berkeley, CA.

Harr, R. D., R. L. Fredericksen, and J. Rothacher. 1979. Changes in streamflow following timber harvest in southwestern Oregon. USDA Forest Service Res. Pap. PNW-249, Pac. Northwest For. Range Expt. Sta., Portland, OR.

Harr, R. D., W. C. Harper, J. T. Krygier, and F. S. Hsieh. 1975. Changes in storm hydrographs after road building and clear-cutting in the Oregon Coast Range. *Water Resour. Res.* 11:436–444.

Harr, R. D. and F. M. McCorison. 1979. Initial effects of clearcut logging on the size and timing of peak flows in a small watershed in western Oregon. *Water Resour. Res.* 15:90–94.

Hart, G. and H. W. Lull. 1963. Some relationships among air, snow and soil frost. USDA Forest Service Res. Note NE-3, Northeastern For. Expt. Sta., Upper Darby, PA.

Haupt, H. F. 1972. Relation of wind exposure and forest cutting to changes in snow accumulation. Paper presented at Intl. Symp. on the Role of Snow and Ice in Hydrology, Session IASH-WMO-5, Banff, Alberta.

Haveraaen, O. 1981. The effect of cutting on water quantity and water quality from an East-Norwegian coniferous forest. (English summary). *Medd. Norsk Inst. Skogforskning* 36.7.

Haverly, B. A., R. A. Wolford, and K. N. Brooks. 1978. A comparison of three snowmelt prediction models. *Proc. Western Snow Conf.* 46:78–84.

Heede, B. H. 1981. Rehabilitation of disturbed watersheds through vegetation treatment and physical structures. pp. 257–268. In: D. M. Baumgartner, Ed., *Interior west watershed management.* Washington State University, Pullman, WA.

Helvey, J. D. 1980. Effects of a north central Washington wildfire on runoff and sediment production. *Water Resour. Bull.* 16:627–634.

Helvey, J. D. and A. R. Tiedemann. 1978. Effects of defoliation by Douglas-fir tussock moth on timing and quantity of streamflow. USDA Forest Service Res. Note PNW-326, Pacific Northwest For. Range Expt. Sta., Portland, OR.

Helvey, J. D., A. R. Tiedemann, and W. B. Fowler. 1976. Some climatic and hydrologic effects of wildfire in Washington State. *Proc. Tall Timbers Fire Ecology Conf.* 15:201–222.

Hetherington, E. D. 1983. A first look at logging effects on the hydrologic regime of Carnation Creek Experimental Watershed. pp. 45–63. In: G. Hartman, Ed. *Proceedings of the Carnation Creek Workshop, a 10 year review.* Dept. Fisheries and Oceans, Fisheries Res. Br., Pac. Biol. Sta., Nanaimo, B.C., Canada.

Hewlett, J. D. 1979. Forest water quality: an experiment in harvesting and regenerating Piedmont forest. *Ga. Forest Res. Pap.*, School of Forest Resources, University of Georgia, Athens, GA.

Hewlett, J. D. and R. Doss. 1984. Forests, floods, and erosion: a watershed experiment in the southeastern Piedmont. *For. Sci.* 30:424–434.

Hewlett, J. D. and J. D. Helvey. 1970. Effects of forest clear-felling on the storm hydrograph. *Water Resour. Res.* 6:768–782.

Hibbert, A. R. 1967. Forest treatment effects on water yield. pp. 527–543. In: W. E. Sopper and H. W. Lull, Eds., *Intl. Symp. For. Hydrol.* Pergamon Press, Oxford.

Hibbert, A. R., E. A. Davis, and O. D. Knipe. 1982. Water yield changes resulting from treatment of Arizona chaparral. pp. 382–389. In: *Dynamics and management of Mediterranean-type ecosystems.* USDA Forest Service Gen. Tech. Rept. PSW-58, Pac. Southwest For. Range Expt. Sta., Berkeley, CA.

Hibbert, A. R., E. A. Davis, and D. G. Scholl. 1974. Chaparral conversion potential in Arizona Pt. I: Water yield response and effects on other resources. USDA Forest Service Res. Pap. RM-126, Rocky Mt. For. Range Expt. Sta., Ft. Collins, CO.

Hickok, R. B. 1967. Water management on semiarid watersheds. *Proc. Arizona Watershed Symp.* 11:9–14.

Hill, L. W. 1960. Forest plantation development influences streamflow. *Proc. Soc. Am. For.*, pp. 168–171, Washington, DC.

Holtan, H. N., G. J. Stiltner, W. N. Henson, and L. C. Lopez. 1975. USDAHL-74 revised model of watershed hydrology. USDA Agric. Research Service Tech. Bull. No. 1518, Washington, DC.

Hornbeck, J. W. 1973. Storm flow from hardwood-forested and cleared watersheds in New Hampshire. *Water Resour. Res.* 9:346–354.

Hornbeck, J. W. 1975. Streamflow response to forest cutting and revegetation. *Water Resour. Res.* 11:1257–1260.

Horton, J. S. and C. J. Campbell. 1974. Management of phreatophytes and riparian vegetation for maximum multiple use values. USDA Forest Service Res. Pap. RM-117, Rocky Mt. For. Range Expt. Sta., Ft. Collins, CO.

Huber, A. L. 1983. A comparison of several snow accumulation and ablation algorithms used in watershed modeling. *Proc. Western Snow Conf.* 51:76–88.

Hutchison, F. A. 1966. A comparison of evaporation from snow and soil surfaces. *Bull. Int. Assoc. Sci. Hydrol.* 11:34–42.

Ingebo, P. A. 1971. Suppression of channel-side chaparral cover increases streamflow. *J. Soil Water Conserv.* 26:79–81.

Isaac, L. A. 1946. Fog drip and rain interception in coastal forests. USDA Forest Service Res. Note No. 34, Pac. Northwest For. Range Expt. Sta., Portland, OR.

Johnson, R. R. and J. F. McCormick. 1978. Strategies for protection and management of floodplain wetlands and other riparian systems. USDA Forest Service Gen. Tech. Rept. WO-12, Washington, DC.

Johnston, R. S. 1984. Effect of small aspen clearcuts on water yield and water quality. USDA Forest Service Res. Pap. INT-333, Intermountain For. Range Expt. Sta., Ogden, UT.

Judson, A. 1965. The weather and climate of a high mountain pass in the Colorado Rockies. USDA Forest Service Res. Pap. RM-16, Rocky Mt. For. Range Expt. Sta., Ft. Collins, CO.

Kelliher, F. M., T. A. Black, and D. T. Price. 1986. Estimating the effects of understory removal from a Douglas fir (*sic*) forest using a two-layer evapotranspiration model. *Water Resour. Res.* 22:1891–1899.

Kerfoot, O. 1968. Mist precipitation on vegetation. *For. Abst.* 29:8–20.

King, J. G. 1984. Ongoing studies in Horse Creek on water quality and water yield. pp. 28–35. In: *Forestry management practices and cumulative effects on water quality and utility.* Tech. Bull. No. 435. Nat. Council for Air and Stream Improvement.

King, J. G. and L. C. Tennyson. 1984. Alteration of streamflow characteristics following road construction in north central Idaho. *Water Resour. Res.* 20:1159–1163.

Kittredge, J. 1953. Influences of forests on snow in the ponderosa-sugar pine–fir zone of the central Sierra Nevada. *Hilgardia* 22:1–96.

Krammes, J. S. and L. W. Hill. 1963. "First aid" for burned watersheds. USDA Forest Service Res. Note PSW-29, Pac. Southwest For. Range Expt. Sta., Berkeley, CA.

Kuusisto, E. 1980. On the values and variability of degree-day melting factor in Finland. *Nordic Hydrol.* 11:235–242.

Langford, K. J. 1976. Change in yield of water following a bushfire in a forest of *Eucalyptus regnans. J. Hydrol.* 29:87–114.

Leaf, C. F. 1975. Watershed management in the Rocky Mountain subalpine zone: the status of our knowledge. USDA Forest Service Res. Pap. RM-137, Rocky Mt. For. Range Expt. Sta., Ft. Collins, CO.

Leaf, C. F. and G. E. Brink. 1973. Computer simulation of snowmelt within a Colorado subalpine forest. USDA Forest Service Res. Pap. RM-107, Rocky Mt. For. Range Expt. Sta., Ft. Collins, CO.

Lewis, D. C. 1968. Annual hydrologic response to watershed conversion from oak woodland to annual grassland. *Water Resour. Res.* 4:59–72.

Love, L. D. 1955. The effect on streamflow of the killing of spruce and pine by the Engelmann spruce beetle. *Trans. Am. Geophys. Union* 36:113–118.

Lovett, G. M., W. A. Reiners, and R. K. Olson. 1982. Cloud droplet deposition in subalpine balsam fir forests: hydrological and chemical inputs. *Science* 218:1303–1304.

Lull, H. W. and H. K. Orr. 1950. Induced snow-drifting for water storage. *J. For.* 48:179–181.

Lynch, J. A., W. E. Sopper, E. S. Corbett, and D. W. Aurand. 1975. Effects of management practices on water quality and quantity: the Penn State experimental watersheds. pp. 32–46. In: *Municipal Watershed Management Symposium Proc.* USDA Forest Service Gen. Tech. Rept. NE-13, Northeastern For. Expt. Sta., Upper Darby, PA.

Mader, D. L., W. P. MacConnell, and J. W. Bauder. 1972. The effect of riparian vegetation control and stand density reduction on soil moisture in the riparian zone. Res. Bull. 597, Ag. Expt. Sta, University of Massachusetts, Amherst, MA.

Martinec, J. and M. R. de Quervain. 1975. The effect of snow displacement by avalanches on snow melt and runoff. *Snow and ice symp.* IAHS-AISH Pub. 104:364–377.

Martinelli, M., Jr. 1959. Alpine snowfields—their characteristics and management possibilities. *Symp. Hann.-Munden., Internat. Sci. Hydrol. Assoc. Pub.* 48(1):120–127.

Martinelli, M., Jr. 1965. An estimate of summer runoff from alpine snowfields. *J. Soil and Water Conserv.* 20:24–26.

Martinelli, M., Jr. 1975. Water-yield improvement from alpine areas: the status of our knowledge. USDA Forest Service Res. Pap. RM-138, Rocky Mt. For. Range Expt. Sta., Ft. Collins, CO.

Meiman, J. R. and C. W. Slaughter. 1967. Long-chain alcohol suppression of snow evaporation. *J. Hydraul. Div. Proc. Am. Soc. Civ. Eng.* pp. 271–279.

Nakano, H. 1971. Effect on streamflow of forest cutting and change in regrowth on cut-over area (English summary). Bull. No. 240, Govt. For. Expt. Sta., Tokyo, Japan.

Oberlander, G. T. 1956. Summer fog precipitation on the San Francisco Peninsula. *Ecology* 37:851–852.

O'Loughlin, E. M. 1986. Prediction of surface saturation zones in natural catchments by topographic analysis. *Water Resour. Res.* 22:794–804.

Oron, G., J. Ben-Asher, A. Isaar, and T. M. Boers. 1983. Economic evaluation of water harvesting in microcatchments. *Water Resour. Res.* 19:1099–1106.

Pase, C. P. and O. D. Knipe. 1977. Effect of winter burns on herbaceous cover on a converted chaparral watershed. *J. Range Mana.* 30:346–348.

Pase, C. P. and A. W. Lindenmuth, Jr. 1971. Effects of prescribed fire on vegetation and sediment in oak-mountain mahogany chaparral. *J. For.* 69:800–805.

Patric, J. H. 1973. Deforestation effects on soil moisture, streamflow, and water balance in the central Appalachians. USDA Forest Service Res. Pap. NE-259, Northeastern For. Expt. Sta., Upper Darby, PA.

Pearce, A. J., L. K. Rowe, and C. L. O'Loughlin. 1980. Effects of clearfelling and slash-burning on water yield and storm hydrographs in evergreen mixed forests, western New Zealand. IAHS-AISH Pub. 130:119–127.

Peterson, T. C. and R. A. Schmidt. 1984. Outdoor scale modeling of shrub barriers in drifting snow. *Agric. For. Meteorol.* 31:167–181.

Plammondon, A. P. and D. C. Ouellet. 1980. Partial clearcutting and streamflow regime of ruisseau des Eaux-Volees experimental basin. IASH Pub. No. 130: 129–136.

Potts, D. F. 1984. Hydrologic impacts of a large-scale mountain pine beetle (*Dendroctonus ponderosae* Hopkins) epidemic. *Water Resour. Bull.* 20:373–377.

Regelin, W. L. and O. C. Wallmo. 1975. Carbon black increases snowmelt and forage availability on deer winter range in Colorado. USDA Forest Service Res. Note RM-296, Rocky Mt. For. Range Expt. Sta., Ft. Collins, CO.

Reinhart, K. G., A. R. Eschner, and G. R. Trimble, Jr. 1963. Effect on streamflow of four forest practices in the mountains of West Virginia. USDA Forest Service Res. Pap. NE-1, Northeastern For. Expt. Sta., Upper Darby, PA.

Reiser, D. W. and T. C. Bjornn. 1979. Habitat requirements of anadromous salmonids. USDA Forest Service Gen Tech Rept. PNW-96, Pac. Northwest For. Range Expt. Sta., Portland, OR.

Rice, R. M., R. P. Crouse, and E. S. Corbett. 1965. Emergency measures to control erosion after a fire on the San Dimas Experimental Forest pp. 122–130. In: *Proc. Fed. Interagency Sedimentation Conf.* USDA Misc. Publ. 970, Washington, DC.

Rice, R. M., R. R. Ziemer, and S. C. Hankin. 1982. Slope stability effects of fuel management strategies—inferences from Monte Carlo simulations. pp. 365–371. In: *Proc. dynamics and management of Mediterranean-type ecosystems.* USDA Forest Service Gen Tech Rept. PNW-58, Pac. Southwest For. Range Expt. Sta., Berkeley, CA.

Rich, L. R. 1965. Water yields resulting from treatments applied to mixed conifer watersheds. *Proc. Arizona Watershed Symp.* 9:12–15.

Rich, L. R. 1972. Managing a ponderosa pine forest to increase water yield. *Water Resour. Res.* 8:422–428.

Rich, L. R. and G. J. Gottfried. 1976. Water yields resulting from treatments on the Workman Creek experimental watersheds in central Arizona. *Water Resour. Res.* 12:1053–1060.

Rich, L. R. and J. R. Thompson. 1974. Watershed management in Arizona's mixed conifer forests. USDA Forest Service Res. Pap. RM-130, Rocky Mt. For. Range Expt. Sta., Ft. Collins, CO.

Robinson, T. W. 1958. Phreatophytes. US Geol. Survey Water-Supply Pap. 1423.

Rosen, K. 1984. Effect of clear-felling on run-off in small watersheds in central Sweden. *For. Ecol. Manage.* 9:267–281.

Rothacher, J. 1973. Does harvest in west slope Douglas-fir increase peak flow in small forest streams? USDA Forest Service Res. Pap. PNW-163, Pac. Northwest For. Range Expt. Sta., Portland, OR.

Rowe, P. B. 1963. Streamflow increases after removing woodland-riparian vegetation from a southern California watershed. *J. For.* 61:365–370.

San Dimas Experimental Forest. 1954. Fire-flood sequences on the San Dimas Experimental Forest. Tech. Pap. No. 6, Calif. For. Range Expt. Sta., Berkeley, CA.

Sartz, R. S. 1969. Folklore and bromides in watershed management. *J. For.* 67:366–371.

Sartz, R. S. and D. N. Tolsted. 1974. Effect of grazing on runoff from two small watersheds in southwestern Wisconsin. *Water Resour. Res.* 10:354–356.

Satterlund, D. R. 1969. Combined weather and vegetation modification promises synergistic streamflow response. *J. Hydrology* 9(2):155–166.

Satterlund, D. R. and A. R. Eschner. 1964. Snowmelt studies on the Tully Forest in central New York. pp. 14–22. In: *Proc. 21st Ann. Eastern Snow Conf.*

Satterlund, D. R. and A. R. Eschner. 1965. Land use, snow, and streamflow regimen in central New York. *Water Resour. Res.* 1:397–405.

Saulmon, R. W. 1973. Snowdrift management can increase water-harvesting yields. *J. Soil Water Conserv.* 28:118–121.

Schneider, W. J. and G. R. Ayer. 1961. Effect of reforestation on streamflow in central New York. US Geol. Survey Water Supply Pap. 1602.

Seyhan, E., A. S. Hope, and R. E. Schultze. 1983. Estimation of streamflow loss by evapotranspiration from a riparian zone. *S. Afr. J. Sci.* 79:88–90.

Silvils, B. E. and J. H. Brock. 1981. Sand dams as a feasible water development for arid regions. *J. Range Manage.* 34:238–239.

Smith, D. R. 1969. *Vegetation, soils and their interrelationships at timberline in the Medicine Bow Mountains, Wyoming.* Sci. Monograph 17, Agric. Expt. Sta., University of Wyoming, Laramie.

Smith, J. L. and H. G. Halverson. 1971. Suppression of evaporation losses from snowpacks. pp. 5–25. In: *Symp. on interdisciplinary aspects of watershed management.* Bozeman, MT, Am. Soc. Civ. Eng.

Solomon, R. M., P. F. Ffolliott, M. B. Baker, Jr., and J. A. Thompson. 1976. Computer simulation of snowmelt. USDA Forest Service Res. Pap. RM-174, Rocky Mt. For. Range Expt. Sta., Ft. Collins, CO.

Sturges, D. L. 1973. Soil moisture response to spraying big sagebrush the year of treatment. *J. Range Manage.* 26:444–447.

Sturges, D. L. 1975. Hydrologic relations on undisturbed and converted big sagebrush lands: the status of our knowledge. USDA Forest Service Res. Pap. RM-140, Rocky Mt. For. Range Expt. Sta., Ft. Collins, CO.

Sturges, D. L. 1977. Soil moisture response to spraying big sagebrush: a seven-year study and literature interpretation. USDA Forest Service Res. Pap. RM-188, Rocky Mt. For. Range Expt. Sta., Ft. Collins, CO.

Sturges, D. L. and R. D. Tabler. 1981. Management of blowing snow on sagebrush rangelands. *J. Soil Water Conserv.* 36:287–292.

Swank, W. T., J. E. Douglass, and G. B. Cunningham. 1982. Changes in water yield and storm hydrographs following commercial clearcutting on a southern Appalachian catchment. pp. 583–594. In: *Proc. Symp. Hydrol. Res. Basins* Sonderh. Landeshydrologie, Bern.

Swindel, B. F. and J. E. Douglass. 1984. Describing and testing nonlinear treatment effects in paired watershed experiments. *For. Sci.* 2:305–313.

Swindel, B. F., C. J. Lassiter, and H. Riekerk. 1983a. Effects of different harvesting and site preparation operations on the peak flows of streams in *Pinus elliottii* flatwoods forests. *For. Ecol. Manage.* 5:77–86.

Swindel, B. F., W. R. Marion, L. D. Harris, L. A. Morris, W. L. Pritchett, L. F. Conde, H. Rierkerk, and E. T. Sullivan. 1983b. Multi-resource effects of harvest, site preparation, and planting in pine flatwoods. *South. J. Appl. For.* 7:6–15.

Tabler, R. D. 1968. Physical and economic design criteria for induced snow accumulation projects. *Water Resour. Res.* 4:513–519.

Tabler, R. D. 1973. Evaporation losses of windblown snow and the potential for recovery. *Proc. Western Snow Conf.* 41:75–79.

Tabler, R. D. 1974. New engineering criteria for snow fence systems. *Transp. Res. Rec.* (NAS-NRC, Washington, DC) 506:65–78.

Tabler, R. D. 1975. Predicting profiles of snowdrifts in topographic catchments. *Proc. Western Snow Conf.* 43:87–97.

Tabler, R. D. 1980. Geometry and density of drifts formed by snow fences. *J. Glaciol.* 26:405–419.

Tabler, R. D. and R. L. Jairell. 1980. Studying snowdrifting problems with small-scale models outdoors. *Proc. Western Snow Conf.* 48:1–13.

Tabler, R. D. and D. L. Sturges. 1986. Watershed test of a snowfence to increase streamflow: preliminary results. pp. 53–61. In: D. L. Kane, Ed. *Cold regions hydrology.* Proc. of a symposium, Am. Water Resour. Assoc., Bethesda, MD.

Tarum, R. D. and J. R. Meiman. 1979. Water yield increases through surface dusting of a permanent ice field. *Water Resour. Bull.* 15:1331–1340.

Tennessee Valley Authority (TVA). 1961. Forest cover improvement influences upon hydrologic characteristics of White Hollow Watershed, 1935–1958. Div. Water Control Plan., Knoxville, TN.

Tennessee Valley Authority (TVA). 1962. Reforestation and erosion control influences upon the hydrology of Pine Tree Branch watershed, 1941–1960. Div. Water Control Plan. Div. For. Devel., Knoxville, TN.

Tiedemann, A. R., C. E. Conrad, J. H. Dieterich, J. W. Hornbeck, W. F. Megahan, L. E. Viereck, and D. D. Wade. 1979. Effects of fire on water. USDA Forest Service Gen. Tech. Rept. WO-10, GPO, Washington, DC.

Toky, O. P. and P. S. Ramakrishnan. 1981. Run-off and infiltration losses related to shifting agriculture (jhum) in northeastern India. *Env. Conserv.* 8:313–321.

Trimble, S. W. and F. H. Weirich. 1987. Reforestation reduces streamflow in the southeastern United States. *J. Soil Water Conserv.* 42:274–276.

Troendle, C. A. 1982. The effects of small clearcuts on water yield from the Deadhorse watershed, Fraser, Colorado. *Proc. Western Snow Conf.* 50:75–83.

Troendle, C. A. 1983. The potential for water yield augmentation from forest management in the Rocky Mountain region. *Water Resour. Bull.* 19:359–373.

Troendle, C. A. and R. M. King. 1985. The effect of timber harvest on the Fool Creek Watershed, 30 years later. *Water Resour. Res.* 21(12):1915–1922.

Troendle, C. A. and C. F. Leaf. 1980. Hydrology. Chapt. III. In: *An approach to water resources evaluation of non-point silvicultural sources.* EPA 60018-80-012, Env. Res. Lab., Athens, GA.

Troendle, C. A. and C. F. Leaf. 1981. Effects of timber harvest in the snow zone on volume and timing of water yield. pp. 231–243. In: D. M. Baumgartner, Ed. *Interior west watershed management.* Coop. Exten., Washington State University, Pullman, WA.

Tschinkel, H. M. 1963. Short-term fluctuation in streamflow as related to evaporation and transpiration. *J. Geophys. Res.* 68:6459–6469.

Ursic, S. J. 1966. Forest hydrology research in Mississippi. pp. 98–103. In: *Proc. Water Resour. Conf.* Water Res. Instit., Miss. St. University, Starkville, MS.

Ursic, S. J. 1970. Hydrologic effects of prescribed burning and deadening upland hardwoods in northern Mississippi. USDA Forest Service Res. Pap. SO-54, Southern For. Expt. Sta., New Orleans, LA.

US Army. 1956. *Snow hydrology.* North Pac. Div., US Army Corps of Engineers, Portland, OR.

US Army. 1960. Runoff from snowmelt. US Army Corps of Engineers, Eng. Manual 1110-2-1406.

Verry, E. S. 1986. Forest harvesting and water: the Lake States experience. *Water Resour. Bull.* 22(6):1039–1047.

Webber, P. J. and J. D. Ives. 1978. Damage and recovery of tundra vegetation. *Environ. Conserv.* 5:171–182.

Weisman, R. N. 1977. The effect of evapotranspiration on streamflow recession. *Hydrol. Sci. Bull.* 12:371–377.

Wesche, T. A. 1974. Relationship of discharge reduction to available trout habitat for recommending suitable streamflow. Ser. No. 53. Wyoming Water Resour. Res. Inst., Laramie, WY.

West, A. J. 1961. Cold air drainage in forest openings. USDA Forest Service Res. Note No. 180, Pac. Southwest For. Range Expt. Sta., Berkeley, CA.

Wight, J. R., E. L. Neff, and F. H. Siddoway. 1975. Snow management on eastern Montana rangelands. pp. 138–143. In: *Proc. Snow Mgt. on the Great Plains Symp.* Great Plains Agric. Council Pub. 73, Agric. Expt. Sta., University of Nebraska, Lincoln, NE.

Ziemer, R. R. 1981. Storm flow response to road building and partial cutting in small streams of northern California. *Water Resour. Res.* 17:907–917.

12 Control of Water Quality

Concern for a quality environment has focused on water as never before. Streamflow issuing from undisturbed wildlands represents the highest standard of quality in the public mind. Despite very wide natural variation, most people think of it as always pure, clear, and cool—and want it kept that way. At the same time, they demand greater levels of the myriad other values produced on wildlands. It is the task of watershed management to reconcile these often conflicting demands insofar as it is possible to do so.

Many characteristics combine to determine water quality. Foremost are those that determine human health and safety. Waterborne diseases were a major scourge of humanity until public water supplies were recognized as potential sources of illness and steps were taken to protect them from contamination by microorganisms, or to treat them to eliminate the danger. Dissolved or entrained chemicals such as pesticides, drilling muds from oil wells, acids from mines, or nitrates from feedlots may damage desirable aquatic ecosystems. Contrary to popular opinion, however, complete chemical purity of water is not natural and attaining it is seldom possible or even desirable. Aquatic life depends on natural organic and inorganic materials moving through the food chain and the taste of water may be improved by low levels of dissolved minerals.

High quality water is usually cool and high in dissolved oxygen. The solubility of oxygen is higher in cool water than in warm, giving cool, well-aerated water a better taste, providing a better habitat for demanding fish such as salmonids, and engendering its properties of self-purification. The decomposition process of organic material is usually that of oxidation, so heavy loads of organic materials exert a *biochemical oxygen demand* (BOD) that can reduce or eliminate the supply of dissolved oxygen in the water. The problem may become acute if waters are not well aerated or if they are warmed by exposure to the sun or discharge of heated water. Warm waters that are low in dissolved oxygen often exhibit undesirable colors, tastes, and odors that result from anaerobic decomposition processes.

Of all the factors that affect water quality, however, few receive more attention than sediment and turbidity. Sediment is far and away the major polluting agent in the world. In terms of volume, it exceeds all other sources combined: domestic wastes, industrial wastes, and others. Most sediment comes from cultivated and urban lands; semiarid rangelands also contribute large quantities, and forested basins the least (Patric et al., 1984).

Nevertheless, sediment still overshadows all other wildland water quality problems. There are several reasons that sediment is the primary problem. For example, turbid water is easily recognized by almost anyone. Furthermore, sediment can be more dramatically changed than any other water characteristic. It is a difficult and exacting task to demonstrate changes in annual yields or regimen resulting from land management activities, for they are readily masked by natural variations and seldom exceed an order of magnitude of change. But sediment production can be increased by 1000 times or more. A study in the mountains of West Virginia where no attention to the water resource was given during logging resulted in a peak turbidity more than 3000 times greater than that from a nearby undisturbed watershed (Reinhart et al., 1963). Though such practices are no longer common today, it is not difficult to appreciate that severe disturbance on a small part of a watershed can result in extensive visible damage to water quality.

Sediment is not independent of other water quality characteristics. Erosion may be accompanied by organic wastes and adsorbed chemicals such as nutrients and pesticides. Sediment deposited in a stream or lake may retain these materials in the system for extended periods. Sediment may also contribute to higher water temperature when it reduces the depth of streams or lakes. The higher temperature and decomposing organic materials may combine to reduce dissolved oxygen. It should be clear that control of erosion may contribute to the solution of many other quality problems in wildland management, as well as solve the problem of sediment directly.

12.1 EROSION AND SEDIMENT CONTROL

All erosion control rests upon the ability to recognize, modify, or prevent the causes of erosion. Since many causes are natural and cannot be controlled by humans, some erosion and sedimentation will always occur. This is true of all kinds of erosion. But where logging, road building, grazing, or other wildland activities take place, erosion can be accelerated beyond the geologic norm. The increase often cannot be completely prevented, but it can be minimized. It is in this sense of keeping accelerated erosion to a minimum that the term erosion control is used. Most lands have the capacity to withstand the impact of use without permanent loss of productivity so long as accelerated erosion is kept within acceptable limits.

12.1.1 Basic Principles

Each land use activity—logging, road construction, grazing, mining, and so on—presents individual problems not shared completely with the others and, therefore, will be discussed separately. However, certain basic elements of erosion control are common to all and are reviewed here before individual problems are discussed.

12.1.1.1 Surface Erosion Control. The most widespread form of erosion subject to control is surface erosion by raindrops and overland flow. In most situations on wildlands, control is achieved by protecting the soil surface from exposure, compaction, and displacement. Many proverbs point out that prevention is easier than cure, but most land uses inevitably result in surface disturbance that initiates some acceleration of erosion. However, if the disturbance is not excessive or is stopped, the site can maintain or regain a stable, dynamic equilibrium condition, whereby erosion rates remain within acceptable limits.

12.1.1.1.1 The Critical Point in Deterioration. A useful principle to illustrate these points is that of the critical point in deterioration, which may be stated as follows. For most sites there likely exists a critical point in deterioration due to disturbance causing surface runoff and erosion, at which further deterioration of the site becomes self-sustaining, with natural stabilizing factors relatively incapable of reversing the process. A diagrammatic representation of this principle on a moderately stable site is shown in Fig. 12.1.

This principle applies because accelerated erosion deteriorates the site to an ever greater degree unless the process is reversed. The more surface runoff occurs, the more soil is removed and the less water and nutrients remain to support the plant growth that would protect it from further deterioration. The microclimate deteriorates and excessive heat may dessicate the surface and frost heaving and scouring by runoff may inhibit establishment of new vegetation. Raindrop compaction and exposure of dense subsoils further reduce infiltration capacities, increasing the proportion of precipitation that becomes surface runoff even more. A vicious cycle is initiated that reduces favorable conditions for plant growth to a minimum, and erosion rates reach their maximum. Once the critical point in deterioration is exceeded, the cycle is carried to completion with the removal of the soil mantle unless people intervene to reverse the process.

The critical point is not fixed for any given site, but rather is influenced by erosive and plant recovery conditions subsequent to disturbance. If weather conditions are favorable, that is, if gentle, frequent rains and mild temperatures occur, healing of even severe disturbances may be rapid and complete, whereas severe storms and extremes of temperature may lead to the breakdown of site stability subsequent to much lesser disturbances. Thus, the critical point in deterioration oscillates over a range of conditions even on the same site.

Nevertheless, sites vary widely in their resistance to deterioration and in their natural stabilization potential. The factors of climate, soils, topography, vegetation, and others are all important in this respect. Thus, we would not expect the same disturbance to cause the critical point to be reached on different sites. Some sites, badlands, for example, lie naturally at or beyond the critical point in deterioration. Erosion can be expected to continue to degrade the site until a new dynamic equilibrium is reached, the rate being limited only by the availability of runoff to detach and transport material away. Other sites,

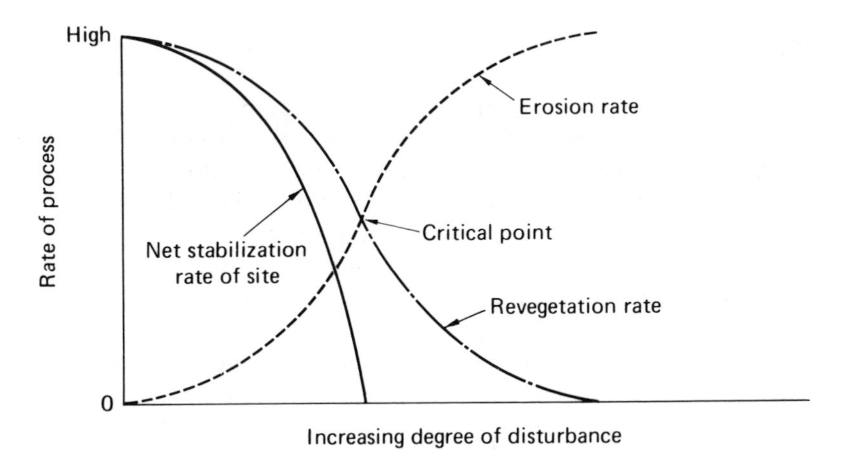

FIG. 12.1 Relationship of deterioration of a site by erosion to the rate of revegetation. At the critical point in deterioration, accelerated erosion becomes self-sustaining and the site cannot recover by natural processes of revegetation. To the right-hand side of the critical point, deterioration continues to completion. To the left, the site recovers its stability at an increasing rate.

such as some of the steep southwest slopes of the Batholith of central Idaho or certain alpine slopes, are in a delicately balanced equilibrium under natural conditions. Geologic erosion rates are relatively modest and slopes are stable as long as the thin line of vegetative cover is not breached. But on such fragile sites, very little disturbance is sufficient to induce accelerated, self-sustaining erosion. On the other hand, there are some sites so stable and characterized by such favorable growing conditions that it is extremely difficult to keep vegetation from forming a dense, protective cover over the site. They reach the critical point in deterioration only under complete and prolonged disturbance that maintains a continuously bare surface. As fines are removed by erosion, certain stony soils may develop a surface armor layer of coarse fragments called *erosion pavement* that helps protect the underlying soil from further erosion (Megahan, 1974a).

It follows from the above that the closer the critical point in deterioration of any site is approached, the greater the amount and frequency of surface runoff and erosion, and the longer it takes before complete recovery is achieved. Conversely, the farther above the critical point, the less damage and more complete and rapid is the recovery from any disturbance.

It is not enough to know the critical point in deterioration for a site. Unacceptable damage, both on- and off-site, is likely to occur long before that point is reached. Managers must be able to classify various sites with which they work according to relative resistance to, and recovery from, natural and human-caused impacts that disturb them.

In many parts of the United States, formal soil–vegetation surveys of public and private lands have been undertaken that provide the type of information needed and, in addition, also provide expert interpretation on specific hazards of use (Gilkeson et al., 1961; Corliss and Dyrness, 1965; Mallory and Alexander, 1965; etc.). Recent soil survey reports of the Soil Conservation Service contain valuable interpretations for wildland use (Barker, 1981). Similar surveys have been produced locally by the US Forest Service and Bureau of Land Management. In some places interpretations of soil characteristics for entire regions are available, as in the Puget Sound Trough Area of Washington (Schlots and Quam, 1966). Aerial photographs, geologic maps, and US Geological Survey Topographic Quadrangle sheets are useful, but require considerable interpretation by the person on the ground (Burroughs et al., 1976).

Even the best of surveys cannot substitute for skilled observers on the ground. No survey can detail every key feature and a given soil or land type does not have the same degree of resistance to damage or potential for recovery everywhere. Good surveys will help indicate what to look for and where to look, but only careful observations on the ground will reveal the varying impacts of past use on a variety of sites. From such information, local guidelines can be developed based on topography, soils, and vegetation. They can be used to train field personnel to recognize limitations and potentials of individual sites that are too small to delineate on most maps, but are often the key to erosion problems. Once these are understood, measures can be taken to prevent unacceptable deteriorations.

12.1.1.1.2 Establishing and Maintaining Vegetation. In many cases, past land use has caused conditions to approach the critical point in deterioration and unacceptable erosion continues. Though the original cause of deterioration may have been long since removed, where natural forces of revegetation are too slow to prevent continuing erosion, people must intervene to establish and maintain a vegetative cover. Furthermore, so long as managers lack perfect knowledge and forecasting skills, some land management activities will inevitably result in unacceptable erosion. Thus, there are always some lands requiring rehabilitation.

The first requirement in revegetation is a careful evaluation of each particular situation to identify the potentials and constraints existing on the site. Any of several factors may be preventing plant establishment, ranging from simple lack of seed to severe physical or chemical conditions. Among the site elements that should be examined are soil physical and chemical conditions. High soil densities may limit water and root penetration. Shallow soils may provide little rooting depth or water holding capacity. Soil crusts may prevent infiltration or inhibit seedling emergence. Silty textures may be associated with susceptibility to frost heaving or high erodibility. Seeds may roll down steep slopes when they are dry or be washed off during rain. Slopes also influence the microclimate, with bare equatorward and west slopes being hotter and drier in summer and exhibiting more freeze–thaw cycles in many regions. Excesses of soil acidity or salts may inhibit plant growth. Toxic substances may be present, or essential nutrients may be deficient. Nitrogen is particularly likely to be in short supply on bare, eroding soils.

Once the limitations and potentials of the site are determined, then vegetation can be selected that is adapted to the situation and appropriate measures of establishment can be taken. Where the site is severely degraded, intensive effort may be necessary to insure plant establishment. In severe cases, as where slopes are steep and severely rilled or gullied, it may be necessary to reshape the land surface, providing drainageways with stable base levels to prevent further cutting. In many situations a suitable seedbed must be established, often by cultivation. Fertilizers and soil amendments may be required. Seedlings or plantings may need protection from washing, dessication, or frost heaving. Mulches often are used to aid establishment. In dry climates, irrigation may be helpful, particularly during the first few years. Where irrigation is impractical, it may be possible to design tiny water harvesting systems to concentrate limited water supplies on planted seedlings. The measures to be taken depend not only on the site involved, but also on the plant species selected.

Each plant species may have a different value in erosion control, may be best adapted to a particular type of site and may require different cultural measures for establishment. The first criterion in choosing vegetation to control erosion is to find a species or combination of species that can be established on the site and will provide a maximum of protective cover. Local species are often, but not always, superior to introduced species. In some situations, as in reclaiming strip mines, local species may be required by law or regulations.

Pioneer species (plants that occur early in succession) are usually better adapted to the severe conditions likely to exist than are climax species. However, some species that are most easily established on disturbed sites may become noxious weeds and should be avoided. Plants that fix atmospheric nitrogen are often desirable. Those having the capacity to spread or reproduce vegetatively, as by stolons, rhizomes, tillers, layering, sprouts, or suckers, may provide a more rapid cover than species that lack such features.

Sometimes there may be several species or combinations of species that will control erosion equally well. In such cases, choice may well revolve upon secondary considerations of capacity for site improvement, aesthetics, forage, wildlife, or timber benefits. However, species highly palatable to domestic or wild animals should be avoided if they are likely to attract damaging use before the site is fully stabilized.

To evaluate site-specific needs, a complete and detailed knowledge of available plant species and techniques of establishment are needed. Fortunately, many very good specialized reports are available that are applicable to almost every situation, except for some of the lesser developed countries. The number of journals treating such topics has expanded in the past decade and include older standbys as *Journal of Soil and Water Conservation, Journal of Range Management, Agronomy Journal, Soil Science Society of America Journal, Arctic and Alpine Research,* and newer ones such as *Environmental Conservation, Reclamation Review, Reclamation and Revegetation Research,* among others. They provide up-to-date information regularly. In the United States, manuals are available from federal agencies, such as the Soil Conservation Service and the Forest Service, and from numerous state agencies such as highway departments, fish and game departments, universities, and county extension agents, and many others. They provide methodology, site-specific recommendations, and key references (e.g.: Stark, 1966; Springer et al., 1967; Plummer et al., 1968; Hafenrichter et al., 1968; Gallup, 1974; USDA Forest Service, 1979, 1980; Cole and Schreiner, 1981; Laycock, 1982; Thornburg, 1982; Long et al., 1984; Barro and Conard, 1987). The Food and Agriculture Organization (FAO) of the United Nations has published guidelines that are useful in many parts of the world (e.g.: Kunkle and Thames, 1976, 1977).

Where large gullies have developed and the land cannot be reshaped, the gully system may be stabilized with check dams that provide a new base level that prevents further cutting. Large gullies may be considered to be ephemeral stream channels. Channel cutting is considered in a separate section.

12.1.1.2 Control Over Mass Movement. The economic control of mass movement on wildlands depends almost completely on recognizing unstable areas and avoiding disturbance. The primary factors influencing slope stability were presented in Chapter 10. They are (a) slope steepness in relation to the angle of internal friction, (b) soil moisture, (c) geology and soil characteristics, and (d) vegetation as it affects soil moisture and rooting depth and strength. Danger signals of potential instability include (Varnes, 1958; Arnold, 1963 a, b; Bailey, 1971; Swanston, 1974; Swanson and James, 1975; Rice, 1977; Pole and Satterlund, 1978; Sidle et al., 1985; Thomas, 1985; Duncan et al., 1987):

1. *Slope Shape, Position, and Steepness.* Most mass failures occur on moist slopes with steepness near their angle of internal friction. Both moisture and steepness are related to slope position. *Headwalls*, the steep concave or straight slopes just above the head of first-order streams may be the most unstable part of small drainages. Bedrock depressions with surface swales having their long axis downslope that collect colluvial fill (and water) are common failure sites. The zone of contact where steep slopes break away from flatter uplands are often unstable, particularly where soil depth is sharply reduced and groundwater is forced to the surface. Highly dissected slopes tend to be less stable than smooth ones. Water tends to concentrate on the lower portions of straight and convex slopes, but tends to occur just below the steepest half of concave slopes and wherever slopes converge (Figs. 12.2–12.4). Other factors equal, stability decreases very rapidly beyond a steepness of about 35% ($\sim 20^\circ$) for most slopes.

2. *Wet Areas Such as Springs, Seeps, or Ponded Water.* Any place with visible wetness near the surface of any slope is likely to be unstable for some distance both above and below the wet area.

3. *The Kind, Amount, and Condition of Vegetation.* Lush understory vegetation on slopes may indicate groundwater near the surface, as do certain tree species. For example, in the Pacific Northwest of the United States, red alder grows vigorously on stream bottoms, but less well on slopes, except where there is seepage water near the surface or where heavy soil disturbance (e.g.,

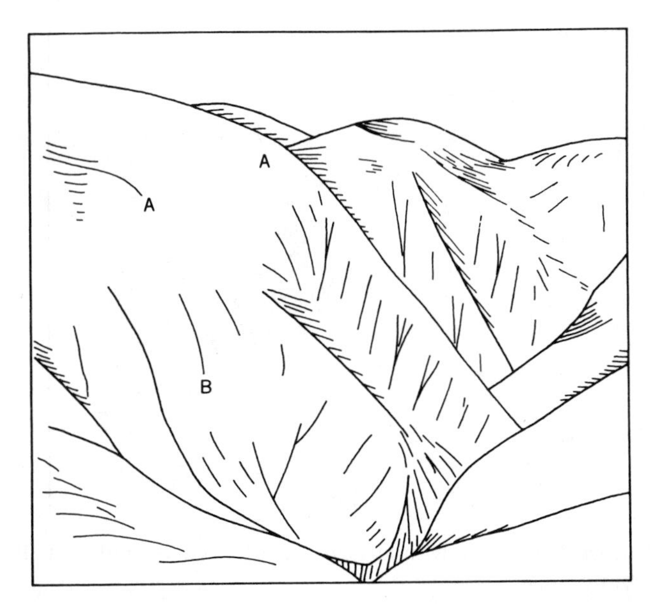

FIG. 12.2 The zone of contact where steep slopes break away from areas of deep soils on flatter uplands (A) are often unstable, as are the steep portions of convex slopes (B) where water concentrations occur.

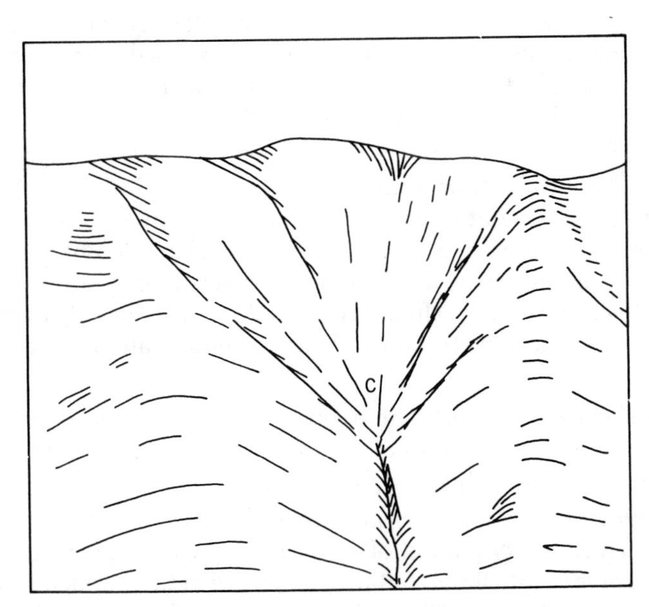

FIG. 12.3 Slope convergence zones (at and below point C) are places of water concentration that are often unstable.

FIG. 12.4 Bedrock depressions with surface swales that collect colluvial fill and water are frequent failure sites.

past mass movement) has occurred. The presence of vigorous alder patches, especially in streaks down through a coniferous slope should alert the manager of a potential instability problem. Similar indicators exist to some degree in every region, so when common streamside or pioneer vegetation is found on slopes, beware. Patches of blowdown often occur where trees are shallow rooted because of excess moisture on exposed slopes. In areas of ultramafic rocks, grassy glades in conifer forests may indicate the presence of unstable serpentine soils (Burroughs et al., 1976).

4. *Leaning or Jackstrawed Trees.* Slow soil creep often causes trees on a slope to tilt gradually outward. Sometimes the lower bole will be tilted, but the tip continues to grow straight upward, giving a broad upward concavity to the stem. On aerial photographs, leaning trees may give a tentlike appearance over streams and swales in unstable areas, whereas they grow upright on nearby stable portions of the slope. Trees tilted or tipped crazily in more than one direction (*jackstrawed*) often indicate recent rotational movement of the slope.

5. *Stress Cracks in the Soil.* These always indicate recent movement and represent development of a partial failure plane. Where vegetation is dense, they may be obscured by litter or ground cover.

6. *Microrelief.* Many unstable areas can be discerned by characteristic microrelief. Hummocky ground often indicates creep, and stepped terraces are evidence of multiple slump failures. The weathered, rounded forms of old slides, slumps and flows are often evident to the careful observer for centuries or more.

7. *Exposed Rock Strata.* Many rock types have weak bedding planes that dip parallel to the slope and are prone to slide failure. They often extend for long distances along a slope of the same dip, so their presence can be inferred from exposed areas even though they may be visible only intermittently. Where strata dip perpendicular to a slope, stability is enhanced.

8. *Other Evidence that May Not Be Visible at the Surface.* These include such characteristics as bedrock type and stratigraphy; location and character of faults, folds and joints; degree of weathering; and soil characteristics of type of parent material, cohesiveness, internal drainage, depth, plasticity, and texture. The presence of these features is often indicated on geological and soil maps.

Sometimes it is possible to prevent accelerated development of shear stress and oversteepened slopes from active cutting of stream channels or gullies by protecting banks, controlling runoff, and building check dams. Proper road location and construction practices can also minimize additional shear stress and reduction in shear resistance, or at least insure that the necessary disturbance is applied to sites capable of withstanding its impact. Timber harvesting can be restricted or modified (e.g., partial cuts) where live roots are known to contribute the critical balance to stability. All these factors are discussed in more detail in the following sections.

12.1.1.3 Control of Channel Erosion. Maintenance of channels in good condition usually means keeping them in, or restoring them to, an equilibrium state so that neither aggradation nor degradation remains a dominant condition. Where channel conditions have deteriorated in the past, either process may dominate until equilibrium is restored, as when fine sediments that clog spawning gravels and pools are flushed away (Megahan et al., 1980) or as when check dams trap and stabilize sediments in gully control (Heede, 1976). Control of aggradation problems usually rests upon limiting the sediment input from uplands, but channel degradation, once initiated, may continue to produce sediment by cutting as the channel adjusts to a new base level. Downcutting may also cause steep slopes to fail, with possible catastrophic channel degradation as debris torrents scour banks and bottom down to bedrock, with debris displaced downstream.

Past land use has left many streams and rivers in a degraded condition. Trapping completely eliminated beaver from many streams throughout North America. Beaver dams were not maintained and their failure released vast quantities of sediment, and new cutting cycles were initiated as stable base levels were removed. In many western streams, later overgrazing eliminated much of the riparian vegetation, preventing repopulation of potentially suitable habitat and increasing runoff from surrounding uplands.

Across the northern parts of the world, the use of streams to transport timber products also resulted in channel degradation. Channels were cleared of all obstructions to prevent jams, and splash dams were often built to release stored water so that logs would be flushed downstream by a surge of high water. The transport network was mostly in third or higher order streams, but sometimes extended into first- and second-order tributaries. In many states, such as Washington, any stream capable of floating logs during freshets, even if it dried up in the dry season, was considered public water and available for transporting logs downstream. Most river driving of logs has ceased in the United States, but past drives have left scoured streambeds and gouged banks that have not yet healed (Sedell and Duval, 1985).

Solutions to the problems of channel instability, as with most others, require first that the cause of instability be removed before any further control efforts can be successful. This means that the manager must understand dynamic equilibrium processes of streams (Heede, 1986; Beschta and Platts, 1986). Where aggradation is the problem, a primary requirement is to prevent excessive sediment from reaching the channel; but where degradation is the problem, a variety of causes and potential solutions may be involved, requiring control of streamflow energy, increasing resistance of the channel to cutting, or both.

12.1.1.3.1 Control of Streamflow Energy. Stream power is a function of stream discharge and flow velocity (cf. Section 10.4.1). Therefore, reducing discharge, reducing velocity, or both will help to reduce the energy available for channel erosion. Furthermore, in a stream not supplied with sediment from outside the channel, more energy is available for cutting than with sediment-laden water where most of the energy is required to transport the existing load. So success in eliminating the entry of sediment into the channel system or trapping it in reservoirs may increase the danger of scouring channel banks or bottom downstream, particularly in alluvial channels.

All channels ultimately come into equilibrium with their characteristic streamflow. For example, many gully systems have developed where land use caused an increase in the amount and frequency of stormflow. Climatic change may also cause channel adjustment; DeWalle and Rango (1972) found that a sustained 15% increase in precipitation would ultimately increase channel width, depth, and area by 3, 4, and 8% in small forested basins of Pennsylvania in response to increased runoff. That is not to suggest that all, or even most, nonpermanent increases in runoff are likely to initiate degradation in stable channels. Most stream systems are naturally subjected to considerable variability in flow that may persist for several decades without noticeable effect. For example, Troendle and King (1985) do not mention any changes in the channel of Fool Creek, Colorado, in the 30 years since treatment increased annual water yields by 40%, and peak flows increased as much as 23% over those prior to treatment. Sediment yields from the entire watershed averaged only 16.3 kg ha^{-1} in the first 15 years since road construction and logging and decreased in the latter one-half of the period (Leaf, 1970). The rate of adjustment to increased flow in most stable systems must be either very slow or dependent on the occurrence of exceptionally high flows that are unlikely to occur before most watersheds recover from disturbance.

Unstable channels are different. They may quickly deteriorate when subjected to increased runoff, particularly if the increases are concentrated as stormflow. Once begun, cutting may continue even after runoff is reduced to normal. Therefore, land management to limit increased stormflow will help to prevent such channel degradation.

The major control over streamflow energy in most streams lies in control of the stream gradient. Base levels must be maintained or reestablished where they have been breached. Stable obstructions such as bars, logs, snags, boulders, and others should be left undisturbed. They not only provide reduction in stream velocity by creating steps, but may increase the Manning roughness coefficient (Eq. 10.4) in the channel (Marzolf, 1978). In forest headwater streams, the manager should leave enough trees to supply a continuing source of large debris to replace that lost to decomposition, disintegration, and removal (Bryant, 1985).

Where large debris is not available, structures may be necessary to establish stable control points. Much structural work has been done in gully rehabilitation projects on semiarid rangelands and in torrent control in alpine regions. *Check dams* are small dams designed to provide a stable, erosion resistant step in a stream channel to reduce the upstream gradient and trap and store sediment. *Sills* provide a low, erosion resistant base level over which water flows, but do not substantially reduce gradients or store sediment. They do prevent further increases in stream gradient. Large boulders have been placed in channels to reduce water velocities and create pools and gravel bars (Reeves and Roelofs, 1982).

Check dams may be porous or nonporous and may be made of loose rock, wire bound rocks, wire mesh gabions, fences filled with rock or brush, concrete, or other solid structures. Most check dams and sills are designed as much to provide resistance to channel cutting as to control stream power (Heede, 1976; Hattinger, 1976).

12.1.1.3.2 Control of Resistance to Channel Cutting. No channel can be stabilized so long as downcutting occurs. The first requirement for a stable channel is the maintenance of a stable base level. Therefore, large forest debris that is relatively stable should not be removed from channels, and any other activity that would disturb or modify a channel in a manner that might initiate downcutting should be avoided. Bryant (1983) and Bilby (1984) provide some guidelines for identifying stable woody debris.

Where downcutting is active, it must be controlled before any other channel stabilization measure can be effective. The most common measures to control downcutting are construction of check dams and sills, which stabilize the channel gradient. Sills constructed of gabions have also been used to retain gravel for spawning habitat for salmonids, as well as to protect against downcutting (Reeves and Roelofs, 1982). Downcutting also occurs as headcuts extend channels into headwater areas. Headcuts usually require structural stabilization before vegetation can be successfully established. Design criteria for check dams may be obtained from several sources (Heede and Mufich, 1973; Heede, 1966, 1976; among others).

Sometimes gullies can be reshaped and rerouted to build a new watercourse with a reduced gradient and broad, shallow cross section that reduces streamflow energy and permits establishment of vegetation (Heede, 1968, 1978). The same principles may be applied to design drainageways in severely disturbed lands such as strip mined lands.

Beaver have also been reintroduced to stabilize deteriorated channels where habitat is suitable to support them (Apple, 1985). However, they should not be introduced where they will become a nuisance by damming irrigation ditches or culverts or where they flood valuable land.

Leaving riparian trees to provide a continuing source of roots and large debris to maintain resistance to channel downcutting is recommended, but there is considerable difference between natural debris entering channels and logging slash that is allowed to enter streams. Natural debris tends to be added in two discrete size classes: small and large. The small material; leaves, needles, fruits, and twigs, is deposited regularly, but is rapidly broken down and lost from the system. The large material enters only intermittently. It tends to consist of long pieces and is frequently anchored to the banks. Natural debris seldom constricts the channel enough to back up a large head of water and tends to become stabilized over time.

If logging occurs along channels, slash tends to be deposited in large quantities, all at once. Small and large pieces are intermixed, and debris jams may be formed that back up a considerable head of water. However, the larger pieces are usually much shorter than natural debris and less well anchored to the banks. Such jams may be strong enough to resist moderate rates of streamflow, but fail when subjected to high flows. Then, streambanks and beds may be subjected to a surge of water mixed with debris and previously stored sediment that can sluice a channel down to bedrock where channel gradients are steep. Once in motion, such a surge may cut through otherwise stable stream segments.

Even where downcutting is controlled, some channels still actively erode by lateral cutting, particularly if streamside cover is poor or banks are high, steep, and undercut. Low banks may be undercut on low gradient streams and remain stable if they are well vegetated and root cohesion is high; in fact, such undercut banks are highly valuable fish habitat. Cattle trampling and grazing can be a major cause of bank erosion on rangelands, particularly during periods of high streamflow (Hunt, 1979; Platts, 1981; Marlow et al., 1987).

Streamside cover is the key to maintaining bank stability under most circumstances. However, where strong currents are directed against steep, high banks, undercutting or bank steepening may jeopardize the entire slope. Slopes may be protected by log or rock jetties anchored or keyed into the bank and angled downstream to deflect the current and protect the bank. Such structures mimic the action of large natural forest debris. Other structural means to protect banks include placement of rock riprap, concrete blocks, retaining walls, or rock-filled gabions from the toe of the slope upward to cover the surface receiving the impact of high flows (Striffler, 1960; Tobiaski and Tripp, 1961; Urie, 1967; Reeves and Roelofs, 1982). The size of rock used for riprap must be sufficient to withstand displacement at the highest velocity of flow expected. Tables are available to determine the size of stone needed (Searcy, 1965). Where high banks are undercut or oversteepened, they must be cut back or otherwise reach a stable angle of repose before any other protective measure can be successful.

Frequently, stones or concrete are too costly or unavailable locally. Then, less effective substitute measures must be taken. If grading oversteepened banks is necessary, a slope flatter than the natural angle of repose may be desirable to help reduce flow impact on structural measures or vegetation used to stabilize the bank. Logs, whole trees, or brush mulches held down with wire netting or fencing may be anchored to protect the banks. For example, whole juniper trees were used to halt erosion on the South Fork of the John Day River in Oregon (Sheeter and Claire, 1981). Pilings are sometimes driven to stabilize slope toes. Vegetation may be established in protected areas and gives long-lasting protection, but it may require continued protection from the current. Organic materials such as logs, trees, and brush mulches decompose and disintegrate with time, and unless natural sources are available to replace them, periodic inspection and maintenance are necessary to insure continuing protection.

Measures needed to establish vegetation on streambanks are similar to those used elsewhere and may include seedbed preparation, fertilization, planting, and protection from scouring. The choice of species may differ from those used on upland sites; for example, plants should be adapted to withstand periodic inundation, scouring, or possible sediment deposition (Logan, 1979). Over a 10-year period, as much as 91 cm of sediment was deposited on the banks of a rehabilitated section of Camp Creek, Oregon (Winegar, 1977). Where eroding low banks occur upon trout streams of low gradient, woody species that develop a prostrate growth form, such as speckled alder or Sitka alder, should be avoided. They frequently weaken streambanks as they tip into the stream and tend to accumulate sediment along the edge of the stream, causing the channel to widen, become shallower, and straighten. Where a fisheries is important, grasses and forbs may be superior for protecting against bank erosion while still providing suitable aquatic habitat next to the bank (Hunt, 1979).

On rangelands, streamside cover may be destroyed by browsing, grazing and trampling. Where such overuse requires rehabilitation to restore bank stability, use must be controlled. Fencing the stream or other means limiting use can be highly effective (Winegar, 1977; Elmore and Beschta, 1987), but requires careful planning, some cost, and continuous maintenance.

12.1.2 Timber Production and Harvesting

Timber production activities contributing to problems of erosion and sediment are primarily those of harvesting, site preparation for regeneration and road building. Roads will be the subject of a separate section, for alone they can be a major source of water quality problems and they are important for nearly all land management activities.

In forestry, most regeneration is initiated by removal, in whole or in part, of the existing stand, so harvesting and regeneration go hand in hand and are appropriately considered together. In fact, *silviculture systems*, the programs of treatment over a forest rotation, are named for the type of regeneration cutting by which a stand is reproduced. Thus, they may include clearcutting, seed tree, shelterwood, or selection methods of regeneration (Smith, 1962; USDA Forest Service 1973). Silviculture systems should not be confused with *logging methods*, which refer to the techniques used to remove trees from the forest and are usually named for the method used to transport the material to landings or roads. Thus, logging methods may include horse logging, tractor (crawler or wheeled) skidding, various cable yarding systems, balloon or helicopter logging, and others (Simmons, 1979; Conway, 1982; Wenger, 1984).

From the standpoint of erosion and sediment production, the logging method and care used in harvesting are usually far more important than the silvicultural system chosen. The only intrinsic differences between clearcutting and partial cutting systems are those related to root cohesion and soil moisture during the period of regeneration and early growth of the new forest. Clearcutting therefore poses a greater hazard than partial cutting only where trees are important for slope stability, whereas any logging method that causes extensive soil exposure and compaction may result in surface erosion regardless of silviculture system and may cause water concentrations that contribute to slope failure as well.

Planning for protection against erosion and sediment transport during timber harvesting and stand regeneration must take into account not only the goals of management but also the capacities and limitations of the site, stand, logging equipment, and techniques. There can be many critical areas that require careful attention (Megahan and King, 1985), but overall, the most important sites are runoff source areas. Local topography and soil characteristics as they determine equipment feasibility, slope stability, soil erodibility, trafficability, and susceptibility to compaction are also important. Climate may influence the timing and degree of site and equipment limitations and determines the probability of erosive conditions. A step-by-step examination of the timber harvesting process will be undertaken in the next sections that will help reveal how logging and site preparation techniques may best be matched to sites to achieve the goal of minimizing erosion and sediment transport from the forest.

12.1.2.1 Timber Felling and Related Activities. In most cases, direct watershed damage from felling, limbing, and buckling of trees into logs by individual loggers using chainsaws is probably negligible. Trees weighing up to many tons inevitably compact the soil surface where they slam to the earth, but the degree and areal extent of compaction in felling operations is likely to be small. For example, in two studies where forests were clearcut and the trees left where they fell, there was no measurable increase in surface runoff and only negligible increases in erosion (Bates and Henry, 1928; Hoover, 1944).

Nevertheless, there are places where felling warrants special care. Trees felled into or across stream channels increase the likelihood of sediment production. Banks may be broken down or weakened, and removal of cover may expose soils directly to the cutting force of water. Logging debris should be kept out of the channel.

It is not enough to avoid damage and debris in perennial streams. Ephemeral channels that are dry during summer may become strong torrents during storms or snowmelt. Unless stored locally, the sediment they carry has no place to go but downstream. No large stream can be kept clear if its tributaries are producing sediment.

Directional felling will keep trees out of stream channels and offers the potential to reduce damage from, and the cost of, skidding operations (Hunt and Henley, 1981; Gerstkemper, 1985). Felling trees toward skid trails will reduce vehicle traffic and skid trail area. Careful placement of fallen trees can reduce the maneuvering by skidders when picking up a turn of logs on short, steep slopes. Fallers tend to take great pride in their skills and, if made aware of the benefits of using them in such ways, can be a strong force in minimizing watershed damage.

The growing trend toward use of mechanized feller-bunchers (Schuh and Kellogg, 1988; Silversides, 1984) carries the potential for much greater soil disturbance in the felling operation than hand methods, and a number of recent studies have evaluated such environmental concerns (Breadon, 1985; Gent and Morris, 1986; Martin, 1988; Miyata et al., 1984; Reisinger et al., 1988; Shetron et al., 1988; Sidle and Laurent, 1986). The maneuvering of the tractor to align itself with each tree and the great amount of travel between trees may leave little of the cutting unit free of disturbance. Stable and fragile areas alike are likely to be traversed with heavy equipment. The impact of this system may be less than skidding with comparable equipment, but will be additive to skidding disturbance where skidding is a separate process. The damage potential increases rapidly with increasing slope, and the machines should be restricted on source areas or unstable sites. Manual felling should be considered where there is some risk of watershed damage.

Tree cutting alone may influence the occurrence of erosion by landsliding on marginally stable slopes if there are significant changes in soil moisture or root strength. Studies have indicated an increase in mass movement following clearcutting (Swanston and Swanson, 1976), although the aerial photo surveys used probably resulted in some bias of the data (Pyles and Froehlich, 1987). Where landslide risks exist, modifying the silvicultural system may be necessary. If even aged management is desired, then shelterwood cuts may be preferable to seed tree or clearcut systems. Uneven aged systems, such as single tree selection, retain a high level of root cohesion at all times. However, even a relatively small amount of tree or shrub cover can effectively reduce landsliding (Hawley and Dymond, 1988; Megahan et al., 1978). Most so-called "protection forests" that are used for timber production are managed under the single tree selection system. Though potentially costly, on a few very risky sites it may be necessary to prohibit all cutting.

Limbing and bucking logs has little direct effect on the watershed, but can influence other operations. Whole tree skidding requires more power and may disturb a greater proportion of the surface than skidding either limbed full tree lengths or logs bucked in the woods. For example, Mace (1970) found that whole tree skidding in a thinned pine stand increased bulk density of the soil twice as much as tree length skidding, and the area disturbed was also greater. However, slash disposal operations and site preparation measures may be reduced or eliminated by whole tree skidding. Decisions on these matters should be made in terms of the total harvest-regeneration operation, not just one segment of it.

12.1.2.2 ***Skidding and Yarding.*** The great variety of techniques and machines used for skidding and yarding provide many alternatives for minimized watershed damage. Many systems permit harvesting on sites that were too hazardous to be considered in the recent past. Improvements in machines and methods can be expected that will further reduce the impact of harvesting operations on the site.

Even the best of equipment and methods are not enough to insure that damage to watersheds is kept below acceptable levels unless each is used where it is most suited. Though many criteria of economics, silviculture and other factors of management enter into the choice of equipment and methods for skidding and yarding, the risk of damage to the site and water quality should serve as an absolute constraint upon acceptable alternatives.

Proper choice of skidding and yarding methods that fit the needs of the operation is only the first step in minimizing watershed damage. By itself, it is not enough. Similar methods used on similar sites have resulted in highly variable damage, depending primarily on the care and skill exercised. Careful operation requires a desire to limit damage combined with a knowledge of erosion processes, and site and equipment capacities and limitations, all applied imaginatively and skillfully to the job at hand. The operator is the key, and to the degree that the above characteristics are evident, the more likely the job will be well done. It may be necessary for the manager to provide training and encouragement for operators, but it must be done diplomatically, for heavy equipment operators tend to take great pride in the manner in which they work and may not accept change readily.

12.1.2.2.1 Methods Compared. In skidding and yarding, one cardinal principle holds: *The less the compaction and disturbance of the forest floor, the less watershed damage is likely to result.* Thus, air transport systems whereby the log is lifted from the ground at the point of cutting and transported to the landing without further contact will result in the least damage, whereas large, heavy, poorly maneuverable crawler tractors dragging whole trees fully on the ground over the entire cutting without regard to areas of special hazard are likely to create maximum damage. Table 12.1 lists disturbance from skidding and yarding for a wide range of methods, sites, cutting systems, and intensity from throughout the United States. Crawler tractors seem to cause more total and heavy disturbance, with wheeled skidders and most cable systems that do not lift logs from the ground close behind. Horses and mules have among the least impact of all ground skidding systems, and a highly mobile mini cable system called the "Radio Horse" (Post, 1986) should be comparable. Animals and the "Radio Horse" have limited power and cannot skid large timber or be used on steep adverse slopes. Skyline, balloon, and helicopter yarding tend to disturb the site least because timber is lifted free of the ground over at least part of the distance from the point of felling to the landing. These yarding systems, however, are considerably more costly than other systems.

It should be noted that the data in Table 12.1 include only three studies (Bradshaw, 1979; Froehlich et al., 1981; Olsen and Seifert, 1984) of the use of designated skid trails to reduce the area of site disturbance from logging vehicles. This type of planned approach to ground-based logging can dramatically decrease the area covered in skid trails (Fig. 12.5), with little or no effect on harvest costs if trail and felling patterns are carefully implemented (Garland, 1983).

There are differences within, as well as between, systems that may affect watershed disturbance. For example, crawler tractor operations tend to cause

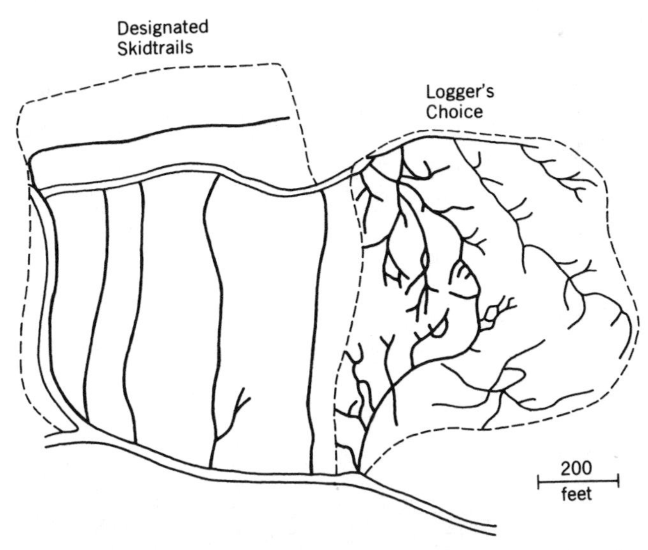

FIG. 12.5 Skid trail patterns: designated trails at various spacings on the left; logger's choice trails on the right (adapted from Froehlich, Aulerich, and Curtis, 1981).

TABLE 12.1 Soil Disturbance by Skidding and Yarding Methods in the United States

Method[a]	Forest Type	Silvicultural System	Slope[a] (%)	Soil Disturbance[b]		References
				Slight	Deep	
Skidder-T	Hardwood	Clearcut	18–22	12	25	McMinn, 1984
Skidder-L	Hardwood	70% Partial cut	21–42	7	9	McMinn, 1984
Skyline-L	Hardwood	70% Partial cut	21–42	1	0	McMinn, 1984
Crawler-W	Douglas-fir	Thinning	To 36	9	11	Murphy, 1982
Skidder-W	Douglas-fir	Thinning	To 18	4	6	Murphy, 1982
Crawler-W	Douglas-fir	Thinning	To 18	7	1	Murphy, 1982
Crawler-W	Douglas-fir	Thinning	To 27	3	1	Murphy, 1982
Crawler-W	Douglas-fir	Thinning	To 36	7	5	Murphy, 1982
Skyline-L	Mixed conif.	Clearcut	50–80	(----5----)		Amaranthus and McNabb, 1984
Crawler-U	Mixed conif.	Salvage-burn	To 30	38	36	Klock, 1975
Crawler (snow)-U	Mixed conif.	Salvage-burn	To 40	24	10	Klock, 1975
Jammer-U	Mixed conif.	Salvage-burn	30–90	45	32	Klock, 1975
Skyline-U	Mixed conif.	Salvage-burn	30–90	22	3	Klock, 1975
Helicopter-U	Mixed conif.	Salvage-burn	40–90	11	1	Klock, 1975
Skyline-L	Spruce–hemlock	Clearcut	Mean 77	(----6----)		Ruth, 1967
Highlead-L	Spruce–hemlock	Clearcut	Mean 77	(---16---)		Ruth, 1967
Skyline-L	Douglas-fir	Clearcut	Mean 63	24	8	Dyrness, 1967
Highlead-L	Douglas-fir	Clearcut	Mean 63	22	19	Dyrness, 1967

Balloon-L	Douglas-fir	Clearcut	60–70	16	4	Dyrness, 1972
Skyline-T, L	Douglas-fir	Thinning	20–40	(----4----)		Aulerich et al., 1974
Crawler-T, L	Douglas-fir	Thinning	20–40	(-16–27-)		Aulerich et al., 1974
Low-pressure, torsion–suspension tractor, L	Douglas-fir & mixed conif.	Clearcut and partial cut	5–50	(--2–20--)		Froehlich, 1978
Low-pressure, torsion–suspension tractor, L	Douglas-fir	Partial cut	To 42	(---12---)		Sidle and Drlica, 1981
Skidder-L	Douglas-fir	Thinning	To 31	(--4–20--)		Froehlich et al., 1981
Tractor-L	Mixed conifer	Partial cut	To 60	(--4–22--)		Bradshaw, 1979
Tractor, skidder-T, L	Mixed conifer	Partial cut	U	(--9–17--)		Olsen and Seifert, 1984
Crawler-L	Pine–Douglas-fir	Selection	U	(---14---)		Dickerson, 1968
Skidder-T	Pine–Douglas-fir	Selection	U	(---11---)		Dickerson, 1968
Mule-L	Pine–Douglas-fir	Selection	U	(----4----)		Dickerson, 1968
Jammer-L	Pine–Douglas-fir	Selection	U	(----5----)		Dickerson, 1968
Helicopter-L	Pine–Douglas-fir	Clearcut	U	25	9	Clayton, 1981
Crawler-L	Pine	Selection	U	5	7	Haupt, 1960
Skidder-U	Aspen	Clearcut	To 25	25	9	Zasada and Tappeiner, 1969
Skidder-L	Mixed conif.	50% Partial cut	To 25	5	7	McDonald, 1971
Crawler-L	Mixed conif.	50% Partial cut	To 25	7	15	McDonald, 1971
Skidder-T	Pine	Clearcut	U	(---23---)		Campbell et al., 1973
Skidder-crawler-W	Conif.–hardwood	Clearcut	5	(---18---)		Hornbeck et al., 1986
Skidder-crawler-W	Conif.–hardwood	Clearcut	10	(---11---)		Hornbeck et al., 1986
Skidder-W	Hardwood	Clearcut	10	(----8----)		Hornbeck et al., 1986

[a] T = tree length, L = log length, W = whole tree, and U = unknown.
[b] Disturbance is in percent. Slight means soil exposed, deep means surface horizon removed or heavily compacted.

a high degree of disturbance, but different machines may differ greatly. Machines come in all different sizes, makes, and models. Those either larger or smaller than required will do more damage than those matched to meet the job at hand. Different manufacturers and models have different track, suspension, and steering systems. Some models are articulated and steered by controlled differentials, while others are rigid and steered by clutches and brakes on each track. Most machines are equipped with winches, but some are not. Lift may be provided by arches, or pull may be direct from the drawbar. Load distributions vary. All these differences combine to create a wide range of ground pressure, vibration, and soil shear stresses. Reduced soil disturbance can result from improvements in design, such as low-ground pressure, torsion-suspension yarding vehicles (Froelich, 1978; Sidle and Drlica, 1981).

The variety of wheeled skidders and cable systems is almost as great. Fortunately, more and more logging systems are being chosen on the basis of their impact on the site, as well as their ability to move timber economically.

12.1.2.2.2 Choice of Method. Timber harvest planning must always consider the nature of the timber stand, the regeneration system, site characteristics of topography, surface and mass erodibility, climate, equipment availability, operator skills, and experience. All must be combined to meet cost constraints, regeneration requirements, site protection, and water quality goals.

The volume and size of the timber removed influences site and water quality damage potential. Other factors equal, the greater the volume removed and the larger the timber, the more site disturbance that is likely. However, differences in timber volume and size tend to cause greater differences in disturbance with ground skidding systems than with aerial yarding. Damage potential is lower with timber of uniform size than with a wide range of sizes because each machine tends to have a limited size range within which performance is optimal.

Timber size, volume removed and cutting system all interact. For a comparable volume of wood removed, there tends to be more disturbance when many small pieces must be handled than when pieces are larger but fewer in number. Similarly, removing a unit volume from a partial cut where the logs are scattered throughout the stand, requires greater travel distance per tree than where the volume is concentrated, as in a clearcut. The lighter the partial cut, the more travel per unit volume removed. Therefore, though less damage may occur at any one time with light cuts, the high frequency of entry over a rotation may result in greater total disturbance than where large volumes are removed at long intervals. It is well to remember that most compaction occurs within the first few passes over the ground and complete recovery may require far longer than the interval between cutting cycles. Much of the opposition to clearcutting derives not from the greater damage it causes to the watershed, but because any damage is clearly exposed to view, not hidden beneath a residual stand of trees and litter and spread over time in smaller increments. Depending on circumstances, either system may be more appropriate than the other, but the partial cutting system will almost invariably look better. Clearly, where a concentrated impact would carry a site past the threshold of the critical point in deterioration, it must not be permitted, but the danger of possible cumulative effects of a series of lesser impacts must be recognized as well. Again, the differences, if any, are greater among ground skidding systems than among aerial systems.

Slope is a major factor determining choice of logging method. Tractors, both crawler and wheeled, are best suited to level or moderate topography. When used on steep slopes they may lose traction, slipping, sliding, and shearing the soil surface. Turning is difficult (and dangerous) and churns the soil. Garrison and Rummel (1951) related damage from tractor skidding on slopes from 0 to 60% and found 2.8 times more damage on slopes exceeding 40% as on lesser slopes. Not only is there more surface disturbance, but the resulting erosion per unit area of disturbed soils increases with slope, as shown in the USLE. Tractor skidding should ordinarily be limited to slopes not exceeding 25–35% if excessive damage to soils is to be avoided (Ruth and Silen, 1950). However, steeper slopes may be negotiated with the right equipment and favorable soils and skidding direction (Fig. 12.6).

Cable systems, particularly skylines, are best adapted to steeper slopes. Cable logging systems generally are more costly than ground-based systems,

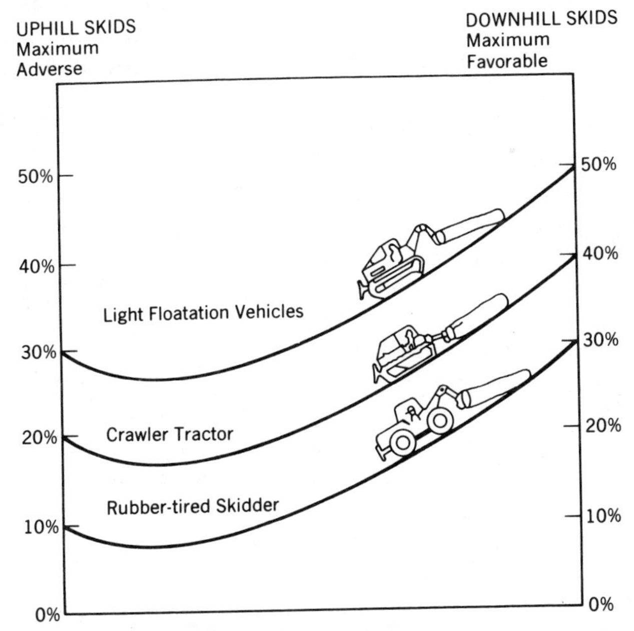

FIG. 12.6 Left and right sides of the graph represent traction on slopes under the best conditions, but soil and weather conditions may reduce gradability (adapted from Johnson and Wellburn, 1976).

but smaller yarders now available have considerably lower operating costs than larger yarders used in the past (Kellogg, 1981). Deflection (sag) of the cable is necessary to maintain load capacity and is difficult to obtain on sites with little slope. On gentle slopes, yarding distances are limited and little lift may be available to keep even the front of the log off the ground. Convex slopes must be steeper than straight or concave slopes to obtain the same degree of deflection. Intermediate cable supports (Kellogg, 1981; Lysne and Armitage, 1983) can be used to deal with terrain features that conflict with line deflection and log suspension.

Most cable logging is done uphill, but sometimes downhill yarding is done. When yarding is done uphill, even with systems such as high lead or Idaho jammer that drag logs in contact with the ground, water that runs down yarding paths is dispersed. Downhill yarding tends to concentrate water along yarding paths that converge on a disturbed landing and generally should be avoided. However, in some cases a well-planned downhill yarding operation with good log suspension and control can help avoid the more risky alternative of road construction on very hazardous, unstable slopes (Patton, 1987).

Some aerial systems, such as balloons and helicopters, can be used on any slope. Their high cost usually restricts use to hazardous or inaccessible sites that cannot be safely yarded with other equipment, although smaller and less costly helicopters have seen increased application.

In most places in the world, tractors tend to be the least costly method, even on steep slopes where they may cause excessive damage, because of their mobility, simplicity of operation, and versatility for use in other operations, such as road building. Extreme care is essential on hazardous sites when their use cannot be avoided.

Regardless of the method chosen for logging, good planning can greatly reduce damage potential. For example, tractors equipped with winches can be used on slopes exceeding 40% without excessive damage where preplanned, permanent skid trails are built, and directional felling is utilized to fell trees for minimum skidding distance to the trails (Bradshaw, 1979; Garland, 1983). Tractors are kept on the trails, and tree length logs are skidded to the trails by cable. Landings are few and widely spaced, for skidding distances and speeds are enhanced by the preplanned skid trail system. When skidding is completed, the skid trails are bladed to repair any ruts and cross drains (e.g., water bars) are maintained to insure free drainage. Compaction and disturbance are concentrated on a well-drained, stable skid trail rather than spread extensively over the forest on stable and unstable sites alike. Such planning is especially needed when soils are erodible and subject to disturbance.

Landings can be critical because they are highly disturbed and compacted, and their location determines skidding and yarding patterns. Where possible, they should be located on stable uplands so water concentrations can be readily dispersed and to avoid cross-drainage yarding.

Above all, it bears repeating that operator skill and care are critical, especially with ground skidding systems on steep, erodible slopes. Even good operators may need training to recognize hazards; for example, to be able to distinguish between source areas and nonsource areas, and to be able to tell highly easily disturbed or erodible soils from those that are less so. Until the operators learn such skills, close supervision and teaching by identifying pertinent examples is desirable. It should go without saying that supervisors must have the necessary skills, but, unfortunately, they sometimes do not. Good loggers, familiar with local conditions, in turn may have much to teach professional land managers, who should be willing to learn from anyone having useful knowledge and experience.

12.1.2.2.3 Effects of Method on Roads. The choice of skidding and yarding methods has a major watershed impact other than direct disturbance in moving logs to landings. It also influences to a high degree the amount and location of haul roads. The distance to which trees can be yarded efficiently varies with the system used, as well as with topography. Animals and Idaho jammers are efficient only out to distances of 100–150 m. High lead and grapple systems commonly reach out 220–325 m. Tractors are highly variable, ranging from 150 to 700 m or so, depending on topography, type of vehicle, presence or absence of designed skid trails, and other factors. Skylines and balloon systems have been used at distances of over 1500 m, and regularly reach out to 700 m. Helicopters have been used at even greater distances, but costs increase rapidly as distances reach beyond 2000 m.

Investigations relating road length to skidding methods confirm the general relations inferred from skidding distances. In north central Washington, Wooldridge (1960) found that a skyline system required only one-tenth as much road as needed when crawler tractors were used. Finding desirable roads, landing, and yarding combinations has been greatly facilitated by the use of desktop computers, with the result that road mileage can be reduced (Sessions and Li Guangda, 1987) and skyline loads may be kept free of the ground for longer distances (Burke, 1975 a, b).

The silvicultural system used may also influence the amount of roads, even with the same skidding or yarding method, particularly when expressed per unit volume of wood removed. For example, in one cable yarding test in the Appalachian Mountains, selection cutting required more than seven times as much road as clearcutting per unit volume removed (Gibson and Biller, 1976).

12.1.2.2.4 Timing of Harvest. It is just as important to recognize the varying susceptibility of watersheds to erosion damage over time as it is to recognize their variation over space. Though mass failure and recovery of stability is not closely related to time of disturbance, surface erosion is. If sediment is to be minimized, the right techniques and equipment must be applied to the right place at the right time.

Several factors influencing watershed damage potential vary systematically with season. There are strong seasonal variations in precipitation amounts, frequency, and intensity in much of the world. Soil moisture often shows a distinct annual pattern, and snow and frost repeat in a regular annual cycle. The manager can take advantage of such variations to minimize adverse effects.

Soil moisture is a major factor in the load bearing capacity (trafficability) of soils, their ability to resist compaction, and susceptibility to frost effects. Coarse, noncohesive soils tend to be well-drained, strong, and seldom exhibit concrete frost or frost heaving. Fine-grained, imperfectly drained, plastic soils tend to be easily compacted, weak, and subject to displacement under loading (Moehring and Rawls, 1970), and develop concrete frost, frost heave and, upon melt, leave great mudholes if subjected to traffic. It follows that operations on the latter should be restricted to periods when soils are solidly frozen or during dry seasons, whereas the former sites might be reserved for wet season operations because they may be the only ones that can be harvested safely at such times. Information about soil conditions also can be used to identify locations and equipment that are less likely to show rutting and related problems (McKee et al., 1985; Olsen and Gibbons, 1983).

Sometimes advantage can be taken of a snow cover to protect a site from damage during harvesting. Thirty centimeters or so of snow can absorb much of the impact of harvesting on many sites. Soils under compacted snow may freeze readily, adding to soil strength in cold weather. Harvesting on winter snow may minimize adverse impacts on many sites, such as runoff source areas, that are very hazardous at other times. With modern equipment, over-snow logging may be more efficient than on bare ground surfaces unless the snow is excessively deep and soft.

The advantages of logging on snow can be quickly lost if work is extended too far into the melt season. Thin snow and soils that may be near saturation (or even supersaturated as soil frost melts) may create an extensive watershed hazard in a very short time.

Some extreme conditions, such as individual intense storms or excessive rainfall, cannot be predicted, but their probabilities can be taken into account in planning allowable disturbance. The higher the probability, the greater the precautions that must be taken to prevent adverse effects. Some operations may be diverted from source areas or, if unusual rains bring soil moisture to hazardous levels, may be closed down entirely until soils dry.

12.1.2.3 Slash Disposal. When forests are thinned or harvested, there is almost always unmerchantable residue, generally termed *slash*, that remains on the site. In some regions, slash may reach several hundred tons per hectare. Slash may create a fire hazard, hinder regeneration, block the movement of wildlife or domestic animals, may provide a source of disease or insect infestation, and may enter streams where it blocks flow, adds leachates and depletes dissolved oxygen. It is also ugly and aesthetically displeasing.

In many situations, particularly thinnings or partial cutting harvest operations, no slash treatment is necessary when quantities are small and scattered and breakdown and decay are rapid. Provided steps are taken to keep slash directly out of streams, few effects should be evident. If slash reaches the stream, it may deplete oxygen and release nutrients and undesirable leachates such as tannins and phenols (Ponce and Brown, 1974; Hart et al., 1980).

Where slash treatment is necessary, the major means include mechanical treatments such as lopping and scattering, chopping, crushing or chipping, removal and disposal, and burning in piles, windrows or broadcast over the surface. Though burial has been shown to be physically feasible and environmentally acceptable (Larson and Wooldridge, 1980), it has never become a widely used practice.

Mechanical treatment may be desirable, undesirable or have no effect on water quality. Slash can be lopped and scattered to provide protection for exposed soil, and be placed to act as effective sediment filters or traps. Chipped residues are often spread over cut and fill slopes of roads, trails, and landings to reduce surface erosion. Chopping and crushing, however, may result in soil exposure and compaction, depending on the kind and amount of slash, soil conditions at the time of treatment, and the size and type of equipment used.

In the western United States where large quantities of large slash may occur on steep erodible slopes, yarding of unmerchantable material (YUM) may be required. The material may be disposed of at the landing by burning, in piles or bins, or it may be chipped and used for pulp or erosion control. Most YUM practices have little net effect on site erosion. Some increased soil exposure may result, but concentrations of slash in draws and intermittent drainages are eliminated where burning may otherwise be intense enough to damage the soil and provide a major surge of ash and nutrients into the streams with the first rains that cause runoff (Brown et al., 1973).

Probably the most widespread means of slash disposal in North America is by burning. Burning always poses some potential hazard to the watershed. Soils can be exposed and infiltration capacities may be reduced by changes in wettability, the loss of organic matter, and, in some cases, by the fusion of clays in the extreme heat, leaving a vitreous surface. Nutrients such as nitrogen and sulfur may be lost by volatilization and others lost in the form of ash that may be washed or blown away (Snyder et al., 1975). Where piling is done by machine before burning, the piling activities may increase soil exposure and compaction.

Broadcast burning may result in a wide range of exposure of mineral soils, from about one-half of the burned area (Tarrant, 1956; Amaranthus and McNabb, 1984) to nearly complete exposure (Morris, 1970; DeByle, 1981). Control of fire intensity is the key to reducing soil exposure and other watershed problems, although this may conflict with other objectives of burning and create more smoke. Vegetation recovery tends to be rapid, but considerable variation exists (Steen, 1966; Dyrness, 1973; DeByle, 1981). Similarly, sediment movement tends to be relatively slight and decreases rapidly, but the potential for severe erosion is always present when bare soils are subject to intense storms, as suggested by observations following wildfires.

In partially cut stands, if burning is done, most slash is piled first. Fires in piled slash are usually more intense than in broadcast burning, but the total soil surface exposed is much less and scattered about. Overland flow and sediment, if any, from the burned areas tends to be trapped and absorbed in the intervening areas unless they have been greatly disturbed and compacted in the piling operation.

Watershed hazards from burning may derive from fireline construction, as well as the burning itself. The effect of firelines is similar to that of skid trails, but is frequently more severe because firelines often involve trenching, have much steeper gradients, and may traverse unstable areas. Though the hazards of firelines may be reduced by such practices as building water bars to divert and disperse surface flows, consideration of their potential effects early in planning the location of cutting boundaries can prevent much of the hazard in the first place.

12.1.2.4 ***Site Preparation.*** Successful regeneration of harvested forests may require preparation of the site to create access for planting or seeding, prepare a suitable seedbed, remove competing vegetation, eliminate habitat for animals that destroy seed or seedlings, and reduce the risk of wildfire that might destroy the next forest. Sometimes past regeneration failures have led to the development of dense shrub fields that must be removed before trees can be established. All site preparation activities carry some risk of damage to water quality even when properly accomplished, but proper choice of method and proper application will reduce the risk.

There are basically four ways to prepare sites for regeneration: (a) hand treatment, (b) mechanical methods, (c) burning, or (d) use of chemicals. Sometimes two or more methods are combined. For example, clearing brush or piling slash may be done by machine or by hand, followed by burning and with chemical treatment of sprouts.

Hand treatment of clearing brush or piling slash tends to result in least site disturbance. Bare soil exposure is limited to scalped seedspots or planting sites. Chemicals applied by backpack sprayers or to cut stumps can be accurately directed to individual spots requiring treatment with little risk of drift or inadvertent application to water supplies. However, hand treatments tend to be limited by high labor costs.

Mechanical methods of site preparation include rolling and chopping or shredding brush or slash, crushing in preparation for burning, clearing and piling material with regular or brush blades on bulldozers, chaining, root plowing (cf. Section 11.4.2.1), flails rotating at high speed, and terracing, discing, ripping, and tilling of compacted or indurated subsoils, bedding, ditching, and plowing furrows (Painter et al., 1974; Beasley, 1976; US Environmental Protection Agency, 1976; Baumgartner and Boyd, 1976; Green, 1977; Swindel et al., 1983; Froehlich and Miles, 1984; McKee, 1985). Some methods, such as rolling, chopping, or flails may leave the soil with a mulch of broken or shredded organic material incorporated into the surface. Others may leave the surface almost as bare as cultivated farmland before seeding.

Guides for predicting erosion hazards from site preparation have been suggested based on the USLE (Curtis et al., 1977; Dissmeyer and Foster, 1980; Dissmeyer and Cost, 1984). Most tests indicate that the USLE can provide useful estimates of erosion (Chang et al., 1982; Chang and Wong, 1983; Chang and Ting, 1986), but does not predict sediment delivery to streams very well. Hewlett and Burns (1983) found that adding an index that reflected both the amount of bare soil and its proximity to the stream channel greatly improved a modified USLE model for predicting sediment yield.

The use of fire to prepare sites for seeding or planting presents the same problems as when it is used to dispose of slash. Fire can be used to remove understory vegetation or brush, as well as to dispose of slash. Sometimes, chemicals are used to dessicate brush fields ("brown and burn") so they can be burned when surrounding vegetation is green, thus limiting the risk of escape or of conflagrations that might erupt when fuels are naturally dry. Brushfields are sometimes crushed before burning to concentrate fuels to insure cleaner burns as well. Once again, however, care must be taken to limit burn intensity and excess exposure of mineral soil.

Chemicals commonly have been used to kill or inhibit vegetation for site preparation. Used alone, site disturbance and erosion is virtually absent. However, problems of potential contamination of water supplies and possible health effects have greatly reduced their use. More will be said about chemicals and water quality in Section 12.4.

Most forest sites prepared for planting by any means tend to recover rapidly with natural vegetation in the absence of repeated treatment.

12.1.3 Roads and Trails

Roads and trails are necessary to wildland management everywhere, for there can be little management without access to the land. Their potential impact upon erosion and sediment often exceeds that of all other activities combined, especially in forests managed for timber.

No other wildland management activity requires such intensive, concentrated soil disturbance. Significant effects of road construction on watersheds are manifest both onsite and downstream (Megahan, 1987). For example, in the northern region of the US Forest Service, studies and observations revealed that as much as 90% of the sediment produced from timber sale areas came from roads (Packer and Christensen, 1964). On rangelands, the mileage of roads necessary for management is much less and the total impact is correspondingly reduced. Nevertheless, per unit length, rangeland roads in rough country may have as great an adverse effect for they are likely to be less well planned, built, and maintained; and the hydrologic environment is often very harsh.

County and town roads in forest and rangeland areas may also contribute substantially to watershed problems, for these levels of government are often underfinanced and their roads may also suffer from a lack of planning, design, and maintenance.

Trails consist of everything from crude footpaths and roads developed by repeated use following natural lines of least resistance to well constructed and designed facilities that fall just short of being suitable for passenger car use in comfort. There is really no strict dividing line between a road and a trail, either in type of use (vehicular, animal, or foot) or in construction. The principles of location, design, construction, and maintenance are common to each, and no distinction will be made in their applications so long as they are deliberately planned and constructed.

At this point it should be mentioned that the problems developed by off-road vehicle (ORV) users have grown explosively. In many wildland areas they have become a major source of watershed deterioration. Few of these problems will be solved by technological progress in the development of roads and trails. Greater opportunities appear to lie with education and cooperation of user groups, and if necessary, restriction or segregation of ORV use (Major, 1987). The control and regulation of ORV use in wildlands is a separate subject that will not be included in this book.

Sediment is the primary problem related to road building, but other characteristics of water yield may be changed as well. Road surfaces are bare of vegetation, compacted as much as possible, and resist infiltration by design to provide maximum bearing strength. This means that virtually all precipitation they receive becomes surface runoff, which is quickly concentrated into drainageways and ditches that may reach nearby streams. Cut and fill surfaces, although generally not so compacted or impervious as road surfaces, are also major sources of surface runoff. In many wildland areas, much storm runoff occurs as subsurface flow through upper layers of soil. Large quantities of such flow may be intercepted by sidehill cuts and concentrated and speeded on its way toward nearby streams through the road drainage system. Concentrated runoff always increases the likelihood of erosion.

All in all, while roads and trails may be necessary to wildland management, they may exert a primary deleterious impact on water quality. The exposed soil, decreased slope stability, and increased and concentrated runoff, all contribute to the problem. Strong efforts must be made to minimize watershed damage. Standards for control must be particularly stringent on watersheds with inherently low resistance to erosion. Roads and trails must, therefore, be located, designed, and constructed with environmental integrity, as well as road use, in mind.

Many good publications present techniques of wildland road construction, some in general terms, others in considerable detail. An excellent example of guidelines useable by nonengineers is the Handbook for Forest Roads, published by the Washington State Departments of Natural Resources and Ecology (Mustard, 1982). Others include Packer and Christensen (1964), US Environmental Protection Agency (1975), Haussman and Pruett (1979), Burroughs et al. (1976), Megahan (1977), Heinrich (1979), Hynson et al. (1982), Garland (1983 a, b, c), Burroughs and King (1989), and Adams and Andrus (1990). Much of the material in these and other sources is summarized in the sections that follow.

12.1.3.1 ***Principles.*** Effective planning and design of roads and trails to meet the needs of wildland management requires skills in many disciplines—engineering, geology, soils, hydrology, ecology, forestry, and range management to name a few. It is not possible to cover each in detail, and only those aspects bearing directly on watershed management will be treated here. Watershed considerations must be integrated with others of variable importance, but in the end no compromise that seriously threatens loss of watershed productivity or function should be permitted.

The first consideration in planning a road system on wildlands is the probable use of the land. Probable use should be projected as far into the future as is reasonably feasible. Once roads and trails are constructed, they tend to become a lasting feature of the landscape. Many roads, originally planned to meet only temporary needs such as timber harvesting, have continued in unforeseen use for which location, construction, and maintenance standards were inadequate, resulting in deterioration of land and water resources. This has proven to be the case in many forested areas where a transition from harvesting old growth timber to sustained-yield, multiple-use management has taken place and on many lands opened to access by the public for recreational activities.

Good roads for wildland management share three characteristics: (a) they represent the least road necessary to meet management needs; (b) they lay as gently as possible upon the land, rather than being cut and blasted out of it (Kraebel, 1936); and (c) they occupy the most stable parts of the watershed. The least road refers not only to road length, but also to width. The alignment, gradient, and width standards of a superhighway are not needed in wildlands. In many cases, narrow, one-lane roads with turnouts, fitted to the natural slope of the land so few cuts and fills are necessary, will minimize the impact on the land. Sometimes compromise of increased road length will be necessary to avoid hazardous sites and obtain the best total result.

Planning well ahead for the entire system needs of small watersheds is necessary to insure the least road. For example, systematic planning of the roads for all of a steep drainage basin in the Douglas-fir region of the western Cascades of Oregon reduced road mileage by 11% and steep grades by 73% over the amount required when they were designed piecemeal to meet only the current cutting requirements of a staggered block system as harvest progressed. Road standards were the same in each case (Silen, 1955).

The high cost of advance road building is often cited as an objection to this procedure and acts as a deterrent. However, it is not necessary to build the entire system in advance of cutting, though the initial construction requirements are likely to be increased. The reduction in road mileage, a more efficient final transport system, increased service life, and the greater watershed protection achieved all tend to offset the costs of a longer investment period and may well prove cheaper in the long run.

The land manager should be alerted to the fact that strong technological and psychological forces combine to support a tendency to overbuild roads. The development of powerful equipment for efficient earth moving and rock removal increases the danger of excessive watershed disturbance. It is easy to think of cost per cubic meter of earth moved rather than minimal requirements for management. And it is difficult for persons operating heavy construction machines, who hold vast power at their fingertips, not to make full use of it. The beauty of smooth alignment, gentle grades, and broad lanes may blind even conscientious engineers to the cost in exposed soil, deep cuts and fills, and sediment in the streams. These are real forces that can cause much unnecessary watershed disturbance. Neither their presence nor their strength should be underestimated, for they are basic to the pride that construction bestows upon its workers—from engineers to heavy equipment operators. It does not help, either, that progress is often measured in terms of rock ripped and earth moved. The potency of these forces can be overcome by constant reminders that roads are only a means of meeting land management objectives, not an end in themselves.

In addition to general principles of planning and design, there are certain rules of location and techniques of construction that will serve as useful guides. Seldom is it desirable to follow them to the letter, for local watershed conditions are usually such as to require some compromise among road length, alignment, grade, drainage, and location to avoid unstable areas or other hazards.

While it is seldom possible to avoid all hazards, it is critically important to recognize them and to compensate for them by modifying design wherever they are encountered. Individual techniques of compensation will be discussed under the topics of design and construction.

12.1.3.2 ***Location.*** All roads produce some sediment. The objective is to minimize erosion and to prevent such sediment as it is produced from reaching a watercourse. Therefore, the major guides in road location are the knowledge that (a) exposed soil and disturbed unstable areas can be the primary source of sediment on the watershed and (b) the easier the path to the stream channel, the more hazardous to water quality it is likely to be. Satisfactory location is a product of both factors. A stable road that produces negligible amounts of sediment can be located safely near a stream, but a large number of obstacles and considerable distance may be needed to trap sediment from an eroding road before it reaches a stream.

Many factors influence road and trail location, including (a) land use, (b) property boundaries, and (c) site characteristics of topography, geology, soils, climate, vegetation, and others. Wildland uses that include timber harvesting usually require the densest network of roads of all standards, from main haul roads and access roads to secondary and temporary haul roads and planned skid trails. At the other extreme, only widely dispersed pack and foot trails are needed to serve users of wilderness areas, but good location is no less important just because they are fewer and have more primitive construction standards. For example, Helgath (1975) found that trail deterioration was more closely related to land form, vegetation habitat type, and grade than it was to amount of use.

Property boundaries seldom serve as a major constraint on road location for large public agencies and industrial owners, but commonly pose severe limitations to small private forest landowners. Cooperation among large owners and public agencies is common, but may be difficult to attain with small private forest owners. The problems include differing management objectives; for example, a significant portion of small private owners have no intention of permitting their timber to be harvested and do not wish to facilitate the harvest of adjoining tracts. Yet, ownership tenure is relatively short compared to the lifetime of a tree, and sooner or later most mature timber in rural areas (in contrast to suburban areas) is harvested; hence, there is little old growth timber in small private ownership. Other problems include the legalities of securing rights of way, determining the degree of benefit to each of several owners, and cost sharing for construction. All in all, property boundaries serve as a major limitation on proper road location in watersheds where land ownership is in small private parcels. Unfortunately, there is no technical solution to the problem, and often no social, political, or legal agreement either. Any solution is likely to be suboptimal under such circumstances.

Topography strongly influences optimum road location. Among important topographic factors are slope steepness and direction, slope form and degree of dissection, and position on a given slope. Generally, steep slopes should be avoided because (a) they are usually (but not always) less stable, and (b) they require excessive amounts of soil exposure from cuts and fills. McCashion and Rice (1983), in a survey of 554 km of roads in northwestern California, found that 52% of the erosion came from 19% of the roads that traversed slopes with a gradient of 60% or more.

The degree of soil exposure from cuts and fills increases rapidly with increasing slope. For example, consider a secondary haul road 4 m in width (about the minimum needed for logging trucks) with a fill slope gradient of 67% and steep cut slopes of 200% gradient. The total horizontal width of disturbed surface increases from about 5 m on a 20% slope through 20 m on a 52% slope, to more than 30 m on a 65% slope when the volume of cuts and fills are equally balanced (Megahan, 1977). On many sites, such steep cut and fill slopes are unstable, but the option for flattening cut and fill gradients is limited, for as either approaches the existing slope gradient, the width of exposed soil approaches infinity. On the other hand, some slope is desirable to provide for ready drainage.

Slopes that are highly dissected also result in more exposed soil from cuts and fills than smoother slopes for it is difficult to fit alignment to sharp, frequent changes in slope direction.

Sometimes slope direction (aspect) can have an important influence on erosion. For example, frost action in temperate regions may be more pronounced on south and west facing slopes than their opposites because they exhibit a greater diurnal range in temperature during freezing periods and thus undergo more frost cycles. On the other hand, such slopes tend to be drier and exhibit greater mass stability in some regions. In humid climates, roads on the sunnier slopes tend to dry more rapidly between rains.

Slope position is important as it influences proximity to stream channels and soil moisture accumulation and drainage. Many times it is possible to take advantage of ridgetops, natural benches and terraces to avoid excessive soil exposure while obtaining the stability of well-drained soils. Care should be taken when crossing the transition where steep slopes with shallow soils break away from rolling uplands with deep soils, for groundwater may approach the surface throughout much of the zone, creating potential instability problems (Bennett, 1975, cf. Fig. 12.2). The inner gorge of actively downcutting streams should be avoided, as such slopes are often delicately balanced near their angle of internal friction. Where the only entry to a watershed is near the mouth of an incised stream channel, road gradients should be increased (with careful attention to drainage design) to get out of the inner gorge and away from the stream as quickly as possible.

Mass instability is a major consequence of poor road location, and locations with such hazards should be avoided. Section 12.1.1.2 presented a long list of indicators of mass instability hazards and should be reviewed to reinforce awareness of this important topic.

Roads should be located not only to avoid obstacles and hazards, but their routes should facilitate good management practices on the rest of the land. For example, in timber harvesting, the location of landings strongly influences the amount and location of disturbance resulting from skidding and yarding operations. Good road location provides for optimum location of landings.

Roads should be located to prevent the sediment they produce from reaching the stream channel. Sediment produced close to a channel is more likely to reach it than sediment produced farther away. The farther sediment must travel before reaching a stream, the more likely it is to be trapped and stabilized before it gets there. The area between a road and a stream can be a protective strip that filters out sediment and absorbs surface runoff, even if the timber on it has been harvested.

Two groups of factors determine the width of protective strip between road and stream needed to trap the sediment from the road: (a) factors determining the amount and concentration of sediment produced, and (b) factors determining the trap efficiency of the strip. Some site characteristics, such as direction of exposure, slope shape, slope gradient, topographic position, and climate, may be common to both groups.

The amount of sediment produced from roads is a function of mass stability, the amount of bare soil exposed, the resistance of exposed soil and roadbed surface materials to erosion, steepness of the road grade, topographic position, climate, steepness of the sidehill above the road, amount of intercepted subsurface flow diverted to surface runoff, and the spacing of cross drains from the road.

The trap efficiency of the strip is a function of the sediment and runoff input (a few large concentrations are much harder to control than many small, well-dispersed ones), and the characteristics of the slope gradient below the road and stream, the distance to the first obstruction, and the density of cover on natural and fill slopes below the road (Packer, 1967). Useful guides for determining safe protective strip width on the basis of all or some of these features are available for several regions of the United States (Haupt, 1959; Kidd, 1963; Packer and Christensen, 1964; Haussman and Pruett, 1979) and should be obtained for detailed directions for application.

It is impossible to keep roads away from streams when they must be crossed. Therefore, stream crossings are particularly hazardous to water quality. Almost total reliance must be placed on the prevention of erosion to maintain sediment-free water because there is not likely to be any place to trap sediment once it is in transport from the road to the channel. The number of stream crossings should be kept to a minimum and located at the most stable feasible sites at right angles to the stream to reduce channel and bank disturbance.

When planning road location, it is useful to classify landscapes with similar geological histories, geomorphic structures, soils, and slope hydrology into land types and map each type. Other maps that are helpful are topographic, soil survey, and vegetation maps. Aerial photos are useful; false IR photos are particularly valuable for locating wet areas if they are taken near the end of a dry period. Wet areas then often appear as distinctive red patches.

Optimum road and trail locations will never be assured by simply following strict rules and criteria of choice alone, no matter how sophisticated they may be. Watersheds are too variable to include all possible situations. The material in the foregoing is designed to help the wildland manager recognize potential problems before they occur. Once the manager learns to see, examination of existing roads on similar nearby land types will provide useful guidance in locating future roads. Probably more information specific to a given locality can be discovered by critical on-site observation than by any other means.

12.1.3.3 ***Design.*** Proper road location needs to be complemented by suitable road design if erosion damage to the land and sediment damage to water is to be prevented. Many of the important characteristics influencing erosion—road gradient, width, drainage, alignment and amount of cuts, and fills—are a function of both design and location. However, good road design can sometimes compensate for hazards that cannot be avoided. An optimum road system can be developed only when appropriate design is coupled with the best feasible location to meet total wildland management needs.

Both mass movement and surface erosion are controllable to some degree by design. Control over mass movement depends mostly on minimizing the impact of roads and trails on the land, but sometimes includes measures to increase or supplement natural stability. Surface erosion control depends primarily on controlling the energy of moving water, but also includes measures designed to increase resistance of soils to movement.

The design of a road depends to a large degree on the amount and kind of use it will receive. Roads may be classed as main, secondary, spur, and skid roads. They may be permanent, as most main and secondary roads are, or temporary, as many spur and skid roads are. This discussion will focus primarily on spur roads, skid roads, and similar trails, for most main and secondary roads are designed by engineers. Nevertheless, the principles involved and many of the techniques are the same for any suitable system.

12.1.3.3.1 Stream Crossings. There are three basic means by which streams can be crossed: bridges, culverts, and fords. No matter which method is used, stream crossings should be kept to a minimum and designed to avoid sediment production because there is no way to trap the sediment before it reaches the stream.

Among the cheapest types of stream crossing is a *ford*, which is simply a shallow section that can be crossed by wading or driving through the water. Undesigned fords obviously pose a high hazard to water quality and are unsuitable in many circumstances, as when streambeds and banks are composed of soft or erodible material, or on larger streams where flow velocity or depth are great. Some states, such as Michigan, prohibit their use (Michigan Dept. of Natural Resources, 1974). Nevertheless, they can be the most appropriate choice where construction and stability criteria are met and use is limited to times of low flow or periods when intermittent and ephemeral streams are dry. Many ephemeral channels in southwestern United States are crossed by fords, even on state and federal highways.

Fords are most suitable where the approach to the streambed is of low gradient, relatively dry, and stable. The approach should be of sufficient grade to prevent diverting the stream down the road during floods. The road surface through the banks and streambed should be perpendicular to the flow and consist of clean, firm stones or solid rock. Gravel or concrete fords can be relatively inexpensive (Milauskas, 1987), but adequate rock riprap or other erosion protection often must be incorporated in the design (Ring, 1987; Warhol and Pyles, 1989). Advantages of a well designed ford include its virtually unlimited capacity to pass high flows and that it is seldom blocked by debris.

Culverts usually are used on streams too small to require a bridge or where fords would be unsuitable. The most critical factor in the use of culverts is choosing a culvert having the capacity to pass the flows likely to occur. Forest practice rules of many states require culverts to pass flows with return periods of from 25 to 100 years, depending on type and useful life of the road. However, some risk of failure exists even using required culvert sizes. For example, if culverts for a given road system are designed to handle a 25-year return period flow, nearly two-thirds of them would be expected to experience a flow larger than a 25-year event within the next 25 years; one-half of them would be expected to experience a flow greater than a 25-year event with 17 years (Adams et al., 1986). Unfortunately, few data are available to determine the 50-year flow or any other design flow for small watersheds on wildlands, except for a few localities (Douglass, 1974; Campbell and Sidle, 1984). The peak rate of runoff usually must be estimated. Rainfall intensity–duration–return period data are available for much of the United States (US Weather Bureau, 1961 a, b, 1962, 1963). With these data, peak flows for corresponding return periods may be estimated using the rational runoff equation:

$$Q = 0.0973\ CIA \tag{12.1}$$

where Q is peak flow in liters per second (L s^{-1}), C is a runoff coefficient ranging from 0 to 1, I is rainfall intensity in millimeters per hour (mm h^{-1}), and A is area in hectares (ha). The critical factor is determining the correct runoff coefficient, which represents the proportion of precipitation that becomes runoff. It should be clear that the runoff coefficient tends to increase as the watershed surface is exposed and compacted by use.

Another means of estimating peak runoff is to examine stream channels for evidence of high flows, such as debris caught in streamside vegetation, scars on trees, silt or mud marks on woody vegetation, scour marks on streambanks, and so on. From such evidence, the cross section of the high channel flow can be determined. Velocity estimates can be made using the Manning equation (Eq. 10.4), and flow rates can then be calculated.

If older culvert crossings downstream or in nearby streams of similar size exist, they may provide some indications of past flow rates and suitable culvert sizes. The size of culvert required to pass a given flow may be determined from manufacturers and suppliers or from capacity charts such as those published by the Bureau of Public Roads (Herr and Bossy, 1965 a, b). Where information is lacking, follow the advice of Mustard (1982, p. 26), "Use good judgment—don't try to force a three-foot wide creek through a 24-in. pipe." Finally, the designer must remember that road construction and timber harvest almost always increase runoff. Peak flows from highly disturbed small watersheds may increase several-fold and be doubled from areas as large as several square kilometers.

Once culvert capacity is determined, the type must be chosen. Concrete culverts tend to be expensive, heavy, and permit high water velocities to develop because they are smooth inside. They should not generally be chosen where fish passage is desired. Round corrugated culverts of metal or plastic are available in a wide range of sizes and are suitable for many small streams. Corrugated pipe arch culverts permit lower fills, and structural steel arch culverts set in concrete footings may be used on larger streams and are most desirable for permitting fish passage (Yee and Roelofs, 1980).

Culverts should generally be placed so as to follow the natural stream course, both as to direction and gradient, and neither above nor below the natural streambed. It is often desirable to protect the approach to culvert inlets with rock or other riprap. Wing walls may be needed to protect the fill over the culvert from the impact of flow. The fill must be well compacted to prevent the development of seepage channels alongside the culvert that can erode away the fill.

Because of the many considerations involved with culvert installations, guidelines have been developed to identify cost-effective designs through an analysis of pipe, installation, and maintenance costs and failure probabilities and costs (Murphy and Pyles, 1989).

Bridges tend to carry the least environmental risk for small stream crossings, provided they have adequate capacity, but may be the most expensive type of structure. On larger streams, bridges are the most frequent choice. Bridges usually require the least amount of fill, are not readily clogged by debris, and their abutments may be kept entirely outside the channel of small streams. There are many kinds of bridges for small streams. Simple timber bridges made of stringers and laminated decks can be relatively economical (Berger et al., 1987), and some modular designs can be quickly erected or dismantled (Parry, 1986), allowing for removal prior to the rainy season. For permanent installations, design features that ensure long structural life become increasingly important (Muchmore, 1986), as do well-protected footings and supports.

Bridges must provide sufficient clearance to pass the design flood. Any fill that may be exposed to flood flows must be protected from scour. The span of the bridge should not represent the lowest point in the road system, as water flowing down the road would then drain directly into the stream. If necessary, the bridge should be raised slightly so the approach is uphill (Mustard, 1982).

12.1.3.3.2 Road Design. Road design includes considerations of *horizontal alignment*, the plan view of a route, including horizontal curvature; *vertical alignment*, the profile of a route, including variations in grade; and *roadway design*, or the cross section through any given section of the right of way (Fig. 12.7). Key features in any design include (a) minimal exposure of bare soil, (b) retention of stable slopes, and (c) provision for control of flowing water.

The horizontal and vertical alignment of any road is a major factor in determining all three features. Overall alignment is determined by site and management features such as management needs, property boundaries, obstacles, and hazards that control the overall route of the road. However, adjusting horizontal and vertical alignment to the topography by using variable radius curves and rolling grades will minimize earth movement. *Dips*, constructed by modifying gradient to create a short road section with the center lower than either end (Fig. 12.8*a*), permit disposal of water from the roadway. *Rolled grades*, which are essentially extended dips, will accomplish the same purpose. Dips are sometimes designed with their low point diagonal to the road center line and outsloped in comparison with the road surface. However, such designs tend to deflect vehicles outward when roads are slick and, in addition, impose twisting stresses on truck frames. Gardner et al. (1978) found that dips of uniform depth across the surface and perpendicular to the center line eliminated such problems and provided good drainage.

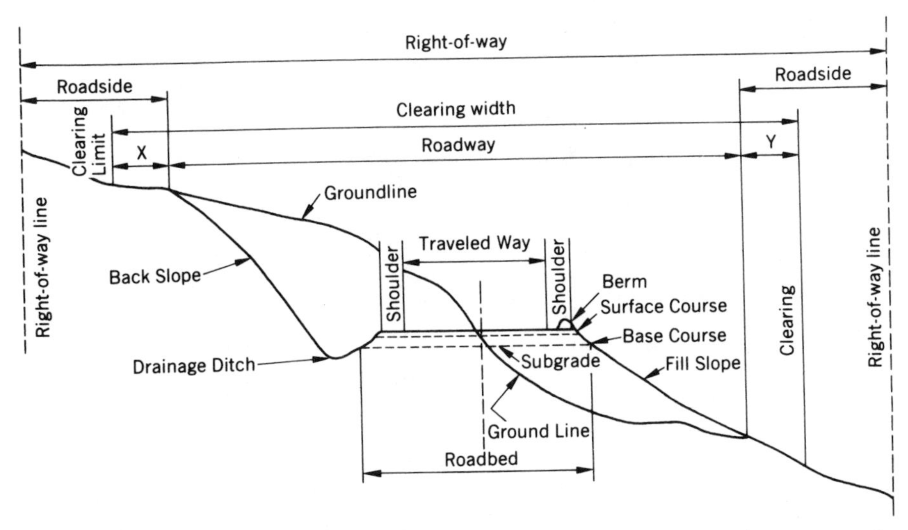

FIGURE 12.7 Illustrated terminology of the elements of a road (from Hynson et al., 1982). *Note:* Shapes and dimensions will vary to fit local conditions. X and Y denote clearing outside of Roadway.

Where hazards cannot be avoided, design becomes critical in maintaining stability and controlling surface erosion. For example, it is not always possible to avoid crossing slopes that have the potential for mass movement. However, it is possible to take certain compensating measures. Keeping the road width narrow across an unstable area minimizes slope disturbance (Gardner, 1979). Where curves are sharp, the road may need to be widened so as to keep the back wheels of long trucks on the roadway (Mifflin and Lysons, 1979). Narrow full bench roads (Fig. 12.9) with disposal of the cut material by depositing it in a stable site (*full-benching* and *end-hauling*), rather than sidecasting or using it as fill, will avoid overloading and oversteepening lower slopes. When combined with steep road grades to access and maintain ridgetop locations, full-bench end-haul construction has been shown to markedly reduce the frequency and size of road-related landslides, when compared to traditional midslope cut and fill (balanced) road construction (Sessions et al., 1987). Steep grades require extra attention not only to surface drainage, but also to potential wheel slippage based on the road surface and vehicles expected (Anderson et al., 1987; Sessions et al., 1986).

Increasing the back slope angle of the cut slope to its natural angle of repose will remove weight from the upper slope and increase its stability. High cut slopes may be built in the form of stepped terraces to reduce the likelihood of failure, and the benches may provide an improved planting site for stabilizing vegetation. Rock buttresses, piling, retaining walls, and gabions may all be used to maintain stability of cut slopes (Yee, 1976). Where windthrow hazard is

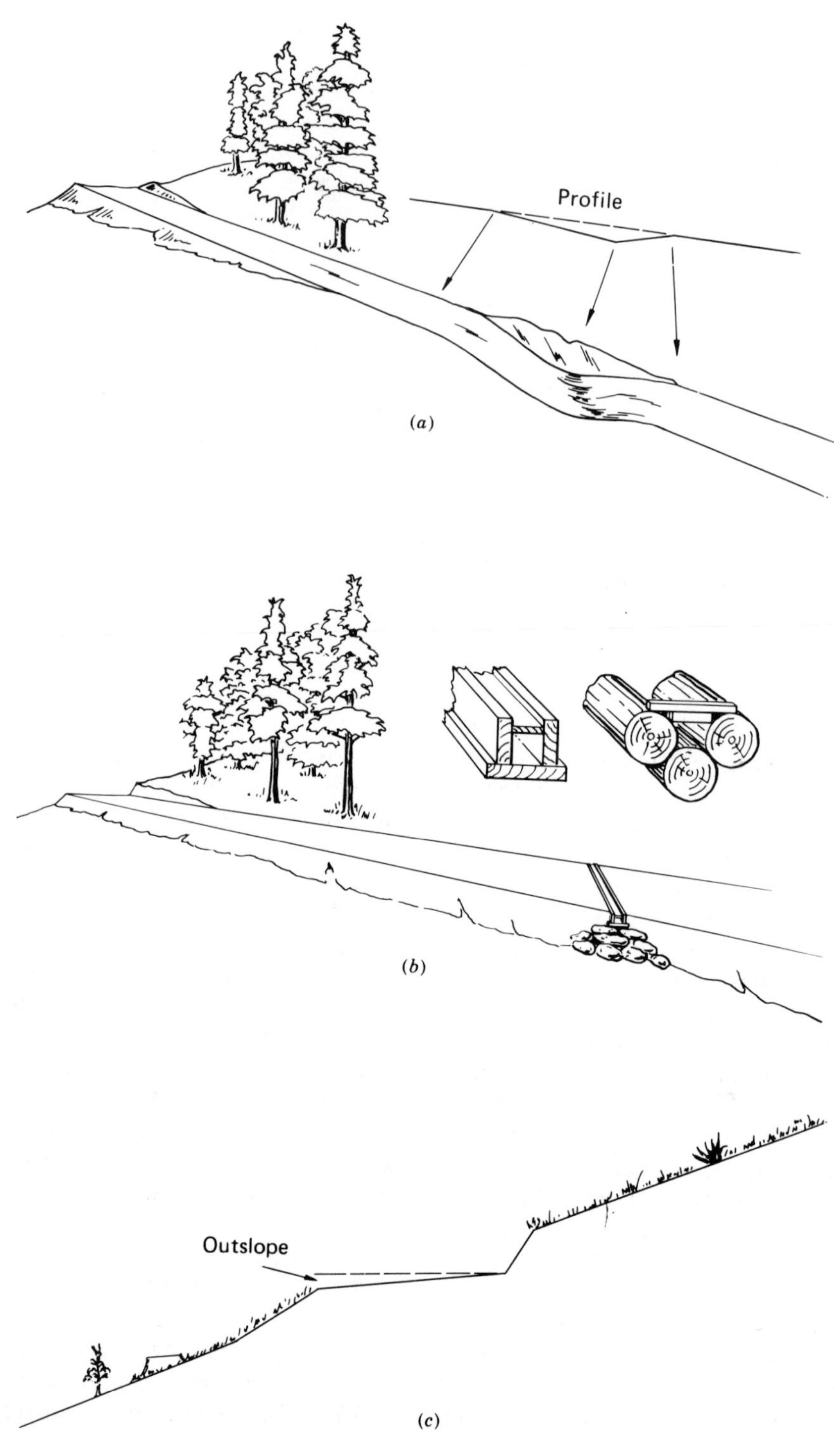
Profile
(a)
(b)
Outslope
(c)

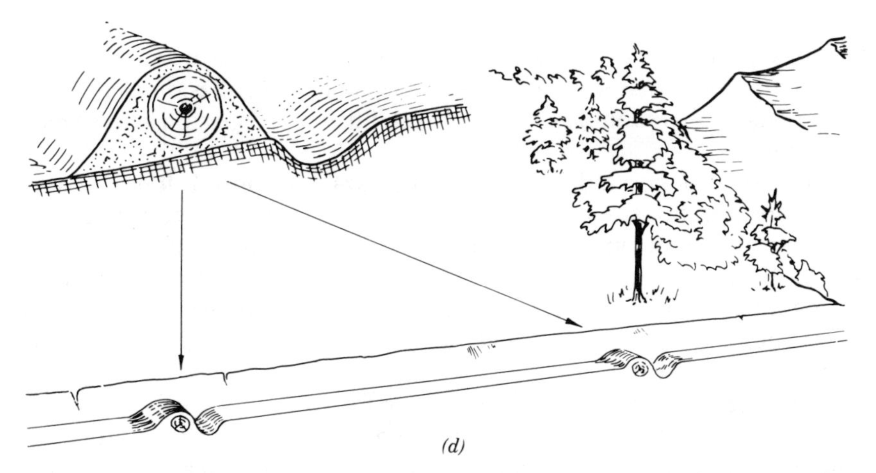

FIG. 12.8 Erosion control structures for roads. (*a*) Dip. There should be a definite adverse grade on both approaches. A rolled grade is a shallow, extended dip. (*b*) An open top culvert. Stones protect fill from erosion. (*c*) Outsloped road. Outslope should not exceed 5% or 5 cm m^{-1}, so even slight ruts will render it ineffective. Best restricted to contour roads. (*d*) Water bars. Short, sharp dips cut into roads. Logs help hold lower grade. Not for roads in use.

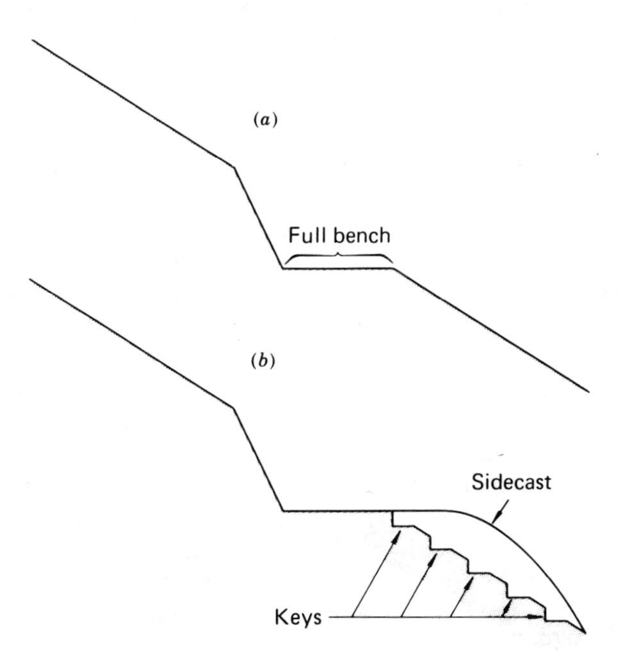

FIG. 12.9 (*a*) Cross section of full benched road. Roadbed is completely on undisturbed material. Cut material has been hauled away. (*b*) Cross section of road cut, showing material sidecast and keyed into lower slope to increase resistance to sliding.

slight or vegetation is in the form of brush or young trees, clearing should not be wider than the roadway to maintain as much root cohesion as possible. Deep rooted plants should be planted to establish root cohesion wherever the site permits.

To maintain slope stability, it may be necessary to control slope moisture. Water concentrations may be avoided by constructing intercepting ditches above unstable slopes to divert water. Drainage wells, sloping slightly upward from the horizontal, may be drilled into wet spots to dewater saturated slopes. The roadbed should not be permitted to block slope drainage. Drainage from roads should be carried across potentially unstable slopes rather than being allowed to infiltrate.

In extreme circumstances, half bridges or full bridges may be used to span short unstable sections or placed on piling over longer ones (Mustard, 1982). Road costs increase rapidly as compensating measures such as bridges, slope support structures, and dewatering systems become necessary. One should question whether roads should be built at all if such measures are required extensively, for the risk–reward ratio can quickly become submarginal on many wildlands.

In far northern areas of permafrost occurrence, areas of ice-rich permafrost pose a special problem. Where cuts are necessary through such material, the road must be protected with insulation. Gabions and rock buttresses may hold insulation against such slopes to reduce thaw depth and prevent slump material from reaching the drainage system. In the long term, vegetation usually must be established to prevent sediment movement (Claridge and Mirza, 1981).

Where the instability hazard is slight, simple measures such as narrow roads with turnouts, and balanced cuts and fills keyed into a natural slope (Fig. 12.9) are often sufficient. In no case should logs, slash, or stumps be buried in the fill, for they will lead to collapse of the fill as they decay. Logs and slash placed just below the fill, however, will help to catch any soil eroded from it. Clearing should extend beyond the roadway only where necessary to increase sight distance along the road or above steep cuts where the extra weight and/or swaying of trees in the wind may initiate slides or slumps.

To reduce exposure of soils, the road should be no wider than needed. Cut slopes should be as steep as stability permits. Many low cuts can be vertical where ditches are not required and soils are cohesive. Vertical cuts should not be used with ditches as even minor slumping will seriously interrupt drainage, but they frequently produce less sediment than backsloped cuts where ditches are not used. Where two different materials are present, the backslope angle should be varied to take advantage of steeper cuts where possible (Mustard, 1982).

It bears repeating that balanced cuts and fills expose the least soil to erosion. Fills should only be as steep as stability requirements permit. Compaction increases the angle of internal friction of fill material. Exposed cuts and fills can be protected from erosion with a variety of mulches, netting, and vegetation (Long et al., 1984; Bethlahmy and Kidd, 1966; Gallup, 1974; Megahan, 1974 b, 1987; US Environmental Protection Agency, 1975; Johnson and Fifer, 1978; Heinrich, 1982), and a summary of the erosion reduction provided by such treatments is shown in Fig. 12.10. Where balanced cuts and fills are not possible, borrow pits and waste dumps can become a source of erosion. They should receive the same protection as cuts and fills and be stabilized as promptly as possible.

Design of the roadbed depends on road use, the load bearing capacity of the soil, the resistance of the surface to displacement and abrasion, and the type of drainage system. Where road use is light or infrequent and subgrade materials are competent, simple outsloped roads without surfacing may be adequate. Preplanned skid trails, built on competent soils and allowed to settle a year before use, have proved successful in limiting soil disturbance from thinning on Crown Zellerbach Corporation operations in Oregon (Anonymous, 1968). Unsurfaced roads are frequently adequate for small private owners, so long as they are not used when wet so as to avoid creating ruts that channel surface water into erosive concentrations. On unsurfaced roads, grass or other herbaceous growth should be encouraged to grow if it does not create a fire hazard. Even wheel tracks may support vegetation if use is infrequent.

Where road use is heavy, soils are easily displaced by traffic, and when wet weather use is likely, the road may have to be surfaced. Gravel or crushed

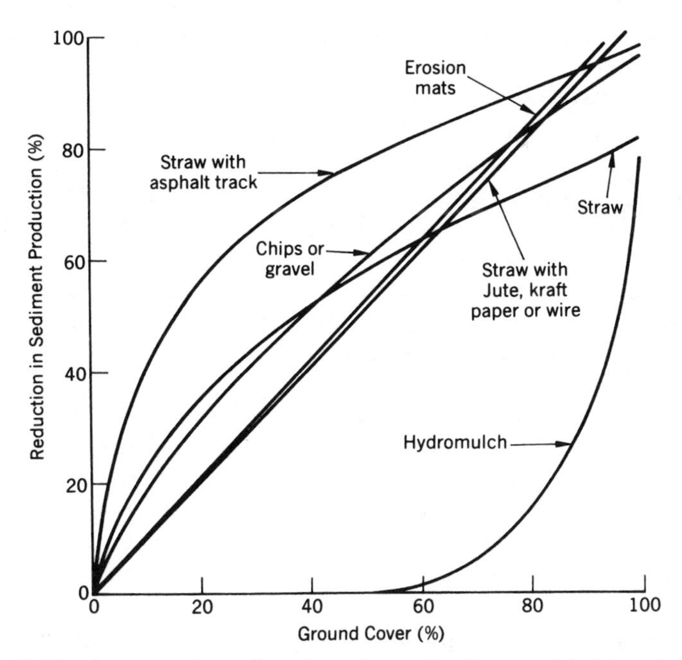

FIGURE 12.10 A comparison of road erosion reduction provided by selected treatments for average site conditions (from Burroughs and King, 1989).

stone is the most common material used, except on main access roads where asphalt is typical. Sometimes surfacing is necessary only on part of the road, such as steep grades where runoff velocities may be high, or on noncohesive, weak or erodible soils that cannot be avoided. Enough rock must be used to avoid deep rutting and mixing of soil and rock (Swift, 1984 a). The amount of rock needed for a given location will vary with expected traffic loads, road grade, and the quality of the road bed and surface rock (Anderson et al., 1986; Berglund and Rowley, 1975; Bilby et al., 1989). A base course of rock, geotextile or polypropylene fabric may be used to distribute loads where underlying soils are weak or easily displaced, as in muskegs (Lively and Vischer, 1977; Mustard, 1982).

To control surface erosion, good road drainage is, without question, the most important design consideration. It should be reiterated that road construction and timber harvesting almost always increase runoff. The traveled way resists infiltration by design, as it must if the road is not to become a mudhole. Surface runoff may be expected to increase in proportion to the area of exposed cuts and compacted fills and with the amount of subsurface flow intercepted by road cuts (Burroughs et al., 1972; Megahan, 1972). Therefore, design must be based on runoff that will occur after road building and logging, not during undisturbed conditions.

The essence of good drainage, whether in ditches parallel to the road or in cross drains, is to remove the water before it becomes sufficiently concentrated to cause serious erosion. The water must then be disposed of in a suitable location, preferably where it can be dispersed and infiltrate into the soil. However, where a surcharge of soil water might create slope instability, the water should be conveyed across the slope rather than induced to infiltrate.

To ditch or not to ditch is a frequent question. Where cuts are unstable and minor slumps and sloughing occur, continual maintenance is necessary to keep ditches functioning properly and the consequences of blocked drainage may be greatly increased erosion. Ditches are risky where all required maintenance cannot be assured. On secondary and temporary haul roads and skid, pack and hiking trails on stable slopes, low vertical cuts, and outsloping or surface drains to divert water are often sufficient. However, ditches are almost mandatory where roads are crowned (higher in the center than on the edges), insloped, or on major haul roads.

Cross drainage is accomplished by a variety of methods. On surfaced, well-maintained roads, the road surface is usually crowned and water drains to each side. Where road location and topography permit, ditches can divert water away from the road without cross drains. However, the opportunity to divert water frequently exists on only one side of the road for long distances. Then culverts are almost mandatory to carry accumulated runoff across a crowned or surfaced road.

Crowned surfaces, outslopes, and inslopes eliminate the possibility of concentrated flow down the road surface when properly designed and maintained. But many wildland roads can only be maintained infrequently or are built with flat cross sections. Water then tends to flow downgrade over the road surface. It must be removed by some sort of cross drain. Methods of cross drainage include shaping the road surface to divert water, as with rolled grades and dips, outsloping or insloping, and water bars, or using structures such as open top drains or culverts (Fig. 12.8).

Each method has its limitations. Road gradient and location determine the feasibility of rolled grades and dips, which cannot be used on many steep slopes or on level areas. A decision-making framework can be useful for comparing road dips versus culverts for cross drainage (Eck and Morgan, 1987). Outsloping should not be used where road surfaces are slippery or on short radius curves for reasons of safety, but insloping may be desirable under such circumstances. Insloping usually requires a ditch and culvert to get cross drainage. Neither insloping nor outsloping should be used on steep grades or on roads where ruts will develop that destroy their effectiveness. Water bars are usually effective only on unused roads, for traffic quickly breaks them down. Open top drains, constructed of poles, concrete or rough lumber, and installed diagonally across uncrowned roads are sometimes used. They offer a good compromise between cost and effectiveness on lightly traveled, seldom maintained roads. For important access roads, plastic pipes (Stjernberg, 1987) may be less costly and easier to transport, install, and maintain than metal or concrete pipes.

Inlets and, especially, outlets of drains may pose a high erosion hazard. Flows are concentrated, gradients often are steep and flow velocities high, and discharge may have to cross fill or other disturbed soil. Therefore, the outlet should not be perched so a waterfall is created. The outfall should be protected with energy dissipating rocks, logs, slash, or other material. Inlets may require similar protection, as well as debris interceptors to avoid clogging. Flow should be carried down and across fills with half culverts or in riprapped channels. Sometimes berms are used to protect fills from cross drainage on crowned or outsloped roads (Swift, 1984 b). Where water is released from the berm, protection as above is also necessary.

Once a suitable type of cross drain is selected, the problem of spacing and location becomes dominant. The spacing should be sufficiently close to: (a) prevent undue erosion from the road surface, ditches, and banks; and (b) insure that the sediment issuing from the outlet can be trapped in the strip between the road and the stream. Published spacing guidelines (e.g., Megahan, 1977; Packer, 1967) provide a starting point, but local experience is essential in refining road drainage planning criteria. Furthermore, on-site reconnaissance is needed so that local cross drain spacing may be modified to avoid locating outlets on fill or other erodible material and to permit dispersion of flow on stable sites.

In many places, once flows become concentrated, it becomes difficult or impossible to disperse them enough to prevent sediment from reaching a watercourse. Under such circumstances, it may be necessary to construct sediment basins to trap the sediment. The sediment carrying capacity is a function of the velocity of flow, so the sediment basin must be large enough to cause a sufficient reduction in velocity so that most of the sediment load is dropped. This can be accomplished with basins having a large ratio of storage to inflow. Where little storage is possible, cross drains must be closely spaced to reduce inflow. Another means of trapping sediment is to construct sediment basins from porous materials such as bales of straw anchored with posts and wire that filter the sediment from the water. Such basins offer only temporary effectiveness, but since erosion from most construction diminishes rapidly with time, they can often meet all needs until erosion control becomes effective.

Research has shown that roads deteriorate rapidly if rills are permitted to erode deeper than about 2 cm. Factors influencing erosion from roads include the resistance of the surface and bank material to erosion and the energy available from flowing water. Resistance is determined largely by the size of exposed surface materials and their degree of cohesion. Thus, drain spacing can be greater where roads are surfaced and soils are cohesive than in fine, noncohesive materials such as silts.

The amount and energy of flowing water is a function of (a) road gradient, (b), topographic position, (c) direction of slope, (d) steepness of sidehill, and (e) seepage from cuts. The velocity increases rapidly as road gradients and the amount of flow increase (cf. Eq. 10.4). The amount of flow tends to be greater on lower than upper slopes, on north and east than on south or west slopes, and steeper sidehills and where deep cuts intercept large amounts of subsurface flow. Where snowmelt is important, south and west slopes may exhibit higher rates of runoff than lower energy slopes. All these factors affect the spacing required to insure that erosion does not exceed acceptable limits.

12.1.3.3.3 Buffer Strips. The criteria of drain spacing interact with the sediment-trapping efficiency of the strip between the road and the stream to determine the amount of sediment reaching the stream. Where many obstructions and good vegetation cover exist to trap sediment, large volumes of flow can be dispersed and sediment caught. Where trap efficiency is low, spacing of drains must be further reduced to prevent concentrated flow from moving directly to the stream.

Several guides are available giving detailed erosion control instructions on the basis of road and buffer strip conditions (Haupt, 1959; Kidd, 1963; Packer and Christensen, 1964; Haussman and Pruett, 1979). Brush barriers at the toe of fills can greatly reduce the downslope extent of sediment movement (Cook and King, 1983; Swift, 1985, 1986; Burroughs and King, 1989). Log, brush, and rock barriers can be designed to spread water over a slope below drainage outlets where natural barriers are lacking.

Buffer strips are most effective where water can be dispersed so that surface roughness features can slow the rate of flow, sediment is deposited, and infiltration is enhanced. But, there will often be circumstances where flow is too concentrated for buffer strips to be effective, and sediment is then carried right through the strip. In such cases, it may be necessary to construct sediment basins by building low head dams, excavating depressions, and constructing basins of logs, straw bales, sandbags, or fabric filter fences (Johnson and Fifer, 1978; Knighton, 1984). Most sediment basins have a short effective life, so any permanent solution requires preventing erosion or continual maintenance.

12.1.3.4 Construction Procedures. There is no way to avoid exposure and disturbance of soils during road construction. If erosion is to be reduced to a minimum, the disturbed surface should be exposed only during the most favorable conditions and for the least amount of time. This means that the major steps of construction should be carefully planned (Garland, 1983 a) so that excavation is avoided during rainy seasons when possible, and protective measures should be installed before adverse weather conditions arrive, such as the fall rains in the Pacific Northwest of North America. A particularly important measure to avoid undue erosion is to schedule operations as tightly as possible to keep the area of highly disturbed surface to a minimum and to complete each phase—clearing, grubbing, earthwork, drainage installation, surfacing, and protective measures such as riprap, mulching, seeding, and revegetation—quickly once construction begins.

Most sediment is produced the first few storms after disturbance. Therefore, it is very important to install proper drainage facilities as soon as possible as construction proceeds. Since there is no way known to avoid erosion when storms strike disturbed sites, provisions must be made to disperse sediment-laden runoff and trap sediment if they should occur during construction. Sediment control features must be built concurrent with drainage facilities as earthwork progresses.

Close control and inspection during construction is needed to see that the above measures are carried out.

12.1.3.5 Use and Maintenance. Heavy traffic, even during dry weather, can increase rates of erosion and sedimentation from wildland roads (Bilby et al., 1989). Measures to limit or disperse traffic may help reduce these problems. If roads must be used in wet weather or other unfavorable conditions, wide tires or lower tire pressures may be effective in reducing damage to road surfaces (Mellgren and Heidersdorf, 1984; Stuart et al., 1987).

Regardless of proper design and construction, roads may fail in their operational and water protective function if they are not maintained (Adams, 1983). Roads should be inspected regularly, especially during wet weather, and any necessary surface grading or culvert or ditch cleaning should be promptly scheduled. The major emphasis must be on maintaining the original drainage capacity. Ditches must be kept clear of obstructions and culverts free of debris or sediment that would reduce their capacity or divert flow over the road (Fig. 12.11). Depressions that permit water to pool on the surface must be filled or drained, and the surface maintained to prevent infiltration that reduces the bearing strength and leads to formation of ruts that concentrate water and keep it from draining off the road. Incipient failures in all drainage facilities

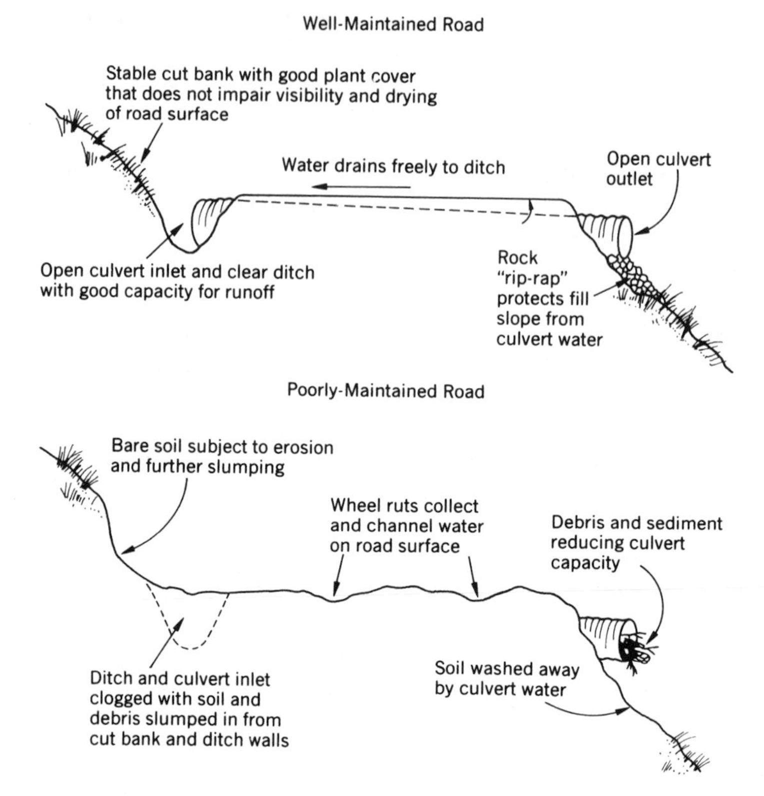

FIG.12.11 Examples of some important differences between well-maintained and poorly maintained woodland roads (from Adams, 1983).

must be detected and repaired before they are subjected to excessive flows. If machinery is used, care must be taken to avoid damage to cross drains (Piehl et al., 1988), and ditch cleaning should not be so aggressive as to eliminate stabilizing vegetation (Bilby et al., 1989).

Incipient mass movement must also be detected and stabilized by reducing shear stress or increasing shear resistance by use of excavation, drainage of soil water, placement of buttresses and retaining walls, or other measures.

Unneeded roads should be retired from use and "put to bed" whenever and wherever feasible. Many spur roads and skid roads used in timber harvest are not needed between harvest or thinning cycles. The high cost of maintaining a stable road bed in some locations also may make road closure economically attractive. Roads can be put to bed by blocking off traffic, and scarification and seeding of compacted surfaces. Sometimes fertilizing and reseeding with wildlife foods can be a fruitful project for sportsman and service organizations, leading to a double return of stability and increased wild game populations.

There are several methods of closure, such as gates or fences, very large water bars, and a particularly effective measure on some slopes is road deconstruction. In road deconstruction, a backhoe is used to pick up sidecast fill and replace it so as to restore the original slope. Revegetation is necessary to prevent erosion of the disturbed slope. The restored slope usually is extended only a short distance back from the road entrance and is easily opened when a road is needed.

The odds of culvert failure through loss of capacity or occurrence of flows that exceed design capacity increase rapidly with time on closed roads. For this reason, culverts are often removed and replaced with water bars or the original cross section of natural drainages is restored when putting roads to bed. Proper water bar function is enhanced when they have an angle greater than 30°, sufficient height to handle flows, and an unobstructed outlet (Yee and Thomas, 1984).

Roads retired from use should be inspected periodically to detect potential problems. Often the only maintenance required will be to clear debris or sediment from drainage structures such as ditches. When the need arises to use a road that has been put to bed successfully, only a small effort is needed to restore trafficability and the risk of erosion is slight.

12.1.3.6 ***Integrated Net Effects.*** Few data are available to demonstrate the effects of improved road location, design, and construction practices, for widespread application in most watersheds has yet to take place. However, a recent study in which 162 km of well-constructed logging roads were built while removing 43% of old growth Douglas-fir in patch cuts from a steep 8082-ha block of land over an 8-year period showed no detectable increase in suspended sediment (Sullivan, 1985). These results are consistent with another study mentioned earlier (Sessions et al., 1987), in which a large area of managed forest land (120,000 ha) was surveyed, and it was found that improved road location and design did indeed reduce landslide erosion. Though two studies cannot be taken as conclusive proof, they do suggest that water quality can be maintained on many intensively managed steep mountain watersheds, provided roads are built with care.

12.1.4 Grazing

Grazing, by both domestic and wild animals, is one of the most widespread uses of wildlands in the world. Furthermore, uncontrolled grazing has been held responsible for severe and widespread watershed destruction that is believed by many to have contributed to the collapse of past civilizations (Lowdermilk, 1953) and the desertification of parts of Africa and the Near East today. However, the great variety of animals and the wide range of sites they utilize—from musk oxen and reindeer in the north; cattle, horses, sheep, and goats throughout the world; buffalo (carabao) in the wet tropics; and various antelopes in seasonally dry tropics—eliminates any possibility of a complete discussion short of encyclopedic length. Therefore, this treatment will be limited to principles of management, mostly as they are applied to grazing lands in North America. A general treatment of range management can be found in Stoddart et al. (1975). Most of the material herein is applicable to semiarid and arid lands in other parts of the world, as well as in North America. Two major publications providing more thorough treatments on range watershed effects are Moore et al. (1979) and Branson et al. (1981).

12.1.4.1 ***Hydrologic Characteristics of Grazing Lands.*** Despite the great variability in grazing lands throughout the world, many share common characteristics and water quality problems.

Most natural rangelands are characterized by relatively low productivity and a harsh physical environment. The more productive and moderate areas tend to be cultivated or occupied by forests. Grazing lands such as natural grasslands, savannas, open woodlands, shrub steppes, and brushlands exhibit limited but intense precipitation, recurrent drought, and extremes of heat. Alpine and arctic areas suffer from short growing seasons, temperature extremes, and often stay wet through much of the snow-free period. Many lands are steep and rocky, have shallow soils, or lack fertility. In short, many grazing lands lie naturally close to their critical point in deterioration.

Important exceptions occur. Some mountain meadows and riparian zones are highly productive, as are most of the prairies and pastures in eastern North America where growing conditions are more favorable. More frequent and greater rainfall, however, increases the danger of disturbance of wet soils, especially in view of the heavy stocking supported by productive sites. Problems are generally fewer in eastern North America (Patric and Helvey, 1986), for woodland grazing of the widespread forests is decreasing, except in parts of the southern pine region that are well adapted to grazing. Most fertile prairie land is now cultivated, with little natural grassland, except in a few areas such as the Flint Hills of eastern Kansas that are too stony to cultivate. However, productive mountain meadows and riparian zones may attract heavy use where they are interspersed among less productive range. Regardless of the differences in environment and productivity, most effects of grazing and principles of erosion control of rangelands are similar everywhere.

12.1.4.2 ***Nature of Grazing Effects.*** Grazing animals influence water quality in several ways. Most important, rangelands are major sources of sediment from wildlands. In western United States it has been said that streamflow is derived mainly from forest and alpine lands, whereas sediment is mainly from the rangelands that lie below them (Branson et al., 1981). In the states of Idaho, Oregon, and Washington, rangelands account for an estimated 28% of total annual sediment production despite some of the highest erosion rates in the nation from cultivated croplands of the Palouse region in the same area (Moore, 1976).

Most of the erosion results from the removal of forage that provides protective cover for the soil surface and from the trailing and trampling that compacts the surface soil, tends to further destroy protective cover, and breaks down streambanks. The effects of trampling and forage removal cannot be separated, except in artificial studies, but it appears that infiltration is affected most by trampling (Stoeckler, 1959; Dadkhah and Gifford, 1980). The effect of trampling varies with soil type, with greater compaction on soils containing higher amounts of silts and clays (Orr, 1960; Van Haveren, 1983).

Erosion and sediment production are more closely related to cover. Where the site will support relatively complete cover, it appears that 50–70% vegetative cover is adequate to prevent significantly accelerated erosion (Meeuwig, 1970 a, b, 1971; Meeuwig and Packer, 1976; Dadkhah and Gifford, 1980; Branson et al., 1981). On sites that support less complete cover, small reductions in cover may be associated with relatively large differences in sediment production (Aldon, 1964; Lusby, 1970; Shown, 1971).

Other effects of grazing that may be important include removal of streamside shade by browsing animals, breakdown of streambanks, and deposition of body wastes in or near streams. Results include increased stream temperatures, increased sediment loads with consequent reduction in spawning and food production areas for fish, loss of cover, eutrophication, and introduction of undesirable microorganisms such as fecal bacteria (Kaufman and Krueger, 1984). Organisms such as liver flukes, *Giardia lamblia*, and others that infect humans can also result. From a public health standpoint, no wildland surface water can be considered free of contamination so long as wild or domestic animals inhabit the watershed.

12.1.4.3 Recognizing Hazards on Rangelands. Range sites, as any other, vary in their ability to withstand the impact of grazing without damage from erosion. Such factors as slope steepness and position, aspect, soil erodibility, moisture content and cover do not require elaboration. However, not all sites are likely to receive similar use. Animal behavior is such that certain site characteristics may strongly influence use, as well as erodibility. For example, steep slopes that tend to increase erodibility also tend to discourage use, particularly by cattle. Thus, the greater hazard of steep slopes may be partially offset by lesser grazing use. On the other hand, certain site factors that contribute to low erosion resistance tend to increase animal use. For example, steep southwest slopes near the South Fork of the Salmon River in Idaho are highly susceptible to erosion. Soils derived from the Idaho Batholith are naturally erodible, and on southwest slopes they are shallow and support only a sparse ground cover. They are also among the first to be exposed by snowmelt in the spring, and the early green forage attracts and concentrates elk and deer. The scanty cover is further depleted by early grazing and trampling when soils are wet. Bare, eroding areas feeding sediment into streams are common. Domestic livestock are not a factor in this example, but they, too, tend to concentrate on similar sites in late winter and spring, both for the early forage and the comfort of a warmer local climate.

A major factor in grazing is the need for most large animals, both wild and domestic, for frequent watering. Furthermore, riparian meadows are highly productive, and their moist soils are often poorly resistant to disturbance. Animals tend to stay in the vicinity of water, and such areas may be heavily overgrazed. Roath and Krueger (1982) found that although riparian meadows occupied only 1–2% of a forest range, they produced up to 20% of the total forage and accounted for 81% of the total herbaceous vegetation removed by the cattle.

Combine heavy use, soil disturbance and proximity to water, and the seriousness of erosion and fecal contamination in streams is compounded. Even minor storms may be sufficient to carry sediment, microorganisms, and nutrients into streams for the riparian transport distance is short and opportunities for trapping and dispersing the loads before they reach the stream are low.

Because many grazing lands are found in harsh environments, they lie naturally close to their critical point in deterioration. It is particularly important to be able to recognize incipient changes in range condition that precede the critical point. The margin of safety before deterioration exceeds acceptable limits for water quality control is small. By the time erosion becomes obvious, it may be too late to achieve rapid, economical control. Early measures to maintain site stability are almost invariably cheaper, faster, and more effective, and may permit continued use of the range.

The recognition of deleterious change at an early stage requires a firm knowledge of range ecology. The first signs almost invariably appear as changes in plant density, composition, and vigor. Such changes precede or accompany changes in surface runoff and erosion. Stoddart et al. (1975) discuss the evaluation of range condition and trend at length, and no attempt will be made here to cover the subject. Analyses of range condition and trend are nevertheless essential first steps if accelerated erosion from grazing lands is to be avoided.

It is also possible to recognize incipient change at an early stage. Evidence is available in the stream channel and streamflow, and on the watershed itself.

Water quality monitoring is one means of detecting changes in nonpoint pollution from rangelands to meet Section 208 requirements of Public Law 92-500 and Public Law 95-217, the Federal clean water legislation mandating nonpoint pollution control. Common monitoring techniques include measurement of turbidity, dissolved solids and salinity, fecal coliforms, and sometimes aquatic organisms. However, most monitoring is done on larger streams and would be prohibitively expensive to conduct routinely on most wildland streams. In addition, natural variability in suspended sediment yield, the primary pollutant of rangeland streams, requires that successful monitoring programs be based on rigorous requirements for instruments, techniques, and sampling that mandate the availability of skilled personnel over a long time period (Thomas, 1985).

Other means of detecting deterioration include observations of channel condition, such as composition of channel sediments, the presence of deposits in pools, and changes in channel morphology toward broad, shallow cross sections from deeper, narrow conditions. Delta desposits where tributaries enter larger streams may help indicate abnormal sediment sources. Streams that are noticeably muddy during stormflow can sometimes be followed to reveal the source of their sediment load, but the short time during which surface runoff is generated and the need to be present and free at unpredictable times and locations limits the feasibility of this procedure.

Fortunately, many kinds of evidence persist between storms (Gleason, 1953). Some have been used to develop resource evaluation systems for uplands (Clark, 1980) and riparian areas and stream channels (Cooper, 1977; Platts et al., 1983; Platts et al., 1987). Included are such indicators as evidence of soil movement and deposition, breakdown of banks, vegetation and litter condition, erosion pavement, and surface runoff.

For example, splash erosion is indicated by one or more of the following conditions:

1. Soil pedestals capped by bits of debris or small stones. The debris or small stones protect the soil beneath from raindrop impact, but unprotected soil is dislodged and carried away by sheet flow. Sometimes the pedestal formation gives the appearance of miniature soil columns scattered over the surface like stumps in a clearcut forest. Living plants may also develop pedestaled bases.
2. Development of puddled spots on the soil surface. The impact of raindrops may disperse aggregates and, as is well known by those who have measured soil texture by use of a hydrometer, the finer particles tend to remain in suspension. The fines may be concentrated in puddles, and when the water infiltrates or evaporates they form an amorphous layer over the surface. If the particles consist of fine clays, a surface glaze may form in every minor depression. As the surface dries, shrinkage cracks often form, creating hexagonal, cupped plates of fine material.
3. The base of plants and stones may be discolored by soil contained in raindrop splash up to a considerable height.

Splash erosion is commonly accompanied by sheet erosion. Early stages of sheet erosion are, therefore, also indicated by the above features, as well as by one or more of the following characteristics:

1. Exposed root crowns and roots indicate loss of the surface soil in which they grew. Following fire, exposed plant material that is uncharred, or charred only on the upper surface, suggests sheet erosion. Rocks are frequently covered by lichens. Where fresh mineral material is exposed along lower surfaces, sheet erosion may be occurring.
2. Miniature terraces, deposits of sediment on the uphill side of bits of debris, stones, or living vegetation, are good evidence of soil movement.

3. Small alluvial cones formed at minor changes of slope indicate sheet erosion. As they become larger, they are easily detected, but close investigation is necessary to see them before rill formation begins. Small cones are easily masked by growing vegetation.
4. The formation of erosion pavement indicates sheet erosion. As the fines are removed by sheet flow, larger particles that are too heavy to move are left behind. They become concentrated over the surface and gradually form an armor that protects the surface from further erosion.
5. Truncated soil profiles. Soils may have their upper soil horizons completely removed by erosion, and lower horizons are exposed at the surface.

The later stages of erosion are usually clearly evident, but in many cases on rangeland, much of the damage has already been done by this time. Rills, gullies, large sediment deposits, and truncated soil profiles may persist and continue to erode for a long time. However, unless evidence of fresh cutting or deposition is present, they may indicate past conditions rather than present condition or trend. It is better to rely on signs of splash or sheet erosion to assess present conditions.

12.1.4.4 Control of Range Use. The severe environment of many rangelands, coupled with their low and variable productivity, makes erosion very difficult to control once it has begun. Semiarid and arid lands often exhibit high rates of geological erosion and naturally lie close to their critical point in deterioration.

To prevent unacceptable damage, it is important to recognize that most rangelands contain certain key areas where the combination of site and animal behavior combine to cause the maximum impact. Such areas include riparian meadows, productive sites having shade, water and salt present or nearby, and others. They are the places where animals tend to congregate and where they will stay until forced to move, even if better forage is available elsewhere. If these areas are identified and managed to avoid overuse, the rest of the range will take care of itself and remain in acceptable condition. Nevertheless, it may not be easy to achieve proper management. Platts and Nelson (1985 a, b) suggest that winter grazing or special riparian pastures may be necessary to protect streamside meadows.

The nature of the range livestock industry, faced with forage productivity that may vary by an order of magnitude from one year to the next, adds to the difficulty. If heavy production of one year is not utilized, it may be largely lost, for nonwoody species tend to die back each year and there is only a limited carryover from one year to the next. Substantial reductions or increases in herd size for optimum utilization of variable forage supplies is biologically and economically problematic. Prices drop when ranchers must sell because of drought and climb when people seek to rebuild herds to take advantage of abundant forage.

The legal status and emotional response to efforts to control populations of large game animals or feral equines provide even greater problems where they are the cause of deterioration of grazing lands. Wild horses, burros, elk, and deer often are "sacred cows" of the American West.

Some wildland users and managers, recognizing these difficulties, have used them as an excuse for permitting continued deterioration of land and water, or for failing to improve them. Though the task is difficult, it must be done and excuses should not be permitted. Much of the range in the western United States is public land in which everyone has an interest. Even private ownership does not carry with it the right to degrade the productive base upon which future generations depend.

These factors do, however, point up the fact that major emphasis must lie in prevention of deterioration at a very early stage before it can become a serious problem. Where deterioration has begun, it must be reversed before it becomes uneconomic or physically beyond control.

The most effective control depends on controlling forage utilization and animal distribution. Repeated studies have shown that most grazing land, deteriorates rapidly under "heavy" use (Cable, 1975; Clary, 1975; Currie, 1975; Martin, 1975; Paulsen, 1975; Thilenius, 1975; Springfield, 1976; Turner and Paulsen, 1976; Moore et al., 1979; Branson et al., 1981). Only a few studies, mostly in the eastern or southern United States (e.g., Duvall and Linnartz, 1967), failed to show adverse changes under heavy grazing. (The reader should be aware that "heavy," "moderate," "light," and other such utilization terms are vague and highly variable by species, region, and productivity. "Heavy" use may be less than 100 kg per ha^{-1} to more than 2000 kg ha^{-1}, depending on circumstances. The proportion of forage used to that produced may range from as little as 40% to more than 75%. Therefore, different studies may not be directly comparable.)

Different grazing systems may also influence the response of plant cover to grazing. Except for continuous grazing, all are designed to promote or maintain plant vigor by taking advantage of protection from grazing at certain times. However, a review of different studies by Gifford and Hawkins (1976) revealed few conclusive results from using different systems.

Many important watershed characteristics, such as infiltration rates, recover during periods of nonuse, but on certain sites where cover was lost, recovery was negligible for years. For example, some sites that lost their cover before 1907 due to excessive sheep grazing were still barren nearly 70 years later despite reduced livestock numbers and improved grazing systems, including nonuse (Strickler and Hall, 1980). In another study, past erosion from heavy use resulted essentially in a permanent change in the vegetation type and density (Reid et al., 1980).

Controlling the distribution of grazing animals can be very important, for animals tend to congregate in preferred areas even after forage is depleted rather than move away from water, shade, or gentle terrain. Useful practices include dispersing the location of salt and water developments, installing fences to control access to key areas such as riparian pastures, and riding or herding. On high hazard, highly productive sites such as riparian meadows, more intensive (and expensive) measures such as fencing may be justified, with potentially dramatic results (Elmore and Beschta, 1987; Chaney et al., 1990). Less intensive measures must be used where economic productivity of the land cannot support heavy investments of capital or labor. In the absence of distribution control, it may be necessary to reduce stocking to that matching the productivity of the key area alone.

Wild game and feral equines are subject to a lesser degree of management control than domestic livestock. Solutions to range deterioration by them are more likely to depend on politics than technology.

12.1.4.5 ***Range Improvements.*** In many cases, deterioration of range vegetation from past misuse has proceeded beyond the point where control of stocking or distribution will insure timely recovery to desirable conditions. Even where such management is ecologically feasible, recovery rates may be unacceptably slow. On ranges that are in relatively good condition, management can sometimes improve productivity so that heavier stocking can be permitted without danger to water quality. In such cases, rehabilitation or improvement measures may be desirable. It should be pointed out, however, that the history of range improvement is replete with failures and many physically and biologically successful measures are not cost effective (Branson et al., 1981). Range improvements should be attempted only after careful analysis of each individual situation.

Range improvements include such measures as removal of competing vegetation to release more desirable species, fertilizing, and seeding or planting. They also include mechanical and structural measures such as soil ripping, terracing, range pitting, contour furrowing or trenching, building check dams and gully plugs, water spreading structures, and others. Actions such as fencing and water developments are often desirable to complement and enhance other rehabilitation or improvement measures. Vallentine (1971) discusses range developments and improvements extensively. Only a brief summary will be presented here.

The most common range improvements are those of removal of undesirable species. Frequently, the problem of range deterioration is accentuated by the invasion of undesirable species that further decrease forage production, yet fail to protect the site. Mesquite, sagebrush, pinyon–juniper, oak savanna, chaparral, and other woody vegetation types have spread widely on deteriorated ranges. Their removal and replacement by more desirable species can sometimes enhance grazing returns while improving watershed protection.

There are many techniques of removing or killing undesirable species. Mechanical methods of cutting, chaining, bulldozing with brush blades, brush chopping, and root plowing are common. Prescribed burning is effective with some species. Herbicides have been widely used in the past, alone or in combination with other methods, but regulatory and social constraints have reduced their use in recent years. Specific techniques are described in numerous publications (Arnold et al., 1964; Bently et al., 1966; Slayback and

Cable, 1970; Vallentine, 1971; Roby and Green, 1976; among others). Most methods involve some risk of reduced water quality in the interval before new vegetation becomes well established.

Sometimes desirable native plants, freed from competition, will recover satisfactorily after the woody vegetation has been killed. Other times, it may be necessary to seed or plant adapted species. Wherever the site is heavily disturbed, seeding or planting often is a necessary follow-up step.

Conversion of type is not desirable or feasible everywhere. On many steep, rocky, or broken sites, those with shallow or infertile soils, and those receiving inadequate rainfall, the chances of successful establishment of a good protective cover may be slight. Many hectares of chaparral in southwestern United States have been converted to grass to increase forage production and to reduce the hazard of wildfire that would lead to the loss of watershed protection. However, conversion of steep slopes of the San Gabriel Mountains of California to grass following wildfire has greatly increased the frequency and extent of soil slips as compared to sites where the brush was allowed to recover naturally (Corbett and Rice, 1966).

No range improvement will be successful if it is not followed up by prudent livestock management. Watershed improvement efforts sometimes go for naught because too much haste to recover investment costs results in subsequent overuse of rehabilitated lands.

Structural and mechanical treatments of the land are less frequently applied than control of undesirable vegetation, but are still common on some sites. Soil ripping and tillage are used to break up compacted soils and increase infiltration. Range productivity is increased and runoff and erosion may be reduced (Hickey and Dortignac, 1964). Contour furrowing and range pitting trap runoff for infiltration and may be justified on favorable sites by increased forage production. Effects of ripping, furrowing, and pitting tend to disappear over 3–20 years, depending on rates of sedimentation (Vallentine, 1971; Aldon, 1972).

Contour trenching and terracing may be effective over longer time periods, but they are very expensive and can be justified only on sites where flood and sediment hazards are severe. They are not feasible where soils are shallow.

Water control structures, such as check dams and gully plugs, are similarly expensive. Proper design is imperative if failures are to be avoided. Design criteria are presented by Heede (1976, 1978) and Heede and Mufich (1973). Enhancement through understanding and use of naturally occurring channel structures, such as log steps and beaver dams, may provide an alternative approach (DeBano and Schmidt, 1989).

Water spreading is a technique for capturing floodwater and dispersing it over rangelands where it can infiltrate and provide supplemental moisture for forage production. It can also reduce gully erosion downstream from the diversion (Miller et al., 1969).

Mechanical and structural methods are not feasible everywhere. Most are unsuited to dry sites with thin soils, and rough, steep, and broken or rocky sites. The manager must recognize that once the critical point in deterioration is reached on such sites, the situation may be essentially irreversible. Money allocated for rehabilitation should be spent elsewhere where results are more promising rather than wasted in quixotic efforts. For example, the more favorable soil and moisture conditions often found in riparian areas may promote relatively rapid and substantial site improvements with proper management (Fig. 12.12).

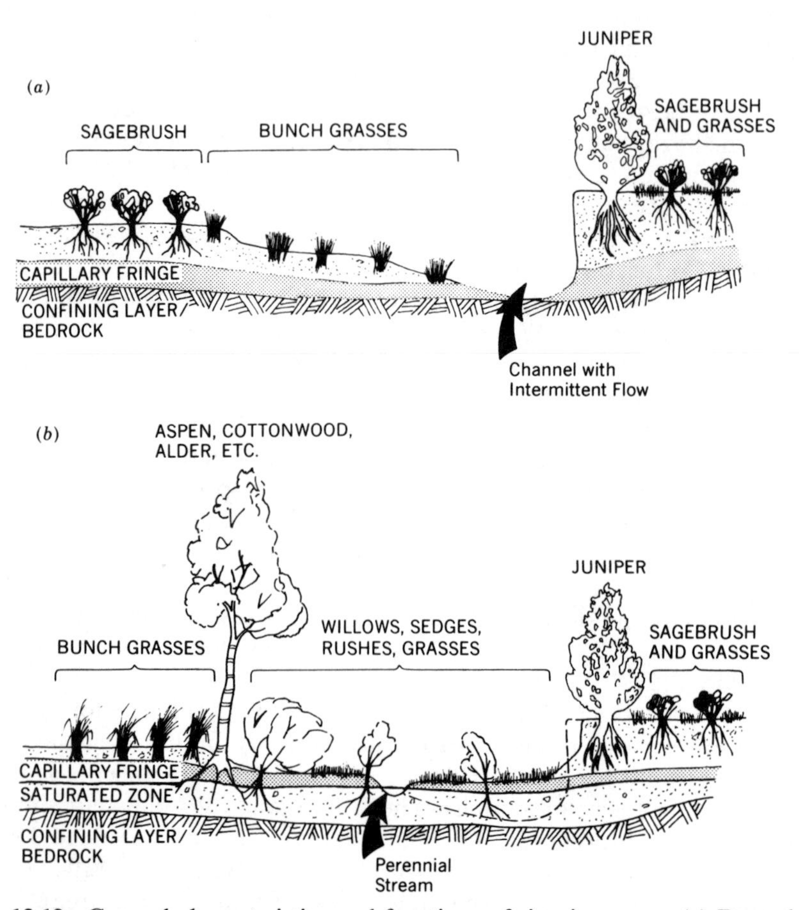

FIG. 12.12 General characteristics and functions of riparian areas: (*a*) Degraded riparian area: Little vegetation to protect and stabilize banks, little shading. Little or no summer streamflow. Warm water in summer and icing in winter. Poor habitat for fish and other aquatic organisms in summer or winter. Low forage production and quality. Low diversity of wildlife habitat. (*b*) Recovered riparian area: Vegetation and roots protect and stabilize banks, improve shading. Elevated saturated zone, increased subsurface storage of water. Increased summer streamflow. Cooler water in summer, reduced ice effects in winter. Improved habitat for fish and other aquatic organisms. High forage production and quality. High diversity of wildlife habitat (adapted from Elmore and Beschta, 1987).

Even where the potential for structural rehabilitation is more favorable, failures are common (Branson et al., 1981). Range lands represent harsh environments where risks are high. All projects must be carefully planned and on a site-specific basis; generalities will not do. Vallentine (1971) and Monson and Shaw (1983) discuss analysis, planning, and design of range improvement projects.

12.1.5 Disturbed Lands

Many lands throughout the world have been laid bare, overturned, mined and quarried, or covered with spoil, tailings, slag heaps, and other dumps. These severely disturbed lands cover an astonishingly large area. In the United States alone almost 2 million ha have been subject to strip mining for clay, coal, gold, oil shale, phosphorus, uranium, and other materials, and the potential deposits of minerals suitable for stripping cover at least 10 times more area. Construction activities result in quarries, sand and gravel pits, pipelines, borrow pits, and land fills. Deep mines, smelting operations, and utilities produce tailings, slag heaps, spoil dumps, and ash. Oil drilling may require the disposal of muds, brines, and other wastes.

These activities are the source or potential source of erosion and sediment, acids or salts, and toxic materials such as heavy metals that contaminate water supplies. Their effects may be highly concentrated and their impact all out of proportion to the area they occupy. Some streams draining relatively small unreclaimed sites are essentially devoid of life for long distances below the source of contamination. Any watershed containing such highly disturbed land or land that may be so disturbed by mining, milling, or construction is at risk until its normal ecologic and hydrologic functioning is restored.

Each environment and type of disturbance combines to create an almost overwhelming number of different problems. Research on the subject is widespread. One bibliography, dealing only with effects of surface mining for minerals other than coal (Richardson and Pratt, 1980), itself contains references to more than 50 other bibliographies, one of which contains 591 annotated references (Czapowskyj, 1976)! The topic has been the subject of many book length treatments, among them Animoto (1978), Hutnick and Davis (1973), Moore and Mills (1977 a, b), Schaller and Sutton (1978), Thames (1977), Vories (1976), and Bradshaw and Chadwick (1980). Alberta, Canada, has provided an exhaustive review of land surface reclamation based on studies conducted throughout the world (Sims et al, 1984). Other reports include Watson et al. (1980) on plants suitable for reclamation, Hermish and Cole (1983) on use of shrubs on oil sand mines, Long et al. (1984), USDA Forest Service (1980 a), Vogel (1981) and Ziemkiewicz (1985) on revegetation methods. Obviously, complete coverage of the whole subject of disturbed lands in a short treatment is impossible. This section will illustrate the nature and some of the range of problems and principles involved in the approach toward solutions. The reader should be aware that most situations are unique and, therefore, each requires a specific prescription tailored to meet the problem involved.

All operations that require drastic disturbance of watershed lands, such as mineral extraction, construction, waste disposal, and others, should be planned from the beginning so that the hydrologic functioning of the land suffers minimal disruption, and restoration is as complete and prompt as possible. Many environmental and economic costs may be avoided by advance planning. A number of skills are needed for rehabilitation or disturbed lands, including those of the geologist, mining engineer, hydrologist, soil scientist, and plant materials specialist. Seldom are all these skills found in the same individual, so conduct of such projects is usually a team effort. The watershed manager is concerned primarily with setting goals for site stability and water yields, and conducting surface restoration activities and monitoring.

Numerous operations in the past were undertaken with no consideration for rehabilitation and were left derelict and so badly devastated that complete restoration of hydrologic functioning is extremely difficult and, in some cases, not possible in anything short of geological time. Nevertheless, ameliorative measures must be taken to reduce the adverse impact as much as possible.

Included in any plan should be (a) the ecologic and hydrologic goals upon completion of the project; (b) methods of appraisal to determine the capacity of the site to support the goals; (c) procedures of operation to insure minimum disruption of the hydrologic function and to facilitate recovery of normal ecologic and hydrologic processes; (d) restoration activities such as land shaping, drainage design, and revegetation of the site; and (e) monitoring activities to uncover any deficiencies and insure that goals are being met.

Ordinarily, the goal of reclamation is to restore the land so it will support the uses and biotic communities existing prior to disturbance. Sometimes it is possible to improve certain characteristics as by adding lakes or ponds to a landscape previously devoid of them when water quality is suitable, thus enhancing the environment. In some situations, goals are prescribed by laws and regulations that limit alternatives available. In many instances, it is not possible to restore previous communities or land uses completely, especially when dealing with derelict lands. Then, goals may be to overcome the worst of the offsite effects, such as acid, alkaline, or toxic contamination of water supplies, and to stabilize the site and establish some type of vegetative cover to improve aesthetics and reduce erosion.

12.1.5.1 Site Appraisal. Determining how best to minimize disruption of hydrologic and ecosystem functions in mining or other operations that drastically disturb the earth's surface requires that original system functioning be known. Thus, the first step in appraisal of a site not yet disturbed is to identify and describe the characteristics of the site and their relations to each other. Where derelict areas remain from the past, inferences should be drawn from nearby sites that were probably similar to the one disturbed.

Among the factors that control the existing ecosystem are climate, soils, and biota. Hydrology is controlled by the above factors plus landform (including drainage systems) and subsurface geology. These factors, plus the nature and degree of change during mineral extraction, largely fix the potential of the succeeding system. The data required will be examined from the surface downward, not necessarily in the order (prospecting, exploration, development, extraction, reclamation, and monitoring) in which most extractive operations are carried out.

Surface data of topography, drainage, soils, and vegetation are assembled from existing maps, such as US Geological Survey quadrangle maps, soils maps of the Soil Conservation Service, Bureau of Land Management, or US Forest Service, vegetation type maps from federal land management agencies, or, for private wildlands in some states such as California, from soil-vegetation surveys. Aerial photos are available for most of the United States and can be interpreted to obtain much of the needed information. Climatic data can be obtained for many areas from the National Climatic Center, Asheville, North Carolina, and other sources (Haines, 1977). Where the data are not available, it may be necessary to undertake local vegetation and soil surveys, and weather stations to collect at least daily temperature and precipitation data should be maintained from the beginning of feasibility study throughout the post-reclamation monitoring and evaluation period. Data on precipitation, growing season length, and chemical properties of spoil materials were found to be important in predicting revegetation success on lands surface mined for coal throughout the interior west (Packer et al., 1982).

At a minimum, the data on vegetation should include species composition and productivity of all communities, and these data should be related to site conditions of soils, topography, and local climate. Observations of possible sources of seed, cuttings, *plugs* (small soil cores containing growing plants) and planting stock may prove useful. If old disturbed areas exist, any natural revegetation should be noted. The observer should particularly notice any plant characteristics that might indicate adaptability to severe site conditions, such as growth form and evidence of tolerance to such stresses as drought, heat, extremes of acidity or salinity, frost and frost heave, blowing snow, shifting soil, submergence, or others.

Soils information should enable the manager to predict its ability to support plant growth, resist erosion, and take in, store, and transmit water. Some soil characteristics, such as texture, influence all of these factors. Among the observations and tests of soil are field observations of horizon thickness, parent material, coarse fragments, color, texture, structure, consistency, root distribution, and carbonates or other soluble salts. Sometimes laboratory tests for dispersion, weatherability, slaking, sodium saturation percentage or sodium adsorption ratio, toxicity, or radioactive tests are conducted in addition to the common pH and fertility testing (USDA Forest Service, 1979 a).

Erodibility of soil or overburden materials is often predicted by the USLE, but predicting the K factor may be difficult. When warranted, the K factor can be estimated by measuring erosion, slope length and steepness, and precipitation on raw spoils and setting

$$K = \frac{A}{RLS} \tag{12.2}$$

where K is the erodibility, A is the erosion in tons per unit area, LS is the slope length and steepness factor, and R is the rainfall factor (C and P, the cover and erosion control practice factors are equal to a value of 1.0 and need not be included). Farmer and Richardson (1976) found K factors that varied from 0.23 to 0.50 in various coal spoils in southeastern Montana.

Information on geology, including overburden materials, may be found from existing geologic maps, observations of nearby operations extrapolated to the site of interest and sample drilling on the site. Wells may need to be drilled to determine ground water geology (USDA Forest Service, 1980 b). Overburden samples are frequently subjected to the same physical and chemical tests as soils. Ground and surface waters may be analyzed for a wide range of constituents, as listed in Table 12.2. Mineralogical analyses for such substances as pyrites are often conducted as well (Barrett et al., 1980). Greenhouse bioassays may reveal mineral deficiencies or toxicities, but the field environment is more severe, and problems may appear that are absent in the greenhouse.

Topographic and *isopach maps* (iso = equal, pach = thickness; lines of equal thickness) showing thickness of overburden, host rock or coal seam, and any interburden between seams are needed to determine potential changes in topography following mining. When overburden materials and rock are broken and removed, they expand in volume. The amount of expansion varies with the degree of breakage and type of material. Depending on the thickness of the material removed in mining, the postmining surface may be raised or lowered as compared to premining conditions. If the mined seam is thick, closed basins may be formed, interrupting normal drainage patterns.

12.1.5.2 ***Mining Operations.*** From the standpoint of the watershed manager, only a few mining activities are of primary interest. They include the develop-

TABLE 12.2 List of Characteristics for Which Analyses May Be Conducted on Surface and Ground Waters in Mining Programs

pH	Arsenic	Nickel
Dissolved O_2	Cadmium	Nitrate (or N)
Total suspended solids	Calcium	Phosphorus
Total dissolved solids	Chromium	Potassium
Electrical conductivity	Copper	Radium
Alkalinity	Flouride	Selenium
Hardness	Iron	Sodium
Carbonate	Lead	Sulfate
Bicarbonate	Magnesium	Vanadium
Redox potential	Manganese	Uranium
Aluminum	Molybdenum	Zinc
Ammonia	Mercury	

Source: Barrett et al., 1980.

ment of a haul road system, the possible segregation and stockpiling of spoil materials and soil, the mass stability of spoil and tailings, and tailings dams and impoundments (USDA Forest Service, 1979 b). Road systems and mass stability have been discussed earlier, and apply to disturbed lands, as well as other situations. Dams and impoundments should be designed by engineers.

Segregation and possible stockpiling of soils and overburden material can be important in two ways: (a) it can make available the most suitable material for plant growth on the reclaimed surface, and (b) it may determine contamination of surface and ground waters by toxic or other undesirable chemicals. The most suitable material for plant growth is usually, but not always, the topsoil from the disturbed area. Sometimes overburden materials may be equal or more suitable for plant growth. Spoils are usually lacking in nitrogen and many microorganisms that enhance growth and may require fertilizers and innoculation with mycorrhizae.

Soils that are stockpiled may lose many of their favorable characteristics during storage and require amelioration, so scheduling application of soils as soon as possible after they are removed may help to retain them.

Contamination of water by undesirable chemicals can be avoided in two basic ways: (a) treating the offending material to remove its effect by detoxifying poisonous salts or neutralizing extremes of acidity or salinity, and (b) by isolating the contaminant from percolating or surface waters. Acid wastes may be treated with lime or mixing with other alkaline materials. Toxic materials should be placed above groundwater and may need to be sealed to prevent percolation of infiltrating water through them, or to prevent oxidation. Lateral flowing soil or groundwater may need to be intercepted and diverted around certain spoils or tailings (Harwood, 1979; USDA Forest Service, 1980 b).

Where spoils from deep mines or tailings dumps must be stored on the land, they should be placed where they will not reach stream channels or occupy source areas. Williams (1982) developed a method to estimate how far material sidecast on steep slopes is likely to extend downslope.

12.1.5.3 Land Shaping and Drainage Design. Current law requires that surface mines be restored approximately to their original contour when feasible. However, that is not always possible where the volume of material removed is great or where spoils or tailings dumps from underground mines or mills are placed on the surface. Nevertheless, if lands are to resist erosion they must not only be as well-vegetated as the climate permits, they must also conform with geomorphic and hydrologic principles of stability (Stiller et al., 1980).

Natural slopes are seldom uniform over any extensive area, but consist of a hierarchy of basins containing streams of different orders. The reclaimed slope should have a pattern similar to natural patterns that have evolved in stream systems at equilibrium in the same climate; first-order basins should be of similar size, stream gradients should be similar entering, within, and leaving the reclaimed area, and base levels should be stabilized with erosion resistant material. During the period after shaping, but before revegetation, above-normal runoff can be expected. Land shaping should direct the runoff to channels protected with rock or other material, and sediment basins should be constructed to trap eroded sediment. Vegetation should be established quickly in riparian zones, as subsidence and settling of the spoil is likely, and good cover may prevent rapid cutting in adjustment to the new equilibrium. It may be necessary to build sills, drop inlets, or check dams if settlement changes stream gradients significantly.

Where spoils or tailings dumps are placed on existing surfaces, they should be shaped to insure mass stability. If they contain toxic materials, a convex surface may direct drainage to move over, rather than through, them. It may be necessary to develop a dense network of surface drains to prevent percolation and leaching of toxic materials from permeable materials. As a rule, it is better to create a channel network that is denser than the natural network than it is to create a more sparse one.

In some regions of limited precipitation, flat or very gently sloping surfaces are desirable. Tillage of compact spoils may further enhance infiltration. Where essentially all precipitation infiltrates, it will help support a better plant cover.

12.1.5.4 ***Revegetation.*** A key element in revegetation of disturbed lands is the nature of the existing substrate for plant growth. Many derelict lands remain bare because they simply cannot support a normal vegetative cover in their present condition. They are too acid or alkaline, are infertile or contain toxic substances, exhibit poor infiltration or drainage, have poor physical structure, form strong crusts, frost heave, exhibit extremes of microclimate, and frequently lack suitable microorganism populations. When they are found in harsh climates such as arid and semiarid areas, alpine areas, or others, the problems are magnified.

The first requirement in revegetation is, therefore, to maintain or improve the substrate. The best material, usually topsoil, but sometimes other strata, should be used for this purpose during mining. As adverse changes in physical, chemical, and biological conditions result from long-term storage, operations should be scheduled so removal is quickly followed by application to the surface. During periods when seeding and planting are not feasible, it may be necessary to apply mulches or other materials to protect the surface from erosion. Where surface materials have been buried by past practices on abandoned lands, it may be necessary to haul in topsoil or treat the existing substrate to create conditions favorable for plant growth.

To neutralize acidity, common treatments include application of lime or physical treatment such as sealing to prevent oxidation of spoil materials. Saline spoils may be treated by application of gypsum. Sewage sludges and other organic wastes can provide nutrients and organic matter. Fertilization may be needed, as well as innoculation of the spoil materials with microorganisms. Soil physical properties may be improved by tillage. Each situation may be somewhat different from others, so the key is analysis of the specific site, followed by a prescription based on that analysis. Many of the references cited in Section 12.15 give specifics of treatment and will not be repeated here.

Sometimes problems arise with certain treatments, such as application of sewage sludges to reclaimed lands. Some of the public consider sludge to be a hazardous waste material. Though sludge from several cities has been shown to be safe and is extensively recycled (Miller, 1986), that from other sources has been shown to cause excessive heavy metal uptake in several forage species grown as cover crops on reclaimed spoils (McBride et al., 1977). Thus, the manager must ascertain that any sludge used is free of danger before recommending such use. Even then, opposition can be expected from some segments of society.

Proper choice of species in revegetation depends on the goals of reclamation, nature of the site, and availability of seed or stock. In many cases, the goal of reclamation is not only to avoid erosion and stream pollution, but to restore natural communities. Such a goal may drastically restrict the choice of species and require special cultural methods. Often, it is not possible to restore climax communities as only earlier successional species are adapted to drastically disturbed sites. Sources of information regarding suitable plant materials were presented in Section 12.1.1.1.2 and throughout Section 12.15 and will not be repeated here.

Even with a suitable substrate and adapted species, successful establishment of vegetation may be difficult on these harsh sites. Arid, semiarid, and alpine regions pose special problems. Careful site preparation and special cultural measures may be necessary. Examples for specific situations are presented by Cook et al. (1974), USDA Forest Service (1979 c), and Thornburg (1982) for western United States, and Brown and Johnston (1978) and Kekerix et al. (1979) for acid alpine mine spoils. Yamamoto (1982) reviewed uranium spoil and mill tailings revegetation in western United States.

Rehabilitated disturbed lands and streams draining them should be monitored during reclamation and until a self-sustaining vegetation is established. Guidelines have been developed for monitoring and evaluation of both revegetation (Chambers, 1983; Chambers and Brown, 1983) and water quality (Kunkle et al., 1987) of disturbed lands. Only by monitoring can one learn what does or does not work so successes can be repeated and failures corrected in future operations.

12.1.6 Fire

Fire is a natural part of the environment in many wildland ecosystems; indeed, many are fire dependent, as the Mediterranean-type shrub ecosystems of North and South America, the Mediterranean Region, South Africa, and Australia. Many coniferous forest ecosystems from taiga to southern pines are characterized by recurrent fire at intervals ranging from decades to several centuries. Fire sweeping through native grasslands of the world is a common occurrence (Chandler et al., 1983; Pyne, 1984).

Natural or not, fire carries with it the potential for sharply degraded water quality (Tiedemann et al., 1978). Whether that potential is realized depends on a wide variety of factors, including the nature and intensity of the fire itself, the nature of the burned community, the nature of the site, the rate of recovery from fire effects, and the particular hydrologic conditions subsequent to the fire.

There are many situations where fire effects are negligible: where light surface fires incompletely remove organic layers, on dormant grasslands that quickly regrow, on level plains with high infiltration capacities, and many others. Even where a major conflagration on steep mountain slopes stripped the ground of all organic material, living and dead, as in the Sundance Fire in northern Idaho in 1967, subsequent mild hydrologic conditions permitted rapid recovery with only minor erosion and damage to water quality. On the other hand, the similar Entiat Burn of 1970 in central Washington resulted in extensive erosion and flooding when rapid snowmelt and severe rainstorms struck before a vegetative cover was reestablished.

Where fire is a normal part of an ecosystem, fire prevention and suppression alone are not viable solutions to prevent erosion and maintain water quality in the long run. The longer fire is absent, the greater the accumulation of fuel and the higher the danger of its ignition. Major wildfires such as those in Yellowstone National Park in 1988 have dramatically demonstrated this fact (Arno and Brown, 1989).

Instead, fire must be managed so that its desirable ecosystem maintenance functions are retained and, wherever possible, its undesirable features are reduced or eliminated. Basically, this means that fire should be applied under climatic and fuel moisture conditions that limit its intensity and that only small portions of a watershed be burned in any one year, with time for recovery before the next portion is burned (Pace and Lindemuth, 1971). Fuel breaks should be constructed to limit wildfire spread (Nord and Green, 1977), and suppression activities should be undertaken to avoid conflagrations.

When unavoidable severe fires do burn large areas, rehabilitation measures may be taken to enhance vegetation recovery (Chandler et al., 1983; Pyne, 1984). However, in some vegetation types, emergency revegetation may not be justified. Research on burned chaparral watersheds in southern California indicate that emergency reseeding of ryegrass can delay recovery of shrubs and has little potential for reducing sediment and may even increase erosion under certain circumstances (Barro and Conard, 1987; Gautier, 1983).

The effects of fire suppression activities sometimes have direct and indirect adverse effects on water quality. For example, fire retardants such as ammonium phosphate and sulfates are fertilizers that may contaminate streams, causing eutrophication or even direct toxicity to some organisms if concentrations are high (Van Meter and Hardy, 1975). In the Arctic, runoff and erosion from burned taiga is often slight, but construction of firelines on permafrost may lead to severe erosion even on very gentle terrain, which may continue for 10 years or more (Viereck and Dyrness, 1979; Viereck, 1982). Firelines should be treated as carefully as roads to avoid erosion, especially since they are almost always more poorly located with respect to potential erosion. The major treatment is closely spaced waterbars, but revegetation, with or without mulching, fertilizing, and other supporting treatments, may be required.

12.2 CONTROL OF DISEASE ORGANISM CONTAMINATION IN WILDLANDS

Surface water from wildlands is a source of domestic supply for urban and rural populations. Furthermore, it is a major attraction for recreational users of wildlands. They build second homes on lakes and streams, and most campgrounds built by public agencies are near water. Such activities carry with them the risk of water-borne disease outbreaks.

It is too much to expect wildland watershed managers to be competent sanitary engineers capable of handling such problems, but it is not too much to expect them to recognize when potential problems exist and seek out expert help when it is needed. In contrast, the wildland manager is expected to deal with potential problems that may arise from users of wildlands such as those who work there and dispersed recreationists. They, too, bear the risk of being infected from contaminated local water, and represent potential sources of contamination.

Among the diseases commonly transmitted by contaminated water are viral infections such as infectious hepatitis (which I contracted, most likely from drinking contaminated water in the field, when I was a young forester—DRS), salmonella and other bacterial infections, giardiasis, and other parasitic diseases. Most are associated with fecal contamination by humans, but many have animal hosts that serve as reservoirs (Meyer and Jarroll, 1980). For this reason, all wildland surface waters should be considered unsafe for drinking unless treated. Boiling water is an effective treatment for all infective agents, whereas one or more types of organisms may be resistant to other treatments.

Management of workers and dispersed recreationists offers the best opportunities to limit water contamination. Forest workers can be provided with portable toilets, as at log landings; in certain situations, such facilities are required by the Occupational Safety and Health Act (OSHA) and various state acts. Dispersed workers and recreationists are not so easily served. Education is the preferred approach, but rules and regulations, regularly enforced, may be necessary.

In most cases, human waste in the form of urine is unlikely to contain pathogenic organisms unless voided by a sick person. Therefore, the problem is primarily that of disposing of feces (Leonard and Plumley, 1979). Soil has long been characterized as a "living filter" (Sopper, 1971). Viruses and bacteria will not move far through well-aerated soil (Aukerman and Springer, 1976;

Gilbert et al., 1976), but material that passes through the anaerobic zone or that which is deposited in a saturated zone can move relatively long distances; hundreds of meters in the course textured materials (Hinesly, 1973; Erickson et al., 1982). Furthermore, shallow burial of feces does not necessarily result in rapid die-off of intestinal bacteria. Fecal bacteria *Escherichia coli* (*E. coli*) and *Salmonella typhinurium* buried in the Bridger Range of Montana persisted for as long as 1 year (Temple et al., 1982).

The implications of the above research are clear. Human feces should be buried carefully on well-drained uplands, well away from water courses (Cole, 1989). In addition, source areas should be avoided, for areas that are dry in summer and fall may be saturated in winter and spring, depending on climate. Unfortunately, studies have shown that even experienced leaders of outings in wilderness areas give low priority to selection of a site for feces disposal, and most disposal areas are close to the campsite (Temple et al., 1982). Individuals and small parties often void feces on the surface, especially if the stay is only overnight or shorter. Under such circumstances, it may be necessary to prohibit camping within a certain distance of back country lakes and streams, and in swales and hollows that may become source areas for runoff in wet weather.

Because of the difficulties in controlling contamination by humans and the virtual impossibility of controlling it from wild animals, it bears repeating that all surface water in wildlands should be considered contaminated. If untreated water is consumed, that issuing from underground springs is least likely to be contaminated, but all water should be treated to insure freedom from infectious disease organisms.

12.3 CONTROL OF STREAM TEMPERATURE

Land management activities, particularly timber harvesting, but also grazing, fire, and road building, can modify the natural temperature regime of many streams. Effects are usually greatest on small headwater and tributary streams that can be completely shaded by riparian vegetation. As streams become larger, their heat capacity and the natural degree of exposure to solar radiation increases, so effects of riparian management on thermal characteristics diminish rapidly. Large rivers, lakes, and reservoirs are unlikely to be noticeably affected.

Temperature changes from wildland management activities are most important from the standpoint of potential adverse effects on the aquatic ecosystem rather than effects on municipal, agricultural, or industrial uses. Most wildland research has been conducted on cold water streams that support trout and salmon (salmonids). Although such fish are known to be sensitive to water temperature, it should be noted that logging-related temperature increases have shown few negative impacts, and in some locations logging has enhanced fish production by increasing primary productivity and food availability (Gregory et al., 1987).

Even the temperature of very small streams can be potentially important. For example, Needle Branch, a small stream in the Coast Range of Oregon, has a minimum flow of about 15 L min^{-1}. At times there is no visible flow between isolated pools, yet the stream is an important rearing area for coho salmon and cutthroat trout (Hall, 1968).

The primary impact on stream temperature results from the removal or reduction of streamside vegetation that exposes the water surface to direct solar radiation by day and the open sky at night. The effect is almost always an increase of maximum temperature during the day in summer, even in the tropics, but the amount of increase varies widely depending upon (a) climate and weather conditions, (b) time of year, (c) amount of water surface exposed, (d) degree and duration of exposure, (e) rate of streamflow, and (f) amount of mixing with groundwater entering the stream. Many studies showed changes in maximum temperatures ranging from a 0.6 °C decrease to increases of more than 11 °C (James, 1957; Eschner and Larmoyeux, 1963; Levno and Rothacher, 1967, 1969; Brown and Krygier, 1967; Brown et al., 1971; Swift and Messer, 1971; Pluhowski, 1972; Burton and Likens, 1973; Neubauer, 1981; Feller, 1981). Temperature increases are likely to persist until significant streamside cover returns, a period that can vary widely, as shown by the range of 5–25 years observed in western Oregon (Beschta et al., 1987).

Greater temperature increases were associated with small, shallow streams completely exposed to the sun for long distances during periods of low flow when solar radiation and air temperatures were high, the sky free of clouds, and little cool groundwater entered the stream. Lesser increases occurred on larger streams during periods of higher flow when copious amounts of cool groundwater entered the channel and the weather was cool and cloudy. Where decreases occurred, they were at night or in cool and cloudy weather. Minimum and winter temperature effects were more variable; both increases and decreases occurred.

Changes in stream temperature can be estimated by applying an energy balance to a stream in its original and modified condition and comparing the two (Beschta and Weatherred, 1984). Brown (1970) developed a simplified temperature equation that yields useful estimates for small streams, but is biased upward where exposures are extensive. It is

$$\Delta T = cAH/D \tag{12.3}$$

where ΔT is the change in temperature in degrees Fahrenheit (°F), c is a constant (0.000267) that converts discharge (D) in cubic feet per second ($ft^3\ s^{-1}$) to pounds of water per minute (lb min^{-1}), A is the exposed stream surface area in square feet (ft^2), and H is the solar radiation in British Thermal Units (BTUs) per minute absorbed per square foot. The constant can be modified to yield ΔT in degrees celsius (°C) when discharge is in liters per second ($L\ s^{-1}$), surface units in square meters (m^2), and heat input in watts per square meter ($W\ m^{-2}$). Beschta and Weatherred (1984) developed an expanded computer model that takes variations in the amount of shade, as well as latent and sensible heat exchange between air and water into account and should provide a more accurate, less biased estimate.

Heated water that enters a shaded stream segment generally cools as it passes downstream. Brown et al. (1971) consider that the major factor involved in such cooling is the inflow and mixing of cool groundwater, but Burton and Likens (1973) estimated that groundwater cooling could account for less than 10% of the temperature drop they observed. Other possible factors involved in cooling include sensible heat flow into the streambed and latent heat flow into the air. Sensible heat flow into the air would only occur when air temperatures were cooler than surface water temperatures. If stream temperatures approach air temperatures, the view factor is positive and the stream is shaded from the side, then radiant cooling could also be involved (DeWalle, 1974).

Protection against temperature changes usually depends on the protection of streamside vegetation and the design of buffer strips. In some situations, the design of buffer strips is specifically prescribed by law or regulation, depending on stream classification and state. In Oregon, for example, 75% of the pre-operation shade must be retained over major streams (Adams, 1988). In addition to the shade provided by trees, shrubs and other smaller vegetation can be an important source of shade, particularly along smaller streams and disturbed areas (Andrus and Froehlich, 1988). Most buffer strips are intended to provide shade, but some design characteristics include consideration of nonthermal factors such as preventing sedimentation, serving as barriers to logging debris and stabilizing stream channels.

Provided that the height of streamside vegetation is known, shadow boundaries are readily determined (Satterlund, 1977). The stream should be shaded during the hours of most intense radiation in seasons of low streamflow and high air temperature, usually from a few hours after sunrise until a few hours before sunset in July and August in the northern hemisphere. On the other hand, view factors should be as high as possible, except toward the sun, to minimize net longwave radiation. Thus, on easterly or westerly flowing streams, it would be most desirable to have no buffer strip at all along the north bank of the stream. However, state laws and regulations often require buffer strips on both sides of streams, regardless of whether both shade the stream.

State laws and regulations may also prescribe buffer strip widths of up to 60 m or more. However, measurements of *angular density* (crown density measured along the path of incoming solar beam radiation) indicate that little shade is gained by requiring strip widths greater than 15–25 m (Brazier and Brown, 1973; Steinblums et al., 1984). Some regulations prohibit removal of any trees from buffer strips, but others allow selective removal of individual trees so long as shading is maintained. Such flexibility helps address one of the inherent costs of buffer strips, that of the economic value of timber foregone.

A major problem with buffer strips in some locations is their susceptibility to windthrow. Many riparian trees are shallow rooted, and soils may be only weakly cohesive, offering little resistance to uprooting. Blowdown is frequently catastrophic, removing complete segments of buffer strips in a single storm. In the Cascade Mountains of Oregon, buffer strip stability increased as exposure to wind decreased and natural resistance of the strip to wind increased (Steinblums et al., 1984). Wind exposure increased with the size of the opening in the direction of damaging wind and the distance to protective ridges, if any. It decreased as valleys become deeper and more protected by nearby ridges. Wind may be channeled along canyons oriented to within 30° or so of the storm wind direction, and strong eddies may develop where canyons change direction and at the mouths of tributary systems (Schroeder and Buck, 1970). Strips bordering clearcuts are more exposed than those adjacent to stands subjected to shelterwood or selection cuttings. Usually, densely stocked strips are more stable than more open strips, as are strips with drier, deeper soils on the edge exposed by cutting. Cutting boundaries should therefore be located on stable sites, and despite the added cost, in some risky situations it may be necessary to have a wider strip than needed for shade to ensure strip stability.

12.4 CONTROL OF CHEMICALS

Public concern for water quality is increasingly focused on possible chemical contamination of water supplies brought on by industrial toxic spills, toxic waste dumps contaminating surface and groundwater, leaching of pesticides, and others. Only a minor fraction of all wildlands are subjected to chemical application, but management activities sometimes require the use of chemicals or affect chemical disposition or cycling. Herbicides may be used to control unwanted vegetation and insecticides, fungicides, rodenticides, and animal repellants may be used to protect resource values. Fertilizers may be applied to enhance desirable growth and fire retardants to fight fire. Oils and greases may be spilled or washed from roads. Heavy metals may leach from mine dumps. Natural processes of weathering result in transport of plant nutrients and other salts into streams and human activities may accelerate the process (Norris et al., 1983). Even remote wildlands may be subject to the influences of acid rain, whose origin may be hundreds of miles away in heavily populated and industrial areas.

The direct effect of chemicals on any organism depends not only on kind, but also on concentration. Some elements or compounds (e.g., selenium), may be essential to an organism for good health. Too little can result in a deficiency disease. As supply is increased, an optimum effect is reached. However, a continued increase in supply may result in toxicity. Other types of chemicals may exert toxic effects in direct proportion to the magnitude and duration of exposure by the affected organism (Horne, 1972; Norris, 1981; Walstad and Dost, 1984). Examples of the *acute toxicities* (expressed as LD_{50} values—the dose that is lethal for 50% of the animals treated in a test) of various chemicals are shown in Table 12.3.

Chemicals may have indirect effects on water quality that may be desirable or deleterious. For example, mechanical site preparation may expose soils to erosion, whereas herbicides may eliminate competition while leaving the pro-

TABLE 12.3 Acute Toxicities of Various Chemicals

Toxicity category (Signal Word on Label)	Chemical Substance	Oral LD_{50} (for rats) ($mg\ kg^{-1}$)	Oral LD_{50} Extrapolation (for 60-kg human)
Very slight		50,000	6.7 L
	Sugar	30,000	($\simeq$1.8 gal)
	Fosamine	24,000	
	Ethyl alcohol	13,700	
	Picloram	8,200	
Slight (caution)	Asulam, Simazine	5,000	670 mL
	Glyphosate	4,300	($\simeq$1.4 pt)
	Table salt	3,750	
	Bleach, dicamba	2,000	
	Atrazine	1,750	
	Aspirin, Vitamin B_3	1,700	
	Hexazinone	1,690	
	Dalapon, Amitrole-T	1,000	
	2,4-DP	800	
	MSMA	700	
	2,4,5-TP, Triclopyr	650	
Moderate (warning)	2,4,5-T	500	67 mL
	2,4-D	370	($\simeq$0.3 cup)
	Caffeine	200	
	Paraquat	150	
Severe (danger—poison)	Nicotine	50	6.7 mL
	Dinoseb	40	($\simeq$1 tsp)
	Strychnine (rodenticide)	30	
	Parathion (insecticide)	13	
		5	0.67 mL
			($\simeq$11 drops)

Source: Adapted from Walstad and Dost, 1984.

tective organic layer of the soil intact. On the other hand, streamside vegetation may be removed by killing with chemicals, as well as by cutting or burning. The indirect effects that result from chemical treatment are not really different from the same effects obtained by other means (Norris, 1981). Usually the matter of primary concern is the direct effect of chemicals reaching streams.

Chemical action on any organism requires that it be present in active form, in sufficient amount, and for a long enough period of time to produce an effect. To assess the hazard of chemicals requires knowledge of the (a) toxicity or biological activity of the chemical and (b) the potential exposure to it. Toxicity alone does not make a chemical hazardous; exposure must also occur. Exposure depends on the behavior of the chemical in the environment, including its initial distribution, and its subsequent movement, persistence, and fate (Fig. 12.13).

The behavior of a chemical, in turn, depends on its properties that determine the ease and method by which it may be transported, stored, or degraded and/or diluted (Norris, 1975, 1984). Transport and storage are determined by its solubility in water, its electrical charge characteristics that determine adsorption and desorption from soil and organic materials, its volatility, and other factors. Pesticide degradation is usually biological, but photolytic and other forms of chemical degradation occur. Some chemicals are highly persistent, whereas others break down rapidly; for example, picloram is relatively long lived, persisting in relatively high concentrations for 6 months or more, whereas carbaryl has a half-life of only 8 days (Norris and Moore, 1971).

Chemicals usually enter water by one or more of the following routes: (a) direct application, (b) drift of aerosols from nearby spray areas, (c) mobilization from ephemeral stream channels and runoff source areas, (d) overland

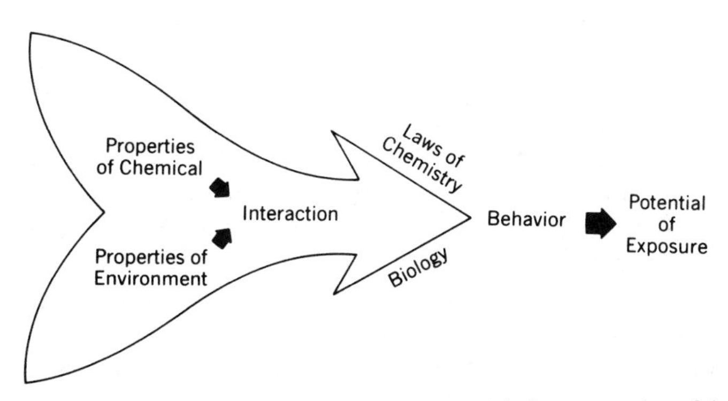

FIG. 12.13 The properties of the chemical interact with the properties of the environment in a manner directed by the laws of nature to produce the movement, persistence, and fate of the chemical—which determine the level and duration of an organism's exposure (from Norris et al., 1983).

flow, and (e) leaching and subsurface flow. The route that is followed depends primarily on the nature of application, the characteristics of the chemical and the nature of the treated area (Norris, 1981; Norris and Moore, 1981).

The most common mode of entry of pesticides is direct application to surface waters, ephemeral channels, and runoff source areas during aerial application, with drift from nearby spray areas another common source of contamination. Overland flow is uncommon (outside of source areas) from watersheds in good condition, but can be a source of entry, especially from highly disturbed areas such as eroding sites, log landings, and others. Leaching is the least likely source of entry by most pesticides used on wildlands, but may be the most common means of entry of nitrogen from fertilizers and fire retardants. It is also the most common route for other plant nutrients, though concentrations are seldom high enough to deteriorate water quality.

Most chemicals used on wildlands, except for fertilizers, are toxic to target organisms and may endanger nontarget organisms as well. Therefore, they must be used according to label directions. Most states require pesticide applicators to be licensed or work under the direct supervision of licensed applicators. But this alone is not enough.

The first principle in using chemicals on any watershed is *If you do not want chemicals in the water, do not put them there.* Most pesticides that enter water do so through direct application or from drift. Therefore, boundaries of treatment areas should be laid out to exclude streams, riparian zones, and runoff source areas. The method of application strongly influences the risk of contamination. For example, direct deposition and/or drift is most likely when using aerial application and may be reduced by ground spraying where feasible. Tree injection of herbicides essentially eliminates such risks.

Where aerial application is the only economic alternative, boundaries of treatment zones must be well marked. Buffer zones may be appropriate and required by law. Drift can be avoided by using appropriate chemical formulations and carriers and by operating under proper atmospheric conditions. Equipment must be properly calibrated and maintained. Operations should be suspended whenever any of these conditions are not met. Buffer zones set up for aerial operations may be treated by ground sprays if treatment of such areas is otherwise desirable.

Roadsides, landings, and other heavily disturbed areas are likely to be subject to surface runoff. If sprayed, chemical formulations that are readily degraded or easily bound to the soil are desirable. These are also likely sites for field maintenance and refueling of equipment, and occasionally pesticides and fuel are spilled, and waste oils, grease, and used containers are disposed of where they or their contents may be washed into streams. Operators must be trained to be aware of such hazards and to use proper disposal techniques for such wastes.

Despite the fact that many of the chemicals used in wildland management are toxic to one or more organisms, most pose little danger to water quality when used with care and with a proper understanding of their properties and behavior in the environment.

Management chemicals are not the only source of concern on wildlands. People have worried about disruption of normal nutrient cycles and stream pollution from forest management activities ever since early studies of clear-cutting and repeated spraying of Bromacil® to completely inhibit vegetation recovery for two years on Hubbard Brook, New Hampshire by Bormann et al. (1969) revealed that nitrate concentrations were increased beyond established pollution levels for more than 1 year, and eutrophication caused summer algal blooms. Although many misunderstood these unusually harsh treatments to be standard practice, related concerns resulted in numerous follow-up studies to elucidate nutrient cycling processes in a wide range of forest conditions, ranging from ecosystems exhibiting no disturbance to complete removal. Reviews of many of the North American studies are contained in Fredriksen et al. (1975), Fredriksen and Harr (1979), Sopper (1975), Corbett et al. (1978), Stone et al. (1978), Vitousek et al. (1979), Berg (1989), Harr and Fredriksen (1988), Hornbeck et al. (1987), Feller (1989), Mann et al. (1987), Martin and Harr (1989), Patric (1980), Verry (1986), Waring and Schlesinger (1985), and Swank et al. (1989), among others.

Results of the above studies indicate that severe disturbances, such as clear-cutting followed by slash burning or severe fire generally results in a flush of nutrients removed in solution. However, few treatments yielded amounts sufficient to be considered polluting, and most yielded relatively small amounts that persisted for only a short time. A notable exception occurred in Arizona where chaparral conversion to grass caused a high level of nitrate pollution that persisted for more than 15 years (Davis and DeBano, 1986).

Provided revegetation is not inhibited, most disturbed sites rapidly revert to a condition of nutrient accumulation (Fig. 12.14). Thereafter, as succession proceeds, the rate of accumulation slows and losses of nutrients in solution again increase until a steady state develops between inputs from weathering, biological fixation, and atmospheric transport and export in streamflow (Marks and Bormann, 1972).

The absolute amount of loss in the pulse of export following cutting is highly variable, depending on soil characteristics, climate, type of vegetation community, the nutrient involved, and the type and degree of disturbance. Deep soils with high exchange capacities tend to retain most of the nutrients that are released by decomposition or fire, whereas nutrients may be readily lost from shallow, coarse soils. Leaching of nutrients tends to be more intense in wet climates than dry, and may be pronounced in the wet tropics.

Some nutrients, such as nitrate nitrogen, are very mobile, even in a drier environment such as the Arizona chaparral community mentioned earlier where conversion to grass caused large quantities of nitrate nitrogen to be released over a long time even though soils were very deep and the site was quickly revegetated with grasses. Other nutrients such as sodium, potassium, and chlorides are also readily mobilized. Still others, such as phosphorus, may be tightly bound in well-drained soils. Burning may release a flush of nutrients, including those that are readily bound, when ash and surface soil particles are washed into streams by surface runoff.

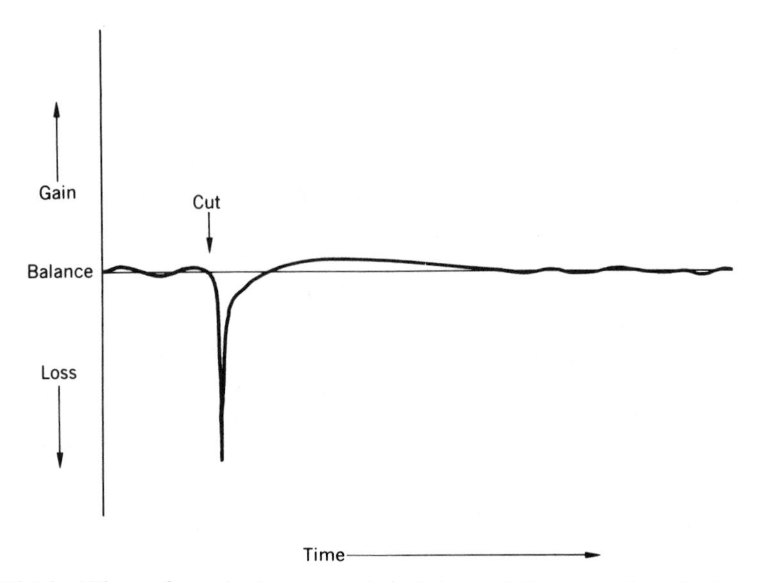

FIG. 12.14 When a forest is clearcut and slash burned, there may be a flush of plant nutrients lost from the site via streamflow. As the site becomes revegetated, conditions rapidly revert to accumulation of nutrients and the site gradually reverts to a steady state where nutrient inputs and exports come into balance.

A unique situation exists in southwestern Australia where chloride deposits cause groundwater to be saline. On forested catchments, leaching of the chlorides has caused chloride output in streamflow to come into equilibrium with water yield and chloride input from rain and dust. But when the forests were cleared for farmlands, the reduced evapotranspiration caused increased ground water discharge, which in turn increased leaching of salts and resulted in degradation of streamwater quality. At present rates of excess chloride output, it is estimated that it would take from 30 to 400 years before cleared farm catchments come into equilibrium between chloride input and output and the increased water yield is as suitable for use as before forest removal (Peck and Harle, 1973).

The evidence reviewed above indicates that in most cases nutrient transport by surface runoff or in solution is not likely to cause water quality problems in wildland management outside the wet tropics. Even in the wet tropics, on-site degradation is more likely than water pollution. It usually takes a combination of extreme circumstances, such as clearcutting and burning slash on shallow, coarse soils, before such problems are likely to arise. However, enough adverse examples have been noted that the land manager should consider the possibility whenever ecosystems are drastically disturbed, even temporarily. On the positive side, land managers also should consider taking advantage of wildland riparian buffers to reduce the water quality impacts of upland agricultural practices (Lawrence et al., 1984, 1985).

Acid precipitation (more commonly referred to as acid rain, but other forms of precipitation also can be acid) has been of widespread public and scientific concern, including its potential effects on water quality. It is created when the gaseous oxides of nitrogen and sulfur are dissolved in precipitation to form nitric and sulfuric acid. Although nitrogen and sulfur gases occur in the atmosphere naturally, it has been the increased use of fossil fuels by automobiles and industry that apparently has led to pH levels of precipitation as low as 3 (roughly the same as vinegar) in some locations in the eastern United States (Hornbeck, 1981). Among the impacts that have been cited are reduced fish populations in areas of upstate New York (Schofield, 1976), probably due to the increased mobility and presence of toxic metals such as aluminum under acid conditions (Cronan and Schofield, 1979). Acid precipitation has been the subject of considerable study, including a major federal research program (Herrmann and Johnson, 1983). However, recent comprehensive analyses indicate that water quality impacts remain relatively uncommon, and that long-term, rather than crisis-level, strategies should be used to address the problem (NAPAP, 1990).

Concerns about chemical water quality on wildlands should extend not only to potential human influences, but natural influences as well. First, it should be emphasized that there is really no such thing as "pure water" (i.e., solely liquid water) in the wildland environment. Even in the absence of acid rain, there are small quantities of gases, aerosols, and dust that dissolve in precipitation. For example, the simple dissolution of carbon dioxide in water results in the formation of the weak acid, carbonic acid (pH = approximately 5.6 at equilibrium. Acid rain is thus generally defined by pH levels <5.6). Precipitation becomes further enriched chemically as it leaches and weathers material from the vegetation canopy, soil, and rocks. Examples of the common chemical constituents found in streamwater are shown in Table 12.4. In most instances, these components pose no problems and may even benefit humans or aquatic organisms, but occasionally they result in brackish or hazardous (e.g., high levels of arsenic) waters. Hem (1970) provides a comprehensive review of chemical, geologic, and hydrologic processes that control the composition of natural waters.

Wildland water bodies also are particularly susceptible to the influences of natural vegetative inputs (leaves, twigs, etc.). These inputs can represent an important energy and nutrient source for aquatic ecosystems (Cummins, 1974; Vannote et al., 1980), but natural waters can sometimes become temporarily overloaded and have reduced quality, particularly for domestic use. For example, leaves from deciduous trees can cause undesirable color, taste, and odor problems during autumn low flow periods (Allen, 1960; Corbett and Heilman, 1975; Slack and Feltz, 1968; Taylor and Adams, 1986). Figure 12.15 illustrates the inverse pattern between streamwater color and dissolved oxygen during autumn leaf fall from riparian trees. Such water quality changes also may cause some more serious problems for water treatment and public health (Carlo and Mettlin, 1980; Le Chevallier et al., 1981; Stevens et al., 1976).

TABLE 12.4 Mean Concentrations of Major Chemical Constituents of Streamwater in Undisturbed Forest Watersheds

Constituent	Western Oregon	New Hampshire
	(mg L^{-1})	
HCO_3-C	3.9	0.3
SO_4-S		1.5
NO_3-N	0.003	0.5
NH_4-N	0.04[a]	0.04
PO_4-P	0.007	
Ca	3.0	2.0
Na	2.5	1.0
K	0.4	0.6
Mg	0.6	0.3

[a]includes dissolved organic N plus NH_4-N

Source: Adapted from Martin and Harr (1989) and Martin (1979).

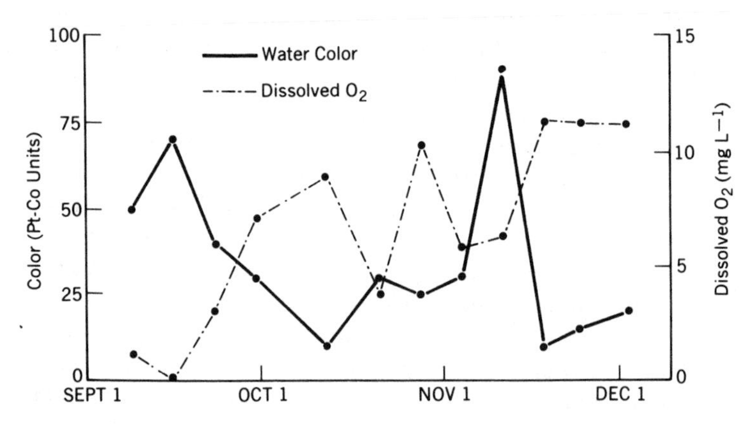

FIG. 12.15 Water color and dissolved oxygen in a pool in a small forest stream in Northwest Oregon in 1981. Annual low flows occurred during September, and maximum leaf inputs from riparian red alder trees occurred about November 1. Flows were increased during storms in early October and again in mid-November (from Taylor and Adams, 1986).

12.5 PROTECTING STREAM FISHERY RESOURCES

Watershed management is predicated on the principle that land management activities may modify the aquatic ecosystem. A valuable component of that ecosystem is the fishery resource. This section will briefly review the relations of stream fisheries to land use and identify measures to protect this valuable resource. A detailed examination of the subject is available in the following references (Reiser and Bjornn, 1979; Swanston, 1980; Chamberlin, 1982; Yee and Roelofs, 1980; Sedell and Duvall, 1985; Everest and Harr, 1982; Platts, 1981; Marcus et al., 1990; Martin and Platts, 1981; Norris et al., 1983; Clark et al., 1985; Schmiege, 1980; Hall and Baker, 1982; Reeves and Roelofs, 1982; Huppert et al., 1985; Salo and Cundy, 1987; and Toews and Brownlee, 1981).

The linkage between the aquatic and terrestrial ecosystems is strongest in first-order headwater streams and diminishes rapidly downstream. Both the physical environment and the biota of the aquatic system are largely determined by conditions on the surrounding land in first-order streams. In low-order streams, the rate of streamflow is largely a direct function of inflow to the channel, water temperatures largely reflect the presence or absence of overhead or lateral shade, and in-stream production is dependent mostly on the entry of organic matter, living and dead, produced outside the stream on the land. (The organic pathway of energy in the stream is *allochthonous*: other-produced.)

Downstream, in higher order streams, inflow directly from the adjacent land makes only a minor contribution to total flow, which represents the accumulated flow from many different tributaries. Whether the banks are forested or not has little effect on stream temperatures. Furthermore, organic productivity is mostly dependent on primary production via photosynthesis within the stream (the pathway is *autochthonous*: self-produced). Ponds and lakes are even less tightly linked to land use than streams. Though the strength of the linkage of land to water diminishes downstream, it never completely disappears. However, this discussion will emphasize relations on small stream systems, fourth order or less, where linkages between land and water are most pronounced.

Stream fish are either resident or migratory. Fish that migrate from the sea to spawn in fresh water are *anadromous*; those that migrate from lakes are *adfluvial*. Even resident fish may be quite mobile, though at certain times adults may be very territorial. Migratory fish are dependent on small streams for only part of their life cycle, whereas resident fish spend their entire life within one stream system.

All fish go through a similar life cycle. They spawn, the eggs incubate and hatch, juveniles grow to adulthood, and the cycle repeats itself. Anadromous and adfluvial fish have two migratory phases: downstream as juveniles and upstream as adults. Resident fish may move within a small stream system, sometimes from tributaries to large streams as they grow and may return to tributaries to spawn.

All stream fish share certain requirements, such as: (a) access to the stream; (b) suitable streamflow; (c) a suitable substrate; (d) cover; (e) suitable temperatures; (f) adequate dissolved oxygen; (g) an adequate food supply; and (h) freedom from excessive sediment and turbidity (Reiser and Bjornn, 1979). Some species are more environmentally demanding than others; among the most sensitive and highly valued are the salmonids (salmon, trout, char, grayling, and whitefish). Other gamefish, such as smallmouth bass, may have a somewhat greater tolerance for higher temperatures, lower dissolved oxygen, and reduced clarity, but still require good quality water. As water quality decreases, only rough fish such as carp may survive. Regardless of the exact species involved, maintenance of any desirable fishery depends on meeting the needs listed above. The discussion that follows will focus primarily on salmonids because of their environmental sensitivity and the availability of many studies relating land management activities to salmonid habitat.

Access to stream segments otherwise suitable for fisheries production can be blocked by natural or human-made features. Effects are primarily adverse to anadromous species. Natural factors include excessively high or low streamflows, falls, debris jams, landslides, and pyroclastic flows as occurred during the eruption of Mt. St. Helens in 1980. Temporary blockages may result from high temperatures or low dissolved oxygen in a segment of a stream. Even temporary blockages can disrupt normal spawning patterns and decrease survival of migrants (Reiser and Bjornn, 1979; Swanston, 1980).

Far more important, however, are blockages due to human activities. Dams have blocked access to huge sections of the Columbia River system. Even where fish ladders permit passage, losses due to nitrogen bubble disease may be high where spillway waters carry entrained air deep below the surface, resulting in waters supersaturated with nitrogen. Many downstream migrants are killed directly by passage through turbines. Losses were so great in the lower Snake and Columbia River system that the US Army Corps of Engineers, now traps and transports millions of smolts hundreds of miles downstream in barges past eight dams from Lower Granite Dam on the Snake River near Pullman, Washington. Thermal blockages occur regularly each hot summer in the reservoir pools, delaying normal upstream migration and causing increased adult mortality. Attempts to mitigate losses by establishing hatcheries, building fish ladders, redesigning spillways, screening turbine entrances, and transporting part of the population downstream in barges have been only partially successful.

Land management, too, has impacted migration of fish. Logging sometimes has resulted in impenetrable debris jams, blockage by landslides, poorly designed road culverts that prevent fish passage, and exposure of streams resulting in temperature increases that cause thermal barriers. Other impacts of land management include changes in the amount and timing of streamflow that may be adverse or beneficial in providing access. Excess sediment from poor land stewardship can also temporarily block migration.

The rate of streamflow strongly influences the suitability of any stream as fish habitat. The major factor influenced by streamflow is the amount of habitat or living space available in the stream segment, such as the food producing areas, spawning and incubation areas, and the amount of cover. Of two streams with the same annual flow, the one with the more uniform regimen tends to provide the better habitat. Wesche (1974) showed that available habitat of a trout stream in Wyoming was not seriously affected until discharge was reduced below 25% of average daily flow.

Other factors influenced by streamflow include migratory blockages due to high flow velocities during storm runoff and shallow water during low flows. Each salmonid species tends also to have a preferred depth and velocity in selecting spawning areas, mostly 15–50 cm depth and from 20 to 90-cm s^{-1} velocity (Reiser and Bjornn, 1979). Incubation success is related to intragravel flow velocities in the *redd* (the "nest" of gravel in which eggs are deposited). Cover is a function of water depth and velocity, as well as channel morphology and structure. High temperatures and low oxygen levels are usually associated with low flows. In general, extreme rates of flow on either the high or low end of the scale are undesirable, but low flows are a more frequent and severe limiting factor to most fish populations.

The substrate of the stream channel is related to spawning success and food production. Most salmonids choose to build redds in gravel 6 mm and larger in size, increasing with the size of the fish. Particles finer than 6 mm reduce the permeability of spawning gravel, inhibiting intragravel flow and water interchange between stream and gravel, resulting in reduced oxygen concentrations in the redds and poorer removal of metabolic wastes. Fry emergence from redds may be sharply inhibited as silts and sands exceed 20–25% of the substrate. Production of stream invertebrates is highest where substrates consist of coarse gravel and rubble (Reiser and Bjornn, 1979).

Cover necessary to protect fish from disturbance and predation is a function of channel morphology, streamflow, large organic debris, and vegetation. Overhead cover consists of riparian vegetation, turbulent water, logs, and undercut banks. Submerged cover consists of rubble and large rocks in the substrate, aquatic vegetation, logs, and other sunken debris.

Most concerns about temperature are related to high summer temperatures, which have already been discussed in Section 12.3. However, low temperatures may also be undesirable, restricting productivity in summer, delaying incubation, and creating ice conditions that eliminate habitat in winter. *Anchor ice* that forms on the substrate, may prevent water interchange between stream and gravel, trap fingerlings seeking cover in rubble, destroy invertebrates, and, with surface ice, may greatly reduce total living space available to fish. The same cover relations that prevent excessive heating, by reducing net longwave radiation losses in winter, help to reduce icing conditions.

Dissolved oxygen can become a problem where streams are heated, contain heavy loads of fine organic debris that exert a high biochemical oxygen demand (BOD), and particularly where restricted flow limits reaeration. The solubility of oxygen in water decreases with increasing temperature. It is only slightly more than one-half as great at 25 as at 0 °C. Substantial oxygen deficits have been observed in small systems heavily loaded with fine logging slash, such as leaves and branches (Ponce, 1974).

Reaeration counteracts oxygen depletion. Reaeration rates depend on stream temperature, active stream width, maximum flow velocity, slope of the stream, the rate of discharge, the oxygen deficit, and BOD loadings. Where little soluble BOD occurs downstream, reaeration of small turbulent forest streams may permit recovery of oxygen deficits within 100 m or so, but where debris loadings are high oxygen may remain depleted for much longer distances downstream (Ice and Brown, 1978). Large organic debris contributes very little to BOD loading after it has been in the stream for a short while.

Most of the food supply in headwater streams with woody riparian vegetation is allochthonous, based on detritus: leaves, twigs, and wood. The amount of sunlight reaching such streams is small and limits production by *periphyton* (organisms such as diatoms attached to underwater surfaces) and *macrophytes* (mosses, flowering plants, filamentous algae, and others). The source of food shifts to autochthonous processes in larger third- and fourth-order streams where sunlight reaches the water in sufficient amounts to stimulate photosynthesis. Stream invertebrates (mostly insects) that feed on these sources also shift in kind and amount downstream, reflecting changes in food supply. These invertebrates are the primary source of food for fish. Terrestrial invertebrates also reach small stream systems in amounts that may contribute importantly to the food supply.

After the basic requirement of an adequate water supply is met, a primary need for a quality fishery is freedom from excessive sediment and turbidity (Everest et al., 1987). High levels of suspended sediment will reduce light penetration and inhibit autochthonous production, abrade and clog fish gills, prevent feeding by sight feeders, may stop migration, and sometimes causes fish to avoid any use of turbid reaches. Combined with high flows, sediment scours away periphyton and many stream invertebrates. Fine sediment deposits may seal rubble and gravel substrates, decreasing spawning area, egg survival, escape of fry, and hiding cover for fingerlings. Invertebrate food supplies are reduced. Pool volumes may be drastically decreased, resulting in direct loss of living space (Reiser and Bjornn, 1979; Toews and Brownlee, 1981).

Most of the management measures required to maintain quality fisheries habitat have already been discussed in some detail in conjunction with previous topics and will only be briefly summarized here. Few land use activities are likely to be either completely deleterious or beneficial, but, as with most activities, it is far easier to damage the fishery resource than to improve it. Even limiting damage takes conscious, deliberate effort based on a thorough knowledge of the interrelations between terrestrial and aquatic systems. In many respects this knowledge and its application remain to be refined, especially when a given activity has both beneficial and adverse effects that interact (Adams et al., 1988).

Another key to maintaining a quality fishery is retention of desirable channel morphology and stability (Bisson et al., 1987; Sullivan et al., 1987). In most cases, any activity in or on streambanks should be avoided; gravel removal should be prohibited, timber should not be felled into the streams or logs yarded across them, except possibly with aerial yarding systems where the logs can be flown entirely in the air. Cattle should not be allowed to break down banks. Protection of the riparian zone by leaving stable buffer strips helps insure the integrity of the stream and its banks, and provides a long-term supply of large woody debris for desirable habitat features. In some instances, however, as where speckled alder encroaches on low gradient streams, removal may improve morphology by decreasing stream width, increasing average depth, and creating stable undercut banks (Hunt, 1979).

Management effects on streamflow may be beneficial or detrimental. Any activity that increases major flood peaks is likely to be undesirable because of increased channel scour and deposition of sediment that often accompanies higher flows. However, most timber harvest activities have a greater effect on increasing the smaller storm flows and improve low summer streamflow, increasing the habitat for fish. Even a small improvement of extremely low flows can have a major beneficial effect on the ability of small streams to support fish.

The maintenance of riparian vegetation provides protection from temperature extremes and may improve cover for fish. On the other hand, a heavy canopy over the stream prevents significant autochthonous productivity and development of instream cover by aquatic macrophytes. As mentioned earlier, fish production may increase when canopy removal increases food availability in light-poor streams (Gregory et al., 1987).

Most land management activities have the potential to increase sediment entry into streams. Nutrient input also may be increased, but the amounts and duration of both sediment and nutrient increases are highly variable, as are fishery impacts. Coarse sediments, such as gravel and rubble provide a desirable substrate, but excessive and persistent fine sediments are undesirable. Most fish can tolerate short-term increases in suspended sediment and some siltation, and even debris torrents may, at times, provide desirable channel complexity and spawning gravels along certain stream reaches (Everest et al., 1987). Except in rare instances, the increased nutrient inputs from wildlands usually increase aquatic productivity, and, unless they are accompanied by excessive heating, BOD loads or sediment, are beneficial or neutral in effect.

There is little information on the integrated net effect of all factors on stream fisheries. Each case seems likely to be different. Clearly, many fisheries have been damaged by careless wildland management that resulted in high stream temperatures, excessive sedimentation, toxic chemical release, and channel degradation. On the other hand, several studies have found increased trout and insect biomass on small forest streams exposed by clearcut logging (Moring and Lantz, 1975; Erman et al., 1977; Gregory et al., 1987; Murphy and Hall, 1981; Murphy et al., 1981; Toews and Brownlee, 1981; Triska et al., 1982; Chamberlin, 1982; Norris et al., 1983). Moreover, anadromous fish are subject to many other influences, such as ocean fishing and marine habitat conditions, that may be of considerably greater importance to population levels than wildland management.

The specific effect on any given stream is probably related to the condition with respect to local factors limiting the fish population. Where there is too little of a factor (heat, nutrients, light, gravel, pools, and other rearing areas), then increases in those factors tend to improve the fishery. However, if such factors are already optimum, further increases may carry the factor to excess and reduce the capacity of the stream fishery. It is the balance of changes in the complex of important factors that determines the integrated net effect of any land management activity. Though past activities have had some adverse effects, evidence is accumulating that careful wildland management need not be detrimental to fish and may even provide opportunities to improve conditions (Adams et al., 1988).

LITERATURE CITED

Adams, P. W. 1983. Maintaining woodland roads. EC 1139. Oregon State University Extension Service, Corvallis, OR.

Adams, P. W. 1988. Oregon's Forest Practice Rules. EC 1194 (revised). Cooperative Extension Service. Oregon State University, Corvallis, OR.

Adams, P. W. and C. W. Andrus. 1990. Planning secondary roads to reduce erosion and sedimentation in humid tropic steeplands. pp. 318–327. In: *Research Needs and Applications to Reduce Erosion and Sedimentation in Tropical Steeplands.* Proc. Fiji Symp., Int. Assoc. Hydrolo. Sci., Wallingford, UK Pub. 192.

Adams, P. W., R. L. Beschta, and H. A. Froehlich. 1988. Mountain logging near streams: opportunities and challenges. pp. 153–162. In: Proc. Intl. Mtn. Log. & PNW Skyline Symp., December, 1988. For. Engr. Dept., Oregon State University, Corvallis, OR.

Adams, P. W., A. J. Campbell, R. C. Sidle, R. L. Beschta, and H. A. Froehlich. 1986. Estimating streamflows on small forested watersheds for culvert and bridge design in Oregon. For. Res. Lab. Res. Bull. 55, College of Forestry, Oregon State University, Corvallis, OR.

Aldon, E. F. 1964. Ground cover changes related to runoff and erosion in westcentral New Mexico. USDA Forest Service Res. Note RM-34. Rocky Mt. For. Range Expt. Sta., Ft. Collins, CO.

Aldon, E. F. 1972. Reactivating soil ripping treatments for runoff and erosion control in the southwestern United States. *An. Arid Zone* 11:154–160.

Allen, J. E. 1960. Taste and odor problems in new reservoirs in wooded areas. *J. Am. Water Works Assoc.* 52:1017–1032.

Amaranthus, M. and D. H. McNabb. 1984. Bare soil exposure following logging and prescribed burning in southwest Oregon. pp. 234–237 In: *New forests for a changing world,* Proc. SAF Natl. Conv., Soc. Am. Foresters, Washington, DC.

Anderson, P. T., M. R. Pyles, and J. Sessions. 1986. Surfacing of steep low volume roads. Tech. Rel. Vol. 8, No. 6. Logging Industry Research Assoc., Rotorua, New Zealand.

Anderson, P. T., M. R. Pyles, and J. Sessions. 1987. The operation of logging trucks on low-volume roads. pp. 104–111. In: Fourth Intl. Conf. Low-Volume Roads, Vol. 2. Trans. Res. Rec. 1106. Trans. Res. Board, Natl. Res. Council, Washington, DC.

Andrus, C. and H. A. Froehlich. 1988. Riparian forest development after logging or fire in the Oregon Coast Range: Wildlife habitat and timber value. pp. 139–152. In: *Symp. Streamside Management: Riparian Wildlife and Forestry Interactions,* University of Washington, Seattle, WA.

Animoto, P. Y. 1978. *Erosion and sediment control handbook.* Div. Mines and Geol., Cal. Dept. Cons., EPA Rept. 4401 3-78-003, Sacramento, CA.

Anonymous. 1968. Skid trails built now await use next year. *For. Ind.* 95(6):48–49.

Apple, L. L. 1985. Riparian habitat restoration and beavers. pp. 489–490. In: R. R. Johnson et al., tech. coords., *Riparian ecosystems and their management: reconciling conflicting uses.* USDA Forest Service Gen. Tech. Rept. RM-120, Rocky Mt. For. Range Expt. Sta., Ft. Collins, CO.

Arno, S. F. and J. K. Brown. 1989. Managing fire in our forests-time for a new initiative. *J. For.* 87(12):44–46.

Arnold, J. F. 1963a. The delineation of field stability hazards. pp. 193–214. In: *Symposium of forest watershed management.* Oregon State University, Corvallis, OR.

Arnold, J. F. 1963b. Road location to retain maximum stability. pp. 215–231. In: *Symposium of forest watershed management.* Oregon State University, Corvallis, OR.

Arnold, J. F., D. A. Jameson, and E. H. Reid. 1964. The pinyon–juniper type of Arizona: effects of grazing, fire, and tree control. USDA Forest Service Prod. Res. Rept. No. 84. USDA, Washington, DC.

Aukerman, R. and W. T. Springer. 1976. Effects of recreation on water quality in wildlands. Eisenhower Consortium Bull. 2, Ft. Collins, CO.

Aulerich, D. E., K. N. Johnson, and H. Froehlich. 1974. Tractors or skylines: what's best for thinning young Douglas-fir? *For. Ind.* 110(11):23–26.

Bailey, R. G. 1971. Landslide hazards related to land use planning in Teton National Forest, northwest Wyoming. USDA Forest Service, Intermountain For. Range Expt. Sta., Ogden, UT.

Barker, R. J. 1981. Soil survey of Latah County Area, Idaho. USDA Soil Cons. Serv. in coop. with University of Idaho and Idaho Soil Conserv. Comm.

Barrett, J., P. C. Deutsch, F. G. Ethridge, W. T. Franklin, R. D. Heil, D. B. McWhorter, and A. D. Youngberg. 1980. Procedures recommended for overburden and hydrologic studies of surface mines. USDA Forest Service Gen. Tech. Rept. INT-71. Intermountain For. Range Expt. Sta., Ogden, UT.

Barro, S. C. and S. G. Conard. 1987. Use of ryegrass seeding as an emergency revegetation measure in chaparral ecosystems. USDA For. Serv. Gen. Tech. Rep. PSW-102. Pac. Southwest For. Range Expt. Sta., Berkeley, CA.

Bates, C. G. and A. J. Henry. 1928. Forest and streamflow experiment at Wagon Wheel Gap, Colorado. US Weather Bureau, Monthly Weather Review Suppl. 30.

Baumgartner, D. M. and R. J. Boyd, Eds. 1976. *Tree planting in the Inland Northwest.* Conf. Proc., Coop. Ext. Serv., Washington State University, Pullman, WA.

Beasley, R. S. 1976. Potential effects of forest management practices on stormflow sources and water quality. pp. 111–117. In: *Proc. Mississippi Water Resour. Conf.*, Water Resour. Res. Inst., Mississippi State University, Starkville, MS.

Bennett, M. W. 1975. Identification of potentially unstable areas on the Clearwater National Forest. MS Spec. Prob. on file at Department of Forestry and Range Management, Washington State University, Pullman, WA.

Bentley, J. R., L. R. Green, and A. B. Evanko. 1966. Principles and techniques in converting native chaparral to stable grassland in California pp. 55–59. In: *Proc. X Intl. Grassland Conf.*, Helsinki, Finland.

Berg, N. H. (tech coord.) 1989. Proc. symp. on fire and watershed management. USDA For. Serv. Gen. Tech. Rep. PSW-109. Pac. Southwest For. Range Expt. Sta., Berkeley, CA. p. 164.

Berger, L., J. Greenstein, and J. Arrieta. 1987. Guidelines for the design of low-cost water crossings. pp. 318–327. In: Fourth Intl. Conf. Low-Volume Roads, Vol. 2. Trans. Res. Rec. 1106. Trans. Res. Board, Natl. Res. Council, Washington, DC.

Berglund, E. R. and M. Rowley. 1975. Rocking woodland roads. EC 859. Oregon State University Extension Serv., Corvallis, OR.

Beschta, R. L., R. E. Bilby, G. W. Brown, L. B. Holtby, and T. D. Hofstra. 1987. Stream temperature and aquatic habitat: Fisheries and forestry interactions. pp. 191–232. In: *Proceedings Symposium on Streamside Management: Forestry and Fisheries Interactions*, University of Washington, Seattle, WA.

Beschta, R. L. and W. S. Platts. 1986. Morphological features of small streams: significance and function. *Water Resour. Bull.* 22:369–379.

Beschta, R. L. and J. Weatherred. 1984. Temp-84. A computer model for predicting stream temperatures resulting from the management of streamside vegetation. WSDG Rept. WSDG-AD-00009, USDA Forest Service Watershed Systems Development Group, Ft. Collins, CO.

Bethlahmy, N. and W. J. Kidd, Jr. 1966. Controlling soil movement from steep road fills. USDA Forest Service Res. Note INT-45, Intermountain For. Range Expt. Sta., Ogden, UT.

Bilby, R. E. 1984. Removal of woody debris may affect stream channel stability. *J. For.* 82(10):609–613.

Bilby, R. E., K. Sullivan, and S. H. Duncan. 1989. The generation and fate of road-surface sediment in forested watersheds in southwestern Washington. *For. Sci.* 35(2):453–458.

Bisson, P. A., R. E. Bilby, M. D. Bryant, C. A. Dolloff, G. B . Grette, R. A. House, M. L. Murphy, K. V. Koski, and J. R. Sedell. 1987. Large woody debris in forested streams in the Pacific Northwest: Past, present, and future. pp. 143–190. In: *Proceedings Symposium on Streamside Mangement: Forestry and Fisheries Interactions*, University of Washington, Seattle, WA.

Bormann, F. H., G. E. Likens and J. S. Eaton. 1969. Biotic regulation of particulate and solution losses from a forested ecosystem. *Bioscience* 19:600–610.

Bradshaw, A. D. and M. J. Chadwick. 1980. *The restoration of land.* University of California Press, Berkeley, CA.

Bradshaw, G. 1979. Preplanned skid trails and winching versus conventional harvesting on a partial cut. For. Res. Lab. Note 62, Oregon State University, Corvallis, OR.

Branson, F. A., G. F. Gifford, K. G. Renard, and R. F. Hadley. 1981. *Rangeland hydrology.* 2nd ed. Range Sci. Ser. No. 1, Soc. Range Manage., Denver, CO.

Brazier, J. R. and G. W. Brown. 1973. Buffer strips for stream temperature control. For. Res. Lab. Res. Pap. 15, School of Forestry, Oregon State University, Corvallis, OR.

Breadon, R. E. 1985. The Kaiser Spyder X5M: Vancouver Island field trials, 1985. FERIC Tech. Rep. TR-63. For. Engr. Res. Inst. Canada, Vancouver, BC.

Brown, G. W. 1970. Predicting the effect of clearcutting on stream temperature. *J. Soil Water Conserv.* 25:11–13.

Brown, G. W., A. R. Gabler, and R. B. Marston. 1973. Nutrient losses after clear-cut logging and slash burning in the Oregon Coast Range. *Water Resour. Bull.* 9:1450–1453.

Brown, G. W. and J. T. Krygier. 1967. Changing water temperatures in small mountain streams. *J. Soil Water Conserv.* 22:242–244.

Brown, G. W., G. W. Swank, and J. Rothacher. 1971. Water temperature in the Steamboat Drainage. USDA Forest Service Res. Pap. PNW-119, Pac. Northwest For. Range Expt. Sta., Portland, OR.

Brown, R. W. and R. S. Johnston. 1978. Rehabilitation of a high elevation mine disturbance. pp. 116–130. In: S. T. Kenny, Ed. *High altitude revegetation workshop* No. 3. Inf. Ser. No. 28. Colorado Water Resour. Inst., Colorado State University, Ft. Collins, CO.

Bryant, M. D. 1983. The role and management of woody debris in west coast salmonid nursery streams. *N. Am. J. Fish. Manage.* 3:322–330.

Bryant, M. D. 1985. Changes 30 years after logging in large woody debris and its use by salmonids. pp. 329–334. In: R. R. Johnson *et al.*, tech. coords., *Riparian ecosystems and their management: reconciling conflicting uses.* USDA Forest Service Gen. Rept. RM-120, Rocky Mt. For. Range Expt. Sta., Ft. Collins, CO.

Burke, D. 1975a. Running skylines reduce access road needs, minimize harvest site impact. *For. Ind.* 102(5):46–48.

Burke, D. 1975b. New tools allow examination of skyline alternatives speedily. *For. Ind.* 102(6):48–50.

Burroughs, E. R., Jr., G. R. Chalfant and M. A. Townsend. 1976. *Slope stability in road construction.* Bur. Land Manage. Oregon State Office, Portland, OR.

Burroughs, E. R., Jr. and J. G. King. 1989. Reduction of soil erosion on forest roads. USDA For. Serv. Gen. Tech. Rep. INT-264. Intermountain Res. Sta., Ogden, UT.

Burroughs, E. R., Jr., M. A. Marsden, and H. F. Haupt. 1972. Volume of snowmelt intercepted by logging roads. *J. Irrig. Drain. Div. Proc. ASCE* 98:1–12.

Burton, T. M. and G. E. Likens. 1973. The effect of strip-cutting on stream temperatures in the Hubbard Brook Experimental Forest, New Hampshire. *BioScience* 23:433–435.

Cable, D. R. 1975. Range management in the chaparral type and its ecological basis: the status of our knowledge. USDA Forest Service Res. Pap. RM-155, Rocky Mt. For. Range Expt. Sta., Ft. Collins, CO.

Campbell, A. J. and R. C. Sidle. 1984. Prediction of peak flows on small watersheds in Oregon for use in culvert design. *Water Resour. Bull.* 20:9–14.

Campbell, R. G., J. R. Willis, and J. T. May. 1973. Soil disturbance by logging with rubber-tired skidders. *J. Soil Water Conserv.* 28:218–220.

Carlo, G. L. and C. J. Mettlin. 1980. Cancer incidence and trihalomethane concentrations in a public drinking water system. *Am. J. Pub. Health* (70(5):523–525.

Chamberlin, T. W. 1982. Influence of forest and rangeland management on anadromous fish habitat in western North America. 3. Timber harvest. USDA Forest Service Gen. Tech. Rept. PNW-136, Pac. Northwest For. Range Expt. Sta., Portland, OR.

Chambers, J. C. 1983. Measuring species diversity on revegetated surface mines: an evaluation of techniques. USDA For. Serv. Res. Pap. INT-322. Intermountain For. Range Expt. Sta., Ogden, UT.

Chambers, J. C. and R. W. Brown. 1983. Methods for vegetation sampling and analysis on revegetated mined lands. USDA For. Serv. Gen. Tech. Rep. INT-151. Intermountain For. Range Expt. Sta., Ogden, UT.

Chandler, C., P. Cheney, P. Thomas, L. Trabaud, and D. Williams. 1983. *Fire in forestry.* Vols. I and II. Wiley, New York.

Chaney, E., W. Elmore, and W. S. Platts. 1990. Livestock grazing on western riparian areas. US Environmental Protection Agency, US Gov. Printing Office, Washington, DC.

Chang, M., F. A. Roth, and E. V. Hunt, Jr. 1982. Sediment production under various forest-site conditions. *IASH Publ. No.* 137:13–22.

Chang, M. and J. C. Ting. 1986. Applications of the universal soil loss equation to various forest conditions. *IASH Publ. No.* 157:165–174.

Chang, M. and K. L. Wong. 1983. Effects of land use and watershed topography on sediment delivery ratio in east Texas. *Beitr. Hydrolog.* 9(1):55–69.

Claridge, F. B. and A. M. Mirza. 1981. Erosion control along transportation routes in northern climates. *Arctic* 34:147–157.

Clark, R. 1980. Erosion condition classification system. USDI Bureau of Land Management Tech. Note 346. Bur. Land Manage., Denver, CO.

Clark, R. N., D. R. Gibbons, and G. B. Pauley. 1985. Influence of forest and rangeland management on anadromous fish habitat in western North America. 10. Influences of recreation. USDA Forest Service Gen. Tech. Rept. PNW-178, Pac. Northwest For. Range Expt. Sta., Portland, OR.

Clary, W. P. 1975. Range management and its ecological basis in the ponderosa pine type of Arizona: the status of our knowledge. USDA Forest Service Res. Pap. RM-158, Rocky Mt. For. Range Expt. Sta., Ft. Collins, CO.

Clayton, J. L. 1981. Soil disturbance caused by clearcutting and helicopter yarding in the Idaho Batholith. USDA Forest Service Res. Note INT-305, Intermountain For. Range Expt. Sta., Ogden, UT.

Cole, D. N. 1989. Low-impact recreational practices for wilderness and backcountry. USDA For. Serv. Gen. Tech. Rep. INT-265. Intermountain Res. Sta., Ogden, UT.

Cole, D. N. and E. G. S. Schreiner. 1981. Impacts of backcountry recreation: site movement and rehabilitation—an annotated bibliography. USDA For. Serv. Gen. Tech. Rep. INT-121. Intermountain For. Range Expt. Sta., Ogden, UT.

Conway, S. 1982. *Logging practices: principles of timber harvesting systems.* Rev. ed. Miller Freeman, San Francisco, CA.

Cook, C. W., R. M. Hyde, and P. L. Sims. 1974. Revegetation guidelines for surface mined areas. *Range Sci. Ser.* No. 16, Colorado State University, Ft. Collins, CO.

Cook, M. J. and J. G. King. 1983. Construction cost and erosion control effectiveness of filter windrows on fill slopes. USDA Forest Service Res. Note INT-335, Intermountain For. Range Expt. Sta., Ogden, UT.

Cooper, J. L. 1977. Technique for evaluating and predicting the impact of grazing on stream channels. USDA Forest Service, Idaho Panhandle Natl. For., Coeur d'Alene, ID.

Corbett, E. S. and J. M. Heilman. 1975. Effects of management practices on water quality and quantity: the Newark, New Jersey, municipal watersheds. In: Municipal Watershed Management Symposium Proceedings, USDA For. Serv. Gen. Tech. Rep. NE-13:47-57.

Corbett, E. S., J. A. Lynch, and W. E. Sopper. 1978. Timber harvesting practices and water quality in eastern United States. *J. For.* 76:484–488.

Corbett, E. S. and R. M. Rice. 1966. Soil slippage increased by brush conversion. USDA Forest Service Res. Note PSW-128. Pac. Southwest For. Range Expt. Sta., Berkeley, CA.

Corliss, J. F. and C. T. Dyrness. 1965. A detailed soil-vegetation survey of the Alsea area in the Oregon Coast Range. pp. 457–483. In: *Forest-soil relationships in North America.* Oregon State University Press, Corvallis, OR.

Cronan, C. S. and C. L. Schofield. 1979. Aluminum leaching response to acid precipitation: effects on high-elevation watersheds in the northeast. *Science* 204:305–306.

Cummins, K. W. 1974. Structure and function of stream ecosystems. *BioScience* 24(11):631–641.

Currie, P. O. 1975. Grazing management of ponderosa pine-bunchgrass ranges in the central Rocky Mountains: the status of our knowledge. USDA Forest Service Res. Pap. RM-159, Rocky Mt. For. Range Expt. Sta., Ft. Collins, CO.

Curtis, N. M., Jr., A. G. Darrach, and W. J. Sauerwein. 1977. Estimating sheetrill erosion and sediment yield on disturbed western forest and woodlands. Tech. Notes, Woodland, No. 10. USDA Soil Conserv. Service, West Tech. Service Ctr., Portland, OR.

Czapowskyj, M. M. 1976. Annotated bibliography on the ecology and reclamation of drastically disturbed area. USDA For. Serv. Gen. Tech. Rep. NE-21. Northeastern For. Expt. Sta., Upper Darby, PA.

Dadkhah, M. and G. F. Gifford. 1980. Influence of vegetation, rock cover, and trampling on infiltration rates and sediment production. *Water Resour. Bull.* 16:979–986.

Davis, E. A. and L. F. DeBano. 1986. Nitrate increases in soil water following conversion of chaparral to grass. *Biogeochemistry* 2:53–65.

DeBano, L. F. and L. J. Schmidt. 1989. Improving southwestern riparian areas through watershed management. USDA For. Serv. Gen. Tech. Rep. RM-182. Rocky Mt. For. Range Expt. Sta., Ft. Collins, CO.

DeByle, N. V. 1981. Clearcutting and fire in the larch/Douglas-fir forests of western Montana—a multifaceted research summary. USDA Forest Service Gen. Tech. Rept. INT-99, Intermountain For. Range Expt. Sta., Ogden, UT.

DeWalle, D. R. 1974. Effect of partial vegetation and topographic shade on radiant energy exchange of streams—with application to thermal loading problems. Res. Pub. No. 82, Inst. for Res. on Land and Water Resour., The Pennsylvania State University, University Park, PA.

DeWalle, D. R. and A. Rango. 1972. Water resources applications of stream channel characteristics on small forested basins. *Water Resour. Bull.* 8:697–703.

Dickerson, B. P. 1968. Logging disturbance on erosive sites in north Mississippi. USDA Forest Service Res. Note SO-72, Southern For. Expt. Sta., New Orleans, LA.

Dissmeyer, G. E. and N. D. Cost. 1984. Multiresource inventories: watershed condition of commercial forest land in South Carolina. USDA Forest Service Res. Pap. SE-247, Southeastern For. Expt. Sta., Asheville, NC.

Dissmeyer, G. E. and G. R. Foster. 1980. A guide for predicting sheet and rill erosion on forest land. USDA Forest Service Tech. Pub. SA-Tp11. Southeastern Area, Atlanta, GA.

Douglass, J. E. 1974. Flood frequencies and bridge and culvert sizes for forested mountains of North Carolina. USDA Forest Service Gen. Tech. Rept. SE-4, Southeastern For. Expt. Sta., Asheville, NC.

Duncan, S. H., J. W. Ward, and R. J. Anderson. 1987. A method for assessing landslide potential as an aid in forest road placement. *Northwest Sci.* 61(3):152–159.

Duvall, V. L. and N. E. Linnartz. 1967. Influence of grazing and fire on vegetation and soil of longleaf pine-bluestem range. *J. Range Manage.* 20:241–247.

Dyrness, C. T. 1967. Soil surface conditions following skyline logging. USDA Forest Service Res. Note PNW-55, Pac. Northwest For. Range Expt. Sta., Portland, OR.

Dyrness, C. T. 1972. Soil surface conditions following balloon logging. USDA Forest Service Res. Note PNW-182, Pac. Northwest For. Range Expt. Sta., Portland, OR.

Dyrness, C. T. 1973. Early stages of plant succession following logging and burning in the western Cascades of Oregon. *Ecology* 54:57–69.

Eck, R. W. and P. J. Morgan. 1987. Culverts versus dips in the Appalachian region: A performance-based decision-making guide. pp. 330–340. In: Fourth Intl. Conf. Low-Volume Roads, Vol. 2. Trans. Res. Rec. 1106. Trans. Res. Board, Natl. Res. Council, Washington, DC.

Elmore, W. and R. L. Beschta. 1987. Riparian areas: perceptions in management. *Rangelands* 9(6)260–265.

Erickson, D. C., H. L. Gary, S. M. Morrison, and G. Sanford. 1982. Pollution indicator bacteria in stream and potable water supply of the Manitou Experimental Forest, Colorado. USDA Forest Service Res. Note RM-15, Rocky Mt. For. Range Expt. Sta., Ft. Collins, CO.

Erman, D. C., J. D. Newbold, and K. B. Roby. 1977. Evaluation of streamside buffer strips for protecting aquatic organisms. Cont. No. 165, California Water Resour. Ctr., University of California, Davis, CA.

Eschner, A. R. and J. Lamoyeux. 1963. Logging and trout: four experimental forest practices and their effect on water quality. *Prog. Fish Culturalist* 25:59–67.

Everest, F. H., R. L. Beschta, J. Charles Scrivner, K. V. Koski, J. R. Sedell, and C. J. Cederholm. 1987. Fine sediment and salmonid production: A paradox. pp. 98–142. In: *Proceedings Symposium in Streamside Management: Forestry and Fisheries Interactions*, University of Washington, Seattle, WA.

Everest, F. H. and R. D. Harr. 1982. Influence of forest and rangeland management on anadromous fish habitat in western North America. 6. Silvicultural treatments. USDA Forest Service Gen. Tech. Rept. PNW-134, Pac. Northwest For. Range Expt. Sta., Portland, OR.

Farmer, E. E. and B. Z. Richardson. 1976. Hydrologic and soil properties of coal mine overburden piles in southeastern Montana. pp. 120–130. In: *Proc. NCA/BCR Coal Conf. and Expo. III,* Louisville, KY.

Feller, M. C. 1981. Effects of clearcutting and slashburning on stream temperature in southwestern British Columbia. *Water Resour. Bull.* 17:863–867.

Feller, M. C. 1989. Effects of forest herbicide applications on streamwater chemistry in southwestern British Columbia. *Water Resour. Bull.* 25(3):607–616.

Fredriksen, R. L. and R. D. Harr. 1979. Soil, vegetation, and watershed management in the Douglas-fir region. pp. 231–260. In: P. E. Heilman, H. W. Anderson, and D. M. Baumgartner, eds. *Forest soils of the Douglas-fir region.* Cooperative Extension, Washington State University, Pullman, WA.

Fredricksen, R. L., D. G. Moore, and L. A. Norris. 1975. The impact of timber harvest, fertilization and herbicide treatment on streamwater quality in western Oregon and Washington. pp. 283–313. In: B. Bernier and C. H. Winget, Eds. *Forest soils and forest land management.* University of Laval Press, Quebec, PQ, Canada.

Froehlich, H. A. 1978. Soil compaction from low-ground pressure, torsion-suspension logging vehicles on three forest soils. For. Res. Lab. Res. Pap. 36. Oregon State University, Corvallis, OR.

Froehlich, H. A., D. E. Aulerich, and R. Curtis. 1981. Designing skid trails systems to reduce soil impacts from tractive logging machines. FRL Res. Pap. 44. Oregon State University, Corvallis, OR.

Froehlich, H. A. and D. W. R. Miles. 1984. Winged subsoiler tills compacted forest soil. *For. Ind.* 111:42–43.

Gallup, R. M. 1974. Roadside slope revegetation. USDA Forest Service Equip. Devel. and Test. Rept. 7700-8. Equip. Test Dev. Ctr., San Dimas, CA.

Gardner, R. B. 1979. Some environmental and economic effects of alternative forest road designs. *Trans. Am. Soc. Agric. Engr.* 22(1):63–68.

Gardner, R. B., W. S. Hartog, and K. B. Dye. 1978. Road design guidelines for the Idaho Batholith based on the China Glenn road study. USDA Forest Service Res. Pap. INT-204, Intermountain For. Range Expt. Sta., Ogden, UT.

Garland, J. J. 1983. Designated skid trails minimize soil compaction. EC 1110. Oregon State University Extension Serv., Corvallis, OR.

Garland, J. J. 1983a. Planning for woodland roads. EC 1118. Oregon State University Extension Serv., Corvallis, OR.

Garland, J. J. 1983b. Designing woodland roads. EC 1137. Oregon State University Extension Serv., Corvallis, OR.

Garland, J. J. 1983c. Road construction on woodland properties. EC 1135. Oregon State University Extension Serv., Corvallis, OR.

Garrison, G. A. and R. S. Rummell. 1951. First-effects of logging on ponderosa pine forest range lands of Oregon and Washington. *J. For.* 49:708–713.

Gautier, C. R. 1983. Sedimentation in burned chaparral watersheds: is emergency revegetation justified? *Water Resour. Bull.* 19:793–802.

Gent, J. A., Jr. and L. A. Morris. 1986. Soil compaction from harvesting and site preparation in the upper gulf coastal plain. *Soil Sci. Soc. Am. J.* 50(2):443–446.

Gerstkemper, J. 1985. Directional felling by tree lining in old-growth Douglas-fir. pp. 136–139. In: *Forest operations in politically and environmentally sensitive areas.* Conference Proceedings, Council on Forest Engineering, Tahoe City, CA.

Gibson, H. G. and C. J. Biller. 1976. Letter to the editor. *J. For.* 74:325.

Gifford, G. F. and R. H. Hawkins. 1976. Grazing systems and watershed management: a look at the record. *J. Soil Water Conserv.* 31:281–283.

Gilbert, R. G., R. C. Rice, H. Bouwer, C. P. Gerba, C. Wallis, and J. L. Melnick. 1976. Wastewater renovation and reuse: virus removal by soil filtration. *Science* 192:1004–1005.

Gilkeson, R. H., A. Starr, and E. C. Steinbrenner. 1961. Soil survey of the Snoqualmie Falls Tree Farm. Weyerhaeuser Co., Centralia, WA, and Washington State University, Pullman, WA.

Gleason, C. H. 1953. Indicators of erosion on watershed land in California. *Trans. Am. Geophys. Union* 34:419–426.

Green, L. R. 1977. Fuel reduction without fire—current technology and ecosystem impact. pp. 163–171. In: *Symposium on environmental consequences of fire and fuel management in Mediterranean ecosystems.* USDA Forest Service Gen. Tech. Rept. WO-3, Washington, DC.

Gregory, S. V., G. A. Lamberti, D. C. Erman, K. V. Koski, M. L. Murphy, and J. R. Sedell. 1987. Influence of forest practices on aquatic production. pp. 233–255. In: *Proceedings Symposium in Streamside Management: Forestry and Fisheries Interactions*, University of Washington, Seattle, WA.

Hafenrichter, A. L., J. L. Schwendiman, H. L. Harris, R. S. MacLauchlan, and H. W. Miller. 1968. Grasses and legumes for soil conservation in the Pacific Northwest and Great Basin States. USDA Soil Conserv. Serv. Agric. Handbook 339, Washington, DC.

Haines, D. A. 1977. Where to find weather and climatic data for forest research studies and management planning. USDA Forest Service Gen. Tech. Rept. NC-27. North Central For. Expt. Sta., St. Paul, MN.

Hall, J. D. 1968. Effects of logging on fish resources. *Loggers Handbook* Sec. 2. pp. 24–28.

Hall, J. D. and C. O. Baker. 1982. Influence of forest and rangeland management on anadromous fish habitat in western North America. 12. Rehabilitating and enhancing stream habitat: 1. review and evaluation. USDA Forest Service Gen. Tech. Rept. PNW-138, Pac. Northwest For. Range Expt. Sta., Portland, OR.

Harr, R. D. and R. L. Fredriksen. 1988. Water quality after logging small watersheds in the Bull Run Watershed, Oregon. *Water Resour. Bull.* 24(5):1103–1111.

Hart, G. E., N. V. DeByle, and R. W. Hennes. 1980. Soil solution chemistry and slash disposal. pp. 318–325. In: *Symposium on watershed management.* Vol. I. Am. Soc. Civ. Eng., New York.

Harwood, G. 1979. A guide to reclaiming small tailings ponds and dumps. USDA Forest Service Gen. Tech. Rept. INT-57, Intermountain For. Range Expt. Sta., Ogden, UT.

Hattinger, H. 1976. Torrent control in the mountains with reference to the tropics. pp. 119–134. In: S. H. Kunkle and J. L. Thames, Eds., *Hydrological techniques for upstream conservation.* FAO Cons. Guide No. 2, FAO, United Nations, Rome.

Haupt, H. F. 1959. A method for controlling sediment from logging roads. USDA Forest Service Misc. Pub. No. 32, Intermountain For. Range Expt. Sta., Ogden, UT.

Haupt, H. F. 1960. Variations in areal disturbance produced by harvesting methods in ponderosa pine. *J. For.* 58:634–639.

Haussman, R. F. and E. W. Pruett. 1979. Permanent logging roads for better woodlot management. Appendix III. In: F. C. Simmons, *Handbook for eastern timber harvesting.* USDA Forest Service, Eastern Region, State and Private Forestry, Broomall, PA.

Hawley, J. G. and J. R. Dymond. 1988. How much do trees reduce landsliding? *J. Soil Water Conserv.* 43(6):495–498.

Heede, B. H. 1966. Design, construction, and cost of rock check dams. USDA Forest Service Res. Pap. RM-20. Rocky Mt. For. Range Expt. Sta., Ft. Collins, CO.

Heede, B. H. 1968. Conversion of gullies to vegetation-lined waterways on mountain slopes. USDA Forest Service Res. Pap. RM-40. Rocky Mt. For. Range Expt. Sta., Ft. Collins, CO.

Heede, B. H. 1976. Gully development and control: the status of our knowledge. USDA Forest Service Res. Pap. RM-169. Rocky Mt. For. Range Expt. Sta., Ft. Collins, CO.

Heede, B. H. 1978. Designing gully control systems for eroding watersheds. *Environ. Manage.* 2:509–522.

Heede, B. H. 1986. Designing for dynamic equilibrium in streams. *Water Resour. Bull.* 22:351–357.

Heede, B. H. and J. G. Mufich. 1973. Functional relationships and a computer program for structural gully control. *J. Environ. Manage.* 1:321–344.

Heinrich, R., Ed. 1979. *Mountain forest roads and harvesting.* Tech. Rept., 2nd FAO/Austria training course. FAO, United Nations, Rome.

Heinrich, R. 1982. Road embankment stabilization with biological and engineering works for forest roads. pp. 81–91. In: R. Heinrich, Ed. *Logging of mountain forests.* FAO Forest Pap. 33, FAO, United Nations, Rome.

Helgath, S. F. 1975. Trail deterioration in the Selway-Bitterroot wilderness. USDA Forest Service Res. Note INT-193, Intermountain For. Range Expt. Sta., Ogden, UT.

Hem, J. D. 1970. Study and interpretation of the chemical characteristics of natural water. USDI Geol. Surv. Water-Supply Pap. 1473. US Govt. Printing Office, Washington, DC.

Hermish, R. and L. M. Cole. 1983. *Propagation study: use of shrubs for oil sands mine reclamation.* Alberta Land Cons. and Reclamation Council Rept. No. OSESG-RRTAC 84-2, Edmonton, Alberta.

Herr, L. A. and H. G. Bossy. 1965a. Hydraulic charts for the selection of highway culverts. Hyd. Eng. Circ. No. 5, Bureau of Public Roads, Dept. Commerce. Govt. Print. Off., Washington, DC.

Herr, L. A. and H. G. Bossy. 1965b. Capacity charts for the hydraulic design of highway culverts. Hyd. Eng. Circ. No. 10, Bureau of Public Roads, Dept. Commerce. Govt. Print. Off., Washington, DC.

Herrmann, R. and A. I. Johnson, Eds. 1983. Acid rain: a water resources issue for the 80's. *Am. Wat. Res. Assoc.*, Bethesda, MD.

Hewlett, J. D. and R. G. Burns. 1983. A decision model to predict sediment yield from forest practices. *Water Resour. Bull.* 19:9–14.

Hickey, W. C. and E. J. Dortignac. 1964. An evaluation of soil ripping and soil pitting on runoff and erosion in the semiarid Southwest. *IASH* Pub. No. 65:22–23.

Hinesly, T. D. 1973. Water renovation for unrestricted re-use. *Water Spectrum* 5(2):1–8.

Hoover, M. D. 1944. Effect of removal of forest vegetation upon water yields. *Trans. Am. Geophys. Union* 6:969–975.

Hornbeck, J. W. 1981. Acid rain: facts and fallacies. *J. For.* 79(7):438–443.

Hornbeck, J. W., C. W. Martin, and C. T. Smith. 1986. Protecting forest streams during whole-tree harvesting. *Northern J. Appl. For.* 3:97–100.

Hornbeck, J. W. and others. 1987. The northern hardwood forest ecosystem: ten years of recovery from clearcutting. USDA For. Serv., NE-RP-596. Northeastern For. Expt. Sta., Broomall, PA. p. 30.

Horne, R. A. 1972. Biological effects of chemical agents. (letter). *Science* 177:1152–1153.

Hunt, D. L. and J. W. Henley. 1981. Uphill falling of old-growth Douglas-fir. USDA Forest Service Gen. Tech. Rept. PNW-122. Pac. Northwest For. Range Expt. Sta., Portland, OR.

Hunt, R. L. 1979. Removal of woody streambank vegetation to improve trout habitat. Tech. Bull. No. 115, Dept. Nat. Resour., Madison, WI.

Huppert, D. D., R. D. Fight, and F. H. Everest. 1985. Influence of forest and rangeland management on anadromous fish habitat in western North America. 14. Economic considerations. USDA Forest Service Gen. Tech. Rept. PNW-181, Pac. Northwest For. Range Expt. Sta., Portland, OR.

Hutnik, R. J. and G. Davis, Eds. 1973. *Ecology and reclamation of devastated land.* 2 vols., Gordon and Beach, New York.

Hynson, J., P. Adams, S. Tibbetts, and R. Darnell. 1982. *Handbook for protection of fish and wildlife from construction of farm and forest roads.* FWS/OBS-82/18, USDI, Fish and Wildlife Serv., Biological Serv. Prog., Washington, DC.

Ice, G. G. and G. W. Brown. 1978. Reaeration in a turbulent stream system. WRRI 58, Water Resour. Res. Inst., Oregon State University, Corvallis, OR.

James, G. A. 1957. The effect of logging on discharge, temperature, and sedimentation of a salmon stream. USDA Forest Service Tech. Note No. 39, Alaska For. Res. Ctr., Juneau, AK.

Johnson, W. and V. Wellburn, Eds. 1976. *Handbook for ground skidding and road building in the Kootenay area of British Columbia.* For. Engr. Res. Inst. Canada, Vancouver, BC.

Johnson, W. N. and R. S. Fifer. 1978. Water quality considerations for highway planning and construction, I-70 Vail Pass, Colorado. USDA Forest Service, White River National Forest, Glenwood Springs, CO.

Kaufman, J. B. and W. C. Krueger. 1984. Livestock impacts on riparian ecosystems and streamside management implications—a review. *J. Range Manage.* 37:430–438.

Kekerix, L. K., R. W. Brown, and R. S. Johnston. 1979. Seedling water relations of two grass species on high-elevation acid mine sites. USDA Forest Service Res. Note INT-262, Intermountain For. Range Expt. Sta., Ogden, UT.

Kellogg, L. D. 1981. Machines and techniques for skyline yarding of smallwood. FRL Res. Bull. 36. For. Res. Lab, Oregon State University, Corvallis, OR.

Kidd, W. J., Jr. 1963. Soil erosion control structures on skid trails. USDA Forest Service Res. Note INT- 1, Intermountain For. Range Expt. Sta., Ogden, UT.

Klock, G. O. 1975. Impact of five postfire salvage logging systems on soils and vegetation. *J. Soil Water Conserv.* 30:78–81.

Knighton, M. D. 1984. Simple runoff control structures stand test of time. USDA Forest Service Res. Note NC-323, North Central For. Expt. Sta., St. Paul, MN.

Kraebel, C. J. 1936. Erosion control on mountain roads. USDA Cir. No. 380, Washington, DC.

Kunkle, S., W. S. Johnson, and M. Flora. 1987. Monitoring stream water for land-use impacts: a training manual for natural resource management specialists. Water Resources Div., National Park Service, Ft. Collins, CO.

Kunkle, S. H. and J. L. Thames, Eds. 1976. *Hydrological techniques for upstream conservation.* FAO Conserv. Guide No. 2, FAO, United Nations, Rome.

Kunkle, S. H. and J. L. Thames, Eds. 1977. *Guidelines for watershed management.* FAO Conserv. Guide No. 1, FAO, United Nations, Rome.

Larson, A. G. and D. D. Woolridge. 1980. Effects of slash burial and stream water quality. *J. Environ. Qual.* 9:18–20.

Laycock, W. A. 1982. Seeding and Fertilizing to improve high elevation rangelands. USDA For. Serv. Gen. Tech. Rep. INT-120. Intermountain For. Range Expt. Sta., Ogden, UT.

Leaf, C. F. 1970. Sediment from central Colorado snow zone. *J. Hydraul. Div. ASCE* 96:87–93.

LeChevallier, M. W., T. M. Evans, and R. J. Seidler. 1981. Effect of turbidity on chlorination efficiency and bacterial persistence in drinking water. *App. Environ. Microbiol.* 42(1):159–167.

Leonard, R. E. and H. J. Plumley. 1979. Human waste disposal in eastern backcountry. *J. For.* 77:349–352.

Levno, A. and J. Rothacher. 1967. Increases in maximum stream temperatures after logging old-growth Douglas-fir watersheds. USDA Forest Service Res. Note PNW-65, Pac. Northwest For. Range Expt. Sta., Portland, OR.

Levno, A. and J. Rothacher. 1969. Increases in maximum stream temperatures after slash burning in a small experimental watershed. USDA Forest Service Res. Note PNW-110, Pac. Northwest For. Range Expt. Sta., Portland, OR.

Lively, M. and W. A. Vischer. 1977. Polypropylene fabric as a reinforcement of muskeg subgrades in road construction. USDA Forest Service Field Notes 9(3):4–10, Eng. Tech. Info. Systems, Washington, DC.

Logan, L. D. 1979. Native vegetation for streambank erosion control. pp. 15–18. In: *Riparian and wetland habitats of the Great Plains.* Proc. 31st Ann. Mtg. Great Plains Agric. Council Pub. No. 91, Rocky Mt. For. Range Expt. Sta., Ft. Collins, CO.

Long, S. G., J. K. Burrell, N. M. Laurenson, and J. H. Nyenhuis. 1984. Manual of revegetation techniques. USDA Forest Service Equip. Dev. Ctr., Missoula, MT.

Lowdermilk, W. C. 1953. Conquest of the land through seven thousand years. USDA Soil Conserv. Serv., Agric. Info. Bull. No. 99, Washington, DC.

Lowrance, R., R. Leonard, and J. Sheridan. 1985. Managing riparian ecosystems to control nonpoint pollution. *J. Soil Water Conserv.* 40(1):87–91.

Lowrance, R., R. Todd, J. Fail, Jr., O. Hendrickson, Jr., R. Leonard, and L. Asmussen. 1984. Riparian forests as nutrient filters in agricultural watersheds. *Bioscience* 34(6):374–377.

Lusby, G. C. 1970. Hydrologic and biotic effects of grazing versus nongrazing near Grand Junction, Colorado. US Geol. Survey Prof. Pap. 700-B, pp. B232–B236.

Lysne, D. H. and S. E. Armitage. 1983. Multispan logging of old-growth timber in southwest Oregon. FRL Res. Note 74. For. Res. Lab., Oregon State University, Corvallis, OR.

Mace, A. C., Jr. 1970. Soil compaction due to tree length and full tree skidding with rubber tired skidders. Minn. Forestry Res. Notes No. 214, School of Forestry, University of Minnesota, St. Paul, MN.

Major, M. J. 1987. Managing off-the-road vehicles. *J. For.* 85(11):37–41.

Mallory, J. I. and E. B. Alexander. 1965. Soils and vegetation of the Lassen Peak Quadrangle. State Coop. Soil-Vegetation Survey: Pac. Southwest For. Range Expt. Sta.,

Div. of Forestry, Dept. Conserv., State of California and University of California, Berkeley, CA.

Mann, L. K., D. W. Johnson, D. C. West, D. W. Cole, J. W. Hornbeck, C. W. Martin, H. Riekerk, C. T. Smith, W. T. Swank, L. M. Tritton, and D. H. VanLear. 1987. Effects of whole-tree and stem-only clearcutting on postharvest hydrologic losses, nutrient capital, and regrowth. *For. Sci.* 34(2):412–428.

Marcus, M. D., M. K. Young, L. E. Noel, and B. A. Mullan. 1990. Salmonid-habitat relationships in the western United States: a review and indexed bibliography. USDA For. Serv. Gen. Tech. Rep. RM-188. Rocky Mt. For. Range Expt. Sta., Ft. Collins, CO.

Marks, P. L. and F. H. Bormann. 1972. Revegetation following forest cutting: mechanisms for return to steady-state nutrient cycling. *Science* 176:914–915.

Marlow, C. B., T. M. Pogacnik, and S. D. Quinsey. 1987. Streambank stability and cattle grazing in southwestern Montana. *J. Soil and Water Conserv.* 42:291–296.

Martin, C. W. 1979. Precipitation and streamwater chemistry in an undisturbed forested watershed in New Hampshire. *Ecology* 60(1):36–42.

Martin, C. W. 1988. Soil disturbance by logging in New England-review and management recommendations. *North J. Appl. For.* 5:30–34.

Martin, C. W. and R. D. Harr. 1989. Logging of mature Douglas-fir in western Oregon has little effect on nutrient output budgets. *Can. J. For. Res.* 19:35–43.

Martin, S. B. and W. S. Platts. 1981. Influence of forest and rangeland management on anadromous fish habitat in western North America. 8. Effects of mining. USDA Forest Service Gen. Tech. Rept. PNW-119, Pac. Northwest For. Range Expt. Sta., Portland, OR.

Martin, S. C. 1975. Ecology and management of southwestern semidesert grass–shrub ranges: the status of our knowledge. USDA Forest Service Res. Pap. RM-156. Rocky Mt. For. Range Expt. Sta., Ft. Collins, CO.

Marzolf, G. R. 1978. The potential effects of clearing and snagging on stream ecosystems. FWS/OBS-78/14. USDI Fish and Wildlife Service, Biol. Serv. Prog., Washington, DC.

McBride, F. D., C. Chavengsaksongkram, and D. Urie. 1977. Sludge-treated coal mine spoils increase heavy metals in cover crops. USDA Forest Service, Res. Note NC-221. North Central For. Expt. Sta., St. Paul, MN.

McCashion, J. D. and R. M. Rice. 1983. Erosion on logging roads in northwestern California: how much is avoidable? *J. For.* 81:23–26.

McDonald, G. A. 1971. Forest soil disturbance in the Blue Mountains from wheeled and crawler skidders. MS thesis on file at Washington State University, Pullman, WA.

McKee, W. H., Jr. 1985. Forestry and forest management impacts on wetlands. pp. 216–224. In: Groman *et al.*, Eds., *Proc. Wetlands of the Chesapeake.* Environmental Law Inst., Washington, DC.

McKee, W. H., Jr., G. E. Hatchell, and A. E. Tiarks. 1985. Managing site damage from logging—a loblolly pine management guide. USDA For. Serv. Gen. Tech. Rep. SE-32. Southeastern For. Expt. Sta., Asheville, NC.

McMinn, J. W. 1984. *Soil disturbance in mountain hardwood stands.* Ga. For. Res. Pap. 54, GA Forestry Comm., Macon, GA.

Meeuwig, R. O. 1970a. Infiltration and soil erosion as influenced by vegetation and soil in northern Utah. *J. Range Manage.* 23:185–188.

Meeuwig, R. O. 1970b. Sheet erosion on Intermountain summer ranges. USDA Forest Service Res. Pap. INT-85. Intermountain For. Expt. Sta., Ogden, UT.

Meeuwig, R. O. 1971. Soil stability on high elevation rangeland in the Intermountain area. USDA Forest Service Res. Pap. INT-94. Intermountain For. Expt. Sta., Ogden, UT.

Meeuwig, R. O. and P. E. Packer. 1976. Erosion and runoff on forest and range lands. pp. 105–116. In: H. Heady et al., Eds. *Watershed management on range and forest lands.* Utah Water Res. Lab., Utah State University, Logan, UT.

Megahan, W. F. 1972. Subsurface flow interception by a logging road in the mountains of central Idaho. pp. 350–356. In: Csallany et al., Eds. *Watersheds in transition,* Proc. Ser. No. 14, Am. Water Resour. Assoc., Urbana, IL.

Megahan, W. F. 1974a. Erosion over time on severely disturbed granitic soils: a model. USDA Forest Service, Res. Pap. INT-156. Intermountain For. Expt. Sta., Ogden, UT.

Megahan, W. F. 1974b. Deep-rooted plants for erosion control on granitic road fills in the Idaho Batholith. USDA Forest Service, Res. Pap. INT-161. Intermountain For. Expt. Sta., Ogden, UT.

Megahan, W. F. 1977. Reducing erosional impacts of roads. pp. 237–261. In: *Guidelines for watershed management.* FAO Cons. Guide Food Agric. Org. of the United Nations, Rome.

Megahan, W. F. 1987. Effects of forest roads on watershed function in mountainous areas. pp. 335–348. In: A. S. Balasubramaniam et al., Eds., *Proc. Symp. Environmental Geotechnics and Problematic Soils and Rocks.* A. A. Balkema Publ., Brookfield, VT.

Megahan, W. F., N. F. Day, and T. M. Bliss. 1978. Landslide occurrence in the western and central northern Rocky Mountain physiographic province in Idaho. pp. 116–130. In: C. T. Youngberg, Ed., *Forest Soils and Land Use.* Proc. 5th N. Am. For. Soils Conf., Colorado State University, Ft. Collins, CO.

Megahan, W. F. and P. N. King. 1985. Identification of critical areas on forest lands for control of nonpoint sources of pollution. *Environ. Manage.* 9(1):7–18.

Megahan, W. F., W. S. Platts, and B. Kuleszo. 1980. Riverbed improves over time: South Fork Salmon. pp. 380–395. In: *Symposium on watershed management.* Vol. 1. Am. Soc. Civ. Eng., New York.

Mellgren, P. G. and E. Heidersdorf. 1984. The use of high flotation tires for skidding in wet and/or steep terrain. Tech. Rep. TR-57. For. Engr. Res. Inst. Canada, Vancouver, BC.

Meyer, E. A. and E. L. Jarroll. 1980. Giardiasis. *Am. J. Epidemiol.* 111:1–12.

Michigan Department of Natural Resources. 1974. Rules and regulations concerning Inland Lakes and Streams Act. Hydrological Survey Div., Lansing, MI.

Mifflin, R. W. and H. H. Lysons. 1979. Glossary of forest engineering terms. USDA Forest Service, Pac. Northwest For. Range Expt. Sta., Portland, OR.

Milauskas, S. J. 1987. Low-water stream crossing options for Southern haul roads. pp. 11–15. In: Fourth Intl. Conf. Low-Volume Roads, Vol. 2. Trans. Res. Rec. 1106. Trans. Res. Board, Natl. Res. Council, Washington, DC.

Miller, M. 1986. Sludge compost recycling: the Philadelphia story. *J. Soil Water Conserv.* 41:292–296.

Miller, R. F., I. S. McQueen, F. A. Branson, L. M. Shown, and W. Buller. 1969. An evaluation of range floodwater spreaders. *J. Range Manage.* 22:246–257.

Miyata, E. S., C. N. Maun, and T. L. Ortman. 1984. Field evaluation of Menzi–Muck feller-buncher on difficult terrain in southeast Alaska. pp. 262–280. In: P. A. Peters and J. Luchak, Eds. *Mountain logging symposium proc.*, West Virginia University, Morgantown, WV.

Moehring, S. M. and I. W. Rawls. 1970. Detrimental effects of wet weather logging. *J. For.* 68:166–167.

Monson, S. B. and N. Shaw (compilers). 1983. *Managing intermountain rangelands—improvement of range and wildlife habitats.* USDA Gen. Tech. Rept. INT-157. Intermountain For. Range Expt. Sta., Ogden, UT.

Moore, E. 1976. *Livestock grazing and protection of water quality.* Working Paper, Environ. Prot. Agency, Washington, DC.

Moore, E., E. Janes, F. Kinsinger, K. Pitney, and J. Sainsbury. 1979. *Livestock grazing management and water quality protection (state of the art reference document).* US Environ. Prot. Agency and USDI, Bur. Land Manage., US Environ. Prot. Agency, Region 10, Seattle, WA.

Moore, R. and T. Mills. 1977a. *An environmental guide to western surface mining.* Pt. 2, Vol. 2, *Impacts, mitigation, and monitoring in the Rocky Mountains.* Ecology Consultants, Inc., Ft. Collins, CO.

Moore, R. and T. Mills. 1977b. *An environmental guide to western surface mining.* Pt. 2, Vol. 3, *Impacts, mitigation, and monitoring in the Intermountain and Southwest.* Ecology Consultants, Inc., Ft. Collins, CO.

Moring, J. R. and R. Lantz. 1975. The Alsea Watershed study: the effects of logging on the aquatic resources of three headwater streams in the Alsea River, Oregon. Pt I - Biological studies. Rept. No. 9, Oregon Dept. Fish and Wildlife, Corvallis, OR.

Morris, W. G. 1970. Effects of slash burning in overmature stands of the Douglas-fir region. *For. Sci.* 16:258–270.

Muchmore, F. W. 1986. Designing timber bridges for long life. pp. 12–17. In: *Timber Bridges.* Trans. Res. Rec. 1053. Trans. Res. Board, Natl. Res. Council, Washington, DC.

Murphy, G. 1982. Soil damage associated with production thinning. *New Zealand J. For. Sci.* 12:281–292.

Murphy, G. and M. R. Pyles. 1989. Cost-effective selection of culverts for small forest streams. *J. For.* 87(10):45–50.

Murphy, M. L. and J. D. Hall. 1981. Varied effects of clear-cut logging on predators and their habitat in small streams of the Cascade Mountains, Oregon. *Can. J. Fish. Aquat. Sci.* 38:137–145.

Murphy, M. L., C. P. Hawkins, and N. H. Anderson. 1981. Effects of canopy modification and accumulated sediment on stream communities. *Trans. Am. Fish. Soc.* 110:469–478.

Mustard, R. H. 1982. *Handbook for forest roads.* DOE Rept. No. 82-5, Washington State Depts. of Natural Resources and Ecology, Olympia, WA.

NAPAP, 1990. Integrated assessment. External review draft. Natl. Acid Precip. Assess. Prog., Washington, DC.

Neubauer, C. P. 1981. The effects of land clearing on a small watershed in southern Guam. Tech. Rept. No. 24, Water and Energy Res. Inst. of the Western Pac., University of Guam.

Nord, E. C. and L. R. Green. 1977. Low-volume and slow-burning vegetation for planting on clearings in California chaparrel. USDA Forest Service Res. Pap. PSW-124, Pac. Southwest For. Range Expt. Sta., Berkeley, CA.

Norris, L. A. 1975. Behavior and impact of some herbicides in the forest. pp. 159–176. In: *Herbicides in forestry.* Proc. of the 1975 John S. Wright Forestry Conf., Purdue University, W. Lafayette, IN.

Norris, L. A. 1981. The behavior of herbicides in the forest environment and risk assessment. pp. 192–215. In: *Weed control in forest management.* Proc. of the 1981 John S. Wright Forestry Conf., Purdue University, W. Lafayette, IN.

Norris, L. A. 1984. Use, ecotoxicology, and risk assessment of herbicides in forestry. pp. 381–393. In: W. Y. Garner and J. Harvey, Jr., Eds., *Chemical and biological controls in forestry.* ACS Symposium series No. 238, American Chemical Society, Washington, D.C.

Norris, L. A., H. W. Lorz, and S. V. Gregory. 1983. Influence of forest and rangeland management on anadromous fish habitat in western North America. 9. Forest chemicals. USDA Forest Service Gen. Tech. Rept. PNW-149, Pac. Northwest For. Range Expt. Sta., Portland, OR.

Norris, L. A. and D. G. Moore. 1971. The entry and fate of forest chemicals in streams. pp. 138–158. In: *Proc., Symp. forest land uses and stream environment.* Forestry Extension, Oregon State University, Corvallis, OR.

Norris, L. A. and D. G. Moore. 1981. Introduced chemicals and water quality. pp. 203–220. In: D. M. Baumgartner, Ed. *Interior west watershed management,* Coop. Extension, Washington State University, Pullman, WA.

Olsen, E. D. and D. J. Gibbons. 1983. Predicting skidding productivity: A mobility model. OSU For. Res. Lab. Bull. 43.

Olsen, E. D. and J. C. W. Seifert. 1984. Machine performance and site disturbance in skidding on designated trails. *J. For.* 82(6):366–369.

Orr, H. K. 1960. Soil porosity and bulk density on grazed and protected Kentucky bluegrass range in the Black Hills. *J. Range Manage.* 13:80–86.

Pace, C. P. and A. W. Lindemuth, Jr. 1971. Effects of prescribed fire on vegetation and sediment in oak-mountain mahogany chaparral. *J. For.* 69:800–805.

Packer, P. E. 1967. Criteria for designing and locating logging roads to control sediment. *For. Sci.* 13:2–18.

Packer, P. E. and G. F. Christensen. 1964. Guides for controlling sediment from secondary logging roads. Intermountain For. Range Expt. Sta. Northern Region, USFS, Missoula, MT.

Packer, P. E., C. E. Jensen, E. L. Noble, and J. A. Marshall. 1982. Models to estimate revegetation potentials of land surface mined for coal in the west. USDA For. Serv. Gen. Tech. Rep. INT-123. Intermountain For. Range Expt. Sta., Ogden, UT.

Painter, R. B., K. Blyth, J. C. Mosedale, and M. Kelly. 1974. The effect of afforestation on erosion processes and sediment yield. *IASH Publ. No.* 113:62–67.

Parry, J. D. 1986. A prefabricated modular timber bridge. pp. 49–55. In: *Timber Bridges.* Trans. Res. Rec. 1053. Trans. Res. Board, Natl. Res. Council, Washington, DC.

Patric, J. H. 1980. Effects of wood products harvest on forest soil and water relations. *J. Environ. Qual.* 9(1):73–80.

Patric, J. H., J. O. Evans, and J. D. Helvey. 1984. Summary of sediment yield data from forested lands in the United States. *J. For.* 82:101–104.

Patric, J. H. and J. D. Helvey. 1986. Some effects of grazing in the eastern forest. USDA For. Serv. Gen. Tech. Rep. NE-115. Northeastern For. Expt. Sta., Broomall, PA.

Patton, V. P. 1987. Downhill yarding on Horse Creek to minimize stream impacts. pp. 85–87. In: Managing Oregon's riparian zone for timber, fish and wildlife. Tech. Bull. No. 514. National Council of the Paper Industry for Air and Stream Improvement. New York.

Paulsen, H. A., Jr. 1975. Range management in the central and southern Rocky Mountains: the status of our knowledge. USDA Forest Service Res. Pap. RM-154, Rocky Mt. For. Range Expt. Sta., Ft. Collins, CO.

Peck, A. J. and D. H. Hurle. 1973. Chloride balance of some farmed and forested catchments in southwestern Australia. *Water Resour. Res.* 9:648–657.

Piehl, B. T., R. L. Beschta, and M. R. Pyles. 1988. Ditch-relief culverts and low-volume roads in the Oregon Coast Range. *Northwest Sci.* 62(3):91–98.

Platts, W. S. 1981. Influence of forest and rangeland management on anadromous fish habitat in western North America. 7. Effects of livestock grazing. USDA Forest Service Gen. Tech. Rept. PNW-124, Pac. Northwest For. Range Expt. Sta., Portland, OR.

Platts, W. S., C. Armour, G. D. Booth, M. Bryant, J. L. Bufford, P. Cuplin, S. Jensen, G. W. Lienkaemper, G. W. Minshall, S. B. Monsen, R. L. Nelson, J. R. Sedell, and J. S. Tuhy. 1987. Methods for evaluating riparian habitats with applications to management. USDA For. Serv. Gen. Tech. Rep. INT-221. Intermountain Res. Sta., Ogden, UT.

Platts, W. S., W. F. Megahan, and G. W. Minshall. 1983. Methods for evaluating stream, riparian, and biotic conditions. USDA For. Serv. Gen. Tech. Rep. INT-138. Intermountain For. Range Expt. Sta., Ogden, UT.

Platts, W. S. and R. L. Nelson. 1985a. Streamside and upland vegetation use by cattle. *Rangelands* 7:5–7.

Platts, W. S. and R. L. Nelson. 1985b. Will the riparian pasture build good streams? *Rangelands* 7:7–10.

Pluhowski, E. J. 1972. Clear-cutting and its effect on the water temperature of a small stream in northern Virginia. US Geol. Survey Prof. Pap. 800-C, pp. C257–C262.

Plummer, A. P., D. R. Christensen, and S. B. Monsen. 1968. Restoring big game range in Utah. Pub. No. 68-3. Utah Div. Fish and Game, Ephraim, UT

Pole, M. W. and D. R. Satterlund. 1978. Plant indicators of slope instability. *J. Soil Water Conserv.* 33:230–232.

Ponce, S. L. 1974. The biochemical oxygen demand of finely divided logging debris in stream water. *Water Resour. Res.* 10:983–988.

Ponce, S. L. and G. W. Brown. 1974. Demand for dissolved oxygen exerted by finely divided logging debris in streams. For. Res. Lab. Res. Pap. 19. School of Forestry, Oregon State University, Corvallis, OR.

Post, I. L. 1986. The radio horse: a new machine for thinning without damage. *Nat. Woodlands* 9(2):9–12.

Pyles, M. R. and H. A. Froehlich. 1987. Discussion: rates of landsliding as impacted by timber management activities in northwestern California. *Bull. Assoc. Engr. Geol.* 24(3):425–429.

Pyne, S. J. 1984. *Introduction to wildland fire.* Wiley, New York.

Reeves, G. H. and T. D. Roelofs. 1982. Influence of forest and rangeland management on anadromous fish habitat in western North America. 13. Rehabilitating and enhancing stream habitat: 2. field applications. USDA Forest Service Gen. Tech. Rept. PNW-140, Pac. Northwest For. Range Expt. Sta., Portland, OR.

Reid, E. H., G. S. Strickler, and W. B. Hall. 1980. Green fescue grassland: 40 years of secondary succession. USDA Forest Service Res. Pap. PNW-274. Pac. Northwest For. Range Expt. Sta., Portland, OR.

Reinhart, K. G., A. R. Escher, and G. R. Trimble. 1963. Effect on streamflow of four forest practices in the mountains of West Virginia. USDA Forest Service Res. Pap. NE-1, Northeastern For. Expt. Sta., Upper Darby, PA.

Reiser, D. W. and T. C. Bjornn. 1979. Influence of forest and rangeland management on anadromous fish habitat in western North America. 1. Habitat requirements of anadromous salmonids. USDA Forest Service Gen. Tech. Rept. PNW-96, Pac. Northwest For. Range Expt. Sta., Portland, OR.

Reisinger, T. W., G. L. Simmons, and P. E. Pope. 1988. The impact of timber harvesting on soil properties and seedling growth in the south. *Southern J. Appl. For.* 12(1):58–67.

Rice, R. M. 1977. Forest management to minimize landslide risk. pp. 271–287. In: S. H. Kunkle and J. H. Thames, Eds. *Guidelines for watershed management.* FAO Conserv. Guide No. 1, FAO, United Nations, Rome.

Richardson, B. Z. and M. M. Pratt, compilers. 1980. Environmental effects of surface mining of minerals other than coal: annotated bibliography and summary report. USDA Forest Service Gen. Tech. Rept. INT-95, Intermountain For. Range Expt. Sta., Ogden, UT.

Ring, S. L. 1987. The design of low water stream crossings. pp. 309–318. In: Fourth Intl. Conf. Low-Volume Roads, Vol. 2. Trans. Res. Rec. 1106. Trans. Res. Board, Natl. Res. Council, Washington, DC.

Roath, L. R. and W. C. Krueger. 1982. Cattle grazing influence on a mountain riparian zone. *J. Range Manage.* 35:100–104.

Roby, G. A. and L. R. Green. 1976. *Mechanical methods of chaparral modification.* USDA Forest Service Agric. Handbook No. 487. Washington, DC.

Ruth, R. H. 1967. Silvicultural effects of skyline crane and high-lead yarding. *J. For.* 65:251–255.

Ruth, R. H. and R. R. Silen. 1950. Suggestions for getting more forestry in the logging plan. USDA Forest Service Res. Pap. PNW-72, Pac. Northwest For. Range Expt. Sta., Portland, OR.

Salo, E. O. and T. W. Cundy, Eds. 1987. *Streamside management: Forestry and fisheries interactions.* Contribution No. 57. College of Forest Resources, University of Washington, Seattle, WA. p. 471.

Satterlund, D. R. 1977. Shadow patterns located with a programmable calculator. *J. For.* 75:262–263.

Schaller, F. W. and P. Sutton, Eds. 1978. *Reclamation of drastically disturbed lands.* Am. Soc. Agron., Madison, WI.

Schlots, F. E. and A. N. Quam. 1966. Soil survey interpretations for woodland conservation. Progress Rept. USDA Soil Conserv. Serv., Spokane, WA.

Schmiege, D. C. 1980. Influence of forest and rangeland management on anadromous fish habitat in western North America. 11. Processing mills and camps. USDA Forest Service Gen. Tech. Rept. PNW-113, Pac. Northwest For. Range Expt. Sta., Portland, OR.

Schofield, C. L. 1976. Acid precipitation: effects on fish. *Ambio* 5:228–230.

Schroeder, M. J. and C. C. Buck. 1970. *Fire weather.* USDA Forest Service Agric. Handbook 360, Washington, DC.

Schuh, D. D. and L. D. Kellogg. 1988. Timber harvesting mechanization in the western United States: an industry survey. *Western J. Appl. For.* 3(2):33–36.

Searcy, J. K. 1965. Design of roadside drainage channels. Hydraulic Design Series No. 4. Bur. Public Roads, US Dept. Commerce, Washington, DC.

Sedell, J. R. and W. S. Duval. 1985. Influence of forest and range management on anadromous fish habitat in western North America. 5. Water transportation and storage of logs. USDA Forest Service Gen. Tech. Rept. PNW-186, Pac. Northwest For. Range Expt. Sta., Portland, OR.

Sessions, J., J. Balcom, and K. Boston. 1987. Road location and construction practices: Effects on landslide frequency and size in the Oregon Coast Range. *Western J. Appl. For.* 2(4):119–124.

Sessions, J. and Li Guangda. 1987. Deriving optimal road and landing spacing with microcomputer programs. *West J. Appl. For.* 2(3):94–98.

Sessions, J., R. Stewart, P. Anderson, and B. Tuor. 1986. Calculating the maximum grade a log truck can climb. *West. J. Appl. For.* 1(2):43–45.

Sheeter, G. R. and E. W. Claire. 1981. Use of juniper trees to stabilize eroding streambanks on the South Fork John Day River. Tech. Note OR-1, USDI Bureau of Land Management, Portland, OR.

Shetron, S. G., J. A. Sturos, E. Padley, and C. Trettin. 1988. Forest soil compaction: effect of multiple passes and loadings on wheel track surface soil bulk density. *Northern J. Appl. For.* 5:120–123.

Shown, L. M. 1971. Sediment yield as related to vegetation on semiarid watersheds. pp. 347–354. In: *Biological effects in the hydrological cycle.* Dept. Agric. Eng., Agric. Expt. Sta., Purdue University, West Lafayette, IN.

Sidle, R. C. and D. M. Drlica. 1981. Soil compaction from logging with low-ground pressure skidder in the Oregon Coast Ranges. *Soil Sci. Soc. Am. J.* 45(6):1219–1223.

Sidle, R. C. and T. H. Laurent. 1986. Site damage from mechanized thinning in southeast Alaska. *North J. Appl. Forestry* 3:94–97.

Sidle, R. C., A. J. Pearce, and C. L. O'Loughlin. 1985. Hillslope stability and land use. Water Resources Monogr. 11. American Geophysical Union, Washington, DC.

Silen, R. R. 1955. More efficient road patterns for a Douglas-fir drainage. *The Timberman* 56(6):82, 85–86, 88.

Silversides, C. R. 1984. Mechanized forestry: World War II to the present. *For. Chron.* 60(4):231–235.

Simmons, F. C. 1979. *Handbook for eastern timber harvesting.* USDA For. Serv., Northeastern Area, State and Private Forestry. US Govt. Printing Office, Washington, DC.

Sims, H. P., C. B. Powler, and J. A. Campbell. 1984. Land surface reclamation: a review of the international literature. Alberta Land Cons. and Reclamation Council Rept. No. RRTAC 84-1, 2 vols. Edmonton, Alberta.

Slack, K. V. and H. R. Feltz. 1968. Tree leaf control on low flow water quality in a small Virginia stream. *Environ. Sci. Tech.* 2(2):126–131.

Slayback. R. D. and D. R. Cable. 1970. Larger pits aid reseeding of semidesert rangeland. *J. Range Manage.* 23:333–335.

Smith, D. M. 1962. *The practice of silviculture.* 7th ed. Wiley, New York.

Snyder, G. G., H.F. Haupt, and G. H. Belt, Jr. 1975. Clearcutting and burning slash alter quality of stream water in northern Idaho. USDA Forest Service Res. Pap. INT-168, Intermountain For. Range Expt. Sta., Ogden, UT.

Sopper, W. E. 1971. Effects of trees and forests in neutralizing waste. Reprint Series No. 23, Inst. for Res. on Land & Water Resour., the Pennsylvania State University, University Park, PA.

Sopper, W. E. 1975. Effects of timber harvesting and related management practices on water quality in forested watersheds. *J. Environ. Qual.* 4:14–29.

Springer, D. K., J. D. Burns, H. A. Fribourg, and K. E. Graetz. 1967. Roadside revegetation and beautification in Tennessee. USDA Soil and Conserv. Serv. and Coll. of Agric., University of Tennessee, SP 162., Knoxville, TN.

Springfield, H. W. 1976. Characteristics and management of southwestern pinyon-juniper ranges: the status of our knowledge. USDA Forest Service Res. Pap. RM-160. Rocky Mt. For. Range Expt. Sta., Ft. Collins, CO.

Stark, N. 1966. Review of highway planting information appropriate to Nevada. Coll. of Agric. Bull. No. B-7, Desert Res. Inst., University of Nevada, Reno, NV.

Steen, H. K. 1966. Vegetation following slash fires in one western Oregon locality. *Northwest Sci.* 40:113–120.

Steinblums, I. J., H. A. Froehlich, and J. F. Lyons. 1984. Designing stable buffer strips for streamside protection. *J. For.* 82:49–52.

Stevens, A. A., C. J. Slocum, D. R. Seeger, and G. G. Robeck. 1976. Chlorination of organics in drinking water. *J. Am. Water Works Assoc.* 68(Pt. 2):615–620.

Stiller, G. M., G. L. Zimpfer, and M. Bishop. 1980. Application of geomorphic principles to surface mine reclamation in the semiarid West. *J. Soil Water Conserv.* 35:274–277.

Stjernberg, E. 1987. Plastic culverts in forest road construction. Tech. Note TN-110. For. Engr. Res. Inst. Canada, Vancouver, B.C.

Stoddart, L. A., A. D. Smith, and T. W. Box. 1975. *Range management.* 3rd ed. McGraw-Hill, New York.

Stoeckler, J. H. 1959. Trampling by livestock drastically reduces infiltration rate of soil in oak and pine woods in southwestern Wisconsin. USDA Forest Service Tech. Notes No. 556. Lake States For. Expt. Sta., St. Paul, MN.

Stone, E. L., W. T. Swank, and J. W. Hornbeck. 1978. Impacts of timber harvest and regeneration systems on stream flow and soils in the eastern deciduous region. pp. 516–535. In: C. T. Youngberg, Ed., *Forest soils and land use.* Colorado State University, Ft. Collins, CO.

Strickler, G. S. and W. B. Hall. 1980. The Standley allotment: a history of range recovery. USDA Forest Service Res. Pap. PNW-278. Pac. Northwest For. Range Expt. Sta., Portland, OR.

Striffler, W. D. 1960. Streambank stabilization in Michigan—a survey. Sta. Pap. No. 84, Lake States For. Expt. Sta. Mich. Dept. Conserv., St. Paul, MN.

Stuart, E., E. Gililland, and L. Della-Moretta. 1987. The use of central tire inflation systems on low-volume roads. pp. 164–168. In: Fourth Intl. Conf. Low-Volume Roads, Vol. 2. Trans. Res. Rec. 1106. Trans. Res. Board, Natl. Res. Council, Washington, DC.

Sullivan, K. 1985. Long-term patterns of water quality in a managed watershed in Oregon: 1. Suspended sediment. *Water Resour. Bull.* 21:977–987.

Sullivan, K., T. E. Lisle, C. A. Dolloff, G. E. Grant, and L. M. Reid. 1987. Stream channels: The link between forest and fishes. pp. 39–97. In: *Proceedings Symposium on Streamside Management: Forestry and Fisheries Interactions*, University of Washington, Seattle, WA.

Swank, W. T., L. F. Debano, and D. Nelson. 1989. Effects of timber management practices on soil and water. pp. 79–106. In: Burns, R. M. (tech. comp.), The scientific basis for silvicultural and management decisions in the National Forest system. USDA For. Serv. Gen. Tech. Rep. WO-55. Washington, DC.

Swanson, F. J. and M. E. James. 1975. Geology and geomorphology of the H. J. Andrews Experimental Forest, West Cascades, Oregon. USDA Forest Service Res. Pap. PNW-188. Pac. Northwest For. Range Expt. Sta., Portland, OR.

Swanston, D. N. 1974. The forest ecosystem of Alaska. 5. Soil mass movement. USDA Forest Service Gen. Tech. Rept. PNW-17. Pac. Northwest For. Range Expt. Sta., Portland, OR.

Swanston, D. N. 1980. Influence of forest and rangeland management on anadromous fish habitat in western North America. 2. Impacts of natural events. USDA Forest Service Gen. Tech. Rept. PNW-104. Pac. Northwest For. Range Expt. Sta., Portland, OR.

Swanston, D. N. and F. J. Swanson. 1976. Timber harvesting, mass erosion, and steepland forest geomorphology in the Pacific Northwest. pp. 199–221. In: D. R. Coates, Ed., *Geomorphology and Engineering.* Dowden, Hutchinson & Ross, Inc., Stroudsburg, PA.

Swift, L. W. 1984a. Gravel and grass surfacing reduces soil loss from mountain roads. *For. Sci.* 30(3):657–670.

Swift, L. W. 1984b. Soil losses from roadbeds and cut and fill slopes in the southern Appalachian mountains. *Southern J. Appl. For.* 8(4):209–215.

Swift, L. W., Jr. 1985. Forest road design to minimize erosion in the southern Appalachians. pp. 141–151. In: B. G. Blackmon, Ed. *Proc. Forestry and water quality: a Mid-South symposium,* Coop. Ext. Service, University of Arkansas, Fayetteville, AR.

Swift, L. W., Jr. 1986. Filter strip widths for forest roads in the southern Appalachians. *Southern J. Appl. For.* 10:27–34.

Swift, L. W., Jr., and J. B. Messer. 1971. Forest cuttings raise temperatures of small streams in the southern Appalachians. *J. Soil Water Conserv.* 26:111–116.

Swindel, B. F., W. R. Marion, L. D. Harris, L. A. Morris, W. L. Pritchett, L. F. Conde, H. Riekerk, and E. T. Sullivan. 1983. Multi-resource effects of harvest, site preparation, and planting in pine flatwoods. *Southern J. Appl. For.* 7:6:15.

Tarrant, R. F. 1956. Effects of slash burning on some soils of the Douglas-fir region. *Soil Sci. Soc. Am. Proc.* 20:408–411.

Taylor, R. L. and P. W. Adams. 1986. Red alder leaf litter and streamwater quality in western Oregon. *Water Resour. Bull.* 22(4):629–639.

Temple, K. L., A. K. Camper, and R. C. Lucas. 1982. Potential health hazard from human wastes in wilderness. *J. Soil. Conserv.* 37:357–359.

Thames, J. L., Ed. 1977. *Reclamation and use of disturbed land in the Southwest.* University of Arizona Press, Tucson, AZ.

Thilenius, J. F. 1975. Alpine range management in the western United States—principles, practices, and problems: the status of our knowledge. USDA Forest Service Res. Pap. RM-157, Rocky Mt. For. Range Expt. Sta., Ft. Collins, CO.

Thomas, B. R. 1985. Uses of soils, vegetation and geomorphic information for road location and timber management in the Oregon Coast Ranges. pp. 68–77. In: D. N. Swanston, tech. Ed., *Proceedings of a workshop on slope stability: problems and solutions in forest management.* USDA Forest Service Gen. Tech. Rept. PNW-180. Pac. Northwest For. Range Expt. Sta., Portland, OR.

Thomas, R. B. 1985. Measuring suspended sediment in small mountain streams. USDA Forest Service Gen. Tech. Rept. PSW-83. Pac. Southwest For. Range Expt. Sta., Berkeley, CA.

Thornburg, A. A. 1982. Plant materials for use on surface mined lands in arid and semiarid regions. USDA Soil Conserv. Serv. SCS-TP-157, Washington, DC.

Tiedemann, A. R., C. E. Conrad, J. H. Dieterich, J. W. Hornbeck, W. F. Megahan, L. A. Viereck, and D. G. Wade. 1978. Effects of Fire on water. USDA Forest Service Gen. Tech. Rept. WO-10, Washington, DC.

Tobiaski, R. A. and N. R. Tripp. 1961. Gabions for stream and erosion control. *J. Soil Water Conserv.* 16:284–285.

Toews, D. A. A. and M. J. Brownlee. 1981. *A handbook for fish habitat protection on forest lands in British Columbia.* Land Use Unit, Habitat Prot. Div., Field Serv. Br., Dept. Fish. and Oceans, Vancouver, BC.

Triska, F. J., J. R. Sedell, and S. V. Gregory. 1982. Coniferous forest streams. pp. 292–332. In: R. L. Edmonds, Ed. *Analysis of coniferous forest ecosystems in the western United States.* US/IBP Synthesis Series 14, Hutchinson Ross Pub. Co., Stroudsburg, PA.

Troendle, C. A. and R. M. King. 1985. The effect of timber harvest on Fool Creek Watershed, 30 years later. *Water Resour. Res.* 21:1915–1922.

Turner, G. T. and H.A. Paulson, Jr. 1976. Management of mountain grasslands in the central Rockies: the status of our knowledge. USDA Forest Service Res. Pap. RM-161, Rocky Mt. For. Range Expt. Sta., Ft. Collins, CO.

Urie, D. H. 1967. Waterline erosion control essential to streambank rehabilitation. USDA Forest Service Res. Pap. NC-29. North Central For. Range Expt. Sta., St. Paul, MN.

USDA Forest Service. 1973. *Silvicultural systems for the major forest types of the United States.* Agric. Handbook 445. US Govt. Printing Office, Washington, DC.

USDA Forest Service. 1979a. User guide to soils. USDA Forest Service Gen. Tech. Rept. INT-68. Intermountain For. Range Expt. Sta., Ogden, UT.

USDA Forest Service. 1979b. User guide to engineering. USDA Forest Service Gen. Tech. Rept. INT-70. Intermountain For. Range Expt. Sta., Ogden, UT.

USDA Forest Service. 1979c. User guide to vegetation. USDA Forest Service Gen. Tech. Rept. INT-64. Intermountain For. Range Expt. Sta., Ogden, UT.

USDA Forest Service. 1980. User guide to hydrology. USDA Forest Service Gen. Tech. Rept. INT-74. Intermountain For. Range Expt. Sta., Ogden, UT.

USDA Forest Service. 1980a. Trees for reclamation: symposium proceedings. USDA For. Serv. Gen. Tech. Rep. NE-61. Northeastern For. Exp. Sta., Broomall, PA.

US Environmental Protection Agency. 1975. *Logging roads and protection of water quality.* EPA910/9-75-007, Water Div., US Env. Prot. Agency, Region X, Seattle, WA.

US Environmental Protection Agency. 1976. *Forest harvest, residue treatment, reforestation, and protection of water quality.* EPA 910/9-76-020. Region X, US EPA, Seattle, WA.

US Weather Bureau. 1961a. Rainfall frequency atlas of the United States for durations from 30 minutes to 24 hours and return periods from 1 to 100 years. Tech. Pap. No. 40, Govt. Print. Off., Washington, DC.

US Weather Bureau. 1961b. Generalized estimates of probable maximum precipitation and rainfall frequency data for Puerto Rico and Virgin Islands for areas to 400 square miles, durations to 24 hours, and return periods from 1 to 100 years. Tech. Pap. No. 42, Govt. Print. Off., Washington, DC.

US Weather Bureau. 1962. Rainfall frequency atlas for the Hawaiian Islands for areas to 200 square miles, durations to 24 hours, and return periods from 1 to 100 years. Tech. Pap. No. 45, Govt. Print. Off., Washington, DC.

US Weather Bureau. 1963. Probable maximum precipitation and rainfall frequency data for Alaska to areas of 400 square miles, durations to 24 hours, and return periods from 1 to 100 years. Tech. Pap. No. 47, Govt. Print. Off., Washington, DC.

Vallentine, J. F. 1971. *Range development and improvements.* Brigham Young University Press, Provo, UT.

Van Haveren, B. P. 1983. Soil bulk density as influenced by grazing intensity and soil type on a shortgrass prairie site. *J. Range Manage.* 36:586–588.

Van Meter, W. P. and C. E. Hardy. 1975. Predicting effects on fish of fire retardants in streams. USDA Forest Service Res. Pap. INT-166. Intermountain For. Range Expt. Sta., Ogden, UT.

Vannote, R. L., G. W. Minshall, K. W. Cummins, J. R. Sedell, and C. E. Cushing. 1980. The River Continuum Concept. *Canad. J. Fish. Aquat. Sci.* 37:130–136.

Varnes, D. J. 1958. Landslide types and processes. pp. 30-47. In: J. B. Eckel, Ed., *Landslides and engineering practice.* Highway Res. Bd. Spec. Rept. 29. NAS-NRC Pub. 544, Washington, DC.

Verry, E. S. 1986. Forest harvesting and water: the Lakes States experience. *Water Res. Bull.* 22(6):1039–1047.

Viereck, L. A. 1982. Effects of fire and firelines on active layer thickness and soil temperatures in interior Alaska. In: *The Roger J. E. Brown Memorial Vol., Proc. Fourth Canadian Permafrost Conf.*, Calgary, Alberta. Natl. Resour. Council of Canada, Ottawa.

Viereck, L. A. and C. T. Dyrness, tech Eds. 1979. Ecological effects of the Wickersham Dome fire near Fairbanks, Alaska. USDA Forest Service Gen. Tech. Rept. PNW-90. Pac. Northwest For. Range Expt. Sta., Portland, OR.

Vitousek, P. M., J. R. Gosz, C. C. Grier, J. M. Melillo, W. A. Reiners, and R. L. Todd. 1979. Nitrate losses from disturbed ecosystems. *Science* 204:469–474.

Vogel, W. G. 1981. A guide for revegetating coal minerals in the eastern United States. USDA For. Serv. Gen. Tech. Rep. NE-68. Northeastern For. Expt. Sta., Broomall, PA.

Vories, K. C., Ed. 1976. *Reclamation of western surface mined lands.* ECO Consultants, Inc., Ft. Collins, CO.

Walstad, J. D. and F. N. Dost. 1984. The health risks of herbicides in forestry: a review of the scientific record. Special Pub. 10. Forest Research Lab, Oregon State University, Corvallis, OR.

Warhol, T. and M. R. Pyles. 1989. Low water fords: An alternative to culverts on forest roads. In: Proc. 12th Ann. Council For. Engr. Meetings, Coeur d'Alene.

Waring, R. H. and W. H. Schlesinger. 1985. *Forest ecosystems: concepts and management.* Academic, New York.

Watson, L. E., R. W. Parker, and D. F. Polster. 1980. *Manual of species suitability for reclamation in Alberta.* Alberta Land Conserv. and Reclamation Council Rept. No. RRTAC 80-5. 2 vols., Edmonton, Alberta.

Wenger, K. F. 1984. *Forestry handbook.* 2nd ed. Wiley, New York.

Wesche, T. A. 1974. Relationship of discharge reductions to available trout habitat for recommending suitable streamflows. Water Resour. Ser. No. 53, Water Resour. Res. Inst., University of Wyoming, Laramie, WY.

Williams, O. R. 1982. A procedure to determine if mine spoils will reach a downslope stream channel. USDA Forest Service WSDG Rept. WSDG-TN-00004. Watershed Systems Dev. Group, Ft. Collins, CO.

Winegar, H. H. 1977. Camp Creek channel fencing—plant, wildlife, soil, and water response. *Rangeman's J.* 4:10–12.

Woolridge, D. D. 1960. Watershed disturbance from tractor and skyline crane logging. *J. For.* 58:369–372.

Yamamoto, T. 1982. A review of uranium spoil and mill tailings revegetation in the western United States. USDA Forest Service Gen. Tech. Rept. RM-92. Rocky Mt. For. Range Expt. Sta., Ft. Collins, CO.

Yee, C. S. 1976. Small rock buttresses—useful for some forest road slump failures. *J. For.* 74:688–689.

Yee, C. S. and T. D. Roelofs. 1980. Influence of forest and rangeland management on anadromous fish habitat in western North America. 4. Planning forest roads to protect salmonid habitat. USDA Forest Service Gen. Tech. Rept. PNW-109. Pac. Northwest For. Range Expt. Sta., Portland, OR.

Yee, C. S. and B. Thomas. 1984. Waterbars—Making them more effective. Calif. For. Note No. 90. California Dept. For., Sacramento, CA.

Zasada, Z. A. and J. C. Tappeiner. 1969. Soil disturbance from rubber-tire skidders during summer harvesting of aspen. Minn. For. Note 204, University of Minnesota, St. Paul, MN.

Ziemkiewicz, P. F., Ed. 1985. *Revegetation methods for Alberta's mountains and foothills.* Proc. of a workshop held 30 April-1 May 1984, Edmonton, Alberta. Alberta Land Cons. and Reclamation Council Rept. No. RRTAC 85-1, Edmonton, Alberta.

13 Watershed Management Policy and Planning

The ultimate justification of this book rests upon its value in stimulating sound watershed management on wildlands. Unless sound management is achieved, all that has gone before will have been only an exercise of the intellect. There is intrinsic value in intellectual exercise, but that value is multiplied many-fold when it is applied to benefit humankind. However, management without a solid foundation of policy and planning offers scant assurance of success. It is the purpose of this final chapter to discuss key considerations and provide a framework for approaching the challenge of watershed management policy and planning.

Management policy and planning are most effective if they proceed from the broad to the specific. Thus, it is first important to remember that watershed management is only one means among many for dealing with water resource needs and issues. While it may be true that for a given erosion and sedimentation problem land management would provide the best long-term solution, in many other situations the problems of water quality, quantity, or regimen may be solved as well or better by end-use conservation, engineering works, institutional approaches, or some combination of methods. The land need not bear the entire burden.

In fact, few watersheds are managed exclusively or even primarily for water, except for some municipal watersheds. Most are managed for some use or combination of uses. On wildlands, these include such goods and services as timber, livestock, minerals, wildlife, recreation, and amenities, as well as water. Equally notable is the fact that, with the exception of some areas of the western United States, few large watersheds are exclusively owned or administered by the same individual or organization. These facts have some profound implications for the nature of watershed management policy and planning.

For example, even where a watershed is managed by a single individual or organization, the person with primary decision making authority most often is a general land manager whose property or administrative area simply happens to include one or more major watersheds. Rarely is this person a technical specialist trained and focused on water resource outputs from the land. Instead, he/she usually collects input from several specialists or other information sources representing a number of key resources, in order to evaluate tradeoffs and make decisions about land management alternatives. Watershed management planning and application are thus conducted within a broader framework of land and resource management. Although management constraints are often noted when more than one resource is of concern, integrated planning that relies on good communication among technical specialists also can reveal some very positive management opportunities (Adams et al., 1988).

In many areas, multiple ownerships or administration within watersheds present some major challenges for watershed management policy and planning. Such situations require close cooperation and coordination to produce consistent behavior among the local landowners and resource managers. In the absence of other forms of leadership, formal policies and regulations increasingly are being used to promote this behavior and apply a watershed framework to land management. These and other approaches to watershed management policy and planning are discussed in Section 13.1.

Before discussing these approaches, however, it is important to point out that resource policies and decision making are shaped heavily by both technical facts and individual or collective values. When all interested parties agree on the facts and values associated with a watershed management issue, policy and decision making can proceed smoothly (Fig. 13.1). However, in recent years, it seems that increasingly there is considerable disagreement among the public or other involved parties about the facts or values associated with natural resource issues. When there is disagreement only about technical facts, resource professionals can play an important role in providing research, education, and informed opinions about resource management options, costs, and benefits for appropriate groups or individuals.

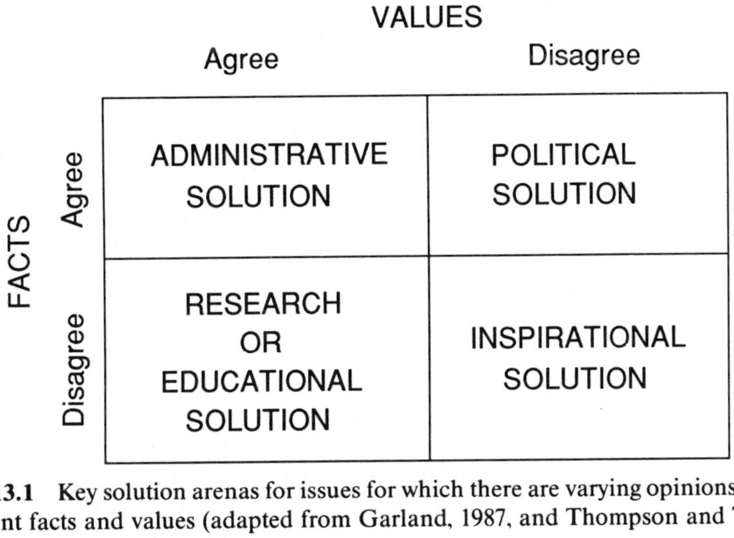

FACTS	VALUES: Agree	VALUES: Disagree
Agree	ADMINISTRATIVE SOLUTION	POLITICAL SOLUTION
Disagree	RESEARCH OR EDUCATIONAL SOLUTION	INSPIRATIONAL SOLUTION

FIG. 13.1 Key solution arenas for issues for which there are varying opinions about relevant facts and values (adapted from Garland, 1987, and Thompson and Tuden, 1959).

The political process, of which compromise is a key component, often is used to formulate policies, laws, and regulations when there is disagreement primarily about the values associated with a natural resource issue. If there is exceptionally wide disagreement about both the facts and values related to a specific issue, a solution may be found only through a charismatic leader or high court (Garland, 1987; Thompson and Tuden, 1959).

13.1 GENERAL APPROACHES TO POLICY AND PLANNING FOR WATERSHED MANAGEMENT

13.1.1 Research and Education

Research establishes much of the knowledge base from which rational watershed policy and planning decisions can be made. Without the considerable watershed research conducted over the last several decades, this book probably would have offered primarily some simple (and possibly erroneous) generalizations and anecdotal examples. General observations and experience certainly can be valuable, but an understanding of broadly applicable scientific principles that are based on carefully designed research will have considerably wider and longer lasting benefits to society.

As mentioned earlier, watershed research can be costly and time-consuming. However, the costs of research should be weighed against the potential costs of policies and practices that have a weak scientific foundation. For example, for many years policies and regulations in the Pacific Northwest supported the removal of large woody debris from forest streams to benefit anadromous fish passage. Research subsequently showed that such practices more likely have degraded fish habitat, and costly habitat restoration and streamside management measures have been implemented to help correct the damage (Adams et al., 1988).

Watershed research programs themselves must be flexible and responsive to our evolving knowledge base and the changing needs and values of society. Brown and Beschta (1985) noted how wildland watershed research in the mid-1980s reflected many, but certainly not all, of the important problems of that time. They also discussed some emerging issues (i.e., cumulative effects, worst-case analyses, monitoring requirements, and riparian management) that they believed deserved greater attention in research. As of this writing, only riparian management has been the subject of a notable amount of research since their paper was published.

Education of policymakers, resource managers, landowners, and operators is essential to the successful application of knowledge gained through watershed research and other means. Although researchers can provide some of this education through research literature, policymakers and resource managers usually have little time to read and interpret such literature, and landowners and operators typically lack the training or interest to do so. These groups are better served by targeted educational materials and approaches that provide not only technical understanding, but other information especially useful for decision making and implementation (e. g., comparing options, equipment, labor, and costs) (Adams, 1991). The need for such "how to do it" materials has been illustrated by the success of the Oregon State University *Woodland Workbook*, which although designed specifically for private, nonindustrial landowners, has been equally popular with professional resource managers (Elwood and Adams, 1989).

Even very basic education and communication can be useful in encouraging appropriate watershed practices. Few landowners, for example, are likely to conduct practices if they are made aware of a high risk of damaging a water supply that they use. Similarly, equipment operators who seek future business and referrals usually will work with landowners and others to meet resource objectives, as long as good two-way communication and understanding is established and maintained. In some cases, knowledgeable operators can help identify potential management problems, and even educate landowners and resource managers about operational limitations or opportunities.

Public education about wildland watershed management has received little attention, yet public opinion heavily influences resource policy and decision making. For the past 25 years or so, the public has shown considerable interest and concern about land use and water resource issues. Although the degree of this concern has varied over the years, population growth during this period has created strong demand for water resources, and increased tourism and outdoor recreation (especially by urban residents) have raised many questions about land management practices and water resources. The result is a renewed interest in protecting or enhancing existing water resources, and a growing need to educate an increasingly urban public about rural land use and natural resources such as water. These trends probably will continue through the turn of the century.

Both research and education in wildland watershed management are inherently interdisciplinary and complex. Innovative approaches, funding, and other incentives may be needed to encourage or strengthen cooperative research and educational programs that will be most effective in advancing management. For example, a "Center for Streamside Studies" recently was established at the University of Washington to conduct interdisciplinary research on many topics with implications for wildland watershed management. In contrast, Klemes (1986) notes that professional education in hydrology remains woefully lacking, and that this has resulted in practicing professionals with biased perspectives and wide knowledge gaps that breed a large variety of misconceptions. To address these problems over the long term, he suggests that hydrology be given a chance to become a broadly based primary discipline in university curricula.

13.1.2 Technical and Financial Assistance

Technical assistance usually refers to various public or private programs that provide help to individual landowners in identifying and dealing with specific resource management problems and opportunities on their property. However, as inferred earlier, organizations that manage sizable areas of land often have a technical staff that internally serves a similar role. For example, the USDA Forest Service and the USDI Bureau of Land Management employ forest and range hydrologists who work in district or regional offices. Private industry also employs some wildland hydrologists and watershed scientists to provide internal technical support (Nutter et al., 1985), as well as to offer consulting services to government and the public.

Essential to the success of technical assistance staff and programs are professionals who are well trained in the broad natural sciences that support watershed management. Still too often, it seems, staff with primary expertise in other areas are given assignments to provide technical assistance related to watershed management, leading to what Klemes (1986) describes as the formidable problems of "dilletantism of which misconceptions are only the symptoms."

Financial assistance and incentives can be useful with private landowners and operators who may lack the resources or inclination to implement watershed protection or enhancement measures. For example, cost-sharing has long been used to promote wildland conservation practices (tree planting, range seeding, windbreaks, etc.) that have watershed benefits. More recently, programs such as Conservation Reserve (CRP) have provided additional funding for focused efforts such as buffer strip planting, and the Stewardship program included in the 1990 Farm Bill provides further resources that can be used to benefit water resources.

Tax incentives have been used much less frequently to provide direct watershed benefits on wildlands. In some cases, income and property tax codes even have encouraged some practices that are now viewed as undesirable, such as wetlands conversion (Guldin, 1989). Oregon does have an innovative tax incentive program that provides property tax exemptions or income tax credits for riparian protection or enhancement activities (Faast, 1983). However, limited funding, management restrictions, and small qualifying acreages for the program have led to relatively minor participation by landowners.

It is important to note that, even when offered free technical assistance or large subsidies to improve management, landowners still need to understand their options and feel that they are making an informed decision about managing their property. A study of woodland owners in Wisconsin, for example, showed that educational programs had the greatest, most enduring influence on management. Thus, continued or increased investments in technical and financial assistance alone are unlikely to attract many landowners to improved management (Bliss and Martin, 1990). This points to a need for close coordination between educational (e.g., extension) and assistance programs for landowners (Adams, 1991), and there may be similar implications for efforts to change behavior within land management organizations.

Assistance programs also may not achieve their potential if they lack a sound scientific or knowledge base. For example, even if cost-share funds are made available for erosion control plantings, funds and efforts may be wasted if plantings fail to survive or are otherwise ineffective against the type of erosion that is common locally (e.g., deep mass soil movement). Thus, close coordination is essential in planning the full spectrum of potential approaches discussed thus far for dealing with watershed management concerns, that is, research, education, and assistance. Figure 13.2 graphically shows how these approaches are intimately linked, the primary leadership for planning and implementing each approach, and how each approach can play a critical role in dealing with key resources issues.

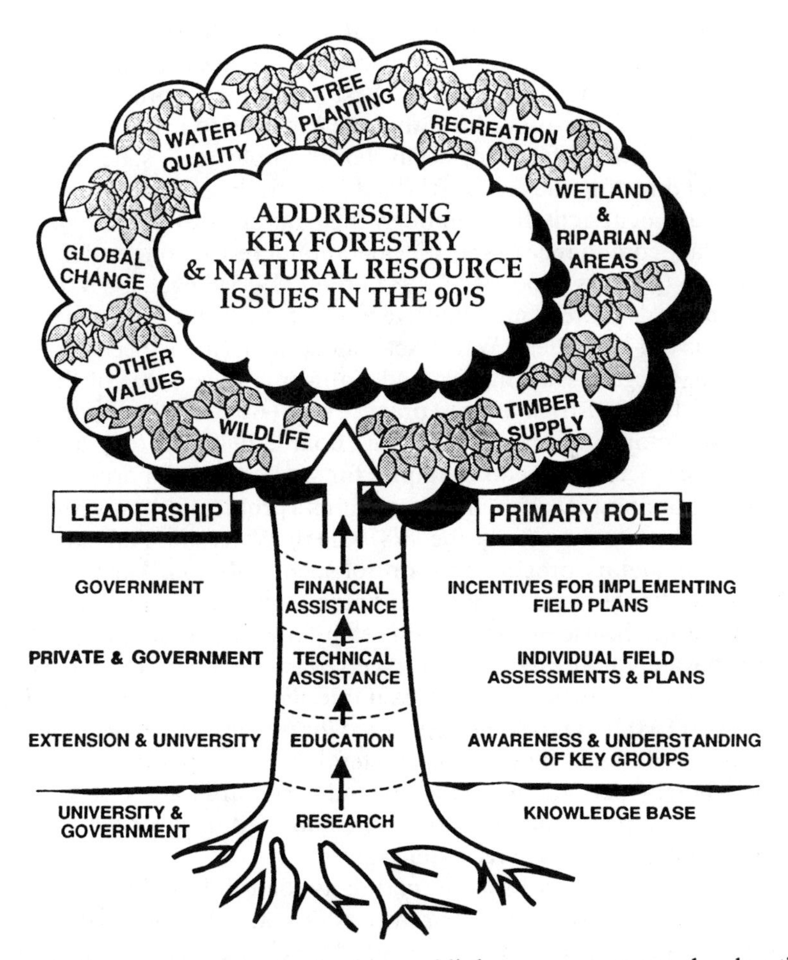

FIG. 13.2 The primary roles, leadership, and linkages among research, education, and assistance programs to address key wildland resource issues (from Adams, 1991).

13.1.3 Regulations

Because of their often punitive nature and the high cost of administration and enforcement, regulations to promote appropriate watershed management could be viewed as a final option after research, education, and assistance programs have failed to achieve the desired actions. Even then, regulations may be ineffective if they cannot be adequately enforced or they may lead to secondary problems if they create unusually high costs for land owners, managers, and users. Regulations are easier to administer and enforce when they are applied broadly and uniformly, but this approach may either do little to control locally serious problems or else result in unwarranted costs where few problems exist. Site-specific regulations and practices usually are most effective for resource protection, but typically these require the greatest knowledge and experience to design and implement.

Regulations to protect or enhance wildland watershed resources can take various forms. Laws requiring the use of "best management practices" (BMPs) are among the most common. Best management practices are those recognized practices or procedures that are expected to avoid or reduce soil and water problems before they occur. For example, most states have some type of laws directing the use of BMPs during timber harvesting and other forest management activities, although their nature varies from guidelines and voluntary compliance to specific operational restrictions and large fines for violations (Cubbage and Siegel, 1985; Irland, 1985; Siegel, 1989). The trend has been towards more comprehensive and specific legal requirements, and laws such as the 1987 Clean Water Act Amendments are expected to further this pattern. Oregon, Washington, and California have some of the oldest, broadest, and most restrictive forest practice laws (Adams, 1988; Spangenberg, 1987; Washington State Forest Practices Board, 1988), so they are recommended as particularly useful examples for study and comparison.

Because nonpoint source pollution remains a problem in many rural areas, efforts have been renewed via the 1987 Clean Water Act Amendments to strengthen regulatory programs to control such pollution. The use of BMP programs continues to be the most common regulatory approach for wildlands. However, the use of water quality standards with monitoring of actual changes in water quality recently has received greater attention in regulating wildland management practices. For example, the concept of "total maximum daily loads" (TMDLs) now is being studied for use on some large watersheds that include wildlands. The approach stems from the point source control method whereby periodic municipal or industrial discharges of waste substances (e. g., nutrients such as N or P) are allocated in accordance with the assimilation capacity of the receiving waters.

Thus, for a given river basin, a total daily load limit (TMDL) would be established for various land uses or ownerships, and management activities on those lands would not be allowed to produce changes exceeding that limit. Establishing appropriate TMDLs probably is one of the most difficult aspects of this approach, because it requires a very thorough understanding of local water quality dynamics, related natural influences, and land use effects. Water quality monitoring to determine and enforce the TMDLs also could be very time-consuming and costly. This approach may therefore be feasible only for a few selected "priority watersheds," like those identified for special agricultural BMP cost-share funding in Wisconsin (Spangenberg, 1987).

This brief discussion highlights only a few examples of the regulatory approaches related to wildland watershed management. Many other federal, state, and local laws and regulations have direct or indirect implications for watershed management. For example, water rights law represents a highly complex and significant area, especially in the western United States (Hutchins, 1971; Tarlock, 1988). Zoning laws and property rights acquisitions normally are used to control urban and suburban land use (Burby et al., 1983), but they also may have some application for wildland watershed management where private ownerships are important.

For federal wildlands, laws such as the National Environmental Policy Act (NEPA) of 1969, the Forests and Rangelands Renewable Resources Planning Act (RPA) of 1974, and the National Forest Management Act (NFMA) of 1976 provide a very important foundation for resource management planning and practice. The reader is referred to other available sources (e.g., Baltic et al., 1989; Canter, 1977; Council of Environmental Quality, 1978; Guldin, 1989; Mills, 1991) for more thorough treatment of these laws and their application.

13.2 FRAMEWORK FOR WATERSHED MANAGEMENT PLANNING

Assuming that there is an individual or organization responsible for wildland management decisions focused on or including watershed concerns, how might management planning proceed? Although broader in scope and currently focused on rangeland concerns, the Coordinated Resource Management Planning (CRMP) process represents a successful conceptual model for interagency, interdisciplinary planning to protect or enhance important natural resources (Cleary, 1988). Similarly, coordinated watershed or basin planning has been implemented in some areas (e.g., Spangenberg, 1987). In the state of Washington, the Timber, Fish, and Wildlife (TFW) program supports cooperative planning, monitoring, research, and regulation efforts that have direct implications for wildland water resources. Various publications also have highlighted conceptual models or other useful information for wildland watershed management planning (e.g., Brooks et al., 1991; Burby et al., 1983; Callaham, 1990; Dissmeyer et al., 1987; Northeastern Forest Experiment Station, 1975; Oregon State Water Resources Board, 1973; Wilford, 1987).

Figure 13.3 shows a simple conceptual model for wildland watershed management planning that should be widely applicable, including public,

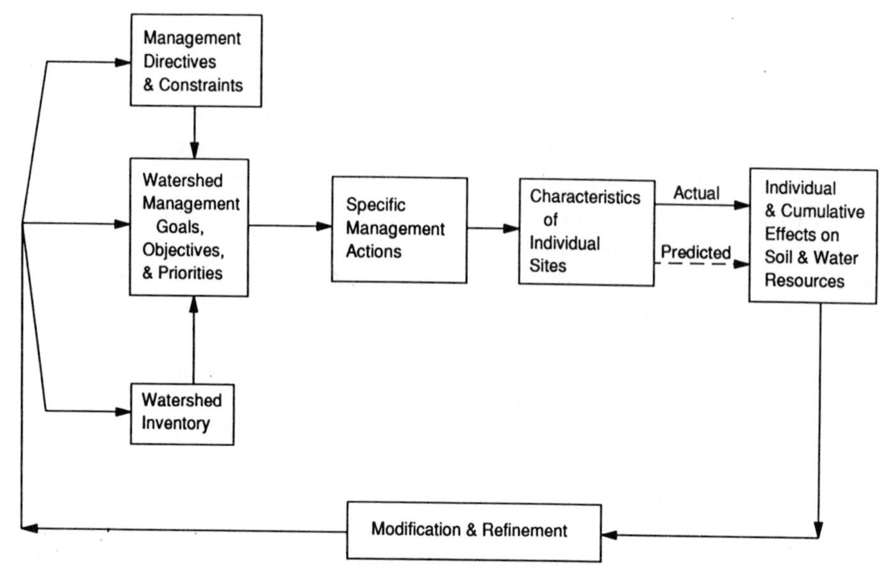

FIG. 13.3 Conceptual model for wildland watershed management planning.

private, and mixed watershed ownerships. At the heart of any management planning effort is the establishment of broad management goals and more specific associated objectives. Because resources for management (time, money, labor, etc.) usually have some limitations, priority setting among the various objectives can be very useful.

Watershed management goals, objectives, and priorities are generated from broader management directives and constraints, as well as from knowledge about the specific watershed of interest. Broad management directives include policy statements and legal mandates (e.g., NEPA, NFMA, RPA) that address social needs and values. On private lands, personal or corporate needs and values would be of similar importance. Constraints often include economic considerations, applicable regulations, and organizational or administrative limitations.

A comprehensive watershed inventory provides key information for both broad and specific management planning activities. Listed and described below are the categories of information that are essential to most watershed inventories. Many of the details behind the significance of these factors, and related references, have been covered in the previous sections and chapters.

Watershed Boundaries. It is important to clearly identify the entire physical watershed, because the water resources from it are influenced by all of this land, from streambanks to ridgetops. Location of important subbasins also may be very useful. Topographic maps or aerial photos most often are used for this work, but subsurface investigation may be needed for accurate watershed definition where impermeable soil or bedrock slopes may route water contrary to surface terrain.

Property Boundaries and Uses. This step identifies key landowners, organizations, and major land uses that can influence the water resources of concern. Such data can facilitate educational, assistance, cooperative, and regulatory efforts. Where private ownerships are of interest, county government offices often maintain the appropriate property records.

Terrain. Characteristics such as slope, aspect, elevation, and drainage configuration can help identify areas of concern for roads and land use. USDI Geological Survey (USGS) topographic maps or stereoscopic pairs of aerial photos (available from government agencies or private companies) are very useful for identifying general terrain features, but the scale and vegetation cover may mask features needed for detailed planning. Private surveyors and aviation companies can provide maps and photos of various, more desirable scales.

Geology and Soils. This information can help identify areas of potential erosion and instability, and clarify hydrologic behavior. The USGS, USDA Soil Conservation Service (SCS), and similar state agencies produce a large variety of useful maps and publications. Consulting firms offer detailed geology and soils mapping and investigation services.

Climate. Data such as rainfall, snowfall, and temperature can help in understanding and predicting streamflow patterns, erosion, influences of vegetation and management, and so on. Information on average, maximum and minimum conditions, and the intensity and duration of extremes are especially useful. Such data are collected for many locations by federal and state agencies, although the lack of sampling in many wildland locations often makes local data collection desirable.

Stream Hydrology. Data on hydrologic behavior are useful for design and operation of water supply, storage, and control systems, as well as for design of stream crossings and roads. Particularly valuable are data about average, peak and low flows, and the frequency and duration of extreme levels. The USGS and state water resources agencies collect flow and other hydrologic data for many larger streams and rivers. These agencies or consulting firms can provide assistance for estimating flows where data are unavailable. Other stream characteristics such as bed and bank materials, large organic debris, channel gradients, and drainage densities can help determine stream stability and expected response to management activities.

Water Quality. Baseline water quality data are essential for understanding existing and potential watershed influences, values, and effects of management. The USGS and other agencies collect some water quality data, but few small wildland watersheds are separately monitored. The seasonal and flow-related dynamics of water quality require careful, and often intensive sampling. Agencies, consulting firms, and commercial laboratories can provide assistance in the collection, analysis, and interpretation of water quality data.

Vegetation. Data on type, age, density, and distribution of major vegetation can help explain watershed behavior and management effects and opportunities. Streamside or source area vegetation may be particularly important to evaluate for such things as shading, bank stabilization, and expected response to disturbance. Although some vegetation data may be available from agencies and landowners, other local inventories may be needed to provide data useful for watershed management planning.

Construction Features. Information on existing roads, bridges, dams, buildings, and other features help in understanding management influences and in planning for access, maintenance, and potential problems. Because roads can be a major influence on wildland water resources, specific data on location, mileages, widths, grades, surfacing, drainage features, stream crossings, traffic control, and problem areas can be especially valuable.

Other Considerations. An inventory of any remaining features, uses and values of the land and environment should be considered, because these also may be significant in watershed management planning, implementation, and outcomes. For example, general ownership and land use surveys may not reveal some potentially important activities such as recreation, hunting, fishing, and grazing, or key user groups such as youth or permittees. Similarly, local fire, wind, ice, insect, and disease hazards may require careful evaluation to clearly establish risks to water resources and possible management strategies to reduce these risks.

To be most useful in planning, information in the watershed inventory should be kept current, organized, and easily accessible to those responsible for planning.

Once the overall watershed management goals, objectives, and priorities are established within the context of broader management considerations and the watershed inventory, planning of more specific actions can begin (Fig. 13.3). This planning should be conducted at two levels, strategic (long-term, area-, or basin-wide actions) and tactical (short-term, local activities) (Garland 1987). Strategic planning is especially important in wildland watershed management to avoid unnecessary road construction or other duplicative action, ensure the continuity and integrity of important riparian management areas, and to understand and deal with potential cumulative effects of management activities (Adams et al., 1988).

Cumulative effects have received considerable attention in wildland management planning in recent years, because federal law now requires an analysis of such effects when environmental analyses are conducted (Council on Environmental Quality, 1978). More specifically, cumulative effects are those environmental effects resulting from an action (e.g., timber harvesting) when added to other past, present, and future incremental actions, regardless of land ownership. Such analyses also need to consider any individually minor, but collectively significant actions occurring over time that may yield such effects. The concept of cumulative effects encompasses a broader spectrum of resources and land uses than typically has been evaluated in watershed research (Sidle and Sharpley, 1991), but a number of studies provide some insights.

Soil compaction from multiple-entry logging may yield cumulative effects, for example, because recovery from the compacted condition may take many years and thus not occur before successive logging entries (Froehlich and McNabb, 1984). Several hydrologic and water quality studies suggest both the occurrence (Beschta and Taylor, 1988; Cristner and Harr, 1982; Lyons and Beschta, 1983) and absence (Duncan, 1986; Sullivan, 1985) of cumulative watershed effects where there has been significant logging and road construction. Physical mechanisms for such hydrologic effects certainly exist, but their occurrence depends on the area and intensity of management activities relative to the size of the watershed of concern (Harr, 1989).

In both strategic and tactical watershed management planning, information about possible actions is combined with site characteristics to predict likely effects on soil and water resources (Fig. 13.3). The expected outcomes are then evaluated relative to management goals and objectives, and if necessary, plans are modified or refined. Both systematic and informal procedures are used to predict potential management effects. Quantitative watershed and hydrologic models often are sought because their systematically generated (usually with computer software), numerical outcomes seem to carry considerable weight with many decision makers. However, using such models to make major decisions about the management of specific watersheds is fraught with potential dangers.

Data used to develop such quantitative models, for example, almost invariably come from watersheds other than those being evaluated for management. Rarely are the key watershed conditions (climate, vegetation, soils, geology, terrain, etc.) comparable enough to confidently allow for such extrapolation. Similarly, management practices and intensities often differ, and much of our quantitative data base for modeling stems from historical research conducted with management planning and practices that now would be considered substandard. Another danger in the development and use of watershed and hydrologic models is the temptation to inappropriately simplify inherently complex physical phenomena (Harr, 1987). The result is user-attractive, but hydrologically indefensible applications (Klemes, 1986), which ultimately could be challenged in court (Garland, 1988).

Quantitive models certainly are not without some value in watershed management planning. However, their use requires a level of professional understanding and expertise that too often is lacking in the organization or administrative level where they are needed. Bunnell (1989) emphasized that teams that include both resource managers and researchers are needed to develop models appropriate for management, and that they must be vigorously evaluated; he even offers "10 commandments" for evaluating models.

In the absence of appropriate quantitative models, a commonly used planning alternative is to conduct a general review of existing research results or other data to develop expectations of management effects. Caution is in order even with this approach. For example, although statistical analyses contained in published research reports usually are considered objective, the specific type of analysis used and the interpretation of the results contain various elements of judgment and subjectivity. This alone may not be a problem, but research reports often fail to provide enough information for the reader to adequately critique the study methods and evaluate the conclusions. Statistical packages for computers also have led to ritualistic use by researchers of relatively few statistical methods, with little question of their suitability in addressing the objectives of specific studies (Warren, 1986). Golbeck (1986) warns that "consumers of research reports must not just accept conclusions, but must investigate the methods used to obtain them," and describes some ways in which conclusions can be evaluated before applying them to managerial or planning decisions.

Parallel to the need for predicting management outcomes for watershed planning is the need for actual monitoring and assessment of management practices that are implemented. In many cases, such monitoring and assessment is required by law (e.g., NFMA), although questions often remain about what should be monitored, when, and with what accuracy (Brown and Beschta, 1985). Because watershed research cannot be conducted for each set of practices and site conditions that will be encountered, monitoring and assessment are essential for testing and validating predictive methods and in modifying and refining practices for specific situations. Guidelines and other information for designing and implementing watershed monitoring programs are found in a variety of references (e.g., Burby et al., 1983; Everett and Schmidt, 1979; Howes et al., 1983; Kunkle et al., 1987; Ponce, 1980; Ponce et al., 1982; USDI Geological Survey, 1983; Waterstone and Burt, 1988).

In both predicting and monitoring management effects on soil and water resources, it should be noted that environmental changes resulting from management are not necessarily synonymous with unacceptable negative impacts. This becomes particularly important as our analytical ability to detect environmental changes increases from new technologies and approaches. Environmental changes can and do occur naturally and dramatically (e.g., via floods, landslides, and wildfire), and soil and water resources can respond in positive, negative, and neutral ways. Resource specialists must evaluate management effects over both short- and long-term time periods, and recommend what is tolerable or desirable for resources of concern. In agriculture, for example, the concept of a tolerable soil loss of up to 10 ton ha^{-1} $year^{-1}$ as a cost of producing crops has been accepted for some time (Johnson, 1987).

Among the most critical features of successful watershed management planning are flexibility and responsiveness to modify plans, based on ongoing review of actual and predicted effects on key resources and changes in technology and management directives and constraints (Fig. 13.3). In modifying and refining management plans, it is essential that proposed management actions be tested for technical, economic, and institutional feasibility (Garland, 1987). For example, cable logging to avoid soil and water impacts may be technically feasible in a given watershed, but local high operating costs or lack of skilled logging personnel or equipment may make such logging economically or institutionally infeasible. Similarly, even seemingly small changes in buffer strip requirements can have substantial effects on logging costs and timber values foregone (Barringer, 1987; Garland, 1987; Olsen et al., 1987), as well as incur significant safety hazards for woodsworkers.

The key to identifying and dealing with such management issues as feasibility limitations are planning activities that regularly seek input from, and encourage communication among, all groups with a stake in watershed management, such as resource users, land owners and managers, resource and operations specialists, field personnel, and contractors. Thus, even the watershed management planning model itself must remain flexible and responsive to changing knowledge and needs.

In some cases, for example, communicating such feasibility limitations and their implications to the public or policymakers may need special emphasis. Often, the immediate responses to wildland watershed concerns are relatively simplistic: ban grazing, stop clearcutting, and so on. As fewer in our society are directly involved with rural land use, there is less understanding of the connection between this land use and our standard of living as represented by relatively inexpensive products of rural lands, for example, food and housing. Also, if demands for such products remain high, yet management restrictions limit domestic outputs, will environmental impacts increase in other countries that act to supply our demands? All of this is not to say that higher costs or other tradeoffs are not appropriate to protect or enhance watershed resources, just that few natural resource issues have simple solutions, and that the public and policymakers should develop *informed* opinions and policy.

Finally, we must remember that wildland watershed management still is a relatively young field, dating essentially since World War II. Though we have learned much, it is only in recent years that a significant part of our wildlands have been evaluated and managed in response to our new knowledge. As we monitor our results and adjust policy, planning, and practices to improve applications, there seems no reason that we cannot soon meet the promise of using all but the most fragile wildlands for many benefits, while maintaining their ability to generate desirable water resources in perpetuity.

LITERATURE CITED

Adams, P. W. 1991. Addressing forestry issues for the future: a vision for coordinated research, education, and assistance programs. Unpublished manuscript.

Adams, P. W. 1988. Oregon's Forest Practice Rules. Ext. Circ. 1194. Oregon State University Ext. Serv., Corvallis, OR.

Adams, P. W., R. L. Beschta, and H. A. Froehlich. 1988. Mountain logging near streams: opportunities and challenges. pp. 153–162. In: Proc. Intl. Mountain Logging and Pacific Northwest Skyline Symp. College of Forestry, Oregon State University, Corvallis, OR.

Baltic, T. J., J. G. Hof, and B. M. Kent. 1989. Review of critiques of the USDA Forest Service land management planning process. USDA For. Serv. Gen. Tech. Rep. RM-170. Rocky Mt. For. Range Expt. Sta., Ft. Collins, CO.

Barringer, J. 1987. Costs and constraints of riparian management of the Wiley Creek cleanup sale. pp. 64–67. In: Managing Oregon's riparian zone for timber, fish and wildlife. Tech. Bull. 514. Natl. Council of Paper Ind. for Air and Stream Improvement, New York.

Beschta, R. L. and R. L. Taylor. 1988. Stream temperature increases and land use in a forested Oregon watershed. *Water Resour. Bull.* 24(1):19–25.

Bliss, J. C. and A. J. Martin. 1990. How tree farmers view management incentives. *J. For.* 88(8):12–18.

Brooks, K. N., P. F. Ffolliott, H. M. Gregersen, and J. L. Thames. 1991. *Hydrology and the management of watersheds.* Iowa State University Press, Ames, IA.

Brown, G. W. and R. L. Beschta. 1985. The art of managing water. *J. For.* 83(10):604–615.

Bunnell, F. L. 1989. Alchemy and uncertainty: what good are models? USDA For. Ser. Gen. Tech. Rep. PNW-GTR-232. Pac. Northwest Res. Sta., Portland, OR.

Burby, R. J., E. J. Kaiser, T. L. Miller, and D. H. Moreau. 1983. *Drinking water supplies: Protection through watershed management.* Ann Arbor Science Pub., Ann Arbor, MI.

Callaham, R. Z. 1990. Guidelines for management of wildland watershed projects. Rep. 23. Wildland Resources Ctr., University of California, Berkeley, CA.

Canter, L. 1977. *Environmental Impact Assessment.* McGraw-Hill, New York.

Cleary, C. R. 1988. Coordinated resource management: a planning process that works. *J. Soil Water Conserv.* 43(2):138–139.

Council on Environmental Quality. 1978. Regulations for implementing the procedural provisions of the National Environmental Policy Act. Fed. Reg. No. 43:55978-56007. US Govt. Printing Office, Washington, DC.

Cristner, J. and R. D. Harr. 1982. Peak streamflows from the transient snow zone, western Cascades, Oregon. pp. 27–38. In: Proc. XXth Western Snow Conf., Colorado State University, Ft. Collins, CO.

Cubbage, F. W. and W. C. Siegel. 1985. The law regulating forest practices. *J. For.* 83(9):538–545.

Dissmeyer, G. E., M. P. Goggin, E. R. Frandsen, S. R. Miles, R. Solomon, B. B. Foster, and K. L. Roth. 1987. *Soil and water resource management: a cost or a benefit? Approaches to watershed economics through example.* Watershed and Air Management Staff, USDA For. Serv., Washington, DC.

Duncan, S. H. 1986. Peak stream discharge during thirty years of sustained yield timber management in two fifth order watersheds in Washington state. *Northwest Sci.* 60(4):258–264.

Elwood, N. E. and P. W. Adams. 1989. Achieving group project success in Oregon: the *Woodland Workbook*. *J. Agron. Educ.* 18(2):119–121.

Everett, L. G. and K. D. Schmidt, Eds. 1979. *Establishment of water quality monitoring programs.* Tech. Pub. TPS79-1. Am. Water Resources Assoc., Minneapolis, MN.

Faast, T. 1983. R_x for riparian zones. *Oregon Wildlife*, July:3–4.

Froehlich, H. A. and D. H. McNabb. 1984. Minimizing soil compaction in Pacific Northwest forests. pp. 159–192. In: E. A. Stone, Ed., *Forest soils and treatment impacts*, Proc. 6th North Am. For. Soils Conf., University of Tennessee, Knoxville, TN.

Garland, J. J. 1987. Aspects of practical management of the streamside zone. pp. 277–288. In: *Proc. Sym. Streamside Management: Forestry and Fisheries Interactions.* University of Washington, Seattle, WA. p. 471.

Garland, J. J. 1988. A modeler's day in court—misuse of poor models. *J. For.* 86(4):57.

Golbeck, A. L. 1986. Evaluating statistical validity of research reports: a guide for managers, planners, and researchers. USDA For. Serv. Gen. Tech. Rep. PSW-87. Pac. Southwest For. Range Expt. Sta., Berkeley, CA.

Guldin, R. W. 1989. An analysis of the water situation in the United States: 1989–2040. USDA For. Serv. Gen. Tech. Rep. RM-177. Rocky Mt. For. Range Expt. Sta., Fort Collins, CO.

Harr, R. D. 1989. Cumulative effects of timber harvest on streamflows. Paper presented at 1989 Soc. Am. Foresters Conf., Spokane, WA.

Harr, R. D. 1987. Myths and misconceptions about forest hydrologic systems and cumulative effects. pp. 137–141. In: R. Z. Callaham and J. J. DeVries, tech. coord., Proc. 1986 Calif. Watershed Manage. Conf., University of California, Berkeley, CA.

Howes, S., J. Hazard, and J. M. Geist. 1983. Guidelines for sampling some physical conditions of surface soils. R6-RWM-146-1983. USDA For. Serv., Pac. Northwest Region, Portland, OR.

Hutchins, W. A. 1971. *Water rights laws in the nineteen western States.* Vols. I–III. USDA Misc. Pub. 1206. USDA Econ. Res. Serv., Washington, DC.

Irland, L. C. 1985. Logging and water quality: state regulation in New England. *J. Soil Water Conserv.* 40(1):98–102.

Johnson, L. C. 1987. Soil loss tolerance: fact or myth. *J. Soil Water Conserv.* 42(2):155–60.

Klemes, V. 1986. Dilettantism in hydrology: transition or destiny? *Water Resour. Res.* 22(9):177S–188S.

Kunkle, S., W. S. Johnson, and M. Flora. 1987. Monitoring stream water quality for land-use impacts: a training manual for natural resource management specialists. Water Resources Div., Natl. Park Serv., Ft. Collins, CO.

Lyons, J. K. and R. L. Beschta. 1983. Land use, floods, and channel changes: Upper Middle Fork Willamette River, Oregon (1936–1980). *Water Resour. Res.* 19(2):463–471.

Mills, T. 1991. Understanding the RPA. *J. For.* 89(5):10–14 and 89(6):17–22.

Northeastern Forest Experiment Station. 1975. Municipal watershed management symposium proceedings. USDA For. Serv. Gen. Tech. Rep. NE-13. Northeastern For. Expt. Sta., Upper Darby, PA.

Nutter, W. L., M. A. Barton, G. E. Hart, and E. L. Miller. 1985. Future roles of forest hydrologists. pp. 153–156. In: Proc. 1985 Soc. Am. For. Natl. Conv., Ft. Collins, CO. Soc. Am. For., Bethesda, MD.

Olsen, E. D., D. S. Keough, and D. K. LaCourse. 1987. Economic impact of proposed Oregon Forest Practice Rules on industrial forest lands in the Oregon Coast Range: a case study. FRL Bull. 61. For. Res. Lab, Oregon State University, Corvallis, OR.

Oregon State Water Resources Board. 1973. Municipal watershed sourcebook: outline for Oregon. State Water Res. Board, Salem, OR.

Ponce, S. L. 1980. Water quality monitoring programs. WSDG-TP-00002. Watershed Sys. Devel. Group, USDA For. Serv., Ft. Collins, CO.

Ponce, S. L., D. W. Schindler, and R. C. Averett. 1982. The use of the paired-basin technique in flow-related wildland water quality studies. WSDG-TP-0004. Watershed Sys. Devel. Group, USDA For. Serv., Ft. Collins, CO.

Sidle, R. C. and A. N. Sharpley. 1991. Cumulative effects of land management on soil and water resources: an overview. *J. Environ. Qual.* 20(1):1–3.

Siegel, W. C. 1989. State water quality laws and programs to control nonpoint source pollution from forest land in the eastern United States. pp. 131–140. In: Proc. 1989 Natl. Symp. Non-point source concerns—legal and regulatory aspects, Am. Soc. Agric. Engr. Southern For. Expt. Sta., New Orleans, LA.

Spangenberg, N. E. 1987. Implementation strategies for agricultural and silvicultural nonpoint source pollution control in California and Wisconsin. *Water Resour. Bull.* 23(1):133–137.

Sullivan, K. 1985. Long-term patterns of water quality in a managed watershed in Oregon: 1. Suspended sediment. *Water Resour. Bull.* 21(6):977–987.

Tarlock, A. D. 1988. *Law of water rights and resources.* Clark Boardman Co., New York.

Thompson, J. D. and A. Tuden. 1959. Strategies, structures, and processes of organizational decision. pp. 195–216. In: J. D. Thompson, P. B. Hammond, B. H. Junker, and A. Tuden, Eds. *Comparative Studies in Administration*. University of Pittsburgh Press, Pittsburg, PA.

USDI Geological Survey. 1983. *National handbook of recommended methods for water-data acquisition.* USDI Geological Survey, Reston, VA.

Warren, W. G. 1986. On the presentation of statistical analysis: reason or ritual. *Can. J. For. Res.* 16:1185–1191.

Washington State Forest Practices Board. 1988. Forest Practices Rules and Regulations. Wash. State Forest Practices Board, Olympia, WA.

Waterstone, M. and R. J. Burt, Eds. 1988. *Water-use data for water resources management.* Am. Water Resources Assoc., Bethesda, MD.

Wilford, D. J. 1987. A watershed perspective in a site specific world. *For. Chron.* August:246–249.

APPENDIX
Units of Measurement with Conversion Factors Used in Hydrology

The Système Internationale d'Unités (SI), a modernized version of the metric system, is replacing the old systems of measurement commonly used in hydrology, meteorology, soil science, and watershed management. However, many of the older systems are still in common use, particularly in the United States, where many data are expressed in such terms as inches of rain or snow, cubic feet per second of streamflow, acre feet of reservoir capacity, or tons of sediment produced per acre per year.

Multiples and subdivisions of base units all use the same prefix in the SI system, which are given below.

Multiples and Subdivisions of Base Units, SI System

			Example	
Prefix	Abbreviation	Multiple	Quantity	Abbreviation
Tera	T	10^{15}	Tetrameter	Tm
Giga	G	10^{9}	Gigameter	Gm
Mega	M	10^{6}	Megameter	Mm
Kilo	k	10^{3}	Kilometer	km
Hecto	h	10^{2}	Hectometer	hm
Deka	da	10^{1}	Dekameter	dam
		10^{0}	Meter	m
Deci	d	10^{-1}	Decimeter	dm
Centimeter	c	10^{-2}	Centimeter	cm
Milli	m	10^{-3}	Millimeter	mm
Micro	μ	10^{-6}	Micrometer	μm
Nano	n	10^{-9}	Nanometer	nm
Pico	p	10^{-12}	Picometer	pm
Femto	f	10^{-15}	Femtometer	fm

Base unit for length—meter (m).

Conversion Factors for Length, Area, and Volume[a]

To convert:			
Length	Multiply by	Length	Multiply by
in. to cm	2.54[b]	cm to in.	0.393701
ft to m	0.3048[b]	m to ft	3.28084
mi to ft	5280[b]	ft to mi	0.000189
mi to km	1.60934	km to mi	0.621371
Area	Multiply by	Area	Multiply by
in.^{2} to cm^2	6.4516[b]	cm^2 to in.^{2}	0.155000
ft^2 to m^2	0.09203	m^2 to ft^2	10.7639
acre to ft^2	43560[b]	ft^2 to ft^2	0.000023
ha[c] to m^2	10000[b]	m^2 to ha	0.0001[b]
ha to acre	2.47105	acre to ha	0.4047
mi^2 to acre	640[b]	acre to mi^2	0.001563
km^2 to ha	100[b]	ha to km^2	0.01[b]
mi^2 to km^2	2.590	km^2 to mi^2	0.3861
Volume	Multiply by	Volume	Multiply by
in.^{3} to cm^3	16.3871	cm^3 to in.^{3}	0.061024
m^3 to L[c]	1000[b]	L to m^3	0.001[b]
m^3 to ft^3	35.3147	ft^3 to m^3	0.028317
acre ft to Mgal	0.32585	Mgal to acre ft	3.06887

[a]Basic SI unit = m
[b]Exactly.
[c]The hectare (ha = dam^2) is sometimes used for land area and the liter (L = dm^3) for liquid measure in the SI system.

Conversion Factors for Weight or Mass and Density[a]

To Convert	to	Multiply by	To Convert	to	Multiply by
oz	g	28.3495	g	oz	0.35274
kg	lb	2.20462	lb	kg	0.453592
ton (t-metric)	kg	1000[b]	kg	t	0.001[b]
ton (t)	lb	2000[b]	lb	t	0.0005[b]
kg m^3	lb ft^3	0.062428	lb ft^3	kg m^3	16.0185

[a]Basic SI unit = kilogram (kg).
[b]Exactly.

Conversion Factors Used in Soils and Climatology

Pressure[a]

To Convert	to	Multiply by	To Convert	to	Multiply by
lb in.^{2}	kPa	6.89465	kPa	lb in.^{2}	0.145040
atm (std)	kPa	101.325[b]	kPa	atm (std)	0.009867
atm	bar	1.01325[b]	bar	atm	0.986924
mbar (mb^{-2})	Pa	100[b]	Pa	mb^{-2}	0.01[b]
dyn cm	Pa	0.1[b]	Pa	dyn cm	10.0[b]
kg m^{-2}	Pa	9.80665[b]	Pa	k gm^{-2}	0.101972

[a]Basic SI unit = pascal (Pa).
[b]Exactly.

Energy[a]

To Convert	to	Multiply by	To Convert	to	Multiply by
calorie (cal)	J	4.1855	J	cal	0.23892
cal cm^{-2}	langley (ly)	1.0[b]	ly	cal cm^{-2}	1.0[b]
ly	J cm^{-2}	4.1855	J cm^{-2}	ly	0.23892
British Thermal Unit (BTU)	k J	1.05507	kJ	BTU	0.94780
kilowatt hour (k Wh)	k J	3600[b]	kJ	k Wh	0.000278

[a]Basic SI unit = joule (J).
[b]Exactly.

Power[a]

To Convert	to	Multiply by	To Convert	to	Multiply by
J s^{-1}	W	1.0[a]	W	J s^{-1}	1.0[a]
ly min^{-1}	W m^{-2}	697.333	W m^{-2}	ly min^{-1}	0.001434

[a]Basic SI unit = watt (W).
[b]Exactly.

Conversion Factors for Water Quantity and Flow Rates[a]

To Convert	to	Multiply by	To Convert	to	Multiply by
US gallon (gal)	lb	8.345	lb	gal	0.1198
gal	kg	3.785	kg	gal	0.2642
ft^3	lb	62.43	lb	ft^3	0.0160
ft^3	kg	28.31	kg	ft^3	0.03532
cm^3	g	1.0	g	cm^3	1.0
L	kg	1.0	g	L	1.0
m^3	kg	1000	kg	m^3	0.001
$m\ s^{-1}$	$ft\ s^{-1}$	3.281	$ft\ s^{-1}$	$m\ s^{-1}$	0.3048
$mi\ h^{-1}$	$ft\ s^{-1}$	1.467	$ft\ s^{-1}$	$mi\ s^{-1}$	0.6818
$m^3\ s^{-1}$	$ft^3\ s^{-1}$	35.31	$ft^3\ s^{-1}$	$m^3\ s^{-1}$	0.2832
$ft^3\ s^{-1}$	$L\ s^{-1}$	28.32	$L\ s^{-1}$	$ft^3\ s^{-1}$	0.03531
$ft^3\ s^{-1}$	$Mgal\ day^{-1}$	0.6463	$Mgal\ day^{-1}$	$ft^3\ s^{-1}$	1.547
$ft^3\ s^{-1}$	$acre\ ft\ day^{-1}$	1.983	$acre\ ft\ day^{-1}$	$ft^3\ s^{-1}$	0.5041
$ft^3\ s^{-1}$	miner's inch	35.7–50[c]	miner's inch	$ft^3\ s^{-1}$	0.020–0.028
$m^3\ s^{-1}$	$Mgal\ day^{-1}$	22.82	$Mgal\ day^{-1}$	$m^3\ s^{-1}$	0.04381
$ft^3\ s^{-1}\ mi^{-2}$	$m\ s^{-1}\ km^{-2}$	0.07335	$m^3\ s^{-1}\ km^{-2}$	$ft^3\ s^{-1}\ mi^{-2}$	13.63
$ft^3\ s^{-1}\ mi^{-2}$	$in.\ h^{-1}$	0.001550	$in.\ h^{-1}$	$ft^3\ s^{-1}\ mi^{-2}$	645.3
$ft^3\ s^{-1}\ mi^{-2}$	$in.\ day^{-1}$	0.03719	$in.\ day^{-1}$	$ft^3\ s^{-1}\ mi^{-2}$	26.89
$ft^3\ s^{-1}\ mi^{-2}$	$m^3\ s^{-1}\ km^{-2}$	0.07335	$m^3\ s^{-1}\ km^{-2}$	$ft^3\ s^{-1}\ mi^{-2}$	13.63
$ft^3\ s^{-1}\ mi^{-2}$	$L\ s^{-1}\ ha^{-1}$	0.1093	$L\ s^{-1}\ ha^{-1}$	$ft^3\ s^{-1}\ mi^{-2}$	9.145

[a] Basic SI units kg, m, $m\ s^{-1}$ (L often used for small quantities)
[b] All volume–weight relations vary slightly with temperature.
[c] Variable by state or province.

INDEX

F

G

H

S